Scientific Data Management

Challenges, Technology, and Deployment

Chapman & Hall/CRC
Computational Science Series

SERIES EDITOR
Horst Simon
Associate Laboratory Director, Computing Sciences
Lawrence Berkeley National Laboratory
Berkeley, California, U.S.A.

AIMS AND SCOPE

This series aims to capture new developments and applications in the field of computational science through the publication of a broad range of textbooks, reference works, and handbooks. Books in this series will provide introductory as well as advanced material on mathematical, statistical, and computational methods and techniques, and will present researchers with the latest theories and experimentation. The scope of the series includes, but is not limited to, titles in the areas of scientific computing, parallel and distributed computing, high performance computing, grid computing, cluster computing, heterogeneous computing, quantum computing, and their applications in scientific disciplines such as astrophysics, aeronautics, biology, chemistry, climate modeling, combustion, cosmology, earthquake prediction, imaging, materials, neuroscience, oil exploration, and weather forecasting.

PUBLISHED TITLES

PETASCALE COMPUTING: Algorithms and Applications
Edited by David A. Bader

PROCESS ALGEBRA FOR PARALLEL AND DISTRIBUTED PROCESSING
Edited by Michael Alexander and William Gardner

GRID COMPUTING: TECHNIQUES AND APPLICATIONS
Barry Wilkinson

INTRODUCTION TO CONCURRENCY IN PROGRAMMING LANGUAGES
Matthew J. Sottile, Timothy G. Mattson, and Craig E Rasmussen

INTRODUCTION TO SCHEDULING
Yves Robert and Frédéric Vivien

SCIENTIFIC DATA MANAGEMENT: CHALLENGES, TECHNOLOGY, AND DEPLOYMENT
Edited by Arie Shoshani and Doron Rotem

Scientific Data Management
Challenges, Technology, and Deployment

Edited by
Arie Shoshani
Doron Rotem

CRC Press
Taylor & Francis Group
Boca Raton London New York

CRC Press is an imprint of the
Taylor & Francis Group an **informa** business

A CHAPMAN & HALL BOOK

Chapman & Hall/CRC
Taylor & Francis Group
6000 Broken Sound Parkway NW, Suite 300
Boca Raton, FL 33487-2742

© 2010 by Taylor and Francis Group, LLC
Chapman & Hall/CRC is an imprint of Taylor & Francis Group, an Informa business

No claim to original U.S. Government works

Printed in the United States of America on acid-free paper
10 9 8 7 6 5 4 3 2 1

International Standard Book Number: 978-1-4200-6980-8 (Hardback)

This book contains information obtained from authentic and highly regarded sources. Reasonable efforts have been made to publish reliable data and information, but the author and publisher cannot assume responsibility for the validity of all materials or the consequences of their use. The authors and publishers have attempted to trace the copyright holders of all material reproduced in this publication and apologize to copyright holders if permission to publish in this form has not been obtained. If any copyright material has not been acknowledged please write and let us know so we may rectify in any future reprint.

Except as permitted under U.S. Copyright Law, no part of this book may be reprinted, reproduced, transmitted, or utilized in any form by any electronic, mechanical, or other means, now known or hereafter invented, including photocopying, microfilming, and recording, or in any information storage or retrieval system, without written permission from the publishers.

For permission to photocopy or use material electronically from this work, please access www.copyright.com (http://www.copyright.com/) or contact the Copyright Clearance Center, Inc. (CCC), 222 Rosewood Drive, Danvers, MA 01923, 978-750-8400. CCC is a not-for-profit organization that provides licenses and registration for a variety of users. For organizations that have been granted a photocopy license by the CCC, a separate system of payment has been arranged.

Trademark Notice: Product or corporate names may be trademarks or registered trademarks, and are used only for identification and explanation without intent to infringe.

Library of Congress Cataloging-in-Publication Data

Scientific data management : challenges, technology, and deployment / editors, Arie Shoshani and Doron Rotem.
    p. cm. -- (Chapman & Hall/CRC computational science series)
Includes bibliographical references and index.
ISBN 978-1-4200-6980-8 (alk. paper)
    1. Science--Data processing. 2. Database management. 3. Supercomputers. 4. Computer simulation. I. Shoshani, Arie. II. Rotem, Doron.

Q183.9.S33 2009
502.85--dc22   2009018535

Visit the Taylor & Francis Web site at
http://www.taylorandfrancis.com

and the CRC Press Web site at
http://www.crcpress.com

Contents

List of Figures ix

List of Tables xvii

Acknowledgments xix

Contributors xxi

Introduction xxvii

I Storage Technology and Efficient Storage Access

1 Storage Technology 3
   Jason Hick and John Shalf

2 Parallel Data Storage and Access 35
   Robert Ross, Alok Choudhary, Garth Gibson, and Wei-keng Liao

3 Dynamic Storage Management 73
   *Arie Shoshani, Flavia Donno, Junmin Gu, Jason Hick,
   Maarten Litmaath, and Alex Sim*

II Data Transfer and Scheduling

4 Coordination of Access to Large-Scale Datasets
   in Distributed Environments 115
   Tevfik Kosar, Andrei Hutanu, Jon McLaren, and Douglas Thain

5 High-Throughput Data Movement 151
   *Scott Klasky, Hasan Abbasi, Viraj Bhat, Ciprian Docan,
   Steve Hodson, Chen Jin, Jay Lofstead, Manish Parashar,
   Karsten Schwan, and Matthew Wolf*

III Specialized Retrieval Techniques and Database Systems

6 Accelerating Queries on Very Large Datasets 183
   Ekow Otoo and Kesheng Wu

7 Emerging Database Systems in Support of Scientific Data 235
   *Per Svensson, Peter Boncz, Milena Ivanova, Martin Kersten,
   Niels Nes, and Doron Rotem*

IV Data Analysis, Integration, and Visualization Methods

8 Scientific Data Analysis 281
   *Chandrika Kamath, Nikil Wale, George Karypis, Gaurav Pandey,
   Vipin Kumar, Krishna Rajan, Nagiza F. Samatova, Paul Breimyer,
   Guruprasad Kora, Chongle Pan, and Srikanth Yoginath*

9 Scientific Data Management Challenges in High-Performance
   Visual Data Analysis 325
   *E. Wes Bethel, Prabhat, Hank Childs, Ajith Mascarenhas,
   and Valerio Pascucci*

10 Interoperability and Data Integration in the Geosciences 369
    Michael Gertz, Carlos Rueda, and Jianting Zhang

11 Analyzing Data Streams in Scientific Applications 399
    *Tore Risch, Samuel Madden, Hari Balakrishan, Lewis Girod,
    Ryan Newton, Milena Ivanova, Erik Zeitler, Johannes Gehrke,
    Biswanath Panda, and Mirek Riedewald*

V Scientific Process Management

12 Metadata and Provenance Management 433
    *Ewa Deelman, Bruce Berriman, Ann Chervenak, Oscar Corcho,
    Paul Groth, and Luc Moreau*

13 Scientific Process Automation and Workflow Management..........467
Bertram Ludäscher, Ilkay Altintas, Shawn Bowers, Julian Cummings, Terence Critchlow, Ewa Deelman, David De Roure, Juliana Freire, Carole Goble, Matthew Jones, Scott Klasky, Timothy McPhillips, Norbert Podhorszki, Claudio Silva, Ian Taylor, and Mladen Vouk

Conclusions and Future Outlook...................................509
Arie Shoshani and Doron Rotem

Index..515

List of Figures

1.1 Data recording onto the ferromagnetic medium and recovery of the information by the read head. 6

1.2 Longitudinal recording vs. perpendicular recording. Perpendicular recording overcomes the challenges of superparamagnetism. 9

1.3 The evolution of area density and price for storage devices (Courtesy of Sun Microsystems). 10

1.4 Graphs showing the performance of disk devices as impacted by the alignment of data transfers relative to the native block boundaries. 14

1.5 Schematic diagrams of emerging new storage technologies: phase-change memory and the MEMS-based Millipede storage system. 29

2.1 In a parallel storage system data is distributed across multiple I/O Servers. Access is via software over an interconnection network. 36

2.2 An example of GPFS deployment. Clients communicate through a virtual shared disk component to access storage attached to I/O servers. 42

2.3 Panasas storage clusters are built of StorageBlades, metadata-managing DirectorBlades; and redundant power, cooling and networking. 44

2.4 Panasas storage uses the DirectFlow file system to enable direct and independent parallel access. 45

2.5 Declustered object-based RAID in Panasas storage, randomizing placement so that all disks assist in every reconstruction. 46

2.6 A diagram of the Blue Gene/P hardware in the Argonne Leadership Computing Facility (ALCF). 51

ix

List of Figures

2.7 The software components in the I/O path on the BG/P. 52

2.8 Los Alamos secure petascale infrastructure diagram with Roadrunner. 54

2.9 Independent I/O vs. collective I/O. Collective I/O opens up opportunities to coordinate storage access, and achieve higher performance. 59

2.10 Comparison of data sheet annual failure rates and the observed annual replacement rates of disks in the field. 64

2.11 Number of disk drives and the number of concurrent reconstructions in the largest of future systems. 65

3.1 The difference in architecture between SRMs and SRB. 76

3.2 SRM interface a to a mass storage system: embedded vs. external implementations. 79

3.3 An architectural diagram of a modular design of an SRM. 80

3.4 Interoperability of SRM v2.2 implementations. 83

3.5 File streaming: stage, use, release, stage next. 86

3.6 Sequence of operations between two cooperating SRMs to support robust file replication. 106

3.7 The Earth System Grid use of SRMs for file streaming to clients. 107

4.1 Storage space management: different techniques. 122

4.2 Storage space management: stork vs. traditional scheduler. 123

4.3 Theoretical TCP behavior on a link with no concurrent traffic. 126

4.4 Whole file transfer vs. remote input and output. 132

4.5 Using Parrot and Chirp for remote I/O. 134

4.6 HARC architecture. NRM = network resource manager. 138

4.7 The EnLIGHTened testbed: (a) geographically; (b) as seen by HARC NRM. 140

List of Figures

5.1	IOgraph example.	161
5.2	DataTap performance, on the left from the Cray XT3, and on the right from Infiniband.	165
5.3	Architectural overview of DART.	167
5.4	Data movement services using IQ-Path.	171
5.5	An autonomic service in Accord.	173
5.6	Self-management using model-based control in Accord.	173
5.7	Overview of the two-level cooperative QoS management strategy.	174
6.1	A taxonomy of indexed access methods.	192
6.2	An example of a single-attribute index.	194
6.3	Illustrative tree-structured index by ISAM organization.	195
6.4	A collection of X-axis parallel line segments.	197
6.5	Shaded region representing a stabbing query.	197
6.6	Shaded region representing an intersection query.	198
6.7	A sample multi-attribute data.	199
6.8	Multi-attribute data in 2D space.	199
6.9	K-D-tree indexing of 2-D dataset.	201
6.10	R-tree indexing of bounded rectilinear 2D dataset.	202
6.11	Examples of two-dimensional space curves.	204
6.12	Hybrid indexing by space-filling curves of 2D dataset.	206
6.13	Spherical indexing with HTM.	207
6.14	Spherical indexing with quaternary/hierarchical triangular mesh.	208
6.15	An illustration of the basic bitmap index.	209

6.16	The effects of compression on query response time.	212
6.17	The query response time of five different bitmap encoding methods with WAH compression.	214
6.18	Time needed to process range queries.	217
6.19	Time needed to answer count queries from the Set Query Benchmark with 100 million records.	223
6.20	Using Grid Collector can significantly speed up analysis of STAR data.	225
7.1	Horizontal vs. vertical organization of tabular (relational) data.	237
7.2	MonetDB architecture: the front end translates queries into BAT Algebra expressions; the back end executes the BAT plan.	258
8.1	Scientific data analysis: an iterative and interactive process.	283
8.2	Performance of indirect similarity measures (MG) as compared to similarity searching using the Tanimoto coefficient.	290
8.3	Growth of sequences and annotations on a log scale using numbers from GenBank and SwissProt.	292
8.4	Comparison of performance of various transformed networks and the input networks.	298
8.5	A three-dimensional PCA plot of a multivariate database of high-temperature superconductors.	301
8.6	Interpreting results by comparing the scoring plot and the loading plot.	302
8.7	Partial least squares (PLS) prediction with lattice constants, space groups, and secondary descriptors.	305
8.8	Evolution of parallel computing with R.	307
8.9	The architecture of pR in use.	308
8.10	Data processing pipeline in mass spectrometry proteomics for proteome quantification.	309

List of Figures

8.11	Scalability of processing task-parallel jobs in ProRata.	310
8.12	Key data analysis steps in ProRata.	311
8.13	Peptide abundance ratio and signal-to-noise ratio estimation via PCA.	313
8.14	Protein abundance ratio and confidence interval estimation via profile likelihood algorithm.	314
9.1	Three partitionings of a visualization pipeline consisting of four stages: data I/O, computing an isosurface, rendering isosurface triangles, and image display.	331
9.2	Illustration of domains that do not have to be read from a disk when creating a particular slice.	332
9.3	Hierarchical Z-ordered data layout and topological analysis components highlighted in the context of a typical visualization and analysis pipeline.	337
9.4	The first five levels of resolution of 2D and 3D Lebesgue's space-filling curves.	340
9.5	The nine levels of resolution of the binary tree hierarchy defined by the 2D space-filling curve applied on a 16×16 rectilinear grid.	341
9.6	Diagram of the algorithm for index remapping from Z-order to the hierarchical out-of-core binary tree order. Example of the sequence of ship operations necessary to remap an index.	343
9.7	Data layout obtained for a 2D matrix reorganized using the hierarchical index.	343
9.8	Comparison of static analysis and real performance in different conditions.	345
9.9	Architectural layout of HDF5_FastQuery.	354
9.10	H5part data loaded into VisIt for comparative analysis of multiple variables at different simulation timesteps.	355
9.11	An example of visual data analysis of an H5Part dataset produced by a particle accelerator modeling code, called ROOT.	356

List of Figures

9.12 Parallel coordinates of particles in a laser Wakefield accelerator simulation using HDF5 FastQuery. 357

9.13 A visualization of flames in a high-fidelity simulation of methane-air jet. 360

9.14 An example of forensic network traffic analysis by examining histograms of suspicious traffic activity at varying temporal resolutions. 361

9.15 Two- and three-dimensional histograms are the building blocks for visual data exploration of network traffic analysis. 362

10.1 Examples of object-based (left) and field-based (right) geospatial data representation. 373

10.2 Illustration of an overlay of themes in a GIS. Geo-referenced and aligned layers include both vector data and field-based data. . . 377

10.3 Open geospatial consortium (OGC) services, operations, and example calls. 382

10.4 Sensor observation service (SOS) sequence diagram for getting metadata and observations, including access to streaming data. . . . 385

10.5 Schematic example of a sensor observation service (SOS) capabilities document. 385

10.6 Conceptual architecture for geospatial data integration and interoperability. 388

10.7 Main stages in the integration of sensor and model-generated data streams. 390

10.8 Schematic SensorML document describing measured ETo. 390

10.9 Example of an O&M observation response. 391

11.1 A leak shows up as additional energy in characteristic frequency bands. 404

11.2 Marmot call detection workflow. 407

11.3 Equivalent WaveScript subquery. 407

List of Figures

11.4 The generic dataflow distribution template PCC for partitioning parallelism of expensive stream query functions. 411

11.5 (a) Central strategy, (b) Round Robin window distribute strategy, (c) FFT-dependent window split strategy. 412

11.6 Parallel window split strategy with tree partitioning of degree four. 413

11.7 FFT performance for parallelism of degree four with various distribution templates. 413

11.8 FFT speedup for parallelism of degree four with different distribution templates. 414

11.9 Stream data flow in the LOFAR environment. 416

11.10 Parallel CQ execution in SCSQ. Wide arrows indicate data streams. 416

11.11 Simulation workflow. 418

11.12 ISAT algorithm. 420

11.13 Updating a local model. 421

11.14 Binary Tree. 424

12.1 The data lifecycle. 434

12.2 An example of image metadata in astronomy. 448

12.3 Earth System Grid project's interface for metadata queries. 450

12.4 Map of sea surface temperature taken February 18, 2009, 10:42 p.m. GMT. 452

12.5 Provenance graph for the provenance challenge workflow. 456

13.1 CPES fusion simulation monitoring workflow (in Kepler). 476

13.2 Kepler supports execution of workflows on remote peer nodes and remote clusters. 485

13.3　An overview of the Pegasus/DAGMan workflow management system. 487

13.4　In VisTrails, workflow evolution provenance is displayed as a history tree, each node representing a workflow that generates a visualization. 492

13.5　Overview of the SDM provenance framework. 493

13.6　SDM dashboard with on-the-fly visualization. 494

List of Tables

5.1 Comparison of GTC run times on the ORNL Cray XT3 development machine for two input sizes using different data output mechanisms. 166

7.1 Elapsed times in seconds for two types of queries against a "small" (1.5 GB) and a "large" (150 GB) dataset. 265

7.2 Main differences between requirements for commercial DBMS and SciDB. 269

8.1 SAR performance of different descriptors. 288

8.2 Types of biological data and analysis techniques used for predicting protein function. 294

9.1 Performance results of visualization processing—slicing, and isocontouring—with and without metadata optimization. 333

9.2 Structure of the hierarchical indexing scheme for a binary tree combined with the order defined by the Lebesgue space-filling curve. 341

9.3 Sample H5Part code to write multiple fields from a time-varying simulation to a single file. 350

9.4 Contents of the H5Part file generated in Table 9.3. 351

Acknowledgments

The editors' work on this book was supported in part by the Director, Office of Science, Office of Advanced Scientific Computing Research (OASCR), of the U.S. Department of Energy, under Contract No. DE-AC02-05CH11231 to Lawrence Berkeley National Laboratory. In particular, the editors wish to acknowledge the support of the Scientific Data Management (SDM) Center by the OASCR. The SDM center is one of several centers and institutions supported under the Scientific Discovery through Advanced Computing (SciDAC) program.* The SDM center was instrumental in exposing its members to real scientific problems, and offered opportunities to explore solutions to such problems. The OASCR supports several large supercomputer facilities that enable very large simulation capabilities; such simulations provide large-scale data management problems that motivated some parts of this book. The editors are grateful to members of the SDM center for several chapters in this book. The editors also wish to acknowledge the contributions of authors from many organizations in the United States and Europe who led and wrote many of the chapters and sections in this book. Many of the authors are supported by their own national research organizations, which are acknowledged in the individual chapters. We are most grateful to all who took time from their busy schedules to contribute to this book. Their conscientious cooperation is greatly appreciated.

---

* http://scidac.org. Accessed on July 20, 2009.

Contributors

Hasan Abbasi
Georgia Institute of Technology
Atlanta, Georgia

Ilkay Altintas
San Diego Supercomputer Center
San Diego, Califonia

Hari Balakrishan
Massachusetts Institute of
  Technology
Cambridge, Massachusetts

Bruce Berriman
California Institute of Technology
Pasadena, California

E. Wes Bethel
High Performance Computing
  Research Department
Lawrence Berkeley National
  Laboratory
Berkeley, California

Viraj Bhat
Rutgers University
New Brunswick, New Jersey

Peter Boncz
Centrum Wiskunde & Informatica
  (CWI)
The National Research Institute for
  Mathematics and Computer
  Science
Amsterdam, the Netherlands

Shawn Bowers
University of California
Davis, California

Paul Breimyer
Oak Ridge National Laboratory
Oak Ridge, Tennessee
and
North Carolina State University
Raleigh, North Carolina

Ann Chervenak
Information Science Institute
University of Southern California
Marina del Rey, California

Hank Childs
High Performance Computing
  Research Department
Lawrence Berkeley National
  Laboratory
Berkeley, California

Alok Choudhary
Northwestern University
Evanston, Illinois

Oscar Corcho
Universidad Politécnica de Madrid
Madrid, Spain

Terence Critchlow
Pacific Northwest National
  Laboratory
Richland, Washington

Julian Cummings
California Institute of Technology
Pasadena, California

David De Roure
University of Southampton
Southampton, United Kingdom

Ewa Deelman
Information Science Institute
University of Southern California
Marina del Rey, California

Ciprian Docan
Rutgers University
New Brunswick, New Jersey

Flavia Donno
European Organization for Nuclear
  Research (CERN)
Geneva, Switzerland

Juliana Freire
University of Utah
Salt Lake City, Utah

Johannes Gehrke
Cornell University
Ithaca, New York

Michael Gertz
Institute of Computer Science
University of Heidelberg
Heidelberg, Germany

Garth Gibson
Carnegie Mellon University
Pittsburgh, Pennsylvania

Lewis Girod
Massachusetts Institute of
  Technology
Cambridge, Massachusetts

Carole Goble
The University of Manchester
Manchester, United Kingdom

Paul Groth
Information Science Institute
University of Southern California
Marina del Rey, California

Junmin Gu
Lawrence Berkeley National
  Laboratory
Berkeley, California

Jason Hick
National Energy Research
  Supercomputing Center
Lawrence Berkeley National
  Laboratory
Berkeley, California

Steve Hodson
Oak Ridge National Laboratory
Oak Ridge, Tennessee

Andrei Hutanu
Louisiana State University
Baton Rouge, Louisiana

Milena Ivanova
Centrum Wiskunde & Informatica
  (CWI)
The National Research Institute for
  Mathematics and Computer
  Science
Amsterdam, the Netherlands

Chen Jin
Oak Ridge National Laboratory
Oak Ridge, Tennessee

Matthew Jones
University of California
Santa Barbara, California

Chandrika Kamath
Lawrence Livermore National
  Laboratory
Livermore, California

George Karypis
University of Minnesota
St. Paul, Minnesota

Martin Kersten
Centrum Wiskunde & Informatica
  (CWI)
The National Research Institute for
  Mathematics and Computer
  Science
Amsterdam, the Netherlands

Scott Klasky
Oak Ridge National Laboratory
Oak Ridge, Tennessee

Guruprasad Kora
Oak Ridge National Laboratory
Oak Ridge, Tennessee
and
North Carolina State University
Raleigh, North Carolina

Tevfik Kosar
Louisiana State University
Baton Rouge, Louisiana

Vipin Kumar
University of Minnesota
Minneapolis, Minnesota

Wei-keng Liao
Northwestern University
Evanston, Illinois

Maarten Litmaath
European Organization for Nuclear
  Research (CERN)
Geneva, Switzerland

Jay Lofstead
Georgia Institute of Technology
Atlanta, Georgia

Bertram Ludäscher
University of California
Davis, California

Samuel Madden
Massachusetts Institute of
  Technology
Cambridge, Massachusetts

Ajith Mascarenhas
Center for Applied Scientific
  Computing
Lawrence Livermore National
  Laboratory
Livermore, California

Jon McLaren
Louisiana State University
Baton Rouge, Louisiana

Timothy McPhillips
University of California
Davis, California

Luc Moreau
University of Southampton
Southampton, United Kingdom

Niels Nes
Centrum Wiskunde & Informatica
  (CWI)
The National Research Institute for
  Mathematics and Computer
  Science
Amsterdam, the Netherlands

Ryan Newton
Massachusetts Institute of
  Technology
Cambridge, Massachusetts

Ekow Otoo
Lawrence Berkeley National
  Laboratory
Berkeley, California

Chongle Pan
Oak Ridge National Laboratory
Oak Ridge, Tennessee
and
North Carolina State University
Raleigh, North Carolina

Biswanath Panda
Cornell University
Ithaca, New York

Gaurav Pandey
University of Minnesota
St. Paul, Minnesota

Manish Parashar
Rutgers University
New Brunswick, New Jersey

Valerio Pascucci
Scientific Computing and Imaging
   Institute
School of Computing
University of Utah
Salt Lake City, Utah

Norbert Podhorszki
Oak Ridge National Laboratory
Oak Ridge, Tennessee

Prabhat
High Performance Computing
   Research Department
Lawrence Berkeley National
   Laboratory
Berkeley, California

Krishna Rajan
Iowa State University
Ames, Iowa

Mirek Riedewald
Cornell University
Ithaca, New York

Tore Risch
Uppsala University
Uppsala, Sweden

Robert Ross
Argonne National Laboratory
Argonne, Illinois

Doron Rotem
Lawrence Berkeley National
   Laboratory
Berkeley, California

Carlos Rueda
Monterey Bay Aquarium Research
   Institute
Moss Landing, California

Nagiza F. Samatova
Oak Ridge National Laboratory
Oak Ridge, Tennessee
and
North Carolina State University
Raleigh, North Carolina

Karsten Schwan
Georgia Institute of Technology
Atlanta, Georgia

John Shalf
National Energy Research
   Supercomputing Center
Lawrence Berkeley National
   Laboratory
Berkeley, California

Arie Shoshani
Lawrence Berkeley National
   Laboratory
Berkeley, California

Claudio Silva
University of Utah
Salt Lake City, Utah

Alex Sim
Lawrence Berkeley National
 Laboratory
Berkeley, California

Per Svensson
Swedish Defence Research Agency
Stockholm, Sweden

Ian Taylor
Cardiff University
Cardiff, United Kingdom

Douglas Thain
University of Notre Dame
Notre Dame, Indiana

Mladen Vouk
North Carolina State University
Raleigh, North Carolina

Nikil Wale
University of Minnesota
St. Paul, Minnesota

Matthew Wolf
Georgia Institute of Technology
Atlanta, Georgia

Kesheng Wu
Lawrence Berkeley National
 Laboratory
University of California
Berkeley, California

Srikanth Yoginath
Oak Ridge National Laboratory
Oak Ridge, Tennessee
and
North Carolina State University
Raleigh, North Carolina

Erik Zeitler
Uppsala University
Uppsala, Sweden

Jianting Zhang
Department of Computer Science
City College of New York
New York City, New York

Introduction

Motivation and Challenges

The tremendous technical advances in computing power and sensory devices have brought about a deluge of data being generated by scientific experiments and simulations. It is common today to have a single simulation on a supercomputer generate terabytes of data, and for experiments to collect multiple petabytes of data. Examples of such data explosions exist in nearly all fields of science in which experiments and large-scale simulations are conducted, including high-energy physics, material science, astrophysics, cosmology, climate modeling, ecology, and biology, to name a few. The data collected represent many challenges in the data generation and collection phase; in the data postprocessing phase; and in the analysis, data mining, and data interpretation phase. These large data volumes represent more detailed and accurate physical, chemical, and biological phenomena, and inevitably lead to new discoveries. Managing, storing, searching, and analyzing such large volumes of data are complex and difficult problems that require innovative and highly efficient data management technology. The data volumes are naturally expected to grow. Already many scientific domains have or are predicting to have in the near future data volumes measured in tens of petabytes, and some are predicting exabyte volumes in less than ten years.

In addition to dealing with extremely large data volumes, scientific data presents three other challenges: multiscale data, diversity of data, and multiple data models and formats. Multiscale data refers to data generated at different scales, such as the modeling of materials at the atomic level, the molecular level, and higher-level composites. Similarly, biological processes can be modeled at the DNA sequence level, the molecular level, or as protein complexes. Having to analyze multiscale data requires special techniques of matching data from multiple levels. Diversity of data arises in scientific projects that involve multiple diverse domain sciences, such as ecology. To understand the ecological effects, it is necessary to model and integrate processes in various domains, including earth and ocean science, climate, biology of plants and animals, behavioral science, economics, and possibly others. Similarly, to model climate, it is necessary to have coupled models of earth, oceans, ice sheets, clouds, vegetation, aerosols, and so forth. Depending on the application domains, different data models and data formats are used. Many application domains use basic regular grid data structures to represent space and time phenomena. For such applications, data formats, such as NetCDF, are used to hold the data. Furthermore, the metadata about the dimensions, their granularity, variables, and their units are stored as headers in such formatted files. Specialized libraries are used by applications to read desired subsets from such formatted files. However, regular grids may not be practical when some areas of the simulation require refined modeling; for example, turbulent regions in combustion modeling. In such cases, more complex data structures are required, such as adaptive mesh refinement (AMR), where some of the regions are modeled by using multiple levels of increasingly finer granularity. For such applications, more complex hierarchical file formats, such as HDF5, are useful. Other mesh structures that adapt better to the objects being modeled are also used; for example, geodesic or icosahedral data structures for modeling the globe. Coupling or combining data from such diverse data models is a challenge that many application domains face. Finally, comparing simulation data with observed data or experimental data as a way to verify the accuracy of scientific models requires complex interpolation techniques.

This volume, complexity and diversity of data cause scientists great hardship and a waste of productive time to explore their scientific problems. When running the simulation models, scientists need to become experts in the peculiarities of the parallel systems they run on, including the parallel file system used, and how to tune their I/O (Input/Output) operations so they do not slow down the computations. In order to access large volumes of data, they typically have to write scripts that manage the data transfers and trace what files are lost because of transient network failures. In order to perform analysis on such large volumes of data, they have to acquire or access large computing facilities and install appropriate software. As a case in point, consider a combustion, astrophysics, or fusion scientist who wishes to run a simulation on a supercomputer. To begin with, the scientist needs to schedule time on the computer based on an allocated quota. Simulation runs need to be monitored in real time to verify that they are progressing correctly. To achieve this, the data generated at each time step has to be summarized or reduced, and perhaps some graphs generated for visualization. Following this step the scientist will want to analyze the data. This requires keeping track of what data was generated and where it is stored or archived, and using effective methods to download the desired subset of the data. Even in communities that are relatively well coordinated and organized because they have developed common coupled models, such as the climate modeling community, many problems of managing, keeping track of, and disseminating data exist. In this community there are a few climate models, such as CCSM[*] and IPCC,[†] that are run on supercomputers in order to share the resulting data from such model runs with

---

[*] http://www.ccsm.ucar.edu/. Accessed on July 20, 2009.
[†] http://www.ipcc.ch/. Accessed on July 20, 2009.

thousands of researchers. How would researchers find what was modeled? How would they select a subset of interest to be downloaded to their sites? What will they use to download the data? The purpose of this book is to address these questions in depth, by describing the problem areas and techniques that address them.

This book presents many important state-of-the-art developments in scientific data management that are intended to answer the challenges listed above. It is intended to serve as a reference book for research scientists and developers of scientific software who need to use cutting-edge technologies for managing and analyzing the huge amount and diversity of data arising from their scientific work. The book can be used as a textbook for graduate and upper-level courses on scientific data management. It can also be used for supplementary material in courses dealing with advanced database systems, advanced I/O techniques, data analysis and visualization, data streams management, scientific workflows, and metadata and provenance management.

Organization and Contents

The book is organized in five parts to provide the reader with a comprehensive understanding of five main aspects and techniques of managing data during the scientific exploration process that scientists typically encounter, from the data generation phase to the data analysis phase. One aspect is the efficient access to storage systems, in particular, parallel file systems, to write and read large volumes of data without slowing a simulation, analysis, or visualization processes. These processes are complicated by the fact that scientific data are structured differently for specific application domains and are stored in specialized file formats. A second aspect is the efficient data movement and management of storage spaces. The efficient movement of large data volumes is one of the more time-consuming and error-prone tasks that scientists face. Failures are induced by temporary network and storage systems breakdowns, slow speed of transfer, firewall and security issues, and storage systems running out of space. The book addresses technologies that support the automation of such tasks. A third aspect addresses the problem that stems from the fact that the data is usually stored in the order it is generated, which often is not a good storage organization for subsequent data analysis. For example, combustion or climate model simulation data is generated one time step at a time over all variables, but later analyzed by requesting a few variables over all time steps. Thus, techniques for automatically optimizing the physical organization of data are necessary for fast analysis. A fourth aspect is how to effectively perform complex data analysis and searches over large datasets. Specialized feature discovery and statistical analysis techniques are needed before the data can be understood or visualized. To facilitate efficient analysis

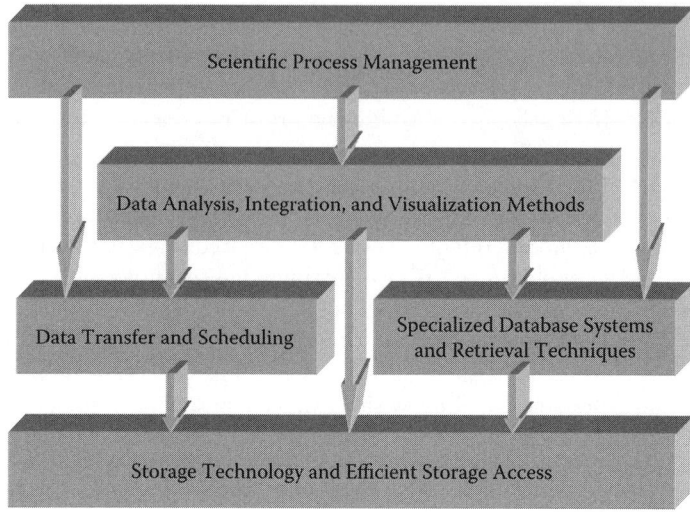

FIGURE I.1 Layered representation of scientific data management technologies.

and visualization, it is necessary to keep track of the location of the datasets, index the data for efficient selection of subsets, and parallelize analysis software. The fifth aspect addressed in this book is the automation of multistep scientific process workflows. Generating the data, collecting and storing the results, data post-processing, and analysis of results is a tedious, fragmented, and error-prone process. The book contains chapters that describe technology for scientific workflow, metadata, and provenance management that ensures executing multiple tasks in sequence or concurrently in a robust, tractable, and recoverable fashion. The five parts of this book cover in detail these aspects of managing scientific data, as well as current technologies to address them, and example applications that use these technologies.

It is convenient to think about the various data management technologies needed and used by scientific applications as organized into four layers, as shown in Figure I.1. This layered view was inspired by work done at the Scientific Data Management Center,[*] supported by the U.S. Department of Energy since 2001. As can be seen in this diagram there are five technology areas, according to which we organized the book parts.

Part I, labeled ***Storage Technology and Efficient Storage Access*** is shown as the bottom (foundation) layer in Figure I.1. It includes three chapters. Chapter 1, *Storage Technology*, describes the evolution of disk technology, redundant arrays of inexpensive disks (RAID) technology, massive arrays of idle disks (MAID), robotic tape technologies, and mass storage systems

---

[*]See http://sdmcenter.lbl.gov. Accessed on July 20, 2009.

(MSS). Chapter 2, entitled *Parallel Data Storage and Access*, describes parallel access to parallel file systems, automating reliability and recoverability, techniques for achieving very fast I/O rates, and uniform access to file systems. Chapter 3, *Dynamic Storage Management*, covers the topics of providing dynamic allocation and reservation of storage space, cleanup of storage for unused replicated data, transparent access to mass storage systems, and interoperability of heterogeneous storage systems.

Two types of technologies are typically built on top of the bottom layer: **Data Transfer and Scheduling** and **Specialized Retrieval Techniques and Database Systems.** Part II, **Data Transfer and Scheduling**, includes two chapters. Chapter 4, *Coordination of Access to Large-Scale Datasets in Distributed Environments,* describes techniques for providing very high wide-area transfer rates, overcoming limitations of firewall security, and setting up and providing transparent authentication/authorization. It also covers techniques for recovery from temporary network and storage system failures. Chapter 5, *High-Throughput Data Movement,* focuses on dynamic storage and transfer of data while they are generated, supporting asynchronous I/O, and dynamic operations on streaming data, such as transposing or summarizing data while they are generated by highly parallel supercomputers.

Part III, **Specialized Retrieval Techniques and Database Systems**, includes two chapters. Chapter 6, *Accelerating Queries on Very Large Datasets,* describes indexing techniques appropriate for scientific datasets and efficient subsetting technologies. Some of the indexing methods described, such as bitmap indexing, take advantage of the fact that scientific data are typically not modified once collected, and by using specialized compression to achieve high indexing efficiency while keeping the index sizes small. Chapter 7, *Emerging Database Systems in Support of Scientific Data*, focuses on techniques of vertical partitioning of datasets, representation of multidimensional data as vertical partitions, and the design of "vertical database systems," such as MonetDB and Vertica. It also reviews recent activities aimed at the development of a scientific database management system referred to as SciDB.

The topic of Part IV is **Data Analysis, Integration, and Visualization Methods**, is represented by the next layer in the diagram of Figure I.1. As can be seen in the diagram, such methods are typically designed to work directly on the storage layer, but can often take advantage of indexing technologies and specialized database systems. Part IV contains four chapters. Chapter 8, *Scientific Data Analysis*, covers techniques for pattern identification, feature extraction, large-scale cluster analysis, and parallel statistical analysis. The chapter contains multiple examples of such techniques applied to different applications domains. Chapter 9, *Scientific Data Management Challenges in High-Performance Visual Data Analysis*, focuses on methods of reducing multi-terabyte data into more compact structures that can reveal patterns and behavior using visualization tools. It describes specialized data structures in support of multi-field and multi-resolution visualization, data models for multidimensional and time-varying data, and parallelization of data structures for real-time visualization of very large datasets. Chapter 10, *Interoperability and Data Integration in the Geosciences,* focuses on integrating geospatial data. This area is an excellent example of having to integrate and couple diverse data formats, to use data exchange tools, to compare simulation and experimental data, to develop standard metadata models and systems, and to federate scientific databases. Chapter 11, *Analyzing Data Streams in Scientific Applications,* focuses on specialized techniques for managing streaming data from sources such as sensors and satellite data. It describes techniques for high-performance, stream-oriented data management, capturing and indexing sensor data, and main-memory real-time query processing of streaming numerical data.

Part V, represented by the top layer in the diagram of Figure I.1, is **Scientific Process Management**. The scientific process often involves multiple steps that require coordination. Unfortunately, such coordination tasks that involve data movement; data transformation; invoking different tools, such as dimensionality reduction; and generating figures, images and movies are usually performed by scientists, causing a waste of time and resources. This part focuses on automating such tasks using scientific workflow tools and automatic collection of provenance information and metadata. As can be seen from Figure I.1, this layer typically uses analysis and visualization technologies from the level below, as well as data transfer and data retrieval technologies from the next lower level. This part includes two chapters. Chapter 12, *Metadata and Provenance Management,* describes newly emerging work on developing schemas for support of metadata, support for ontologies and hierarchical classification structures, metadata collection tools, and support for data provenance (history of data generation). Chapter 13, labeled *Scientific Process Automation and Workflow Management,* describes workflow management and execution tools for the orchestration of complex, multidisciplinary, parallel tasks that involve computation, data movement, data summarization, and data visualization tasks.

Part I

Storage Technology and Efficient Storage Access

Chapter 1

Storage Technology

Jason Hick and John Shalf

National Energy Research Supercomputing Center, Lawrence Berkeley National Laboratory, Berkeley, California

Contents

- 1.1 Introduction 4
- 1.2 Fundamentals of Magnetic Storage 5
  - 1.2.1 Storage Medium 7
  - 1.2.2 Superparamagnetism 7
  - 1.2.3 Magnetoresistive Reading 8
  - 1.2.4 Giant Magnetoresistance (GMR) 8
  - 1.2.5 Perpendicular Recording 8
  - 1.2.6 Future Trends 9
- 1.3 Disk Storage 11
  - 1.3.1 Fundamentals 12
  - 1.3.2 Performance Characteristics 13
  - 1.3.3 Enterprise and Commodity Disk Technology 15
  - 1.3.4 Future Trends 16
- 1.4 Tape Storage 16
  - 1.4.1 Fundamentals 17
  - 1.4.2 Enterprise-Class Tape Technology 18
  - 1.4.3 Commodity Tape Technology 19
  - 1.4.4 Future Trends 20
- 1.5 Optical Storage 20
  - 1.5.1 CDs 20
  - 1.5.2 DVDs 21
  - 1.5.3 Blu-Ray and HD-DVD 21
  - 1.5.4 Future Trends 21
- 1.6 Composite Devices 22
  - 1.6.1 RAID 22
  - 1.6.2 Virtual Tape Libraries (VTLs) 24
  - 1.6.3 Redundant Arrays of Independent Tape (RAIT) 24
  - 1.6.4 Data Integrity 25

1.7 Emerging Technologies 25
    1.7.1 Massive Array of Idle Disks (MAID) 26
    1.7.2 FLASH 26
    1.7.3 MRAM 28
    1.7.4 Phase-Change Technology 28
    1.7.5 Holographic Storage 28
    1.7.6 Direct Molecular Manipulation 30
1.8 Summary and Conclusions 30
Acknowledgments 33
References 33

1.1 Introduction

Computational science is at the dawn of petascale computing capability, with the potential to achieve simulation scale and numerical fidelity at hitherto unattainable levels. However, harnessing such extreme computing power will require an unprecedented degree of parallelism both within the scientific applications and at all levels of the underlying architectural platforms. Power dissipation concerns are also driving High-Performance Computing (HPC) system architectures from the historical trend of geometrically increasing clock rates toward geometrically increasing core counts (multicore),[1] leading to daunting levels of concurrency for future petascale systems. Employing an even larger number of simpler processor cores, operating at a lower clock frequency, is increasingly common as we march toward petaflop-class HPC platforms, but it puts extraordinary stress on the Input/Output (I/O) subsystem implementation.

I/O systems for scientific computing have unique demands that are not seen in other computing environments. These include the need for very high bandwidth and large capacity, which requires thousands of devices working in parallel, and the commensurate fault resilience required to operate a storage system that contains so many components. Chapter 2 will examine the software technology required to aggregate these storage building blocks into reliable, high-performance, large-scale file systems. This chapter will focus on the characteristics of the fundamental building blocks used to construct parallel file systems to meet the needs of scientific applications at scale.

Understanding the unprecedented requirements of these new computing paradigms, in the context of high-end HPC I/O subsystems, is a key step toward making effective petascale computing a reality. The main contribution of this chapter is to quantify these tradeoffs in cost, performance, reliability, power, and density of storage systems by examining the characteristics of the underlying building blocks for high-end I/O technology. This chapter projects the evolution of the technology as constrained by historical trends to understand the effectiveness of various implementations with respect to absolute performance and scalability across a broad range of key scientific domains.

1.2 Fundamentals of Magnetic Storage

Magnetic recording technology was first demonstrated in 1898. Magnetic storage has been closely associated with digital computing since the dawn of the industry. Its performance and cost-effectiveness ensure that it will continue to play an important role in nonvolatile storage for some time to come. The primary factor that consistently drives down the cost of storage media is the ability to double storage density consistently on an annual basis as shown in the storage trends in Figure 1.3. As we reach physical limits of the media, the storage density trends can no longer continue and the market may change dramatically. Until then, magnetic storage will continue to play a leading role in nonvolatile storage technology. This section elucidates the physics underlying the magnetic storage technology and the current challenges to improving storage density.

The underlying media for magnetic recording are ferromagnetic materials. During the write operation, a magnetic field is applied to the recording medium, which aligns the magnetic domains to the orientation of the field. The write head uses an electrical coil to induce the magnetic field in the head to magnetize regions of the medium, and the read head senses the polarity of magnetization in the underlying media using either magnetic induction or the magnetoresistive effect. Figure 1.1(a) shows how the current is applied to the write head to record bits onto the medium and how the read head subsequently recovers the data when the magnetically polarized media induces electrical impulses into the read head as it passes over the magnetically polarized media. The green lines shown under the read head in Figure 1.1(a) show the magnetic flux lines. The current is induced in the read head when the flux lines change direction, thereby making the boundaries between bits detectable.

Figure 1.1(b) shows the hysteresis curve of typical magnetic media, which enables the magnetic flux of the write head (H) to induce a reversible change in the magnetization (M) of the underlying medium. The coercivity of the media refers to the amount of magnetic flux that must be applied to the media to change its magnetization. Higher coercivity media improves the stability and density of the storage (e.g., reducing the instabilities caused by superparamagnetism). The magnetic recording industry employs a number of technologies to improve the recording density of the media. These include improvements to the coercivity of ferromagnetic materials that comprise the media, improved read-head technology (magnetoresistive recording), and improvements to the recording technology (perpendicular recording). In the next

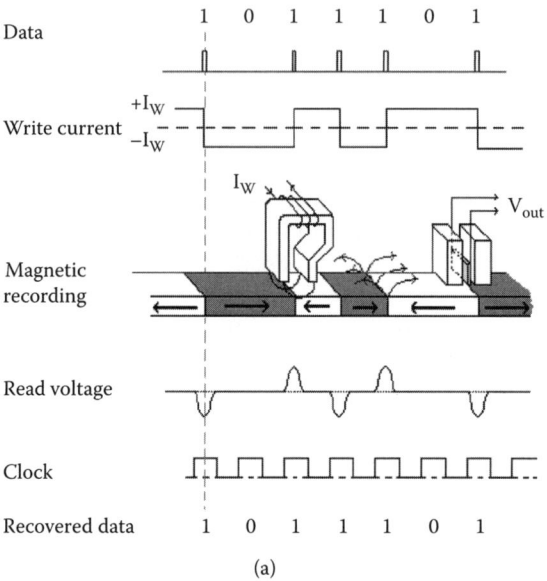

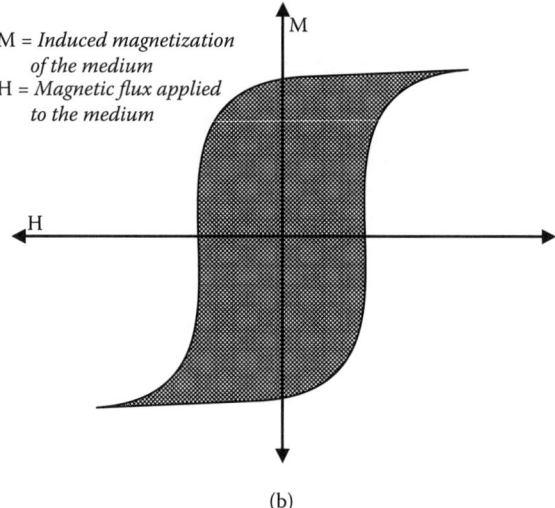

Figure 1.1 The figure (a) shows how data is recorded onto the ferromagnetic medium and how the read head subsequently recovers the information. The green lines shown on the figure are magnetic flux lines. It is the reversal in the magnetic flux lines that the read head detects to determine bit boundaries. The figure (b) shows the typical hysteresis curve of typical magnetic media.

subsections, we will go through a subset of these improvements, describe their inherent physical limitations in terms of bit density, and discuss future approaches to maintaining historical trends in storage density, which will be discussed in Figure 1.3.

1.2.1 Storage Medium

The performance of this technology is strongly dependent on the material properties of the recording medium. The medium is composed of ferromagnetic particles suspended in a binder matrix (typically a polymer or other flexible plastic material). The ferromagnetic particles contain domains that align to the magnetic field if sufficient flux is applied. Coercivity is a measure of a material's resistance to magnetic reversal. As mentioned above, materials with a high coercivity also are better able to hold their magnetization and also tend to maintain a higher magnetic flux, which improves the signal-to-noise ratio (SNR) when reading their magnetization. The first digital storage devices employed iron oxide particles, but chromium dioxide particles and thin-film alloy materials have substantially improved the coercivity of the medium. More recently, thin-film layering of antiferromagnetically coupled (AFM) layers have enabled order-of-magnitude improvements in coercivity over the best available homogeneous thin films.

Improvements in storage density are achieved by reducing the thickness of the magnetic medium to maintain a high aspect ratio of the recorded bits, but this is done at the expense of the SNR of the medium. Even if the lowered SNR can be compensated by using more sensitive read-head technology, the superparamagnetic effect (explained below) sets a lower limit on bit size for conventional longitudinal recording as bit stability becomes increasingly subject to changes due to random temperature fluctuation.

1.2.2 Superparamagnetism

Magnetic media are composed of discrete crystaline grains of ferromagnetic material embedded in some form of binder matrix (polymer or similar materials). The magnetic bit value is determined by the magnetic polarity of a population of about 1,000 of these magnetic "grains" in order to maintain a reasonable SNR. Continued improvements in areal density require proportional decreases in magnetic grain size for the underlying magnetic material. However, as these grains get smaller, they are more susceptible to random changes in their magnetic polarity due to random thermal fluctuations in the medium, in a behavior known as the superparamagnetic effect.[2] Improvements in the coercivity of storage media has kept the superparamagnetic effect at bay since the mid 1990s, but many believe these improvements will reach their limit at 150–200 Gb/mm^2, forcing a move toward alternative approaches to recording, including magneto-optical, heat-assisted recording to reduce coercivity for writes, and perpendicular recording technology.

1.2.3 Magnetoresistive Reading

Another aspect that can limit the density of recorded data is the sensitivity, size, and consequent flying height of the read heads. In early devices, the read operation employed an inductive head that depended on the magnetized medium inducing a current in coiled wire on the read head as the medium moves under the read head. Magnetoresistive materials change their resistance in response to the magnetization of the substrate. Magnetoresistive read heads offered order of magnitude improvements in sensitivity compared with inductive heads, enabling increased recording densities. The performance of a magnetoresistive head is independent of the rate of movement of the underlying medium, which is particularly useful for tape heads where the rate of the medium is very low or subject to velocity changes. The magnetoresistive effect cannot be used to induce magnetization in the medium, so the conventional inductive heads must remain, but magnetoresistive heads are substantially smaller than the typical inductive heads and can often be embedded within the structure of a typical thin-film recording head. The magnetoresistive heads are typically narrower than the recorded track in the medium, which can dramatically reduce the cross-talk between neighboring tracks during the read operation.

1.2.4 Giant Magnetoresistance (GMR)

The "Giant Magneto-Resistive" (GMR) effect was discovered in the 1980s by Peter Gruenberg and Albert Fert, and led to their 2007 Nobel Prize in physics. In their research on materials that are comprised of alternating layers of magnetic and nonmagnetic materials, they observed that sensitivity to magnetization improved between 6% and 50% in comparison to conventional magnetoresistive materials. Further refinement of the GMR technology led to read heads that were an order of magnitude more sensitive than its conventional magnetoresistive approach. The improvements in read sensitivity enabled further reductions in bit sizes and enabled continued annual doubling of storage densities from 1997 to approximately 2005.

The GMR effect also gave birth to a new form of magnetism-based computer logic referred to as "spintronics,"[3] that refers to the underlying mechanism of the effect, which is spin-dependent scattering of electrons. The GMR effect is also being used for new solid-state devices such as MRAM, which we will discuss in the Emerging Technologies section.

1.2.5 Perpendicular Recording

Superparamagnetic effects limit recording density to 150 gigabits/mm^2 for conventional longitudinal recording where the polarization is parallel to the magnetic storage medium. In such an arrangement, the areas between oppositely polarized regions tend to demagnetize their neighbors. As shown in Figure 1.2, perpendicular recording technology induces the magnetic field

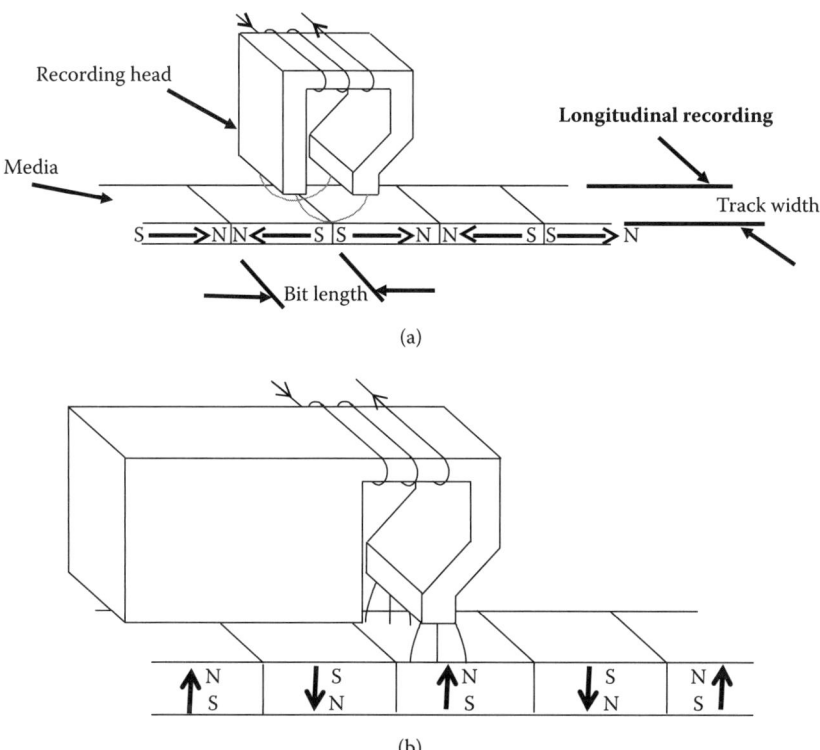

Figure 1.2 Longitudinal recording devices (a) currently dominate the market. However, the superparamagnetic effect limits areal densities of this technique to 150 Gbits/mm^2. Perpendicular recording (b) promises to support continued scaling of areal density up to 1 Tbit/mm^2 because neighboring domains complement rather than counteract each other's polarity. This improves the stability of the recording and thereby overcomes the challenges of superparamagnetism.

perpendicular to the storage medium where the demagnetizing fields tend to have less interaction between neighboring bits. Perpendicular recording is expected to support continued improvements to storage densities that top out at about 1 terabit/mm^2. The first commercial products incorporating vertical recording technology were introduced in 2005 with bit densities of 170 gigabits/mm^2, with density doubling approximately every two years.

1.2.6 Future Trends

Figure 1.3 summarizes trends in storage density and cost per bit. The cost trends depicted in Figure 1.3(b) are strongly dependent on the ability to improve bit densities in Figure 1.3(a) at exponential rates. When a storage

10 Scientific Data Management

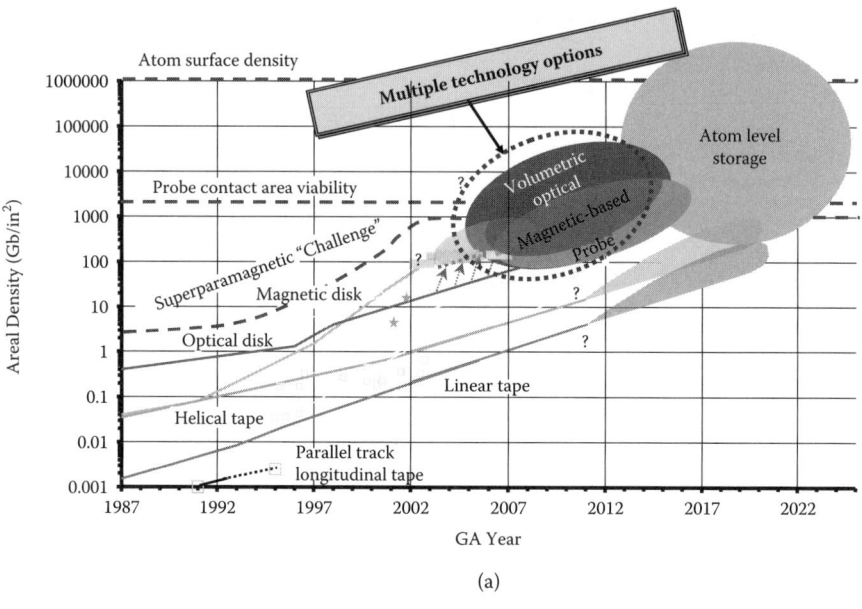

(a)

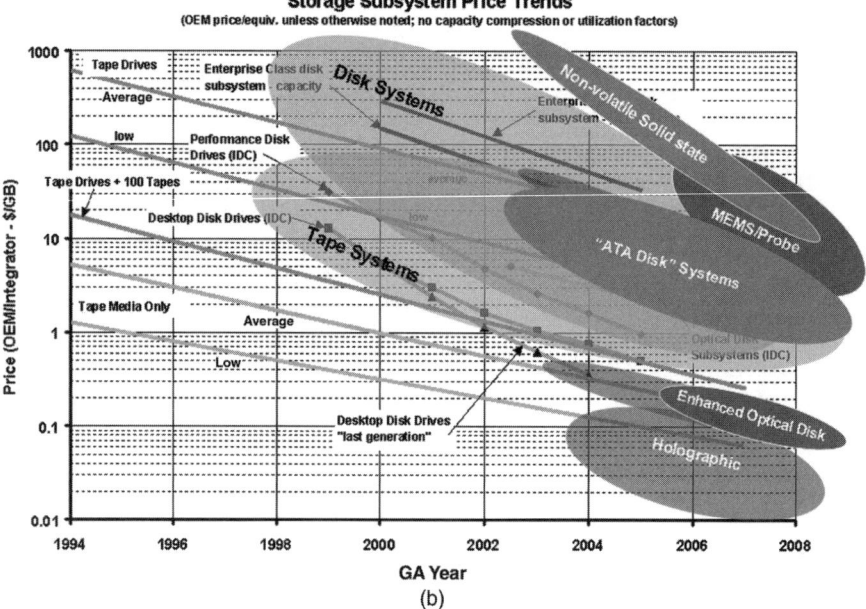

(b)

Figure 1.3 (See color insert following page 224.) This figure summarizes the evolution of area density and price for storage devices (Courtesy of Sun Microsystems). Item (a) summarizes the trends in the areal density for different classes of storage media. Graph (b) shows the cost per bit for different classes of storage media.

media reaches the limits imposed by physics, it opens opportunities for alternative technology to take over. For example, magnetic disk storage density was eclipsed by optical storage in the late 1980s, but was able to improve performance at a far faster pace and achieve higher bit-densities than competing optical solutions. However, as we edge closer to the limits of perpendicular recording technology (1 terabit/mm^2), multiple technology options appear to compete well. At atomic scales, existing approaches to magnetic storage technology, including perpendicular recording, are clearly not viable. However, spintronic devices are able to push past the superparamagnetic limit, and may yet reset the clock again for magnetic technology.

Notice that tape is far below the storage density curve traced by the disk technology in Figure 1.3. Helical tape technology will not reach the superparamagnetic limits to areal storage density until nearly 2020 if densities improve at historical rates. The cost per bit of tape storage remains far superior and differentiated from the cost of low-end disk storage in Figure 1.3, so the tape systems are unlikely to jump to the kinds of storage densities seen in the disk storage systems because there is currently little economic incentive to do so. It also indicates that if there is any market pressure on tape storage, the technology has a lot of margin to improve storage density relative to the physical limits imposed by the underlying recording media were it to be challenged by holographic or enhanced optical storage media in the future.

In the area of desktop disk storage, the storage trends have followed a very consistent slope on the exponential graph except for the nonvolatile solid state storage, such as FLASH. The consumer electronics market applications for FLASH storage have created such rapid increases in manufacturing volume, that the cost per bit has been decreasing far faster than any of the other technology options. The costs of such devices are now within striking distance of the upper end of the consumer grade advanced technology attachment (ATA) disk systems, and may well be a direct competitor to that market segment in the coming years if this accelerated trend can continue. If FLASH becomes competitive with mechanical disks, it will dramatically change the landscape of storage systems—affecting fundamental design decisions for the parallel file systems described in Chapter 2. However, much work still needs to be done to characterize the failure modes of nonvolatile storage solutions to understand how they will behave for large-scale, high-bandwidth scientific storage subsystems before progress can be made in this area.

1.3 Disk Storage

The very first rotating disk storage device was the IBM 350 RAMAC (Random Access Memory Accounting system), which was introduced in 1956. It contained 50 platters spinning at 1200 RPMs and had a total storage capacity of 5 megabytes. Whereas modern disk technology can pack 150 billion bits per mm^2, the RAMAC supported a storage density of approximately 4 bits per mm^2 (100 bits per inch on each track, with 20 tracks per inch). In order to improve the signal-to-noise ratio for the recordings, early disk technology attempted to reduce the flying height of the head to the disk surface; this feature made its debut in the 1960s. Approximately 15 years after the RAMAC, IBM introduced the 3340 using what was termed "Winchester" disk technology, which is considered the ancestor of modern disk technology. The 3340 introduced a thin-film inductive head that flew just 18 um from the surface on a cushion of air. The bernoulli effect enabled the head to fly that close to the disk surface. Modern hard drives continue to employ this same basic flying-head approach of these early washing machine-sized disk units. The flying height for modern disks is now on the order of 15 nm (just 15 atoms of air between the head and the disk surface) with the medium flying at 50–150 MPH beneath the head for a typical 7,500–15,000 RPM rotational rate.

Disks continue to be the preferred technology for secondary storage on computing devices from desktop computers to the largest-scale supercomputing systems. In the following subsections, we will discuss the organization of disk storage devices, technology trends, and their ramifications on design considerations for storage devices for scientific computing systems.

1.3.1 Fundamentals

Ever since the very first RAMAC disk device, disks have been organized as a spinning magnetic media on a platter — with multiple platters in the disk often referred to as a *spindle*. On the platter are concentric *tracks* of recorded data that are often referred to as *cylinders* for tracks located at the same radius on different platters on the same spindle. The tracks are in turn subdivided into "sectors" that contain a fixed number of bytes comprising a disk *block*. The most common block size is 512 bytes. Consequently, disks are typically referred to as *block storage devices* due to this aspect of their organization.

Because the circumference of tracks on the inner diameter of the disk unit are much smaller than for the outer tracks, *zonal* recording packs more sectors on the outer tracks of the disk unit in order to maintain uniform bit density on all tracks. Zonal recording maximizes the storage density of the device, but the sustained bandwidth of data transfers from outer tracks is much higher than that of the inner tracks of the device. Consequently, algorithms for disk file systems preferentially pack data on the outermost tracks in order to sustain maximum performance. Read and write operations that are smaller than the native block size of the devices will waste bandwidth because the device works with data only at the granularity of blocks. For write operations, it is necessary to read an entire block, update a subset of that block, and then write the complete aligned block back to the disk subsystem. Consequently, unaligned accesses can suffer, and write operations consume both the read and the write bandwidth of the disk devices. Composite devices, such as RAID, also have a much larger logical block size, requiring much larger transaction sizes to maintain full performance.

Because disks are mechanical devices, there are a number of latencies involved in access. These latencies include: rotational latency, head-switch latency, track-to-track latency, and seek latency. The disk heads are mounted on mechanical arms that position them over the disk platters. Most disk units operate the heads on all platters in tandem. After the heads are positioned over the correct cylinder or track, the disk must wait for the platters to rotate so that the correct sector is located under the read heads. The latency of the platter rotation is the rotational latency of the disk, and affects the rate at which the sector can be read or written to the cylinder. Read and write operations can only engage one head at a time, so the head-switch time (closely related to the rotational latency of the disk) refers to the amount of time required to switch between heads on different platters. The latency of repositioning the head differs greatly depending on whether it is between neighboring tracks or it involves random accesses. To speed write operations, many disks employ buffer caches to hide these latencies, but only hide latencies for small transactions. Such buffers cannot mask the reduced throughput that results from the time spent moving the heads. Consequently, sequential (append-only writes or streaming reads) offer the best performance for disk devices. Random access (seeking) presents the worst case read and write performance due to the mechanical latencies involved in repositioning the disk heads. Solid state nonvolatile storage devices may offer considerable advantages for such access patterns.

A common misconception about disk performance is that the device interface performance defines the expectations for the disk's performance. In fact, the sustained media transfer performance is often an order of magnitude less than the performance of the device interface. For example disks that adhere to the ATA-100 standard boast 100 MB/s transfer rates, but in practice one is limited by the transfer rate of the raw device, which is typically on the order of 20 MB/s or less internally. Composite devices such as RAID can share a common bus (such as SCSI or Fiberchannel) to saturate its available bandwidth, and are termed "spindle limited" if the performance of the underlying disk media is less than the available bandwidth of the device interface.

1.3.2 Performance Characteristics

Although there is significant work at the file system and operating system levels to hide some of the performance characteristics of block devices, most of the characteristics of the underlying devices end up percolating up through the entire software stack and must become first-order design issues for high-performance scientific I/O applications.

The performance of disk devices is substantially affected by the size of the I/O transaction. The behavior is partly due to the disk device's organization, but also due to the POSIX I/O sequential consistency semantics for

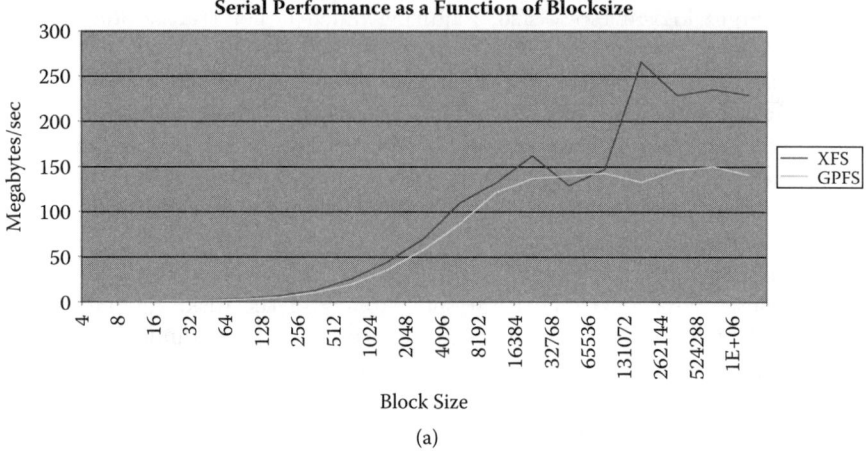

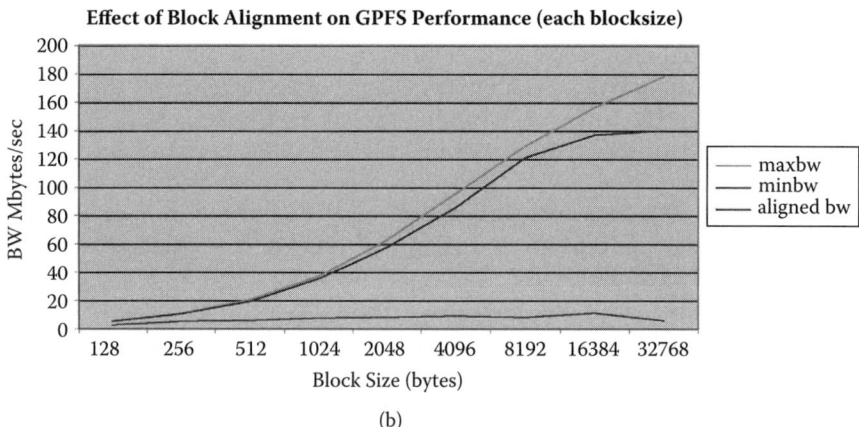

Figure 1.4 This figure (a) compares a cluster file system (GPFS) to the local file system performance (XFS) for varying transaction sizes for contiguous, append-only streaming writes. Figure (b) shows that the performance of disk devices, even for relatively large transaction sizes, is substantially impacted by the alignment of the data transfers relative to the native block boundaries of the disk device.

large-scale systems, which will be described in much more detail in Chapter 2 of this book. Figure 1.4(a) compares a cluster file system (GPFS) to the local file system performance (XFS) for varying transaction sizes for contiguous, append-only streaming writes using POSIX I/O. Given the huge performance disparity between best and worst performance shown on this plot, even a single small transaction interspersed with large transactions can substantially lower the average delivered performance. Small I/O transactions, even for linear operations, can be extremely costly because the entire disk-block must be transferred even when a small subset of the data is updated.

Furthermore, I/O transactions that are aligned to the native block size of the disk offer the best performance, whereas unaligned accesses can severely impact the effective performance delivered to applications as shown in Figure 1.4(b). The block alignment issues manifest themselves at all layers of the storage hierarchy, and even the file system. GPFS is used in this example, but this is common to many different file system implementations. In addition, streaming write operations tend to be slower than reads by up to 50% because of the need to read the disk block into the system memory, modify its contents, and then write it back to disk. Since read and write operations are mutually exclusive for the mechanical device (disk heads are in effect half-duplex), the read-modify-write can impact delivered bandwidth. However, the impact can be modest (<10%) because it is amortized by the track-to-track seek latencies.

1.3.3 Enterprise and Commodity Disk Technology

The disk storage market is typically divided into three market segments—enterprise, consumer, and handheld. The enterprise class storage market emphasizes performance (reduced failure rates, reduced seek times, and increased transfer bandwidth) above all other considerations. Enterprise-class disk interfaces tend to employ host interfaces such as SCSI and Fiber Channel that emphasize reliability, support for multiple outstanding transactions, more robust error checking and correction, as well as support for addressing a larger number of devices on the same loop or channel interface. The spindle speeds tend to be higher than consumer-grade devices, leading to higher power consumption. A typical enterprise class device will have a lower storage capacity per unit than the consumer class devices, but will spin at up to 15,000 RPMs, offer seek times of less than 5 milliseconds, and consume 12–20 watts.

The consumer class devices emphasize low cost and high capacity for volume-driven consumer electronics markets. A typical consumer-grade disk will offer double the storage capacity of the enterprise-class storage devices, but spin at a lower spindle rate of 5,400–7,200 RPMs. The drive interface for these devices tends to rely more on the host computer interface to control the disks than the enterprise-class devices. Consumer-class devices tend to support fewer devices on the same interface, and are less tolerant of faults. For example, IDE devices perform parity checking on the data bus, but not on the control interface. Interface technology is typically derived from integrated device electronics (IDE), ATA and Serial ATA (SATA) standards. The power consumed by such a device is typically limited by applications to less than 10 watts.

The handheld devices favor a design point that emphasizes reduced size and power consumption for devices such as MP3 players, ultraportable computers, and handheld movie cameras. The typical capacity for such devices is an order of magnitude smaller than for enterprise storage devices, but power consumption is typically on the order of 0.5–2 watts. These devices are also being explored as a component for ultraefficient composite (RAID) storage devices such as Massive Arrays of Idle Disks (MAID).[4]

While enterprise-class storage devices have enjoyed a niche market in large-scale storage solutions, they have been gradually supplanted over time by consumer-grade devices that depend on assembly into hierarchical RAID devices to match the reliability and performance of the enterprise-class devices. Recent studies[5,6] have cast doubt on claims that enterprise-class disks offer lower failure rates than their consumer-grade counterparts. Consequently, enterprise-class RAID storage subsystems composed of consumer-grade SATA disks are rapidly overtaking the enterprise market segment.

1.3.4 Future Trends

Magnetic disk technologies are faced with the theoretical limitation of superparamagnetism which proposes that as bit density is increased so must power to the device in order to prevent spontaneous changes of data (e.g., a bit flip) from occurring. Because this is a theoretical limitation, the actual limit is not known and continues to change each time a manufacturer produces a higher capacity disk. Several manufacturers are developing solutions to the problem, but the main concern with the limitation is that solutions will come at a significant increase in cost. This would have ramifications for the high-performance computing sites that rely on squeezing ever increasing amounts of storage into smaller spaces at a fixed cost.

1.4 Tape Storage

Tape has been the primary medium for offline storage since the 1980s. Magnetic disks or hard disk drives (HDD) are its primary competitor in terms of capacity and data transfer capabilities. Through the early 1990s tape remained slow and well behind disk drives in terms of performance, but well ahead of disk in terms of capacity. Disk storage was still in the MBs of space, and disk drives were a premium in terms of cost. In the early 1990s disk storage became commonplace and extremely affordable as the number of personal computers soared.

The key principles or advantages of tape storage that will continue to make it a viable mass storage solution if not the primary mass storage solution well into the future are that they are removable, extremely power efficient, maximize GB/sq ft, and continue to offer a competitive price/GB. Like optical storage, it is removable media. This fact allows tape to be an ideal solution for offsite data requirements. Tape is also primarily used for system backups because data can easily be exchanged between systems or stored separate from the system to prevent risk from colocation with the primary data source. Tapes do not require power or a conditioned computer room environment to retain data or be prepared for use. Tape libraries requiring very little power continue to keep tape access times reasonable for sites with requirements to keep vast amounts of data accessible to users. Archival life of tape is very good, with most tapes capable of retaining data anywhere from 15–30 years.

1.4.1 Fundamentals

Tape media has been made for about a decade and is composed of a durable substrate, mylar in most media, with single- or dual-layer magnetic alloys. Tape cartridges have tracks with data blocks or sections written to the tracks. Depending on the tape drive or application using the tape, file marks are written to the tape to serve as markers for moving to different points or data blocks on the tape. The primary down side of tape is its sequential access limitations. Tape does not lend itself to random access because tape is a continuous medium that must be moved forward and backward until the desired position is reached. The further down the tape the desired data resides, the longer it takes to reach the data.

Cartridges can be single or multireel. Single-reel tapes are fed through the drive with rollers where the tape passes over the recording head of the drive. Dual-reel tapes are loaded into the drive but keep the tape internal to the cartridge with the recording head of the tape drive positioning itself over the tape between the two reels. Tape drive speed and tape recording head capabilities determine the tape's data transfer rate. Tape can be damaged given the high velocity and mechanical nature of reading and writing data. However, data loss is normally limited to a small part of a single tape cartridge.

There are several types of tape technology in use in the high-performance computing industry. Tape drives utilize SCSI and Fiber Channel protocols. New tape drive models usually demand new tape media formats. Commodity tape drives typically don't plan for media reusability with multiple versions of drives, whereas enterprise tape drives plan for media reuse and backward compatibility. Most tape drives are capable of and automatically compress data being written to tape. The largest tape formats allow nearly 1 TB of data uncompressed to fit on a single tape. The cost of storing data on tape is typically an order of magnitude less than the cost of storing the same data on disk, irrespective of upkeep, maintenance, or power costs.

Access times for tape depend on the tape model used, but for the highest performing tape libraries, they typically range from 30 seconds to several minutes to mount the tape and deliver the first byte of data. The highest performing tape drives are capable of transferring uncompressed data at 180 MB/s and achieve about 380 MB/s with compression. Access time is highly dependent on the tape drive's seek capabilities. Tape is fundamentally sequential media, and tape marks are essential to finding the exact beginning and end of user data. Tape marks are a unique pattern written to tape by the drive when requested by the application in order to separate user data on the tape. This separation is application specific; some use tape marks to designate the end of a file, and others use them to separate blocks of a file. Tape mark processing is expensive and defeats the drive's ability to stream data. Enterprise tape drives include features to quickly seek to a particular place on the tape by avoiding the use of tape marks. One enterprise tape drive utilizes a special track on the tape called the servo track and a special index at the beginning of the tape to quickly position the tape to a particular point. The slowest tape seek times occur when applications move down the length of the tape counting tape marks to locate the data. Some applications, such as the high-performance storage system (HPSS) software, are careful to allow placement of small files on tape with better access speed and larger files on slower access, high-capacity tape.

In terms of durability, tape is at least as durable as other data storage technologies. Depending on the amount of reuse, format and density of the tape, and the speed and handling of the drive, tape can be used for several decades to store data.

1.4.2 Enterprise-Class Tape Technology

Enterprise tape drives are designed and manufactured under the strictest specifications to ensure extra reliability and durability of user data. Features include redundant components internal to the tape drives to eliminate any single point of failure to the greatest extent possible. At least one tape drive available in the market has dual tape heads, the most expensive component of the drive, to ensure that tape head failure does not interrupt tape drive data transfer capabilities.

Enterprise-class drives provide ruggedized components to handle extra wear and tear they receive from high utilization. Tape media wear is also a primary consideration in designing enterprise quality tape drives. Enterprise-class tape drives are typically capable of adjusting tolerances, such as tension on the tape or recording head positions, and are an order of magnitude more accurate than commodity tape drives.

Another consideration in differentiating between enterprise and commodity tape drives is the bit error rate capabilities of the drive. Commodity tape drives have bit error rate on the order of 10E-17. Currently, enterprise tape drive manufacturers are able to design the recording head and the pattern of data written to media to reduce bit error rate by up to two orders of magnitude compared with commodity tape drives. One other feature that enterprise class drives provide are multiple control and data path connections to the drive.

The main purpose of tape libraries is to provide storage for tape cartridges not in use and to load or unload tapes in or out of tape drives to the tape library. Enterprise class tape libraries provide redundant components such as robotic arms, grippers, visioning systems, and power feeds to prevent a single point of failure. When failures occur, hot swappable capabilities (such as replacing robotics while the library remains available for use) become another important feature of the industry-leading tape libraries. There is currently only one tape library manufacturer that provides an interchange mechanism between libraries to allow cartridges to be mounted in tape drives existing in separate libraries connected via the passthrough port. This is an extremely useful feature in a mixed media multitape library environment in that it prevents the need to directly manage the usage of tapes in drives.

Experiences at most High-Performance Computing centers with both commodity- and enterprise-class tape drives support the claim that enterprise-class tape drives do actually exhibit fewer problems in terms of placing user data at risk than the commodity drives.

1.4.3 Commodity Tape Technology

For as long as computer technology has existed in the home, there has existed some form of commodity tape intended for general consumer use. There are several types of tape used in electronics, with the most common being magnetic tape.

In the 1980s and 1990s commodity magnetic tape was primarily built around 8 mm tape that was mostly used to handle backups of the machine to which they were attached. In the mid-to late 1990s the commodity tape drive market opened up with the definition of a new kind of tape drive called LTO that defined a new market sector called midrange tape.

The linear tape-open (LTO) tape technology has a strong hold in the marketplace today primarily because of its low cost and relatively good performance capabilities and large capacities. It was primarily intended for use with backup applications where absolute assurance of no single point of failure or data loss was not critical. This is due to the fact that backup data is a copy of the original data and in general loses value over time. However, because of its low cost and relatively good performance and capacity, the technology has started to emerge as a viable option for the most data-intensive high-performance computing centers especially for consideration in dual-copy or the lowest tiers of storage in a hierarchical system. If used in a mass storage system with archival storage requirements, most centers simply make multiple copies of data retained on this technology.

The LTO specification has plans to release six generations or form factors of tape. For LTO-6, the final version plans to provide 3.2 TB of data and a speed of 270 MB/s. Currently, the LTO-4 drive is available and delivers 800 GB of data on a single cartridge and up to 120 MB/s data transfer speed.

In the last few years, several midrange tape libraries have emerged in the market that are capable of handling the full spectrum of drives and tape technologies available. Several of these mid-range libraries are starting to adopt features of the enterprise-class libraries, specifically redundant components, ruggedized robotics, and expansion capabilities. They are excellent for applications that do not require scalability beyond the limits of the library.

1.4.4 Future Trends

Current tape media technology has a physical limitation relative to the particular process or magnetic coating in use for the last 10 years or so. It is believed that a new media formulation will be required to allow tape to exceed 16 TB per cartridge. Tape capacity is increasing at a rate of about 40% per year.

1.5 Optical Storage

Optical storage looked very promising in the 1980s and 1990s. The new method of storage promised very large capacities and fast access methods in its early years. After two decades, it is apparent that the technology has a definite niche with its most popular formats, the compact disc (CD) and digital versatile disc (DVD), primarily suited for storing music and videos. The major benefit of optical storage today is its wide acceptance in the commercial marketplace and low cost. Optical drives are found on nearly every new computer and function particularly well as read-only media. Unfortunately, the technology was never able to achieve the write access rates or capacities that would allow it to compete with magnetic storage. Optical storage media is an order of magnitude behind magnetic storage in available capacity and transfer rates.[7]

1.5.1 CDs

The smallest unit of data on compact discs is called a frame, and discs capable of storing 650 MB have tracks with pitches (distance between tracks) of 1.6 μm. As the track pitch gets closer together, the CD can hold more data, but the ability of drives to read the CD decreases. The highest-capacity CDs today hold 700 MB or slightly more and have track pitches of 1.5 μm. Data rate in writing to CDs is determined by the speed of the drive used, which is represented by the factor of improvement over a baseline (1x) performance required for digital audio playback at 44 KHz. Data rates of modern CDs have leveled off around 48x–52x on writing (48–52 times faster than the baseline rate), which yields a sustained data transfer rate of around 7–9 MB/s. The archival life of CDs is not comparable to magnetic storage either inasmuch as each CD lasts typically 5–7 years before data is at risk. Due to the relatively short archive life, slow transfer speeds, and small capacity in comparison with magnetic storage, the technology is not in use in high-performance computing or mass storage system environments.

1.5.2 DVDs

The digital versatile discs can hold anywhere from 4–16 GB of data currently, depending on whether they are single or double sided and single or double layered. Pitch between tracks for DVD discs is 0.73 μm. Access rates are determined by the speed of the drive used, but can achieve over 20 MB/s on writes for a 16x drive with double-layered discs. Archival life of the media is similar to CD in that a disc is expected to last 5–7 years before data may become unreadable. Speed, capacity, and archival life are still only achieving what the oldest and least capable magnetic storage devices can deliver.

1.5.3 Blu-Ray and HD-DVD

Blu-ray disk and high-definition DVD (HD-DVD) are the newest multilayer optical storage mediums and are currently capable of holding 25 or 50 GB of data per Blu-ray disc, depending on whether they are single or dual layer, and 15 GB per HD-DVD disc. Blu-ray has been locked in a multiyear battle with HD-DVD for the digital video market, but the battle has largely been won by Blu-ray at this point in time. The much higher storage capacity of this technology is largely enabled by the availability of solid-state blue lasers (hence the Blu-ray name). The much shorter wavelength of blue light can be focused to a much smaller point size on the disk medium. The specification and design for this technology includes plans to expand a single disk to 100–200 GB by simply defining new tracks. Current drives are capable of 1x–16x speeds providing anywhere from 6 MB/s to 16 MB/s. The specification allows for up to 35 MB/s. These are substantial improvements over previous generation optical storage, but still not competitive with magnetic storage.

1.5.4 Future Trends

Traditional optical storage technologies have fundamental limits imposed by the physics of far-field diffraction. This means that the limits of the amount of data stored on a two-dimensional disc are determined by the wavelength of light used in the optical drive. In order to increase the amount of data on the disc, new optics with smaller wavelengths (blue and ultraviolet light) are necessary, and there are physical limitations on how small optics can get. Holographic storage promises solutions that reduce or eliminate the far-field diffraction limitations by using several different techniques capable of using the entire volume of media available. However, holographic storage continues to struggle with bringing a product to market, so it is not clear that the physical limitations of optical media will be overcome.

1.6 Composite Devices

Scaling up the performance of disk technology for server-class systems is increasingly expensive. Composite technologies embody the same spirit as commodity clusters to achieve the performance and reliability goals of specialized proprietary hardware using clusters of much simpler consumer-grade building blocks.

Composite devices also play a role in ensuring protection against media. In particular, the probability of a single-bit error in typical disk storage media is one error in 10^{14} to 10^{15} bits, and this rate has been relatively consistent over time. However, with continued exponential increases in the capacity of storage devices, the likelihood of encountering uncorrectable errors is dramatically increased. Composite devices employ various forms of redundancy and error correcting codes (ECC) to improve media integrity.

1.6.1 RAID

The nomenclature for RAID configurations was formalized as RAID "levels" by Chen et al. in their 1994 paper entitled "RAID: High-Performance Reliable Secondary Storage."[8] RAID was originally introduced with five levels, but over time new configurations have been introduced that include nested configurations as well as some proprietary configurations. RAID employs some combination of striping for performance with various approaches to redundancy for fault resilience. The standard RAID levels are as follows:

RAID 0: This configuration, which was not part of the original five RAID levels, offers improved bandwidth at the expense of reliability by "striping" data across parallel disks comprising the volume.

RAID 1: This configuration implements mirroring to improve reliability by maintaining an exact copy of data on each disk comprising the RAID volume, but offers no improvement in performance over a single disk.

RAID 2: Implements hamming codes to support ECC using a dedicated parity disk to store bit-interleaved parity information, much as ECC-corrected dynamic random access memory (DRAM) works. The approach achieves the reliability of mirroring with fewer redundant disks per protected bit, but suffers from poor performance for small transactions relative to mirroring. This form is rarely seen in practice.

RAID 3: Implements byte-level striping with a dedicated parity disk, and relies on the disk controller to identify the failed disk rather than strictly depending on the parity information as is the case for level 2. Consequently, it offers the same performance characteristics as level 2, but employs fewer disks to achieve fault resilience.

RAID 4: Implements block-level striping with a dedicated parity disk, which achieves the same fault resilience as level 3, but matches the transfer unit sizes to the native sector-size for the device. This organization greatly improves performance by ensuring that the entire disk block is utilized on each read. This also greatly improves transfer performance for small transactions, in comparison to the lower levels of RAID.

RAID 5: Uses block-level striping like RAID 4, but distributes the parity across the disks that comprise the RAID set. When a disk fails, the RAID 5 can continue to operate at full performance. Some RAID systems allow the failed disk to be replaced and rebuilt while the unit is running. However, the rebuild process is typically very costly and can greatly reduce the performance of the array while the rebuild is in progress. A second failure while the RAID 5 is in degraded state will render the volume unusable, thereby motivating more robust approaches to encoding the parity information used to detect disk failures.

RAID 6: RAID level 6 was not part of the original RAID configurations, but is now commonly considered a standard RAID configuration. RAID 6 introduces an addtional parity block to handle the increased failure rates that are anticipated with extremely large disk configurations. In contrast, RAID 5 uses the simplest case of Reed-Solomon ECCs, which enables it to handle loss of a single disk. However, technology trends and large data centers have increased the probability of seeing multiple disk failures within the same RAID block. RAID 6 extends the Reed-Solomon error correction field so that it can accommodate multiple simultaneous disk failures. The extended error correction enables RAID 6 to continue to operate in the presence of more than one simultaneous disk failure.

RAID can be implemented in either hardware or software. However, the parity checking and automatic volume rebuild process for RAID 2 and higher typically benefit from dedicated hardware. This has led to a robust hardware controller market for RAID systems. Given that RAID 0 and RAID 1 have less-demanding requirements for fault detection and correction, many operating systems incorporate logical volume managers that support concatenation of volumes, mirroring, or striping of volumes in software.

As large deployments are becoming more common, hierarchical implementations of RAID technology have become more common. The terminology for this hierarchical structure is RAID M + N, or "RAID MN," where N is the baseline building block that is composed together in RAID M fashion. So, for instance, a RAID 05 (also known as RAID 0+5) is a RAID 5 array comprised of a number of striped RAID 0 arrays, whereas a RAID 50 is a RAID 0 array striped across RAID 5 elements. RAID 50 and RAID 60 arrangements are typically the most commonly employed hierarchical RAID implementations.

1.6.2 Virtual Tape Libraries (VTLs)

Virtual tape libraries are a somewhat new concept in storage and primarily came about due to the cheap cost and prevalence of disk. The concept of a VTL is to use disk to mimic tape such that the application using the VTL does not know that it is manipulating disk rather than tape. These are primarily used in backup applications to eliminate dependence on streaming to tape for good performance and to increase the speed of recovery. Testing with at least one VTL in an archive application that writes small files or blocks of data to tape showed only a 2x improvement over a high-capacity tape drive. The thought was that the VTL would do much better given that the tape drive used did not handle small files with many tape marks well. In testing, the site discovered that disk is not that much faster or more efficient than tape at processing tape marks. One major problem with VTLs is that they have a cost somewhere between tape and high-end disk arrays. For mass storage systems, there seems to be little need or use for VTLs when most are already hierarchical storage systems with both disk and tape being used for both performance and cost reasons.

There have been other attempts at placing magnetic disk drives in tape cartridges so that they could fit in tape robotic libraries and mimic tape cartridges. However, there are few applications that can make use of removable disk, and the disk drives must be ruggedized for extra wear and tear. None of these solutions has come to market.

1.6.3 Redundant Arrays of Independent Tape (RAIT)

Many high-performance computing sites could take advantage of redundant tape systems or redundant arrays of inexpensive tape (RAIT). There are several hardware or software solutions on the market, but none that satisfy the needs of providing extreme bandwidths and parity protection. The hardware solution from Ultera provides mirroring or parity protection with an SCSI hardware controller card that connects SCSI tape drives together to perform RAIT. However, the solution is limited in bandwidth for what high-performance storage would require. The software solutions available, primarily HPSS, Veritas, or Legato, generally meet high-performance bandwidth requirements but do not provide parity protection to reconstruct data in the event that a tape within the stripe is lost. At the time of this writing, HPSS is currently working to design and develop a RAIT solution for its customers.

One of the most promising RAIT systems, developed by StorageTek, made it to the prototype stage demonstrating the loss of two tape drives during a read while continuing to stream data to the application.[9] During the final phase of development of turning the prototype into a production system, the company halted the project because the market demand for the product was believed to be too limited.

1.6.4 Data Integrity

Media integrity is often confused as being equivalent to data integrity, but the important and consequential differences are often underappreciated. Composite devices only protect against data-integrity problems with the disk media, but that does not ensure the integrity of data produced by a scientific application. From the application standpoint, any intermediate step along the complex pathway to storage can compromise data integrity before it arrives at the storage device, which will dutifully record the incorrect values to permanent storage. Unfortunately, commonly deployed storage hierarchies (for example, memory to disk to tape) provide no such end-to-end data integrity checking, so it is left to the application developer to employ various forms of checksums and other tests to verify the integrity of their data.

CERN (the European Organization for Nuclear Research) recently performed a study of data integrity issues on their own systems[10] by writing 3,000 special 2 GB files of a predefined pattern every two hours, and then reading them back to check for errors every five weeks. What they found was that even with RAID to protect against media errors, they observed 300 uncorrected and unreported errors in the 2.4 petabytes of data that were stored on the data volumes. In all, after examining 8.7 TB of user data for corruption (33,700 files), they found 22 corrupted files that had not been detected by any part of the storage infrastructure. In some cases, the corruption remained undetected because it was caused by errors in various tiers of the software infrastructure rather than random bit-flips. Ultimately, CERN found an overall byte error rate of one in $3*10^7$ bytes, which is considerably higher than the media error rate of one byte in 10^{14}. Without some form of end-to-end monitoring of data integrity it is clear that the larger disk-based storage systems will observe data inconsistencies. Focusing on media alone for data integrity protection is insufficient.

There are emerging storage technologies such as zettabyte file system (ZFS),[11] that are capable of ensuring end-to-end data integrity. However, until such systems are widely deployed, it is important for scientists to incorporate their own checksum checks (such as cyclic redundancy check [CRC] or message digest algorithm #5 [MD5] and error detection mechanisms into their data storage practices to protect against silent data corruption.

1.7 Emerging Technologies

This section describes technology that is poised to compete with mainstream tape and disk technology, but for various reasons has not yet overtaken these technologies. The reasons include manufacturing economics in a volume market and technologies that are still in development. Many of these technologies are already available, but have had a limited impact due to market economics or a narrow band of applications where the technology offers superior price/performance. Others are still in development; it remains to be seen if an effective volume-manufacturing process can be found to bring them to market.

1.7.1 Massive Array of Idle Disks (MAID)

Given current technology trends for rotating disk storage, data centers are increasingly reliant on larger numbers of disk spindles to maintain I/O bandwidth growth rates that match performance improvements of the computing infrastructure. However, as commensurate power consumption of such system grows, there is increasing concern for reining in the power consumed by the storage subsystem. The concept of a massive array of idle disks (MAID) is to power down disks that are not in use so that power savings may be realized. A secondary benefit of powering down the disks is that they have higher reliability for lack of use. MAID technologies normally use SATA disks, which allow the overall system to present an extremely dense array of disks in a small footprint for a very competitive price.

There are several MAID technologies on the market today, and new products that are not explicitly MAID systems continue to adopt similar power management features. MAID systems are typically close to the cost of enterprise tape drives with the reduced benefit of not being removable or scalable as tape technology provides. There is also concern over the increased wear and tear of the disks due to powering them on and off. In many of the mass storage systems in production, the disk cache normally has high utilization, and taking advantage of the power down feature would not be possible.

MAID systems are novel and timely now that power management is becoming a larger and larger part of storage and compute centers' planning and design.

1.7.2 FLASH

FLASH memory is the evolution of electronically erasable programmable read-only memory (EEPROM) technology. A FLASH memory cell, like its EEPROM predecessors, is based on a field effect transistor (FET). Normally, a FET's on/off state is controlled by charging the gate with electrons to apply an electric field to the transistor channel. In the case of FLASH memory cells, the FET contains a floating gate that is completely surrounded by insulating silicon oxide. The gate is programmed by applying a high (>12V) voltage to the electrically connected FET gate, which causes electrons to tunnel through the insulator and into the floating gate in a process known as *hot electron injection* or *avalanche injection*. The residual electric charge in the floating gate provides enough energy to maintain the gate's programmed state. The gate can be deprogrammed by exposing it to UV light in the case of earlier electronically erasable programmable read-only memories (EPROM), or applying a large inverse voltage in the case of EEPROMs and FLASH memory cells. Whereas EEPROM were designed to be programmed and erased as a whole, FLASH extended the technology to allow finer-grained block-level reprogramming of storage cells.

Most digital FLASH storage devices use single-level cell (SLC), which store a binary 1 or 0 in each cell of the chip. Another FLASH technology referred to as multi-level cell (MLC) can store multiple values per cell as different voltage levels. This can greatly increase storage density but is more susceptible to failure than the SLC device.

Initially, FLASH technology was very low density and high cost, which relegated it to niche applications that were extremely power constrained. However, with improvements to the technology and silicon chip lithography, FLASH prices have made it increasingly popular for storage in consumer electronics devices such as MP3 players and digital cameras. The enormous volumes supported by these applications has brought FLASH memory prices down to the point that they are price competitive with high-end disk solutions and may drop further still. Already, storage devices in laptops and other portable computers are poised to be replaced by FLASH as an alternative.

From the standpoint of scientific applications, FLASH memory can be read in random access fashion with little performance impact. The typical read access latencies are less than 0.1 ms, which makes them considerably higher performance than mechanical disk units, which offer latencies on the order of 7 ms. However, writing data to FLASH takes considerably longer than the read operation due to the much longer latencies required to program the cells. Whereas read rates can be achieved that approach 200 MB/s, the write performance is typically more on the order of tens of megabytes per second or less. Prior to 2008, state-of-the-art NAND-FLASH-based storage devices were typically limited to one megabyte/second peak write performance. New high-performance double data rate (DDR) interfaces, and improvements in the cell organization to reduce effective cell size, are enabling FLASH to push performance past 100 MB/s write and 200 MB/s read.

One of the main problems with FLASH memory is that the cells wear out after a limited number of writes. For a typical NAND-FLASH, 98% of the blocks can be reprogrammed at least 100,000 cycles before they fail. As FLASH densities increase through improvements in chip lithography, the problem of preventing cell wear-out becomes more challenging. Solid state disks attempt to mitigate the cell wear-out problem by using load leveling algorithms. The load-levelers attempt to spread the write operations evenly across the device so as to reduce the chance of cell wear-out. As a result, given the practical bandwidths available for accessing, the device would require five years of continuous access before the device will encounter cell wear-out—which is on par with the mean time between failures (MTBF) of mechanical disk storage devices. However, occasionally the load-leveling algorithm encounters degenerate cases that result in unexpected access delays as data blocks are remapped to maintain even distribution of the writes. The cell wear-out issues with FLASH leave the door open for competing technologies such as magnetoresistive random access memory (MRAM) to step in.

1.7.3 MRAM

MRAM[12] is the first commercially viable solid state device to use the principles of "spintronics" that emerged from the discovery of the giant magnetoresistive (GMR) effect (see Section 1.2.4). MRAM uses electron spin to store information in its cell array. It offers many of the benefits of FLASH memory, such as nonvolatile storage and low power, without suffering from the cell wear-out that is inherent to NAND-FLASH technology. In addition to nonvolatile storage, it also promises to match the performance of SRAM (typically used for CPU cache memory). The bit densities of MRAM are still an order of magnitude below that of leading-edge FLASH memory implementations, but the technology is maturing rapidly. It may prove to be a strong competitor, and possibly the heir-apparent to FLASH memory in portable consumer electronics devices. It is likely to compete with FLASH densities and cost-competitiveness in the 2010 time frame, given current trends in the improvement of this technology. In the short term, MRAM competes against FLASH for lower-density applications that require DRAM-like write bandwidths.

1.7.4 Phase-Change Technology

Phase-change memory is another form of nonvolatile storage that is still in developmental stages. Phase-change storage devices rely on using current-induced heating to reversibly change the chemical composition of Chalcogenide (GeSbTe) material between the cross-points of a wire mesh that forms the storage array[13] shown in Figure 1.5(a). The technology is further from commercial introduction than MRAM at this time, but it has the potential to scale much faster to high-bit densities using commercial manufacturing processes.

1.7.5 Holographic Storage

Holographic storage promises to resolve the limitations that current optical storage contends with in terms of increasing the capacity of a single disc. Currently, optical storage is focused on decreasing the pitch of tracks or the size of the data format in order to fit more data in the same form factor or limit of a two-dimensional space. Holographic storage uses a volumetric approach to storing data, such that data is not strictly limited by the two-dimensional size of the disc. Holographic storage has been in development since at least the 1990s. No products are available in the market today, but the two companies working on this technology have roadmaps that plan to deliver media roughly the same as the current DVD that will hold 300 GB of data and be capable of transferring data at a rate of 20 MB/s. The roadmap extends the media

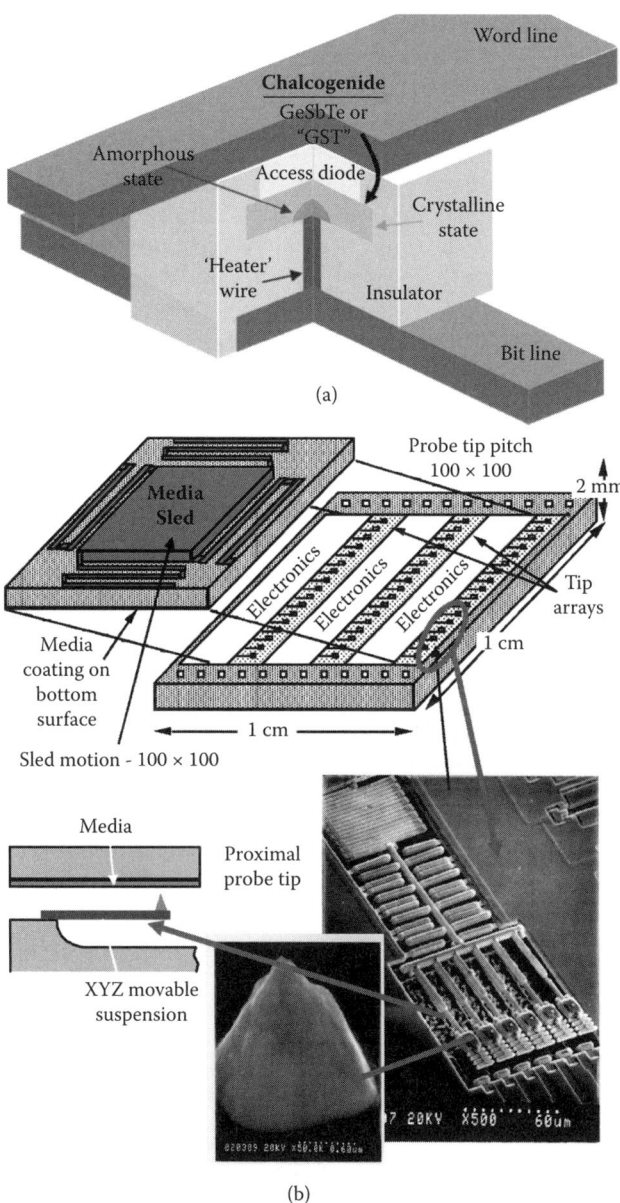

Figure 1.5 (See color insert following page 224.) Part (a) shows a schematic of a single cell of a phase-change memory cell that uses electrically induced heating to induce phase changes in the embedded Chalcogenide (GeSbTe) material. Part (b) shows the architecture of the Micro-Electro-Mechanical system (MEMS)-based Millipede storage system that employs small STM heads to directly manipulate a polymer medium for storage.

to a planned maximum capacity of 2 TB. Initial products are expected to be write-once read-many with future plans to handle rewritable media.

1.7.6 Direct Molecular Manipulation

Continued improvements in cost-effectiveness of devices is closely related to storage densities of the medium. As we press toward areal densities that approach the atomic scale, there has been increased interest in technologies that can encode data by directly manipulating atomic structures. Direct mechanical manipulation attempts to make novel technology cost-competitive with current storage devices by leapfrogging their areal densities. Some examples include direct manipulation of atoms using a scanning tunneling microscope (STM), nanotube devices that depend on the van der Waals interaction between crossed tubes, and IBM's Millipede device (shown in Figure 1.5) that uses many parallel micromechanical (MEMS) styli to encode data into a polymer medium. Each of these devices promises storage densities that exceed the current magnetic limit of 1 Tbit/mm2. However, such devices are still in their infancy.

1.8 Summary and Conclusions

Mechanical magnetic storage devices such as disk and tape have been the dominant technology for secondary storage for the past 30 years. Although solid state devices such as FLASH have been gaining ground over the past few years, areal density and cost trends ensure that disk will remain competitive for at least the next decade.

One important trend that may complicate future storage for the highest-end scientific computing system is the growing gap between disk capacity and the delivered bandwidth of these devices. Storage trends continue to show 40–60% per year compounded growth rate for storage capacity, thanks to continued improvements in areal density. However, the bandwidth delivered by these same disk subsystems is only growing by 17–20% per year. The performance of such systems for random access has become nearly stagnant, which favors linear streaming read or append-only write operations. If these trends continue unchanged, HPC systems will be forced to purchase larger numbers of disk spindles over time that are accessed in parallel in order to maintain existing balance ratios between the HPC system performance and storage subsystem bandwidth. As such, the disk subsystem will likely consume a larger fraction of the area and power budget for future systems without some technology change.

As a result of requiring more disks to achieve performance requirements, HPC centers are deploying file systems that are for the first time eclipsing the size of archival storage systems. Considering the sustainability paradox, presented in Section 3.2.1, it is likely that users will need to more carefully consider how best to use archival storage systems to manage their most important data.

In addition, as the number of spindles continues to increase, the likelihood of device or even RAID system failures occurring increases, as does the rebuild time for RAID systems that are able to recover. These are challenges that have recently been addressed with innovations in offering new levels of RAID. This problem is discussed in some detail in Chapter 2. However, at some point storage devices will not be able to handle error detection or recovery on their own and will require innovation in other technologies to solve (e.g., file system checksums).

Several technologies emerged to keep power consumption under control and to fill the widening gap between primary storage (DRAM) performance and secondary storage (disk) performance, such as MAID and FLASH.

Solid state non-volatile random access memory (NVRAM) technologies such as FLASH are becoming cost competitive with the high end of disk technology and may soon reach parity with consumer disk storage due to dramatic rise of mass-market applications. NVRAM technologies address issues of poor response to random accesses due to lack of mechanical "seek" cost, power dissipation, and bandwidth scalability of conventional disk devices. However, cell wear-out and poor write performance of FLASH relative to the mechanical devices keep impact on high-end scientific computing storage marginal. While load-leveling technology offers some protection against wear-out, the algorithms are still subject to edge cases where intensive recopying of data is required. In the interim, FLASH may offer advantages for read-intensive storage applications such as data-mining applications, but for write-intensive applications (such as data output from time-evolution simulation codes or checkpoint/restart) we may need to wait for commercialization of alternative NVRAM technology that doesn't suffer from cell wear-out such as MRAM, or phase-change devices.

Tape subsystems are also seeing new demands that run counter to their original performance characteristics. Formerly, there was emphasis on streaming performance of tape systems for moving small numbers of very large files. However, over time archival storage, even at scientific computing facilities, has been increasingly dominated by large numbers of small files. For a sequential access medium, managing many small files using current tape technology poses daunting technical challenges, especially as older, smaller capacity devices are replaced by media that hold more and more data. Tape speed is stable, which is desirable as increased speed means increased wear-and-tear on the tape. User wait time to first byte of data will continue to increase linearly as tape media capacity increases. Current high-performance archival storage management software, such as the high-performance storage system (HPSS), are historically geared toward handling large files where data can be streamed to tape. HPSS is planning to deliver a feature to enable aggregation of small files when they are migrated to tape to address part of this problem. However, such systems will need to be refactored to better handle small files by providing policies that optimize aggregation based on the access patterns to the tape.

Every few years, there have been white papers questioning the long-term viability of tapes and the likelihood of their being replaced by spinning disk storage. Despite this, the cost performance and power performance of tapes continue to maintain order of magnitude benefit over disks and even NVRAM devices. Due to their enormous surface area and the efficient storage layout offered by helical scan heads, tapes maintain high storage density despite being far below leading-edge areal densities offered by leading-edge magnetic storage technology, and thus provide very little pressure on vendors to push the limits on the technology. Were a competing technology to emerge that put pressure on the tape market, the tape technology vendors have significant headroom to improve densities and price performance. However, such a competitor has yet to emerge, and current power, density, and storage trends for disk make it unlikely to be the likely successor to tape.

It is not clear that there is a viable competitor to the tape market for the lowest tiers of storage in mass storage systems primarily due to the cost and power savings that tape continues to provide. However, a potential future competitor to the tape market, namely holographic storage, continues to remain on the horizon. Holographic storage has been "a year away from production" for at least two decades, with issues that continue to challenge its ability to achieve commercial production. Holographic storage is much like optical storage in that it will likely have its niche applications and uses within the storage industry when it arrives, but will take a while to develop characteristics or features that would make it competitive with the demands that high-performance computing centers require of disk or tape systems. Likewise, other passive technologies such as direct molecular manipulation (IBM's Millipede) are unlikely to compete with tape on the basis of cost, density, power, or streaming performance alone—but they do offer higher performance for random accesses, which would be more appropriate for storage of large numbers of small files.

In summary, there exist many new and exciting emerging storage technologies with interesting and unusual storage characteristics. In the near term, none of these are likely to disrupt the current high-performance computing industry trends of primarily using disk and tape to store data. As the petascale computing age begins, the storage industry is likely to see another significant increase in the amount of data stored and retrieved from these larger and more capable systems. The leading challenges to disk storage are power and reliability or data integrity. However, tape will be challenged with increases in access times and the amount of data at risk as the size of a single cartridge increases. These are problems that the storage industry is working to solve, but at some point or scale will cease to be reasonable for any storage device to handle. This demands solutions, potentially by data management software to improve on the reliability and access to ever-increasing amounts of data.

Acknowledgments

The authors were supported by the Office of Advanced Scientific Computing Research in the Department of Energy Office of Science under contract number DE-AC02-05CH11231. We also thank Stephen Cranage from Sun Microsystems for his contribution of technology roadmap diagrams to this document.

References

[1] K. Asanovic, R. Bodik, B. C. Catanzaro, J. J. Gebis, P. Husbands, K. Keutzer, D. A. Patterson, W. L. Plishker, J. Shalf, S. W. Williams, and K. A. Yelick. The Landscape of Parallel Computing Research: A View from Berkeley. Technical Report UCB/EECS-2006-183, EECS Department, University of California, Berkeley, 2006. http://www.eecs.berkeley.edu/Pubs/TechRpts/2006/EECS-2006-183.html.

[2] S. H. Charap, P. Ling Lu, and Y. He. Thermal stability of recorded information at high densities. In *IEEE Transactions on Magnetics*, 33, 978–983, January 1997.

[3] S. P. Parkin. The spin on electronics. In *ICMENS*, pages 88–89, 2004.

[4] D. Colarelli and D. Grunwald. Massive arrays of idle disks for storage archives. In *Supercomputing02: Proc. ACM/IEEE Conference on Supercomputing Los Alamitos, CA, USA: IEEE Computer Society Press*, pages 1–11, 2002.

[5] E. Pinheir, W. D. Weber, and L. A. Barroso. Failure trend in a large disk drive population at Google Inc. In *In Proc. of the 5th USENIX conf. (FAST07)*, 2007.

[6] B. Schroeder and G. A. Gibson. Disk failure in the real world: What does an MTTF of 1,000,000 hours mean to you. In *Proc. of 5th USENIX conf. (FAST07)*, 2007.

[7] D. Saird and B. H. Schechtman. A roadmap for optical data storage applications. *Optics and Photonics News.*, April 2007.

[8] P. M. Chen, E. K. Lee, G. A. Gibson, R. H. Katz, and D. A. Patterson. RAID: High-performance, reliable secondary storage. *ACM Comput. Surv.*, 26(2): 145–185, 1994.

[9] J. Hughes, C. Milligan, and J. Debiez. High performance rait. *Storage Technology Corporation funded by DOE ASCI Path Forward*, 2002.

[10] Bernd Panzer-Steindel. Data integrity. *CERN Technical Report Draft 1.3, CERN/IT*, April 8, 2007.

[11] E. Kusarts. Zfs: The last work on filesystems. http://www.opensolaris.org/os/community/zfs/, 2007.

[12] W. J. Gallagher and S. S. P. Parkin. Development of the magnetic tunnel junction MRAM at IBM: From first junctions to a 16-mb MRAM demonstrator chip. *IBM J. Res. Dev.*, 50(1):5–23, 2006.

[13] Y. C. Chen et al. Ultra-thin phase-change bridge memory device using gesb. *International Electron Devices Meeting (IEDM) of the IEEE International Solid-State Circuits Conference*, February 2007.

Chapter 2

Parallel Data Storage and Access

Robert Ross,[1] Alok Choudhary,[2] Garth Gibson,[3] and Wei-keng Liao[2]

[1] *Argonne National Laboratory*
[2] *Northwestern University*
[3] *Carnegie Mellon University*

Contents

2.1 From Disk Arrays to Parallel Data Storage 36
    2.1.1 Data Consistency and Coherence 37
    2.1.2 Fault Tolerance 39
2.2 Parallel File Systems 40
    2.2.1 Block-Based Storage Systems 40
        2.2.1.1 General Parallel File System 41
    2.2.2 Object-Based Storage Systems 43
        2.2.2.1 PanFS 43
        2.2.2.2 PVFS 47
        2.2.2.3 Lustre 49
2.3 Example Parallel File System Deployments 50
    2.3.1 PVFS on the IBM Blue Gene/P 50
        2.3.1.1 Argonne Blue Gene/P System 51
        2.3.1.2 Blue Gene/P I/O Infrastructure 51
        2.3.1.3 Tolerating Failures 52
    2.3.2 PanFS on Roadrunner 53
2.4 Interfacing with Applications 55
    2.4.1 Access Patterns 55
    2.4.2 POSIX I/O 57
    2.4.3 MPI-IO 58
    2.4.4 High-Level I/O Libraries 60
        2.4.4.1 Parallel netCDF 60
        2.4.4.2 HDF5 61
2.5 Future Trends and Challenges 63
    2.5.1 Disk Failures 63
    2.5.2 Solid State and Hybrid Devices 66
    2.5.3 Extreme-Scale Devices 66
2.6 Summary 68
Acknowledgments 69
References 69

2.1 From Disk Arrays to Parallel Data Storage

The preceding chapter covered how disk arrays can combine many disks into a unit that has a high-aggregate input/output (I/O) performance. This technique enables the construction of high-performance file systems that are accessible from single systems. When combined with a networked file system such as NFS and high-performance networking hardware, many client nodes may have access to this resource, enabling I/O to a shared storage system from a parallel application.

However, disk arrays are not the end of the story. Vectoring all I/O operations through a single server can create a serious bottleneck because I/O is limited by the bandwidth between the network and the server and the communication paths internal to the server. To attain the aggregate bandwidths required in today's parallel systems, we need to eliminate this bottleneck as well.

Parallel data storage systems combine multiple network links and storage components with software that organizes this hardware into a single, coherent file system accessible by all the nodes in a parallel system. The software responsible for performing this organization is commonly called a *parallel file system*. Figure 2.1 illustrates a parallel file system. On the left we see a simple directory hierarchy, with one astrophysics checkpoint file and two protein sequence files stored in different directories. On the right we see how this directory structure is mapped onto hardware. I/O servers (IOSs) store components of the file system, including directories and pieces of files. Distribution of data across IOSs is managed by the parallel file system, and the distribution policy is often user tunable. In this example, data is split into stripes that are referenced by *handles* (e.g., H01, H02). The large checkpoint file has been split across four servers to enable greater concurrency, while the smaller

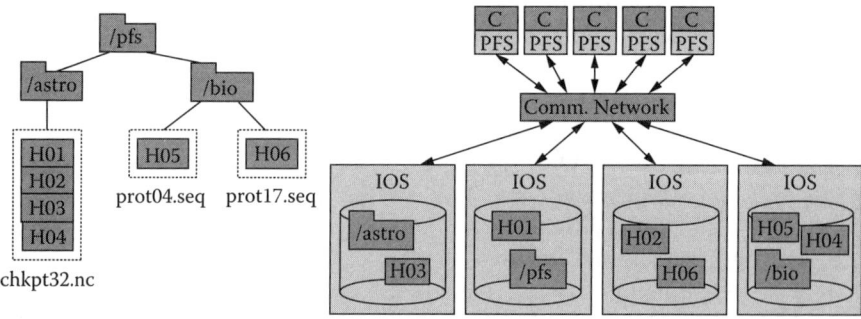

Figure 2.1 In a parallel storage system, data is distributed across multiple I/O servers (IOSs), allowing multiple data paths to be used concurrently to enable high throughput. Clients access this storage via parallel file system (PFS) software that drives communication over an interconnection network.

bioinformatics files were stored each on a single IOS to minimize the cost of metadata management. Compute nodes run a file system component that allows them to access data distributed in this system. The communication network provides network paths between clients and servers, enabling clients to take advantage of the storage hardware at the servers and, in particular, allowing for very high *aggregate* performance when multiple clients access the parallel file system simultaneously.

The location of file data, owner, permissions, and creation and modification dates for a file must also be maintained by the file system. This information, called *metadata*, might be stored on the same IOSs holding data or might be kept on a separate server. In our example system we show the directory structure distributed across the IOSs; in such a case the file metadata is likely distributed as well.

2.1.1 Data Consistency and Coherence

Data consistency and cache coherence problems have long been studied since storage systems became sharable resources. Consistency and coherence define the outcomes of concurrent I/O operations on a shared file. The problems occur when at least one operation is a write and the file system must ensure consistent results. While different levels of data consistency have been defined, the best known is *sequential consistency*, which is also adopted by most UNIX file systems. It requires the results to be as if the multiple I/O operations happened in some sequential order. For example, given two write requests overlapping at a certain file location, the contents of the overlaps must come entirely from either the first write or the second. No interleaved result is allowed. It is relatively easy for a file system with one server and one disk to guarantee sequential consistency, but it is difficult for parallel file systems with more than one server because coordination between servers becomes necessary. Currently, the most popular solution uses a locking mechanism to enforce data consistency. Locking provides exclusive access to a requested file region. Such access, however, also means operation serialization when conflicts exist. As the number of compute processors in current and future parallel machines grows to thousands and even millions, guaranteeing such consistency without degrading the parallel I/O performance is a great challenge.

Since disk drives are currently the most popular storage media, it is important to understand how their data access mechanism impacts the file systems' consistency control. Disk drives can be accessed only in fixed-size units, called disk sectors. File systems allocate disk space for a file in blocks, which consist of a fixed number of contiguous sectors. The disk space occupied by a file is thus a multiple of blocks, and files are accessed only in units of blocks. Under this mechanism, file systems handle an I/O request of an arbitrary byte range by first allocating block-size system buffers for disk access and then copying data between the user and system buffers. The same concept applies to RAID, in which a block consists of sectors distributed across multiple disks. In order to increase the data throughput, parallel file systems employ multiple IOSs, each containing one or more RAID storage devices, and files can be striped across servers.

Traditional file locking uses a single lock manager, but this approach is not scalable on large parallel machines. Various distributed lock protocols have been adopted in modern parallel file systems, such as IBM's general parallel file system (GPFS)[1] and Lustre.[2] In the GPFS distributed lock protocol, lock tokens must be granted before a process can perform any I/O operation. Once a lock is granted, a token holder becomes a local lock manager for granting any further lock requests to its granted byte range. A token also allows a compute node to cache data that cannot be modified elsewhere without revoking the token first. The Lustre file system uses a slightly different distributed locking protocol in which each IOS manages locks for the stripes of file data it owns. If a client requests a conflicting lock held by another client, a message is sent to the lock holder asking for the lock to be released. Before a lock can be released, *dirty* cached data — data that has been changed on the client and has not been updated on storage — must be written to the servers.

Because files are accessed only in units of file blocks, the file system's lock granularity is the size of a block. In fact, the lock granularity on both GPFS[1] and Lustre[2] is usually set to the file stripe size, which can be a multiple of block size. Thus, two I/O requests must be executed sequentially if they access parts of the same block, even if they do not overlap in bytes. Typically, an application's parallel I/O requests are not aligned with such lock boundaries. Once nonaligned accesses occur, I/O parallelism can be drastically degraded owing to the file system's consistency control.

Another I/O strategy used in many file systems that has a significant impact on the parallel I/O performance is client-side file caching. Caching places a replica in the memory of the requesting processors such that successive requests to the same data can be carried out locally without going to the file IOSs. However, storing multiple copies of the same data at different clients introduces coherence problems. Many system-level solutions for coherence control involve bookkeeping of cache status at IOSs and invoking client callbacks as necessary to flush dirty data. Such mechanisms require a lock as part of each read/write request to ensure atomic access to cached data. While forcing a lock request for every I/O call guarantees the desired outcome, it can easily limit the degree of I/O concurrency. The block-level access conflicts that happen in the sequential consistency control can also occur in file caching as *false sharing*. False sharing is a situation that occurs when the granularity of locks in a system forces multiple clients accessing different regions of the file to attempt to obtain the same lock, such as when clients attempt to write to different parts of the same block in a parallel file system. Therefore, it is not wise to directly apply the traditional file caching strategy in a parallel environment, even though caching has been demonstrated as a powerful tool in distributed systems. If not used carefully, caching can seriously hurt the shared-file parallel I/O performance.

2.1.2 Fault Tolerance

Distributing data over a large number of servers and disks creates a wide variety of possible failure scenarios, including failure of disks or disk enclosures, loss of servers, loss or partitioning of the network between clients and servers, and loss of clients during file system operation.

In the preceding chapter we discussed RAID and its use in tolerating disk failures. RAID is typically used in the context of a collection of drives attached to a single server in order to tolerate one or more disk failures in that collection. With drive failures covered, we can begin to consider how to handle server failures.

Servers can, and do, fail. Two approaches can be taken for handling server failures: providing an alternative path to the storage that the server manages, or maintaining a replica of or means of reconstructing the data managed by that server. IBM's GPFS[1] is typically configured to use the first approach. Multiple servers are connected to a single storage unit, often using Fibre Channel or InfiniBand links. When a server becomes inoperable, another server attached to the same storage can take its place. In an *active–passive* configuration, an additional server is attached and remains idle (passive) until a failure occurs. Often this configuration shows no performance degradation after the new server has taken over (the service has *failed over*). In an *active–active* configuration, another active server takes over responsibility for the failed server's activities. In this configuration no resources ever sit idle, but performance degradation is likely when running after a failure.

The Google file system (GFS)[3] takes the second approach. In GFS, data is replicated on multiple servers. When one server dies, another takes over responsibility for coordinating changes to data for which the dead server was previously responsible. A server overseeing the storage as a whole will, on detecting the death of the server, begin replicating its data using the remaining copies to avoid future failures that could cause permanent data loss. This approach can be implemented in commodity hardware, allowing for much lower cost. Additionally, the system can be configured to make more or fewer replicas of particular data, depending on user needs. This capability allows for fine-grained control over the trade-off between performance (more copies cost more time to write) and failure tolerance. On the down side, implementing such a system in software can be complicated, and few enterprise solutions exist.

Clients can also fail. In fact, most systems have many more clients than servers and are more likely to see client failures than server failures. Client failures may have a number of different impacts on file systems. In a system using PVFS (parallel virtual file system) or NFSv3, a client failure has no impact on the file system, because clients do not maintain information (state) necessary for correct file system operation. In a system such as Lustre or GPFS where locks might be cached on the client, those locks must be reclaimed by the servers before all file system resources will be accessible. If a file system cache writes on clients, then dirty data might be lost, or operations might have to be replayed from a log when the client returns.

Network failures can cause two problems. The most obvious is that they can cause a single server (or client) to become inaccessible. This situation is typically handled as if the server (or client) has died. A more complicated situation arises when a network failure *partitions* the network, causing one set of servers and/or clients to be unable to contact another set. This situation is dangerous because if both groups continue to operate, the state in the two groups could become out of synchronization. In order to avoid this situation, file systems often use *quorum* techniques to prevent small groups of partitioned servers or clients from continuing to operate. The quorum approach involves passing messages routinely to establish that a given server can communicate with at least half of the other servers (putting it in a majority). If it cannot, then it shuts down, ensuring that it does not cause file system errors.

2.2 Parallel File Systems

While parallel file systems vary widely in the specifics of their implementation, at a high level these systems can be split into two groups based on how clients access data. In the first group, block-based storage systems, clients perform I/O in terms of (real or virtual) disk blocks in a manner similar to a locally attached disk drive. In the second group, object-based storage systems, clients operate on regions of more abstract "objects," containers holding data and associated attributes. In this section we examine examples of both approaches.

2.2.1 Block-Based Storage Systems

Prior to the 1980s a popular data organization on disk was count-key-data formatting, where applications decided how much data to put into each record and what searchable key is associated with each record. In the early 1980s, SCSI (small computer system interface) created a standard that kick-started competition, leading to a much larger market with much better pricing; responded to a need for smaller form-factor disks for mini-, micro-, and personal computers; and cost-effectively moved enough independent controller function to the device to allow large collections to operate in parallel. In particular, SCSI popularized the now common fixed-block disk format, in which every disk record is the same size and is addressed by its index number in a linear space of N blocks—a memory abstraction. With this simple abstraction, a linear block address space concealing the actual location of each block on disk, came the opportunity to change the physical organization of blocks on a disk, for example, to skip over a damaged block and use a spare block at some other place whenever the damaged block's index is requested.

The combination of small form-factor, cost-effective SCSI disks for small computers and linear block address space led to *disk arrays* in the early 1990s. A disk array is a set of disks grouped together, usually into a common physical box called the array, and representing itself as a much larger "virtual" disk with a linear block address space that interleaves the virtual disk blocks across the component physical disks. Arrays promised higher aggregate bandwidth, more concurrent random accesses per second, and cost- and volumetric-effective large storage systems. But with many more mechanical disk devices in an array, the component failure rates also rise. In a paper called "A Case for Redundant Arrays of Inexpensive Disks," Patterson, Gibson, and Katz described a taxonomy of "RAID" levels showing different ways disk arrays could embed redundant copies of stored data.[4] With redundant copies of data, the failure of a disk could be transparently detected, tolerated, and, with online space disks, repaired. The leading RAID levels are level 0, nonredundant; level 1, duplication of each data disk; and level 5, where one disk stores the parity of the other disks so that a known failed disk can be reconstructed from the XOR of all surviving disks.

SCSI is still with us, and its lower-cost competitors advanced technology attachment (ATA) and serial ATA (SATA) share the same linear block address space and embedded independent controller. RAID has been relabeled Redundant Arrays of Independent Disks because the expensive, large form-factor disks have been displaced by relatively inexpensive, smaller form-factor disks. RAID is a core data management tool in all large data systems. And most important, the linear block address space abstraction is the basic and central storage virtualization scheme at work today.

2.2.1.1 General Parallel File System

IBM's general parallel file system (GPFS) grew out of the Tiger Shark multimedia file system, developed in the mid-1990s. Variants of GPFS are available for both AIX and for Linux. GPFS is one of the most widely deployed parallel file systems today.

GPFS implements a block-based file system, with clients either directly accessing disk blocks via a storage area network or indirectly accessing disk blocks through a software layer (called virtual shared disk [VSD] or network shared disk) that redirects operations over a network to a remote system that performs access on the client's behalf. In large deployments, the cost of connecting all clients to the storage area network is usually prohibitive, so the software-assisted block access is more often employed.

Since GPFS is a parallel file system, blocks move through multiple paths, usually multiple servers, and are striped across multiple devices to allow concurrent access from many clients and high aggregate throughput. To match the network bandwidths of today's servers, disks used in GPFS deployments are typically combined into arrays. Files are striped across all RAIDs with

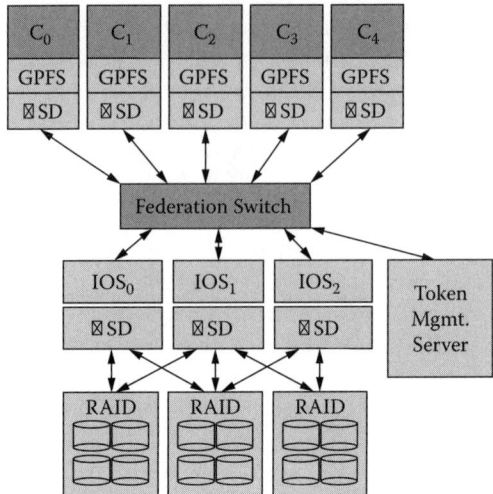

Figure 2.2 GPFS deployments include software on clients and servers. Clients communicate through a virtual shared disk (or network shared disk) component to access storage attached to I/O servers. Redundant links connect I/O servers to storage so that storage is accessible from more than one I/O server. A token management server coordinates access by clients to the shared storage space.

a large stripe unit, such as 4 MB. This allows for the maximum possible concurrency, but it is not efficient for small files.

For reliability, these arrays are often configured as RAIDs. GPFS also supports a replication feature that can be used with or without underlying RAID to increase the volume's ability to tolerate failures; this feature is often used just on metadata to ensure that even a RAID failure doesn't destroy the entire file system. Moreover, if storage is attached to multiple servers, these redundant paths can be used to provide continued access to storage even when some servers, or their network or storage links, have failed. Figure 2.2 shows an example of such a deployment, with clients running GPFS and VSD software and servers providing access to RAIDs via VSD.

This diagram also shows the token management server, an important component in the GPFS system. GPFS uses a lock-based approach for coordinating access between clients, and the token management server is responsible for issuing and reclaiming these locks. Locks are used both to control access to blocks of file data and to control access to portions of directories and allow GPFS to support the complete POSIX I/O standard. Clients obtain locks prior to reading or writing blocks, in order to maintain consistency, and likewise obtain locks on portions of directories prior to reading the directory or creating or deleting files.

Because GPFS is often used in enterprise settings that have limited concurrent write access to shared files, GPFS incorporates a special optimization to reduce lock acquisition overheads for single writers. Specifically, when the token manager sees a single writer accessing a file, it grants a lock to that writer that covers the entire file, allowing that writer to operate on the file without further interactions with the token manager. However, if another writer appears for that file, the lock is revised so that it covers only the region before, or the region after, the region requested by the new client (depending on where it requested its lock originally). This process of breaking up locks is continued if more writers appear, allowing greater concurrency at the expense of increased lock traffic.

Because GPFS has a robust locking system, GPFS is able to extensively cache both data and metadata on clients. Caching is performed by the GPFS page pool. Data can be prefetched and held in page pool buffers on the client, and writes can be buffered and written after control is returned to the client process. Likewise, directory contents may be cached to speed directory listings. These techniques, like lock optimization, are particularly helpful in environments where sharing is limited.

2.2.2 Object-Based Storage Systems

More recently, parallel file systems using *object-based* storage have emerged. Object-based storage devices differ from traditional block storage devices in that they present their storage as a collection of "objects" rather than a collection of blocks. These objects can store data of variable size, as well as a set of attributes on that data. Object-based storage devices also provide functionality for searching based on the attributes of objects and for grouping objects into collections. Overall these features shift the burden of local storage management away from the file system and into the device, allowing file system implementers to focus on other design challenges. Unfortunately, while the interfaces for object-based storage devices have been specified, few products currently provide these capabilities. For this reason, parallel file system designs relying on object-based storage capabilities provide these capabilities by layering software or firmware on top of traditional block storage devices (disks).

2.2.2.1 PanFS

Panasas provides a distributed, parallel file system and highly available, clustered metadata and storage server hardware to large-scale, Linux-based compute clusters used in government, academic, and commercial high-performance computing.[5-7] Panasas is the primary storage used by the world's first petaFLOP computer, Los Alamos National Laboratory's Roadrunner. Panasas systems are constructed of one or more shelves, shown in Figure 2.3. Each shelf is a bladeserver with 11 slots that can contain storage servers,

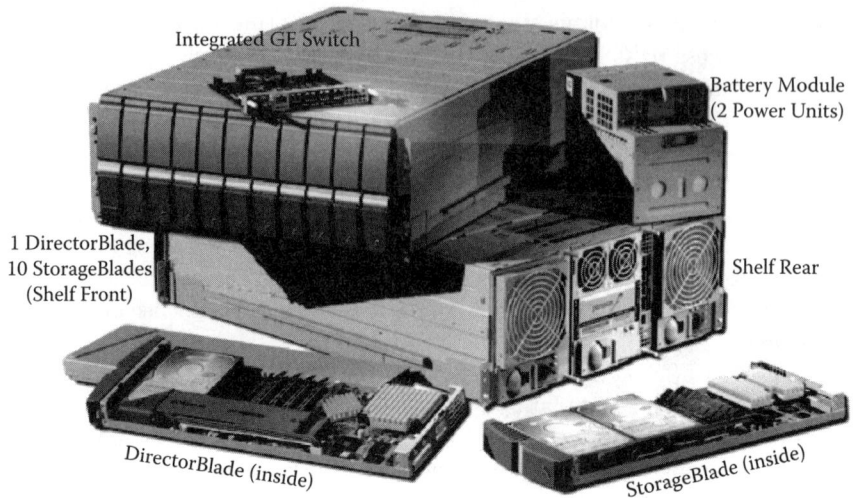

Figure 2.3 (See color insert following page 224.) Panasas storage clusters are built of bladeserver shelves containing object-serving StorageBlades; metadata-managing DirectorBlades; batteries; and redundant power, cooling, and networking.

called StorageBlades, or metadata managers, called DirectorBlades. All slots share redundant gigabit or 10 gigabit Ethernet networking, redundant cooling, redundant power supplies, and an integrated uninterruptable power supply (UPS) large enough to cleanly shut down all blades on power failure. StorageBlades contain two large-capacity commodity disk drives and an embedded controller with a variable amount of cache memory and implement an object storage protocol,[5,8] standardized as the SCSI Object Storage Device (OSD) protocol,[40] transported over IP-based Ethernet using the iSCSI storage area network. DirectorBlades contain a small local disk for code and journals and a server-class processor and memory, and implement network file system (NFS) and common Internet file system (CIFS) protocol servers to access Panasas storage from legacy non-Linux clients as well as the PanFS parallel file system metadata manager. When additional shelves are powered up in the same domain as an existing Panasas system, the new blades are recognized and incorporated into the existing system; and, under administrator control, the existing storage data can be transparently rebalanced over the existing and new blades.

Panasas storage was designed to be accessed in parallel from each client in one or more Linux compute clusters, as illustrated in Figure 2.4. Files are striped over objects stored on different storage blades so that a client can increase performance by accessing multiple objects at the same time in parallel, and multiple clients can be accessing the same or different files in parallel at the same time, up to the bandwidth limitations of the network between

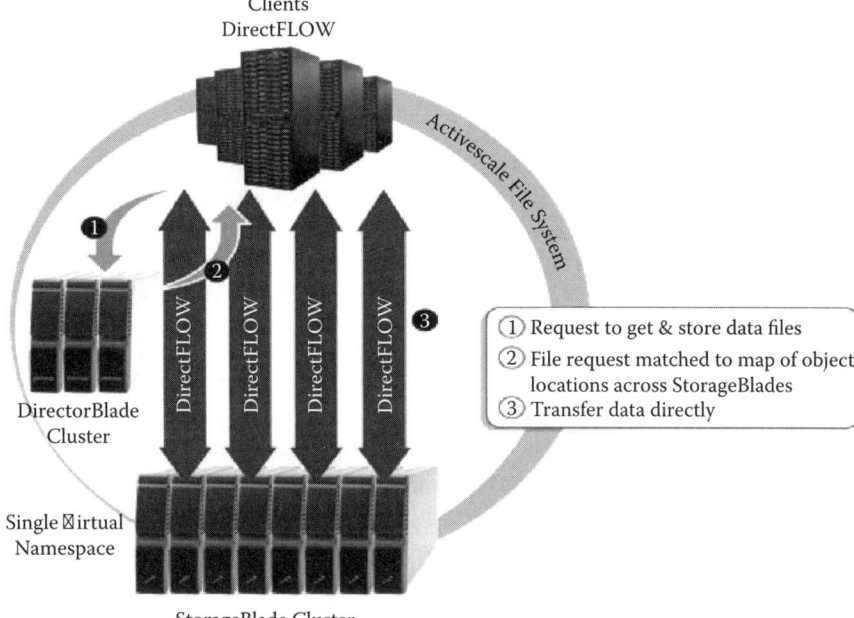

Figure 2.4 Panasas storage uses the DirectFlow installable file system to enable clients to consult clustered metadata servers for maps to the location of data on storage that can be directly and independently accessed in parallel.

storage and clients. Because this object storage protocol is not yet distributed with every release of Linux, Panasas offers a file system client module, called DirectFlow, that can be installed onto each client in the compute cluster. All of Panasas storage then shows up in the namespace of each DirectFlow client at /panfs. Accessing a file in /panfs for the first time causes DirectFlow to communicate with the metadata managers to request permission to access the file, the reply to which will contain a map of the object locations of the file on storage servers. With a map, and the security capabilities that come with it, the client can directly access object data in parallel and rarely needs help from metadata managers, which could not support nearly as high a total bandwidth if all of the data had to pass through the metadata managers as it does in traditional distributed file systems like NFS or CIFS. Since, however, not all machines at a site are in the Linux cluster, the metadata managers in Panasas also implement NFS and CIFS servers to enable other machines to get to the same data. Because Panasas clients, including the metadata's NFS and CIFS servers, have consistent client caches and use a distributed locking protocol to maintain correctness during concurrent access, all NFS and CIFS servers offer the same exported data, a property called clustered network-attached storage (NAS).

Panasas storage clusters are widely available.[6,7] Shelves include redundant networking, power supplies, cooling fans, and an integrated UPS for hardware fault tolerance. Software fault tolerance is provided by three different layers: a replicated global state database and directory; failover mirroring of metadata manager state; and a novel declustered, object-based RAID implementation. A quorum consensus voting protocol is employed on a subset of metadata managers to coordinate changes in configuration and blade states. These changes are rare and need to be consistent across all blades. The manager that implements this global state also implements a directory service so that all blades and clients can find all other services through it. The domain name system (DNS) name of the entire cluster resolves to the addresses on which this directory service is available, so a client's mount table needs little more than the storage cluster's DNS name to bootstrap all services. Metadata managers, however, change state far too often and without needing global synchronization. Metadata manager state changes are mirrored on a backup metadata manager, and each manager journals changes against reboot failures to avoid file system checking.

The RAID implementation of Panasas is unique in several ways. First, each file has its own RAID equation, so small files can be mirrored for efficient small file update, and large files can be parity protected for low-capacity overhead and high large-transfer bandwidth. Then all files are placed in objects spread over all object servers in such a manner that the failure of any object server will engage a small fraction of all other object servers in its reconstruction, as shown in Figure 2.5. This distributed RAID scheme, known as declustering, enables much higher read bandwidth during reconstruction and lower interference in user workload relative to traditional RAID that reconstructs a failure with the help of all of the storage in a small subset of the surviving disks. Panasas RAID also reserves spares differently from most RAID implementations; it reserves a fraction of the capacity of all blades

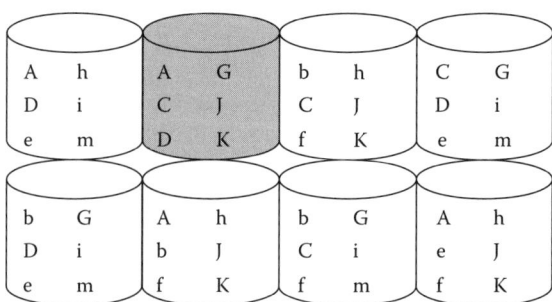

Figure 2.5 Declustered object-based RAID in Panasas storage defines a different RAID equation for every file, randomizing placement so that all disks assist in every reconstruction.

instead of a few dedicated idle spare blades, so that reconstructed data can be written in parallel to all surviving blades much faster than to a single replacement blade. Because each file is an independent RAID equation, PanFS also distributes the rebuild work to all metadata managers and incrementally puts rebuilt files back into normal mode, rather than waiting until all data is recovered. Collectively parallel reconstruction of declustered per-file RAID into reserved distributed spare space yields reconstructions that get faster in bigger systems, whereas traditional RAID reconstruction does not get faster, especially because the amount of work gets larger as the disks get bigger.

A second unique feature of Panasas RAID is that the resistance each disk provides against sectors being unable to be read after they are written is made much stronger by another layer of correcting code in each disk. This protects against media read errors failing a reconstruction by repairing the media error before the data leaves the storage blade.

A third unique feature of Panasas RAID is that clients can be configured to read the RAID parity (or mirror) when reading data to verify the RAID equation. Since PanFS clients compute RAID parity on writing, this allows end-to-end verification that the data has not been damaged silently, on the disk or in network or server hardware. Similar to disk checksums, this end-to-end parity provides against a much wider range of potential silent failures.

The Panasas parallel file system and storage cluster is a high-performance computing storage technology embodying many new technological advances. But it is also an integrated solution designed for ease of use and high availability.

2.2.2.2 PVFS

The parallel virtual file system (PVFS) project began at Clemson University in the early 1990s as an effort to develop a research parallel file system for use on cluster computers. Since then, the project has grown into an international collaboration to design, build, and support an open source parallel file system for the scientific computing community. The project is led by teams at Argonne National Laboratory and Clemson University. PVFS is widely used in production settings in industry, national laboratories, and academia. PVFS is freely available under an LGPL/GPL license from http://www.pvfs.org, and it has served as a starting point for many research projects. Currently Linux clusters and IBM Blue Gene/L and Blue Gene/P systems are supported, with preliminary support for Cray XT series systems. TCP/IP, InfiniBand, Myrinet GM and MX, and Portals networks are natively supported for PVFS communication.

PVFS was designed before the "object-based" nomenclature became popular, but the approach is similar to that used by Lustre and other object-based file systems. File data is striped across multiple file servers, with the stripe units belonging to a particular server being stored together on a local disk at that server. In PVFS the container for the stripe units is called a "datafile," and the current implementation stores these datafiles in a directory structure on a local file system (e.g., XFS). A set of parameters, stored with the metadata for a file, defines how the bytes of a particular file are mapped into the datafiles stored on different servers. Clients access files by requesting reads or writes of byte regions of these datafiles, and accesses do not need to be aligned on block or page boundaries.

PVFS servers can be configured to manage data, metadata, or both. At configuration time the administrator chooses which servers will be responsible for what types of data. This approach allows the administrator to take advantage of special hardware, such as solid state disks, or to place data based on reliability of the underlying hardware.

Because PVFS is freely available, it is often installed on commodity hardware as a scratch file system. In these configurations PVFS is not fault tolerant: the loss of a server holding metadata will likely cause the file system to become unusable until that server is returned to service, and the loss of a disk is likely to lose data permanently. However, PVFS may also be configured as a fault-tolerant system when appropriate hardware is available. Using a RAID volume for underlying storage will allow a PVFS file system to tolerate disk failures. Using hardware with multiple paths to the underlying storage, in conjunction with heartbeat software, will allow PVFS to tolerate server and network link failures as well. (Such a configuration is discussed in greater detail in the next section.)

One characteristic of PVFS that separates it from most other parallel file systems is its lack of critical state stored on clients. By eliminating this type of data on clients, client failures have no significant impact on a running file system. In contrast, systems like GPFS must correctly detect client failures and recover from these failures, for example, by reclaiming locks that the client might have held, in order to ensure that other clients can access files and directories that were being accessed by a dead client. This ability to ignore client failures becomes increasingly important as systems are built with ever larger numbers of clients, because of the increasing chance of client failures.

Because PVFS does not implement locks (those would be critical state on a client), coherent caching and buffering on the client are not implemented. Hence, all client I/O operations must pass over the network, and as a result small file operations can be slower than on systems such as GPFS where caching and buffering can hide this latency. For input datasets that will not change, PVFS does allow the user to mark these files as *immutable*, and PVFS will cache data read from these files on clients for performance purposes, providing some of the benefits of caching without the overhead and consistency complications of locking. Likewise, metadata operations such as creating a new file are performed atomically, so locks are not needed.

2.2.2.3 Lustre

Lustre is a shared file system for clusters originally developed by Cluster File Systems, Inc.[2] In the June 2006 TOP500 list, over 70 of the 500 supercomputers used Lustre technology.[9] The Lustre architecture is made up of file system clients, metadata servers (MDSs), and object-storage servers (OSSs). File system clients handle the requests from user applications and communicate with MDSs and OSSs. MDSs maintain a transactional record of high-level file information, such as directory hierarchy, file names, and striping configurations. OSSs provide file I/O service, and each can be responsible for multiple object storage targets (OSTs). An OST is an interface to a single, exported backend storage volume.

Lustre is an object-based file system targeting strong security, file portability across platforms, and scalability. An object is a storage container of variable length and can be used to store various types of data, such as traditional files, database records, and multimedia data. Lustre implements object-based device drivers in OSTs to offload block-based disk space management and file-to-block mapping tasks from the servers. The object-based device functionalities are built on top of the ext3 file system, with objects stored in files. Similar to traditional UNIX file systems, Lustre uses inodes to manage file metadata. However, the inode of a file on the MDSs does not point to data blocks but instead points to one or more objects associated with the files.

In order to improve I/O performance for large file accesses, a file can be striped into multiple objects across multiple OSTs. The file-to-object mapping information along with other metadata for the file, is stored on an MDS. When a client opens a file, the MDS returns the file's inode so that the client uses this information to convert file access requests to one or more object access requests. Multiple object access requests are performed in parallel directly to multiple OSTs where the objects are stored. Lustre's metadata server software is multithreaded to improve metadata performance. Substantial modifications to the ext3 and Linux VFS have been made to enable fine-grained locking of a single directory. This optimization proves very scalable for file creations and lookups in a single directory with millions of files.

The Lustre failover technique protects the file metadata through MDS replication. This is done by directly connecting the MDSs to a multiport disk array in an OST. When one MDS fails, the replicated MDS takes over. In addition, MDSs can be configured as an active/passive pair. Often the standby MDS is the active MDS for another Lustre file system, so no MDSs are idle. To handle OSS failure, Lustre attaches each OST to different OSSs. Therefore, if one OSS fails, its OSTs can still be accessible through the failover OSSs. Lustre also provides journaling and sophisticated protocols to resynchronize the cluster in a short time.

Lustre is a POSIX-compliant file system. In particular, Lustre enforces the atomicity of read and write operations. When application threads on different compute nodes read and write the same part of a file simultaneously, they see consistent results on the overlapped part of the file. Atomicity for both file data and metadata is achieved by a distributed lock mechanism. In this mechanism, each OST runs a lock server and manages the locking for the portions of files that reside on that OST. Having multiple lock servers removes the potential communication bottleneck on a single lock server. Thus, the lock server capacity scales with the number of OSTs.

The distributed lock mechanism is also used to maintain the globally coherent client-side file cache. When a client requests a lock that conflicts with a lock held by another client, a message is sent to the lock holder asking for the lock to be dropped. Before that lock is dropped, the client must write back any dirty data and remove all data from its cache. A configurable time-out value is used in case of a client's failure. If a client does not drop its lock before the time-out, it is evicted from that OST and will not be allowed to execute any operations until it has reconnected. Locks for an arbitrary byte range are allowed; however, OSTs align the granted locks to file system block boundaries.

Lustre adds a lock manager extension to optimize the performance of reading file metadata. Reading an object's attributes is commonly seen when users track the progress of a job by checking the listing of output files being produced. In order to obtain the most updated file attributes, a lock on the metadata must be obtained. Since the file data is not required, however, such locks can be costly. Lustre adopts the *metadata intent locking method* that bundles up the file attributes as the return of a lock request. By having all of the information available during the initial lock request, all metadata operations can be performed in a single remote procedure call (RPC). If the file is not being actively modified, then the server will grant a lock so that the client can cache the attributes.

2.3 Example Parallel File System Deployments

Having discussed some example parallel file systems, we now will examine how two of these, PVFS and PanFS, are deployed in leadership computing environments.

2.3.1 PVFS on the IBM Blue Gene/P

The IBM Blue Gene series of supercomputers is designed to deliver extreme-scale performance in conjunction with low power consumption and very high reliability. The Blue Gene/P (BG/P) system is the second in this series and was designed in conjunction with the U.S. Department of Energy to ensure that the design best meets the needs of the computational science community. A BG/P rack consists of 1,024 quad-core nodes with 2 terabytes of memory

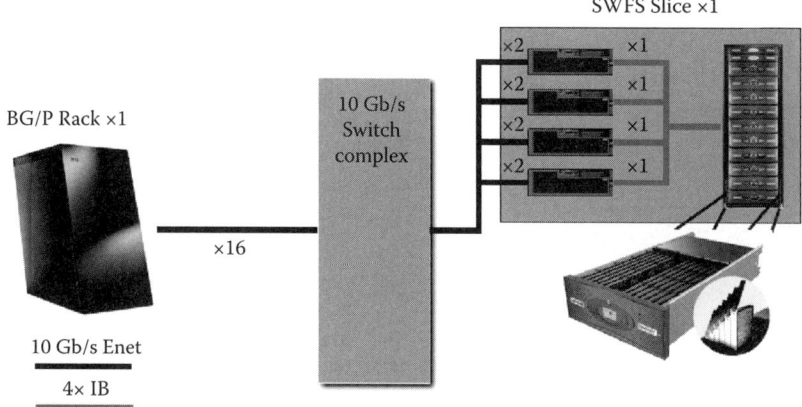

Figure 2.6 (See color insert following page 224.) Each rack of Blue Gene/P hardware in the ALCF system is connected to storage via 16 10-gigabit-per-second links. On the storage side, servers are grouped into units of 4 servers attached to a single rack of disks via InfiniBand. There are 17 of these storage "slices" providing storage services to the ALCF system.

and a peak performance of 13.9 teraflops. BG/P compute nodes perform point-to-point communication via a 3-D torus network, while separate networks for global barriers and collective operations are provided to enable even higher performance for specific common group communication patterns. An additional tree network is used for booting and for offloading system call servicing. This network connects the compute nodes to separate I/O nodes that serve as gateways to the outside world, using 10 gigabit Ethernet for communication.

2.3.1.1 Argonne Blue Gene/P System

The largest Blue Gene/P in operation at the Argonne Leadership Computing Facility (ALCF) is a 40-rack system, providing 80 terabytes of system memory and a peak performance of 556 teraflops (Figure 2.6). The ALCF BG/P is configured with one I/O node for every 64 compute nodes. A Myricom Myri-10G switch complex is used to attach the BG/P to storage and provides redundant connections to each of 128 file servers. Those servers supply the system with an aggregate of 4.3 petabytes of storage using 17 data direct network (DDN) storage devices with a peak aggregate I/O rate of 78 gigabytes per second.

2.3.1.2 Blue Gene/P I/O Infrastructure

Figure 2.7 shows the software involved in the I/O path on the BG/P. Because there is no direct connection between Blue Gene compute nodes and the outside world, the BG/P system uses *I/O forwarding* to accomplish I/O

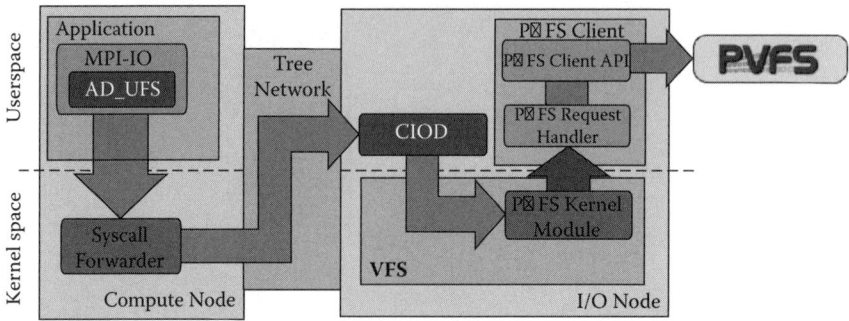

Figure 2.7 I/O accesses from the Blue Gene compute node are serialized and forwarded to a Blue Gene I/O node, where they are passed through the VFS layer for service. On the ALCF system, high-performance I/O is provided via the PVFS file system, which in turn performs file system access through a user-space process.

for applications. The compute kernel (a custom kernel written by IBM) marshals arguments from I/O calls and forwards a request for I/O over the BG/P tree network to the I/O node associated with that compute node. The I/O node, running Linux, processes this request and performs the I/O on behalf of the compute node process by calling the appropriate system calls on the I/O node. This approach allows for I/O to any Linux-supported file system, such as an NFS-mounted file system, a Lustre file system, a GPFS file system, or a PVFS file system. The ALCF system provides PVFS as a high-performance file system.

The PVFS file system is mounted on I/O nodes using the PVFS kernel module. The Linux kernel vectors I/O operations to the mounted PVFS volume to this kernel module. This module, in turn, forwards operations through a device file to a user-space process. This user process interacts with server processes on the file servers, which in turn perform block I/O operations on the DDN storage devices.

2.3.1.3 Tolerating Failures

The combination of PVFS and Linux-HA (high availability) heartbeat software makes an excellent parallel file system solution for Blue Gene with respect to tolerating failures. The Blue Gene I/O nodes are the real PVFS clients in this system; and if an I/O node fails, only the compute nodes attached to that I/O node lose connectivity. Because PVFS clients do not hold file system state, the failure of an I/O node has no impact on other I/O nodes, allowing jobs associated with other I/O nodes to continue operation.

If a server fails, the heartbeat software will detect this failure through the use of quorum. The failed server will be forcefully shut down using intelligent platform management interface (IPMI) power controls, ensuring that the server is no longer accessing storage. Once this shutdown is accomplished, the storage for that file system is mounted on one of the other active servers and a new PVFS server process is started, filling in for the failed server until the original server is brought back online. Using heartbeat clusters of eight nodes, up to three server failures can be tolerated per group of eight servers before the PVFS file system would become unavailable.

2.3.2 PanFS on Roadrunner

On June 9, 2008, the Department of Energy (DOE) announced that Los Alamos National Laboratory's Roadrunner supercomputer was the first computer to exceed a petaFLOP, or 1,000 trillion operations per second, of sustained performance according to the rules of the top500.org benchmark. Roadrunner will be used to certify that the U.S. nuclear weapons stockpile is reliable, without conducting underground nuclear tests. Roadrunner will be built in three phases and will cost about $100 million. It is the first "hybrid" supercomputer, in that it achieves its performance by combining 6,562 dual-core AMD Opteron (x66 class) processors and 12,240 IBM Cell Broadband Engines, which originated from the designs of Sony Playstation 3 video game machines. It runs the Linux operating system in both Opteron and Cell processors, contains 98 TB of memory, is housed in 278 racks occupying 5,200 square feet, and is interconnected node to node with 10 Gbps InfiniBand and node to storage with 10 Gbps Ethernet. One of the most surprising results is that the world's fastest computer (at that time) was also the third most power efficient, according to the green500.org list of supercomputers; it achieves 437 million operations per watt consumed, whereas computers previously topping the speed list were 43rd and 499th on the green list.

Roadrunner is organized as subclusters called compute units, each with 12 I/O nodes routing storage traffic between the clusters and Panasas storage clusters. All Roadrunner compute nodes are diskless; their operating system runs from a RAMdisk with external storage access using Panasas DirectFlow and NFS. The attached Panasas storage contains over 200 shelves of Panasas storage shared over Roadrunner's phases and older Los Alamos supercomputers Lightning and Bolt. Each shelf contains 10 StorageBlades and one DirectorBlade, for a total of over 3 PB, 2,000 object servers, 4,000 disks, and 200 metadata managers.

Figure 2.8 shows the storage architecture of the Panasas storage connected to Roadrunner and shared with other supercomputers in Los Alamos's "red"-level secure computing facility. All supercomputers have access to all Panasas storage in this facility, with the amount of bandwidth available to each cluster determined by the I/O node resources on the supercomputer. While most supercomputers have dedicated storage directly attached to the cluster's I/O nodes, Los Alamos shares storage resources to reduce copying overhead and reduce the cost of scaling storage bandwidth with each new cluster. Their storage architecture, called PaScalBB (parallel and scalable server I/O backbone

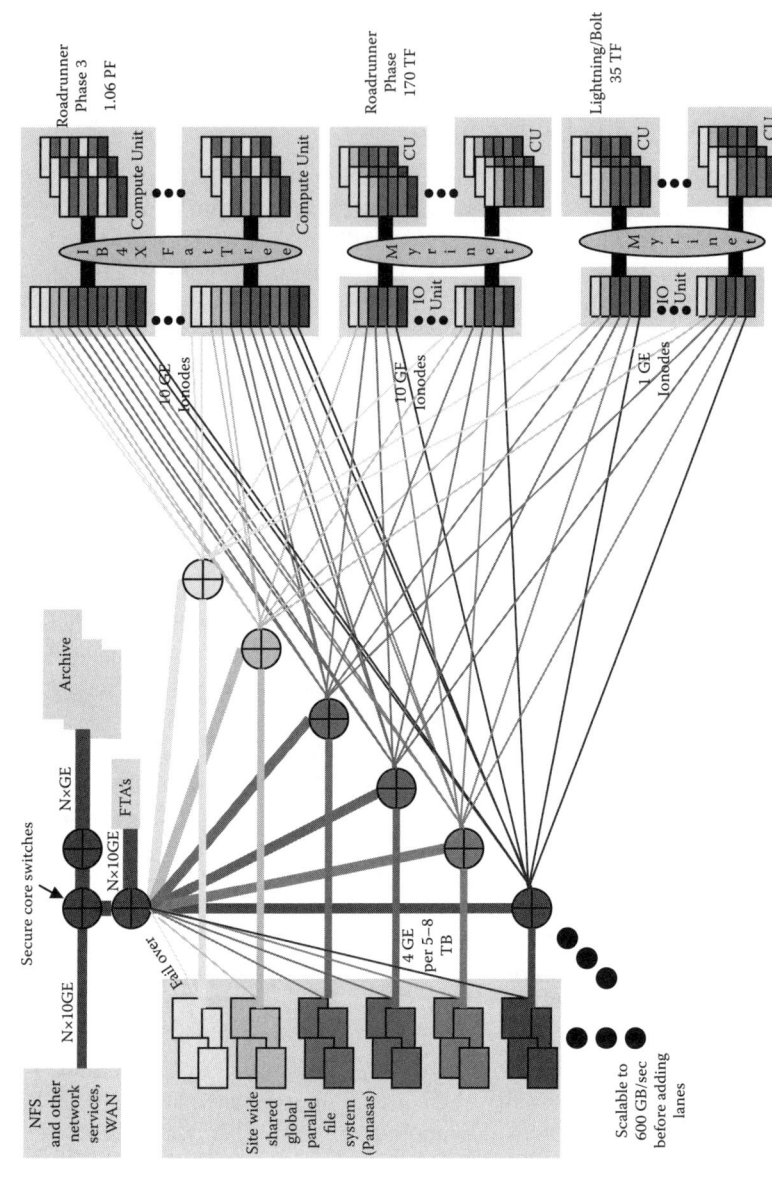

Figure 2.8 (See color insert following page 224.) Los Alamos secure petascale infrastructure diagram with Roadrunner.

network), is novel in the way it scales.[10] Panasas storage shelves are added one "lane" at a time. A lane is a collection of shelves sharing a (redundant) lane switch that is connected to a subset of the I/O nodes in each cluster, with a much lower bandwidth route between shelves for Panasas internal traffic. Most of a cluster's I/O nodes are not routed to each lane, so that the storage switches do not have to get larger as the number of lanes increase. Thus, routing to a storage device is done primarily in the compute nodes when they select an I/O node to route through. In contrast, many supercomputers have a full bisection bandwidth storage network between I/O nodes and storage devices so that compute node to I/O node routing can be based only on the compute node address. With this arrangement, however, the storage network may become very expensive as the storage scales. Additionally, PaScalBB I/O nodes are redundant and load balanced by using standard IP routing software in the Linux client software and Panasas devices.

2.4 Interfacing with Applications

Computational science applications are complex software systems that operate in terms of their own data models. When the time comes for these applications to interact with the storage system, these internal data models must be mapped onto structures that the storage system understands. Additional software layers—an "I/O software stack"—are layered on top of the underlying storage hardware. The interfaces that these layers provide help applications use the available storage efficiently and easily.

The I/O software stacks being deployed today consist of three layers. At the lowest layer, parallel file systems maintain a global name space and provide efficient and coherent access to data. These file systems present a POSIX or POSIX-like interface[11] for applications as well as richer interfaces for use by upper layers. The second layer is the I/O middleware layer. This layer is responsible for mapping I/O into the programming model being used by the application, with MPI-IO[12] being a good example. The top layer is the high-level I/O library layer, with software such as Parallel netCDF[13] and HDF5[14] providing structured interfaces and data formats to help bridge the gap between application data models and the storage system.

In this section we discuss some of the common patterns of access in applications and then cover the interfaces that the I/O software stack makes available for applications to interact with underlying storage.

2.4.1 Access Patterns

As a result of the number of processes involved and the kinds of data being stored, computational science applications have access patterns that in many cases are different from patterns seen in enterprise settings. Two access pattern studies were performed in the mid-1990s, the CHARISMA project[15] and the Scalable I/O Initiative Applications Group study.[16] Although applications and systems have changed some since these studies were performed, many of their conclusions about application access patterns are still valid today. We summarize those findings here.

The CHARISMA study was conducted on two systems, a Thinking Machines CM-5 and an Intel iPSC/860, both with numerous scientists running on them. The team was hoping to discover some commonalities between parallel applications on the two machines in terms of the number of files read and written, the size of the files, the typical read and write request sizes, and the way these requests were spaced and ordered.

The team found it important to differentiate between *sequential* and *consecutive* requests, where sequential accesses begin at a higher file offset than the point at which the previous request from the same process ended, but consecutive ones begin at exactly the point where the last request ended. Almost all write-only files were accessed sequentially, and many of the read-only files were as well. Most of the write-only files were written consecutively, probably because in many applications each process wrote out its data to a separate file. Read-only files were accessed consecutively much less often, indicating that they were read by multiple applications. Overall, about a third of the files were accessed with a single request.

Examining the sizes of the intervals between requests, the team found that most files were accessed with only one or two interval sizes. The request sizes were also very regular, with most applications using no more than three distinct request sizes. Tracing showed that a simple strided access pattern was most common, with a consistent amount of data skipped between each data item accessed. Nested strided patterns were also common, indicating that multidimensional data was being accessed within the file, but occurred about half as often as the simple strided pattern.

The CHARISMA project team concluded that parallel I/O consists of a wide variety of request sizes, that these requests often occur in sequential but not consecutive patterns, and that there is a great deal of interprocess spatial locality on I/O nodes. They believed strided I/O request support from the programmer's interface down to the I/O node to be important for parallel I/O systems because it can effectively increase request sizes, thereby lowering overhead and providing opportunities for low-level optimization.

The Scalable I/O Initiative Applications Group study was performed using three I/O-intensive scientific applications designed for the Paragon. The goal of this work was to observe the patterns of access of these applications, determine what generalizations could be made about the patterns in high-performance applications, and discuss with the authors the reasons for choosing certain approaches.

The group found that the three applications exhibited a variety of access patterns and requests sizes; no simple characterization could be made.

Small I/O requests were common; although programmers can aggregate requests in the application, they often did not.

From conversations with the application programmers, the group found that the I/O capabilities of the system did not match the desired ones. This mismatch resulted in complications in application code and also reduced the scope of feasible problems. For example, in two of the applications the programmers found it easier to read data into a single node and then distribute to the remaining nodes rather than use parallel access modes, because the available modes on the Paragon did not allow for the particular pattern of access they needed.

Since these studies, many advances have been made in I/O middleware and high-level I/O libraries to help applications more effectively access storage. In particular, efforts have been made to make it easier to describe strided access patterns and the relationships between the accesses of different processes. However, we still see applications that write a file from each process and other applications that perform all I/O through one process, and neither of these I/O approaches scales well. As these applications attempt to execute at larger scale, we hope that better I/O tools will be adopted.

2.4.2 POSIX I/O

The standard interface for file I/O is defined as part of the POSIX standard.[11] The interface was developed at a time when file systems were typically local to a single operating system instance, and concurrent access occurred only between multiple processes running on the same computer.

The interface provides a simple set of functions for accessing contiguous regions of files as well as functions for creating and removing files and directories and traversing the contents of directories. It further defines a strict set of semantics for these operations, guaranteeing sequential consistency of I/O operations in all cases, meaning that all operations appear to have occurred one at a time in some sequential ordering that is consistent with the ordering of the individual operations. So, if two processes simultaneously write to the same region, it will appear as if one wrote first, and then the other second. Likewise, if one process reads while the other writes, the reader will see either all or none of the writer's changes. This API provides a convenient mechanism for serial applications to perform a variety of common tasks, and in the UNIX environment it is also the mechanism through which many system activities are performed (by treating devices as special types of files). The limitations of the POSIX interface become apparent only when a file system is made available over a network or simultaneously made available to more than one operating system instance.

When an operating system is interacting with a file system that is attached locally, the latency of communication with that file system is very short. In this environment, the simple building blocks that POSIX provides for performing I/O are more than adequate, because the cost of assembling more complex operations from many small ones is minimal. When a network link is introduced between the file system and the operating system, however, the costs change. The complex access patterns of parallel applications described previously require many individual POSIX operations, each incurring a round trip time at the minimum. Performance drops quickly. Network file system developers combat this problem by enabling read-ahead, caching, and write-back on file system clients (the operating system instances accessing the file system). These allow the client to avoid network communication when predictable patterns are present and adequate memory is free for caching.

When caching is introduced, the consistency semantics of POSIX then become a significant challenge. Because operations must be sequentially consistent, the file system must strictly manage concurrent access to file regions. In parallel file systems, access management typically is accomplished with a single-writer, multiple-reader, distributed-range locking capability that allows concurrent reading in the absence of writers but retains sequential consistency in the presence of writers. This is a proven technique, but maintaining locks introduces communication again, and the locks become a state that must be tracked in case of client failure.

In parallel applications, developers are accustomed to synchronizing processes when necessary and rarely overwrite file regions simultaneously. In order to enable higher-performance application I/O than is possible through the POSIX API, a richer I/O language was needed that enables developers to describe and coordinate access across many application processes, and that enables the greatest degree of concurrency possible.

2.4.3 MPI-IO

The MPI-IO interface is a component of the MPI-2 message-passing interface standard[12] and defines an I/O interface for use in applications using the MPI programming model. The model for data in a file is the same as in POSIX I/O: a stream of bytes that may be randomly accessed. The major differences between POSIX I/O and MPI-IO lie in the increased descriptive capabilities of the MPI-IO interface. Language bindings are provided for C, C++, and Fortran, among others.

One trend that emerged from access pattern studies was that applications often access data regions that are not consecutive, that is, *noncontiguous*, in the file. The POSIX interface forces applications exhibiting these patterns to perform an access per region[1]; this constraint makes this type of access very inconvenient to program, and performing many small operations often

---

[1]POSIX does define the lio_listio call that may be used to access noncontiguous regions, but it places a low limit on how many regions may be accessed in one call (often 16), limits concurrency by forcing the underlying implementation to perform these accesses in order, and requires that memory regions and file regions be of the same size. These constraints make lio_listio of limited use in scientific applications.

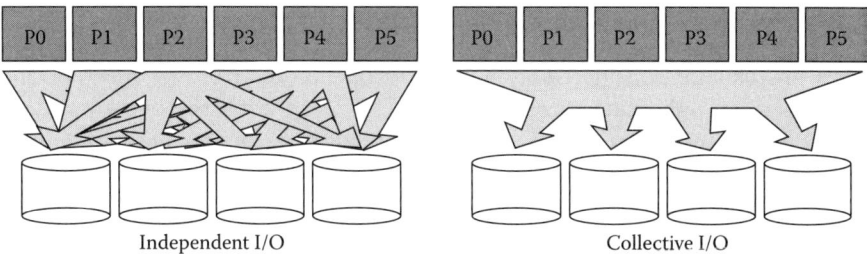

Figure 2.9 Independent I/O (left) results in an uncoordinated stream of accesses to the parallel data storage system, often leading to poor access patterns and low performance. By describing accesses as collective ones, application programmers open up opportunities for software at the MPI-IO layer to reorganize and coordinate storage access, leading to more efficient access and higher performance.

leads to very poor performance. The MPI-IO interface addresses noncontiguous I/O support through using the MPI datatype facility, familiar to MPI programmers, to describe regions in memory and file. This allows MPI-IO implementations to perform optimizations behind the scenes that enable efficient noncontiguous I/O as well.

Computational science applications written in the MPI model execute as a large number of communicating processes oriented at a single goal, such as simulating global weather patterns over time. Because the processes are working in a coordinated manner, MPI applications often use *collective* communication. Collective communication calls require that all processes in a group participate, and these calls provide MPI implementation opportunities to optimize communication between all the participants, such as utilizing special hardware support or scheduling communication to best use the underlying network. Because collective communication operations are so useful in applications, the MPI-IO specification includes collective I/O operations that fill a similar role in I/O (Figure 2.9). These calls are particularly appropriate for applications with phases of computation followed by I/O. Collective I/O calls provide a critical optimization opportunity, describing the relationship between the accesses of many processes. This allows the underlying MPI-IO implementation to schedule I/O traffic to the underlying file system or to reorganize data between processes in order to combine data and align accesses to best match file system needs.[41]

Support for collective and noncontiguous I/O together provides a rich language for describing application I/O patterns, and implementations such as ROMIO[39] leverage the additional information passed through this interface to optimize I/O to great effect. From the application programmer's perspective, however, MPI-IO does not significantly improve the usability of the storage system. This situation has led to the development of high-level I/O libraries.

2.4.4 High-Level I/O Libraries

Files are usually considered as a linear sequence of bytes by most of the file systems. Applications are responsible for interpreting the bytes into logical structures, for instance a two-dimensional array of floating-point numbers. Without metadata to describe the logical data structures, a program has difficulty telling what the bytes represent. Therefore, in order to ensure portability, a file's metadata must accompany the file at all times. This requirement is particularly important for scientific data because many scientific data libraries, such as for visualization and data mining, manipulate data at a higher level than byte streams.

This section describes two popular scientific data libraries, parallel netCDF and HDF5. Both libraries store metadata along with data in the same files. In addition, both define their own file formats and a set of APIs to access the files, sequentially as well as in parallel.

2.4.4.1 Parallel netCDF

The network common data form (netCDF) was developed at the Unidata Program Center.[17,18] The goal of netCDF is to define a portable file format so that scientists can share data across different machine platforms. Atmospheric science applications, for example, use netCDF to store a variety of data types that encompass single-point observations, time series, regularly spaced grids, and satellite or radar images.[19] Many organizations, including much of the climate community, rely on the netCDF data access standard for data storage.[20] However, netCDF does not provide adequate parallel I/O methods. For parallel write to a shared netCDF file, applications must serialize access by passing all the data to a single process that then writes to the file. The serial I/O access is both slow and cumbersome for the application programmer. A new set of parallel programming interfaces for netCDF files, parallel netCDF (PnetCDF), therefore has been developed.[13]

The netCDF file format follows the common data form language (CDL) suitable for interpreting data for human readers. It divides a netCDF file into two parts: file header and body. The header contains all information about dimensions, attributes, and scalar variables, followed by the body part containing arrays of variable values in binary form. The netCDF file header first defines a number of dimensions, each with a name and a length, which can be used to describe the shapes of arrays. The most significant dimension of a multidimensional array can be unlimited for arrays of growing size. Global attributes not associated with any particular array can also be added to the head. This feature allows programmers' annotation and other related information to be added to increase the file's readability. The body part of a netCDF file first stores the fixed-size arrays followed by the variable-sized arrays. For storing a variable-sized array, netCDF defines each subarray comprising all the fixed dimensions as a record, and the records are stored interleaved. All offsets of fixed-size and variable-size arrays are properly saved in the header.

One noticeable limitation of netCDF is the 2 GB file size due to the 4-byte integers used in the CDL format. PnetCDF lifts this limitation by adopting CDL version 2 format. However, even though the file size can grow beyond 2 GB, CDL version 2 still limits any single array size to 2 GB. The next generation of CDL format is under development and will remove this limitation.

The netCDF API is divided into five categories: dataset functions—create/open/close a file, switch to define/data mode, and synchronize changes to the file system; define mode functions—define array dimensions and variables; attribute functions—manage adding, changing, and reading attributes; inquiry functions—return metadata; and data access functions—read/write data in one of the five access methods (single element, whole array, subarray, subsampled array, and mapped strided subarray). Parallel netCDF retains the file format of netCDF version 3, and its implementation is built on top of MPI-IO, allowing users to benefit from existing I/O optimizations adopted in the underlying MPI-IO library, such as data sieving and two-phase I/O strategies in ROMIO[21–23,41] and data shipping in the IBM's MPI-IO library.[24,25] In order to seemingly integrate with the MPI-IO functions, PnetCDF APIs borrow a few MPI features, such as MPI communicators, info objects, and derived data types. An MPI communicator is added to define the participating I/O processes between the file's open and close scope. Adding MPI info objects allows users to pass access hints for further optimizations. The PnetCDF interface's define mode, attribute, and inquiry functions are collective in order to guarantee data consistency among the participating processes. There are two sets of parallel data access APIs. The high-level API closely follows the original data access functions where the read-write buffers must be contiguous in memory but file access can be noncontiguous. The flexible API provides a more MPI-like style of access to permit noncontiguous I/O buffers through the use of MPI datatypes. Similar to MPI-IO, PnetCDF data access functions are split into collective and noncollective modes. netCDF users with MPI background will find PnetCDF easy to adopt because of the many features inherited from MPI.

2.4.4.2 HDF5

Hierarchical data format version 5 (HDF5),[14] developed at the National Center for Supercomputing Applications (NCSA), is a widely used high-level I/O library that serves the same purposes as PnetCDF. HDF5 is a major revision of HDF version 4.[26] Similar to netCDF, HDF4 allows annotated multidimensional arrays and provides other features such as data compression and unstructured grid representation. HDF4 does not support parallel I/O, and file sizes are limited to 2 GB. HDF5 was designed to address these limitations.

As a major departure from HDF4, HDF5's file format and APIs are completely redesigned. An array stored in an HDF5 file is divided into header and body parts. There are four essential classes of information in a header: name, data type, dataspace, and storage layout. The name of the array is a text string. The data type describes the numerical type of array elements, which can be atomic, native, compound, or named. Atomic data types are the primitive data types, such as integers and floats. Native data types are system-specific instances of atomic data types. Compound data types are collections of atomic data types. Named data types are either atomic or compound data types that can be shared across arrays. A dataspace depicts the dimensionality of an array. Unlike netCDF, all dimensions of an HDF5 array can be either fixed or unlimited. The storage layout specifies the way a multidimensional array is stored in a file. The default storage layout format is contiguous, meaning that data is stored in the same linear way that it is organized in memory. The other storage layout is called chunked, in which an array is divided into equal-sized chunks, and chunks are stored separately in the file. Chunking has three important benefits. First, it provides the possibility to achieve good performance when accessing noncontiguous subsets of the arrays. Secondly, it allows large array compression. Third, it enables the dimension extension of an array in any direction. The chunking is also applicable to headers. Therefore, the HDF5 headers can be dispersed in separate header blocks for each object, not limited to the beginning of the file. Another important feature of HDF5 is called *grouping*. A collection of arrays can be grouped together, and a group may contain a number of arrays and other groups that are organized in a tree-based hierarchical structure.

HDF5 APIs are divided into 12 categories: general purpose, attributes, datasets, error handling, file access, grouping, object identifiers, property list, references, data-space, data type, and filter/compression. Writing an HDF5 file comprises the following steps: create a file, create groups, define dataspaces, define data types, create arrays, write attributes, write array data, and close the file. The parallel data access support in HDF5 is built on top of MPI-IO to ensure the file's portability. However, HDF5 does not separate its file access routines into collective and independent versions as MPI-IO does. Parallel I/O is enabled through the setting of properties passed to the file open and data access APIs. The properties tell HDF5 to perform I/O collectively or independently. Similar to PnetCDF, HDF5 allows accessing a subarray in a single I/O call, and it is achieved through defining *hyperslabs* in the dataspace. HDF5's chunking allows writing subarrays without reorganizing them into a global canonical order in the file. The advantage of chunking becomes significant when the storage layout is orthogonal to the access layout, for example, storage layout of a two-dimensional array being in the row major and the access pattern being in the column major. However, this high degree of flexibility in HDF5 can sometimes come at the cost of high performance.[13,27]

2.5 Future Trends and Challenges

High-performance computing systems continue to grow in computational capability and number of processors, and this trend shows no signs of changing. Parallel storage systems must adapt to provide the necessary storage facilities to these ever larger systems. In this section we discuss some of the challenges and technologies that will affect the design and implementation of parallel data storage in the coming years.

2.5.1 Disk Failures

With petascale computers now arriving, there is a pressing need to anticipate and compensate for a probable increase in failure and application interruption rates and in degrading performance caused by online failure recovery. Researchers, designers, and integrators have generally had too little detailed information available on the failures and interruptions that even smaller terascale computers experience. The available information suggests that failure recovery will become far more common in the coming decade and that the condition of recovering online from a storage device failure may become so common as to change the way we design and measure system performance.

The SciDAC Petascale Data Storage Institute (PDSI, www.pdsi-scidac.org) has collected and analyzed a number of large datasets on failures in high-performance computing (HPC) systems.[28] The primary dataset was collected during 1995–2005 at Los Alamos National Laboratory (LANL) and covers 22 high-performance computing systems, including a total of 4,750 machines and 24,101 processors. The data covers node outages in HPC clusters, as well as failures in storage systems. This may be the largest failure dataset studied in the literature to date, in terms of both the time period it spans and the number of systems and processors it covers. It is also the first to be publicly available to researchers (see Reference 29 for access to the raw data). These datasets and large-scale trends and assumptions commonly applied to future computing systems design have been used to project onto the potential machines of the next decade and derive expectations for failure rates, mean time to application interruption, and the consequential application utilization of the full machine, based on checkpoint/restart fault tolerance and the balanced system design method of matching storage bandwidth and memory size to aggregate computing power.[30] If the growth in aggregate computing power continues to outstrip the growth in per-chip computing power, more and more of the computer's resources may be spent on conventional fault recovery methods. Highly parallel simulation applications may be denied as much as half of the system's resources in five years, for example. New research on application fault-tolerance schemes for these applications should be pursued; for example, process pairs[31] mirroring of all computation is such a scheme that would halt the degradation in utilization at 50%.[28]

PDSI interest in large-scale cluster node failure originated in the key role of high-bandwidth storage in checkpoint/restart strategies for application fault tolerance.[32] Although storage failures are often masked from interrupting applications by RAID technology,[4] reconstructing a failed disk can impact storage performance noticeably.[33] If too many failures occur, storage system recovery tools can take days to bring a large file system back online, perhaps without all of its users' precious data. Moreover, disks have traditionally been viewed as perhaps the least reliable hardware component, due to the mechanical aspects of a disk. Datasets obtained describe disk drive failures occurring at HPC sites and at a large Internet service provider. The datasets vary in duration from one month to five years; cover more than 100,000 hard drives from four different vendors; and include SCSI, fibre channel, and SATA disk drives. For more detailed results see Reference 34.

For modern drives, the datasheet MTTFs (mean times to failure) are typically in the range of 1–1.5 million hours, suggesting an annual failure and replacement rate (ARR) between 0.58% and 0.88%. In the data, however, field experience with disk replacements differs from datasheet specifications of disk reliability. Figure 2.10 shows the annual failure rate suggested by the datasheets (horizontal solid and dashed line), the observed ARRs for each of the datasets, and the weighted average ARR for all disks less than five years old (dotted line). The figure shows a significant discrepancy between the observed ARR and the datasheet value for all datasets, with the former as high as 13.5%. That is, the observed ARRs are a factor of 15 higher than datasheets would indicate. The average observed ARR over all datasets (weighted by the number of drives in each dataset) is 3.01%. Even after removing all COM3

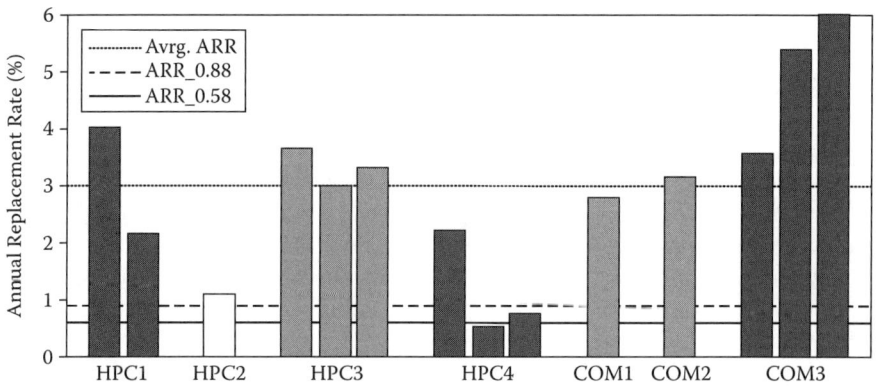

Figure 2.10 Comparison of data sheet annual failure rates (horizontal dotted lines) and the observed annual replacement rates of disks in the field.

data, which exhibits the highest ARRs, the average observed ARR was still 2.86%, 3.3 times higher than 0.88%.

With this cluster and disk failure data, various projections can be developed. First, integrators are projected to deliver petascale computers according to the long-standing trends shown on top500.org;[9] that is, the aggregate compute performance will double every year. Second, integrators will continue to build balanced systems; that is, storage size and bandwidth will scale linearly with memory size and total compute power.[30] As a baseline, projections model the Jaguar system at Oak Ridge National Laboratory after it is expanded to a petaFLOP system having approximately 11,000 processor sockets (dual-core Opterons), 45 TB of main memory, and a storage bandwidth of 55 GB/s.[35] While the architecture of other 2008 petascale machines, such as LANL's Roadrunner,[36] differs from Jaguar in its use of hybrid nodes employing vector/graphics coprocessors, predictions for its failure rates are little different from Jaguar, so they are not included in the following graphs.

Further, individual disk bandwidth will grow at a rate of about 20% per year, which is significantly slower than the 100% per year growth rate that top500.org predicts. In order to keep up, the number of disk drives in a system will have to increase at an impressive rate. Figure 2.11a projects the number of drives in a system necessary simply to maintain balance. The figure shows that, if current technology trends continue, by 2018 a computing system at the top of top500.org chart will need to have more than 800,000 disk drives. Managing this number of independent disk drives, much less delivering all of their bandwidth to an application, will be extremely challenging for storage system designers.

Second, disk drive capacity will keep growing by about 50% per year, thereby continuously increasing the amount of work needed to reconstruct a failed drive, and the time needed to complete this reconstruction. While other trends, such as decrease in physical size (diameter) of drives, will help to limit the increase in reconstruction time, these are single-step decreases limited by the poorer cost-effectiveness of the smaller disks. Overall it is

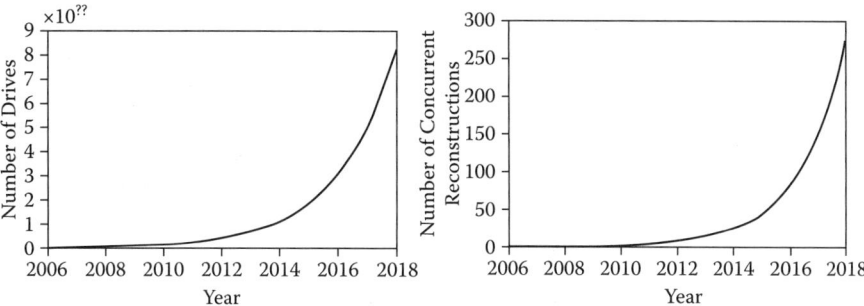

Figure 2.11 (a) Number of disk drives in the largest of future systems. (b) Number of concurrent reconstructions in the largest of future systems.

realistic to expect an increase in reconstruction time of at least 10% per year. Assuming that today reconstruction times are often about 30 hours and that 3% of drives in a system fail per year on average (as shown in Figure 2.10), the number of concurrent reconstructions can be projected to rise in future high-performance computing systems, as shown in Figure 2.11b. This Figure indicates that in the year 2018, on average, nearly 300 concurrent reconstructions may be in progress at any time.

Clearly, designers of petascale storage systems will be spending a large fraction of their efforts on fault tolerance inside the storage systems on which petascale application fault tolerance depends.

2.5.2 Solid State and Hybrid Devices

Rotational delays in disk drives limit our ability to treat these devices as truly "random access." To attain the highest possible performance from storage systems, software developers must spend a great deal of effort to organize disk accesses to maximize sequential block access at the disk level. Although this is trivial for simple, serial, contiguous access patterns, the complex and concurrent access patterns of HPC applications do not lend themselves to this type of optimization. The result is that few HPC applications ever see the full I/O potential of the parallel data systems they are accessing.

Internally, parallel file systems also need very fast, truly random access storage for use in managing metadata and for write-ahead logging. Lowering access latency for these two categories can significantly speed small accesses, file creates and removals, and statistics gathering.

One technique that has been employed in many enterprise products is the use of battery-backed RAM. In this technique battery power is used to allow time for committing data from RAM to storage in the event that power is lost. In normal operation, the RAM serves as a low-latency space for storing small amounts of data. However, the cost and complexity of this approach limit its use.

The cost of solid state disk drives (SSDs) has recently dropped to the point where this technology is becoming a viable component in storage systems. With latencies of 0.1 ms, these devices have latencies as much as two orders of magnitude lower than traditional hard drives. By integrating these devices alongside traditional hard drives in a parallel storage system, parallel file systems can dramatically improve latency of common operations without a significant impact on reliability. Hybrid disk drives are also appearing: traditional drives with hundreds of megabytes of NAND flash storage in the same enclosure.

2.5.3 Extreme-Scale Devices

Modern large-scale parallel computers contain tens to hundreds of thousands of processor cores. This level of concurrency has posed enormous challenges for today's I/O systems for supporting efficient and scalable data movement between disks and memories. With the increasing number of cores in future computer systems, the task for designing scalable I/O systems will be even tougher. Hence, it is important to understand the applications, I/O requirements and reexamine the I/O strategies used by the current file systems.

A recent survey conducted by Shan and Shalf[37] studied the current practice and future requirements of I/O systems from the user community of scientific parallel applications. The results revealed several interesting I/O characteristics commonly appearing in parallel applications. First, the majority of I/O patterns are append-only writes. Parallel applications often have very long execution time and perform data checkpointing at a regular interval for restart purposes. Many applications also keep the checkpoint files for postrun data analysis, such as visualization and feature extraction. However, the data analysis tasks are usually carried out on different machines, because their computation power requirement is not as critical as the parallel applications themselves. The read operations occur mostly at the beginning of the run for data initialization. Compared to the write, the read amount is much smaller and the cost is negligible.

The survey also indicated that the most popular I/O method uses one-file-per-process strategy. This method is easy to program and often gives satisfactory performance when applications run on a small number of processes. However, the immediate drawback is that a restart must use the same number of processes as the run that produced the checkpoint files. Note that if a different number of processes were used in a restart, the I/O is no longer one-file-per-process. A more serious problem is that this method can create a management nightmare for file systems when an application runs on a large number of processes. Thousands of processes can produce hundreds of thousands or millions of files. Because modern parallel file systems employ one of only a small number of metadata servers, serious network traffic congestion can form at the metadata servers when thousands of files are being created simultaneously. Furthermore, after a parallel job exits, millions of newly created files immediately become a challenge for file systems to manage. An alternative solution is the parallel shared-file I/O.

In a parallel application, the problem domain is often represented by a set of global data structures, for example, multidimensional arrays. These global data structures are partitioned among multiple processes so that each process operates on the data in its subdomain. During a data checkpoint, it makes more sense to store the data in global canonical order. Such file layouts often make postrun data analysis and visualization easier. However, shared-file I/O may result in poor performance if the requests are not handled carefully. In order to address this concern, the MPI-IO standard defines a set of programming interfaces for concurrent shared-file access. MPI-IO consists of collective and independent function calls. The purpose of collectives is to enable the MPI-IO implementation to collect together processes to generate the I/O requests that perform faster. An example is the two-phase I/O strategy.[21,23] Another factor causing poor I/O performance is the file system overhead on data consistency control. A POSIX-compliant file system, such as GPFS, Lustre, and Panasas, must guarantee the I/O atomicity, data-sequential consistency, and cache coherence. These requirements are commonly enforced through a lock mechanism. Atomicity needs a lock for every I/O call, but locks can easily degrade the degree of I/O parallelism for concurrent file operations. Meeting these POSIX requirements has been reported as a major obstacle to parallel I/O performance.[38]

Many fundamental problems of parallel I/O not being able to achieve the hardware potential lie in the file systems' obsolete protocols, not suitable for today's high-performance computing systems. File systems have long been designed for sequential access and have treated each I/O request independently. This strategy works well in the nonshared or distributed environments, but poorly for parallel applications where the majority of I/O accesses data that are part of global data structures. For instance, when a program reads a two-dimensional array and partitions it among all running processes, each process makes a request of a subarray to the underlying file system. However, each of these requests will be considered separately by the file system for atomicity, consistency, and coherence controls. In order to address this issue, future file systems must support parallel I/O natively. New programming interfaces will allow applications to supply high-level data access information to the file systems. For example, a new interface could tell the file system that a parallel write by a group of processes should be considered as a single request. Thus, a file system's atomicity and consistency controls could skip checking the internal conflicts of the parallel I/O requests.

Another interesting result from the survey is that most of the programmers use customized file formats rather than the standardized self-describing formats such as netCDF or HDF5. The one-file-per-process is popular because of its simplicity; users are reluctant to adopt complex I/O methods if no significant performance gains or other benefits are promised. As the data size generated from today's parallel applications reaches the scale of terabytes or petabytes, scientists are more willing to consider alternative I/O methods to meet the requirements such as portability and ease of management. However, tradeoffs between performance and productivity will always exist.

2.6 Summary

In this chapter we have discussed at a high level the hardware and software that work together to provide parallel data storage and access capabilities for HPC applications. These technologies build on disk array and RAID techniques discussed in the preceding chapter, and they provide a critical infrastructure that allows applications to conveniently and efficiently use storage on HPC systems. In Chapters 3 and 4 we will see how additional tools build on these parallel data storage facilities to provide additional features and capabilities, such as storage reservations and high-performance storage access across wide area networks.

Acknowledgments

This work was supported in part by the U.S. Department of Energy under Contract DE-AC02-06CH11357, the Department of Energy SCIDAC-2 Scientific Data Management Center for Enabling Technologies (CET) grant DE-FC02-07ER25808, NSF HECURA CCF-0621443, NSF SDCI OCI-0724599, the Department of Energy SCIDAC-2 Petascale Data Storage Institute grant DE-FC02-06ER25767, and the assistance of Los Alamos National Laboratory and Panasas, Inc.

References

[1] F. Schmuck and R. Haskin. GPFS: A shared-disk file system for large computing clusters. In the Conference on File and Storage Technologies (FAST'02), pp. 231–244, January 2002.

[2] Lustre: A scalable, high-performance file system. White paper. Cluster File Systems, Inc., 2003.

[3] S. Ghemawat, H. Gobioff, and S. Leung. The Google file system. *ACM SIGOPS Operating Systems Review* 37, no. 5, pp. 29–43, 2003.

[4] D. Patterson, G. Gibson, and R. Katz. A case for redundant arrays of inexpensive disks (RAID). In Proceedings of the ACM SIGMOD International Conference on Management of Data, pp. 109–116, June 1988.

[5] D. Nagle, D. Serenyi, and A. Matthews. The Panasas ActiveScale storage cluster delivering scalable high bandwidth storage. In Proceedings of the 2004 ACM/IEEE Conference on Supercomputing (SC '04), November 2004.

[6] B. Welch. Integrated system models for reliable petascale storage systems. In Proceedings of the Petascale Data Storage Workshop, Supercomputing (SC'07), Reno, NV, Novemeber 2007.

[7] B. Welch, M. Unangst, Z. Abbasi, G. Gibson, B. Mueller, J. Small, J. Zelenka, and B. Zhou. Scalable performance of the Panasas parallel file

system. In FAST'08: Proceedings of the 6th USENIX Conference on File and Storage Technologies, 2008.

[8] G. A. Gibson, D. F. Nagle, K. Amiri, J. Butler, F. W. Chang, H. Gobioff, C. Hardin, E. Riedel, D. Rochberg, and J. Zelenka. A cost-effective, high-bandwidth storage architecture. In Proceedings of the Eighth International Conference on Architectural Support for Programming Languages and Operating Systems (ASPLOS-VIII), San Jose, CA, October 1998.

[9] Top 500 supercomputing sites. http://www.top500.org, 2007. Accessed July 9, 2009.

[10] G. Grider, H. B. Chen, J. Nunez, S. Poole, R. Wacha, P. Fields, R. Martinez, P. Martinez, S. Khalsa, A. Matthews, and G. Gibson. PaScal—a new parallel and scalable server IO networking infrastructure for supporting global storage/file systems in large-size Linux clusters. In Proceedings of the 25th IEEE International Performance Conference. Computing and Communications Conference (IPCCC), April, 2006.

[11] Portable Operating System Interface (POSIX) Part 1: System Application Program Interface {(API)} [{C} Language]. IEEE/ANSI Std. 1003.1. 1996.

[12] Message Passing Interface Forum (1997) MPI-2: Extensions to the Message-Passing Interface.

[13] J. Li, W. Liao, A. Choudhary, R. Ross, R. Thakur, W. Gropp, R. Latham, A. Siegel, B. Gallaghar, and M. Zingale. Parallel NetCDF: A high-performance scientific I/O interface. In the ACM/IEEE Conference on High Performance Networking and Computing (SC03), pp. 15–21, November 2003.

[14] Hierarchical Data Format, Version 5. The National Center for Supercomputing Applications. http://www.hdfgroup.org/HDF5. Accessed July 9, 2009.

[15] N. Nieuwejaar, D. Kotz, A. Purakayastha, C. Schlatter Ellis, and M. Best. File-access characteristics of parallel scientific workloads. IEEE Transactions on Parallel and Distributed Systems 7, no. 10, pp. 1075–1089. October 1996.

[16] P. E. Crandal, R. A. Aydt, A. A. Chien, and D. A. Reed. Input/output characteristics of scalable parallel applications. In Proceedings of Supercomputing '95, December 1995.

[17] NetCDF. http://www.unidata.ucar.edu/content/software/netcdf. Accessed July 9, 2009.

[18] R. Rew, G. Davis, S. Emmerson, and H. Davies. NetCDF User's Guide for C. Unidata Program Center, http://www.unidata.ucar.edu/packages/netcdf/ guidec, June 1997.

[19] R. Rew and G. Davis. The Unidata NetCDF: Software for scientific data access. In the Sixth International Conference on Interactive Information and Processing Systems for Meteorology, Oceanography, and Hydrology, February 1990.

[20] Where Is NetCDF Used? Unidata Program Center. http://www.unidata.ucar.edu/software/netcdf/usage.html. Accessed July 9, 2009.

[21] J. del Rosario, R. Bordawekar, and A. Choudhary. Improved parallel I/O via a two-phase run-time access strategy. In the Workshop on I/O in Parallel Computer Systems at IPPS '93, pages 56–70, April 1993.

[22] R. Thakur, W. Gropp, and E. Lusk. An abstract-device interface for implementing portable parallel-I/O. interfaces. In the Sixth Symposium on the Frontiers of Massively Parallel Computation, pp. 180–187, October 1996.

[23] R. Thakur, W. Gropp, and E. Lusk. On implementing MPI-IO portably and with high performance. In the Sixth Workshop on Input/Output in Parallel and Distributed Systems, pp. 23–32, May 1999.

[24] J. Prost, R. Treumann, R. Hedges, A. Koniges, and A. White. Towards a high-performance implementation of MPI-IO on top of GPFS. In the Sixth International Euro-Par Conference on Parallel Processing, August 2000.

[25] J. Prost, R. Treumann, R. Hedges, B. Jia, and A. Koniges. MPI-IO/GPFS, an optimized implementation of MPI-IO on top of GPFS. In *Supercomputing*, November 2001.

[26] Hierarchical Data Format, Version 4. The National Center for Supercomputing Applications. http://www.hdfgroup.org/hdf4.html. Accessed July 9, 2009.

[27] R. Ross, D. Nurmi, A. Cheng, and M. Zingale. A case study in application I/O on Linux clusters. In *Supercomputing*, November 2001.

[28] B. Schroeder and G. A. Gibson. Understanding failures in petascale computers, *Journal of Physics: Conference Series* 78 (2007), SciDAC 2007.

[29] Data is available at http://www.lanl.gov/projects/computerscience/data/.

[30] G. Grider. HPC I/O and File System Issues and Perspectives. Presentation at ISW4, LA-UR-06-0473. Slides available at http://www.dtc.umn.edu/disc/isw/presentations/isw4_6.pdf, 2006. Accessed July 9, 2009.

[31] D. McEvoy. The architecture of tandem's nonstop system. In ACM 81: Proceedings of the ACM '81 conference, ACM Press, p. 245, New York, 1981.

[32] E. N. M. Elnozahy, L. Alvisi, Y.-M. Wang, and D. B. Johnson. A survey of rollback-recovery protocols in message-passing systems. *ACM Comput. Surv.* 34, no. 3, pp. 375–408, 2002.

[33] M. Holland, G. A. Gibson, and D. P. Siewiorek. Architectures and algorithms for on-line failure recovery in redundant disk arrays. *J. Distributed & Parallel Databases* 2, no. 3, pp. 295–225, July 1994.

[34] B. Schroeder and G. A. Gibson. Disk failures in the real world: What does an MTTF of 1,000,000 hours mean to you? In FAST'07: Proceedings of the 5th USENIX Conference on File and Storage Technologies, 2007.

[35] P. C. Roth. The path to petascale at Oak Ridge National Laboratory. In Petascale Data Storage Workshop *Supercomputing* '06, 2006.

[36] K. Koch. The new roadrunner supercomputer: What, when, and how. Presentation at ACM/IEEE Conference on Supercomputing, November 2006.

[37] H. Shan and J. Shalf. Using IOR to analyze the I/O performance of HPC platforms. In Cray Users Group Meeting, May 2007.

[38] W. Liao, A. Ching, K. Coloma, A. Nisar, A. Choudhary, J. Chen, R. Sankaran, and S. Klasky. Using MPI file caching to improve parallel write performance for large-scale scientific applications. In the ACM/IEEE Conference on Supercomputing, November 2007.

[39] R. Thakur, W. Gropp, and E. Lusk. Data sieving and collective I/O in ROMIO. In Proceedings of the Seventh Symposium on the Frontiers of Massively Parallel Computation, pp. 182–189, February 1999.

[40] Object-Based Storage Device Commands (OSD), ANSI standard INCITS 400-2004.

[41] R. Thakur and A. Choudhary. An extended two-phase method for accessing sections of out-of-core arrays. *Scientific Programming*, 5(4):301–317, 1996.

Chapter 3

Dynamic Storage Management

Arie Shoshani,[1] Flavia Donno,[2] Junmin Gu,[1] Jason Hick,[1] Maarten Litmaath,[2] and Alex Sim[1]

[1] Lawrence Berkeley National Laboratory
[2] European Organization for Nuclear Research (CERN), Switzerland

Contents

- 3.1 Introduction and Motivation 74
- 3.2 Support for Mass Storage Systems 77
    - 3.2.1 Management of Files in Today's Mass Storage Systems 77
    - 3.2.2 Making Mass Storage Systems Available through SRM Interfaces 78
- 3.3 SRM Functionality 80
    - 3.3.1 Background 80
    - 3.3.2 SRM Concepts 82
    - 3.3.3 Conclusion 90
- 3.4 Data Management in WLCG and EGEE 91
    - 3.4.1 The WLCG Infrastructure 91
        - 3.4.1.1 The Tiers Model 92
        - 3.4.1.2 The WLCG Data Management Services and Clients 92
    - 3.4.2 High-Energy Physics Use Cases 93
        - 3.4.2.1 Constraints for Distributed Computing and Storage 94
    - 3.4.3 High-Level Physics Use Cases 94
        - 3.4.3.1 Reconstruction 94
        - 3.4.3.2 Mainstream Analysis 95
        - 3.4.3.3 Calibration Study 96
        - 3.4.3.4 Chaotic Analysis 97
    - 3.4.4 Storage Requirements 97
        - 3.4.4.1 The Classic Storage Element and SRM v1.1 97
        - 3.4.4.2 The Storage Element Service 99
        - 3.4.4.3 Beyond the WLCG Baseline Services Working Group 100

              3.4.4.4 The Storage Classes 101
              3.4.4.5 The Grid Storage Systems Deployment
                       Working Group 103
      3.4.5 Beyond WLCG: Data Management Use Cases in EGEE 104
3.5   Examples of Using File Streaming in Real Applications 105
      3.5.1 Robust File Replication in the STAR Experiment 105
      3.5.2 The Earth System Grid 107
3.6   Conclusions and Future Work 108
Acknowledgments 109
References 109

3.1 Introduction and Motivation

Dynamic storage space allocation is a feature that was not available for scientific applications. Therefore, scientific researchers usually assume that storage space is preallocated and that the application will have enough space for the input data required and the output data generated by the application. However, in modern computer systems that support large scientific simulations and analysis, this assumption is often false, and application programs often cannot complete the computation as a result of lack of storage space. Increases in computational power have only exacerbated this problem. Although the increased computational power has created the opportunity for new, more precise and complex scientific simulations that can lead to new scientific insights, such simulations and experiments generate ever-increasing volumes of data. The ability to allocate storage dynamically (on demand), to manage the content of the storage space, and to provide sharing of data between users are crucial requirements for conducting modern scientific explorations.

Typically, the scientific exploration process involves three stages of data-intensive activities: data generation, postprocessing, and data analysis. The data generation phase usually involves data collection from experimental devices, such as change-coupled devices measuring electric fields or satellite sensors measuring atmospheric conditions. However, increasing volumes of data are now generated by large-scale simulations, the so-called third pillar of science, which now complement theory and experiments. At the data generation phase, large volumes of storage have to be allocated for data collection and archiving.

Data postprocessing involves digesting the large volumes of data from the data-generation phase and generating processed data whose volume can often be as large as the input data. For example, raw data collected in high energy physics (HEP) experiments are postprocessed to identify particle tracks as a result of collisions in the accelerator. The postprocessed data in such cases is of the same order of magnitude as the raw data. However, in many applications postprocessing usually generates summary data whose volume is smaller. For example, climate-modeling simulation variables such as temperature, which are generated using time granularity of a day or even an hour, are typically summarized in the postprocessing phase into monthly means. Another activity at the postprocessing phase is to reorganize the large volume of data to fit better projected analysis patterns. For example, a combustion model simulation may generate a large number of variables per space–time cell of the simulation. The variables may include measures of pressure or temperature as well as many chemical species. In the analysis phase it is typical that not all variables are needed at the same time. Therefore, organizing the data by variable for all time steps prevents getting the entire volume of data in order to extract the few variables needed in the analysis. This process is often referred to as the "transposition" of the original data. In this phase, it is necessary to ensure that enough space is allocated for the input as well as for the output. Because of the potential large volume of space required, postprocessing tasks are often performed piecewise or delegated to multiple processors, each dealing with part of the data volume. That introduces the issue of reliable data movement between sites and processors.

The data analysis phase typically involves exploration over part of the data, such as climate analysis involving sea-surface temperature and wind velocity in the Pacific Ocean. The most reasonable thing to do is to extract the needed data at the site where the data is stored and only move to the analysis site the needed subset of the data. However, such data extraction capabilities are not always available, and scientists end up copying and storing more data than necessary. Alternatively, the data analysis could be performed in the site or near the site where the data resides. Here again, such analysis capabilities are not usually available. Furthermore, many scientists are only comfortable with their own analysis environments and usually get the data they want to analyze to their local site. In general, at the data analysis phase, storage has to be allocated ahead of time in order to bring a subset of the data for exploration and to store the subsequently generated data products. Furthermore, storage systems shared by a community of scientists need a common data access mechanism that allocates storage space dynamically, manages its content, and automatically removes unused data to avoid clogging the shared data stores.

When dealing with storage, another problem facing the scientist today is the need to interact with a variety of storage systems. Typically, each storage system provides different interfaces and security mechanisms. For these reasons, several activities emerged over time in order to standardize and streamline the access to the storage systems through common interfaces and manage dynamically the storage allocation and the content of these systems. The goal is to present the scientists or software utilities with the same interface regardless of the type of storage system is used. Ideally, the management of storage allocation should become transparent to the client.

The problems described above can be summarized with the following four requirements:

- The ability to reserve storage resources (disk, archival storage) dynamically.
- The ability to request data to be moved into or copied between storage resources in a uniform fashion.
- The ability to dynamically release storage when no longer necessary.
- The ability for storage to be assigned for a given lifetime, and then released automatically in order to avoid clogging storage or rendering it useless as it gets full.

We will describe in some detail such a standard approach, referred to as storage resource management (SRM) in the subsequent sections. The approach taken is to provide a standard specification that can be implemented with any storage system. Such implementations of SRMs can target various storage systems with the goal that all such systems interoperate.

In contrast to the approach of permitting multiple implementations according to a standard specification, another approach is a centralized one, such as the storage resource broker (SRB).[1] SRB provides a single front end, which interfaces to a variety of back-end storage systems, giving the client the illusion that all data is stored in a single monolithic system. This distinction is illustrated in Figure 3.1. Another important difference is that SRB is not designed to provide storage space reservations or manage space quotas. However, since it manages multiple data stores centrally, it can provide the capability of a metadata catalog for all its holdings. SRB has been used very successfully

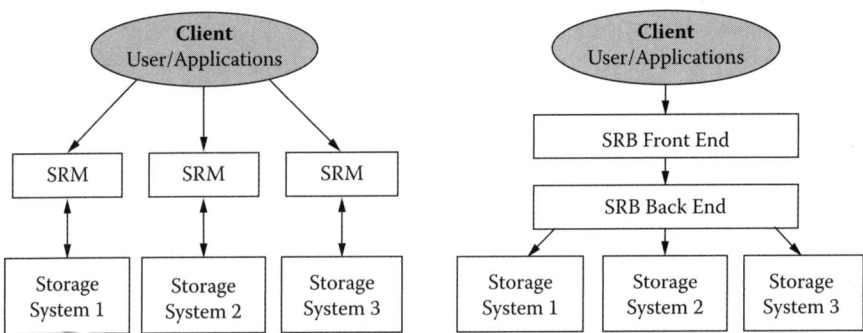

Figure 3.1 The difference in architecture between SRMs and SRB. SRMs permit independent implementations to various distributed storage systems that facilitate space reservation and management. SRB takes a centralized approach that provides a uniform view of all the data in storage systems, but no storage space management.

in many projects where storage space management and dynamic allocation of storage is not a concern.

Another approach to space allocation is to provide a software layer that can operate directly on a variety of hardware, such as a raw disk or a RAID (redundant array of independent disks) system, or directly on a local file system. This approach was taken by the network storage (NeST) project,[2] which provides an abstract interface to disks. The idea is to have a relatively simple software layer that exposes a storage appliance interface. NeST provides dynamic storage allocation in a form of "lots" that can be allocated to a client. It can manage guaranteed and "best effort" type lots and provides multiple protocols for file access. However, unlike SRMs, NeST is not designed to interact with mass storage systems (MSSs) or complex storage systems that have multiple components.

3.2 Support for Mass Storage Systems

3.2.1 Management of Files in Today's Mass Storage Systems

Today's MSSs, including the largest file and archival storage systems, are designed with high performance and reliability as their key design principles. The main goal for performance is the rate at which a user can read or write data to or from the system. Reliability is the degree to which users trust data that exists in the system through many decades and technology changes. In addition, most large file and archival storage systems adhere to standards that contribute to their portability but ultimately limit their features down to a set of known commands or functionality. For instance, many file or archival storage systems comply with the portable operating system interface for unix (POSIX) standard for I/O and provide very few special or custom commands outside the standard set of commands for common Unix file systems.

Given the constraints of designing for reliability as opposed to performance and self-imposed limitations in favor of portability, MSSs do not compete with the data performance of high-performance file systems. High-performance file systems make design and implementation trade-offs that would affect reliability or not be acceptable to an MSS in order to achieve the higher-performance data rates. An acceptable target for one commercial MSS is to stay within 10% or an order of magnitude of what a file system is capable of handling in terms of data rates. Depending on their ability to stripe transfers across multiple disk or tape devices, most MSSs are capable of extremely high data rates which are comparable to high-performance file system data transfer capabilities.

Another trend in MSS evolution is the average size of user files. The average size of files has not appreciably increased over the past few decades as predicted. Instead, most MSSs holding several decades worth of data still have an average file size measured in megabytes. The amount of metadata increases as users store more and more rather small files into MSSs, with minimal regard to managing the data. Metadata includes file names, description of file content, versions of files, and information to help relate files to each other (e.g., create datasets). The poor metadata rates are a hindrance in managing a large number of files. This phenomenon contributes to a paradox in data storage called the sustainability paradox, which states, as the world continues to generate more data, we are more in danger of losing the data we really wish to keep. This is due to the fact that the important datasets continue to be a smaller and smaller portion of a user's total amount of data retained.

As MSSs amass data over the years, users often find the limited metadata file management capabilities and their lackluster metadata performance a burden in terms of managing ever-increasing amounts of data. To accommodate this, users often resort to rudimentary data-management techniques like establishing file-naming conventions or by using aggregation techniques to limit the number of files and grouping-related files into a single file. More advanced data-management techniques include establishing a database external to the MSS to keep track of additional metadata. The most advanced techniques of data management that are emerging are systems that hide the complexities of dealing with tape and disk and hierarchical storage management (HSM). One example of this approach is the SRM standard.[3] There are several implementations of SRM software in use today, which generally focus on providing a common single view of a user's data regardless of the specific MSS–file system. These implementations can be part of the HSM or external to the HSM. Examples of HSM systems that have embedded SRM implementations are CASTOR, an HSM developed at CERN;[4] and Enstore, developed at Brookhaven National Laboratory.[5] Examples of HSM systems that have external SRM interfaces are Berkeley Storage Manager (BeStMan),[6] developed at Lawrence Berkeley National Laboratory for HPSS,[14] and storage resource manager (StoRM),[7] developed in Italy for GPFS. SRM software provides a common set of commands to help manage data between systems that may not share command sets or provide similar functionality. However, SRMs currently do not expose the performance expectation from an MSS (see Conclusions and Future Work section).

3.2.2 Making Mass Storage Systems Available through SRM Interfaces

Given the complexity of MSSs, sometimes it is not straightforward that the SRM functionality be made part of such systems. In practice, there are two possible approaches: extend the MSS to support the SRM functionality or provide an external layer that fronts the MSS. It turns out that both approaches are valid. If the developers of the MSS are involved in SRM projects, they

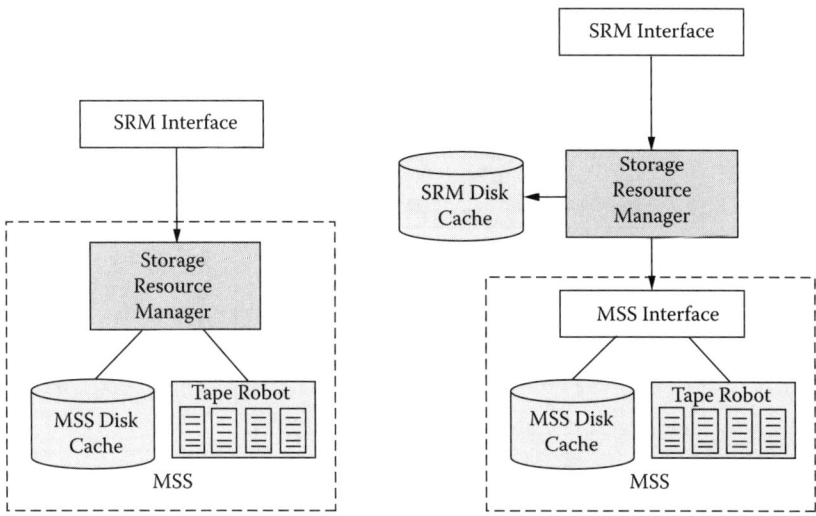

Figure 3.2 SRM interface to a mass storage system. On the left the SRM is embedded in the MSS; on the right it is external to the MSS, managing its own disk cache.

can choose to take the first approach of extending the MSS to support the SRM functionality. This is the case with several systems as described in the next section (JASMine, dCache, and Castor). On the other hand, the MSS system can be treated as a "black box," and a front-end SRM accesses that system through its API. This was the case with HPSS as well as with a legacy MSS system in the National Center for Atmospheric Research (NCAR) as described in Section 3.5.2.

The difference between these two approaches is illustrated in Figure 3.2. On the left the SRM software is embedded in the MSS, thus exposing the SRM interface (in addition to the previously supported interfaces). On the right the SRM software is external to the MSS and communicates with the MSS through its interface. In the case of HPSS the interface is either PFTP (parallel FTP) or HSI (hierarchical storage interface). As can be seen in the case of an external SRM, the SRM manages its own disk cache and files that go into and out of the MSS get copied to the SRM cache. This extra "hop" can add to access latency when getting a file from the MSS. However, if the file is on tape, the time to transfer it from one disk to another is a small fraction of the total latency. On the other hand, having the SRM cache can help, since frequently accessed files by the SRM clients are kept in its cache. This can lead to better access rates, as clients do not have to compete with other MSS clients. A similar advantage can be achieved while putting files into the MSS since the files can be written quickly into the SRM disk cache and then moved lazily by the SRM to the MSS.

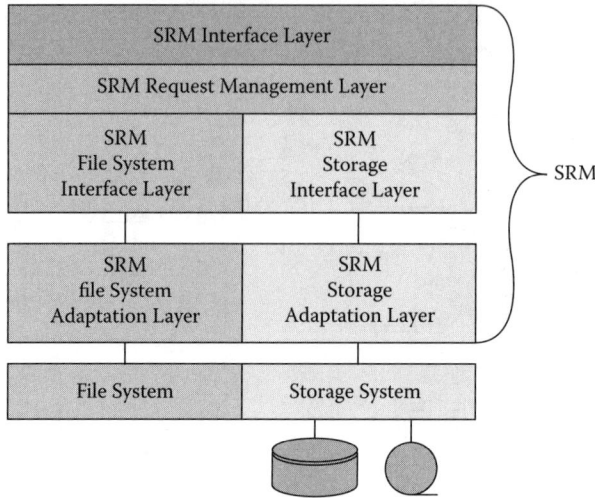

Figure 3.3 An architectural diagram of a modular design of an SRM.

In general, SRMs that front MSSs can be adapted to work with various file systems. The SRM developed by LBNL (called BeStMan) was adapted to front HPSS, the NCAR MSS, L-Store, and Lustre. This is shown in Figure 3.3. It was also adapted to provide an SRM gateway to the XROOTD[20] system used by HEP projects. Similarly, StoRM developed in Italy fronts GPFS. There is also a version of SRM developed in Taiwan to front SRB.

3.3 SRM Functionality

3.3.1 Background

The SRM technology was inspired by the emergence of the grid and the ability to use distributed compute and storage resources for common tasks. The grand vision of the grid is to provide middleware facilities that give a client of the grid the illusion that all the compute and storage resources needed for their jobs are running on their local system. This implies that a client only logs in and gets authenticated once and that some middleware software figures out where the most efficient way to run the job is; reserves compute, network and storage resources; and executes the request. Initially the grid was envisioned as a way to share large computing facilities by sending jobs to be executed at remote computational sites. For this reason, the grid was referred to as a *computational grid*. However, very large jobs are often data intensive, and in such cases it may be necessary to move the job to where the data is located in order to achieve better efficiency. Thus, the term *data grid* was used to emphasize applications that produce and consume large volumes of data. In some applications the volume of data is very large (in the order of hundreds of gigabytes to terabytes), and it is impractical to move entire datasets on demand to some destination. In such cases, the computation is moved to where the data is or partial replication of the data is performed ahead of time to sites where the computation will take place.

In reality, especially in the scientific domain, most large jobs that require grid services involve the generation of large datasets, the consumption of large datasets, or both. Whether one refers to the grid as a computational grid or a data grid, one needs to deal with the reservation and scheduling of storage resources when large volumes of data are involved, similar to the reservation and scheduling of compute and network resources.

In addition to storage resources, SRMs also need to be concerned with the data resource (or files that hold the data). A data resource is a chunk of data that can be shared by more than one client. For the sake of simplifying the discussion, we assume that the granularity of a data resource is a file. In some applications there may be hundreds of clients interested in the same subset of files when they perform data analysis. Thus, the management of shared files on a shared storage resource is also an important aspect of SRMs. In particular, when a file has to be stored in the storage resource that an SRM manages, the SRM needs to allocate space for the file and remove another file (or files) to make space for it, if necessary. Thus, the SRM manages both the space and the content of that space. The decision about which files must be kept in the storage system is dependent on the cost of copying the file from some remote system, the size of the file, and the usage level of that file. The role of the SRM is to manage the storage space under its control in a way that is most cost beneficial to the community of clients it serves.

The desired SRM functionality evolved over time, based on experience of its use and practical requirements. The SRM concepts were first introduced by researchers at LBNL in References 8 and 9, but soon after (as early as 2002) collaborators from Thomas Jefferson Lab National Accelerator Facility (TJNAF), Fermi National Accelerator Laboratory (FNAL), and CERN, European Organization for Nuclear Research, have seen the value of evolving this into a standard that all will adhere to. This resulted in the first stable version, referred to as SRM version 1.1 (or SRM v1.1). Each of the institutions built SRM software based on this specification on top of (or as part of) their own storage systems. LBNL developed an SRM for accessing the MSS HPSS; FNAL for accessing dCache, a system that is capable of managing multiple storage subsystems including an MSS called Enstore; TJNAF for accessing their MSS, called JASMine; and CERN for accessing their MSS called Castor. It was subsequently demonstrated that all these implementations can interoperate.

SRM v1.1 included concepts of keeping a file in disk cache for a period of time, referred to as *pinning*, requesting to get or put files into the SRM space, and managing user quotas internally. However, the experience with SRM v1.1 revealed functionality that was missing: in particular, the need to explicitly reserve space through the interface and manage its lifetime, explicitly creating and managing directories through the interface and providing access control. This led to SRM v2.2, which was stabilized to a great degree based on practical requirements from the Large Hadron Collider (LHC) SRM development team (see Section 3.4). By now several more implementations resulted, including an SRM for a disk pool at CERN, called DPM; an SRM for the parallel file system GPFS in Italy, called StoRM; and an SRM that can be deployed to disk systems or various MSS system at LBNL, called BeStMan. A brief description of each of these implementations is provided in Reference 10 as well as large-scale daily testing of their interoperation. The interoperation of SRM implementations that adhere to SRM v2.2 specification and locations of their use is illustrated in Figure 3.4.

The historical discussion above was intended to show that the SRM specification was developed over time by a large number of international collaborators. SRM v2.2 was submitted and approved as a standard by the Open Grid Forum (OGF).[3] The practical requirements of the SRM usage for the LHC are discussed in some detail in Section 3.4 as an example of a successful deployment of SRM technology in a large international project. The change from SRM v1.1 to SRM v2.2 illustrates the difficulties of evolving standards. The SRM deployment in the Earth Systems Grid (ESG) is discussed in Section 3.6.

We discuss next the main concepts that emerged over time for SRM regardless of their versions. We focus only on the latest version.

3.3.2 SRM Concepts

To illustrate the functionality of SRMs, we start with a simple example of having to run a job on a local machine; the files that are needed for the job are at remote locations including remote MSSs. Assume that the job requires a large number of files (100–1000). A user will have to perform the following actions:

- allocate space
- bring in all input files or some fraction of the files if they do not fit in the space
- ensure correctness of files transferred
- monitor and recover from errors
- if files don't fit space, remove files to make space for more files
- save output files, possibly moving them to archival storage
- if files come from different storage systems (including MSSs), different protocols may need to be used

Dynamic Storage Management 83

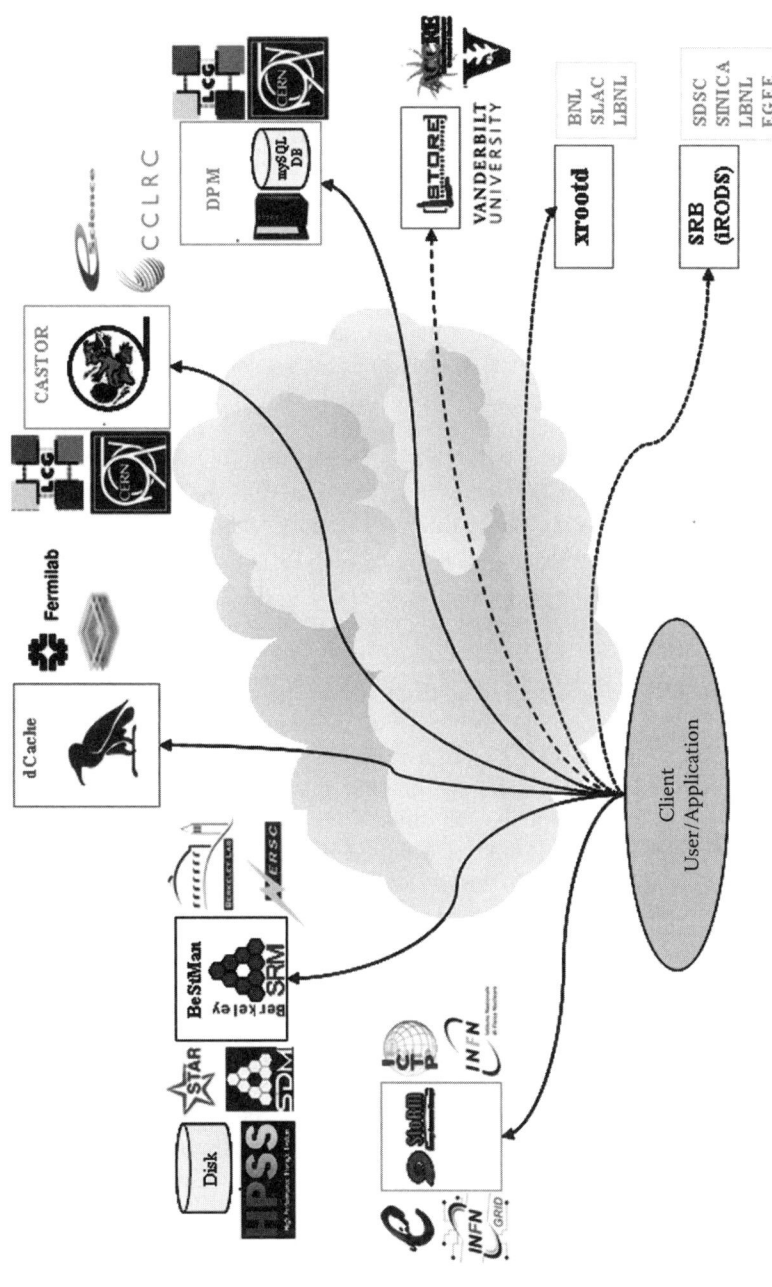

Figure 3.4 (See color insert following page 224.) Interoperability of SRM v2.2 implementations.

This is a lot of work for each user to manage. One can imagine writing specialized scripts to manage this task. SRMs are designed to provide this functionality through appropriate interfaces. Furthermore, SRMs are designed to provide this functionality on shared storage resources, such as a shared disk cache by a community of users, by allocation of space quotas to users, and to ensure fairness of space allocation and scheduling. In addition, SRMs provide uniform access to all storage systems including MSSs. The concepts involved in fulfilling this scenario by SRMs are explained next.

1. **Logical file name for files at a site.** As mentioned above, an SRM may front multiple storage systems at a site. A file may reside on any one or more of these systems. Furthermore, it may be advantageous for the SRM at that site to copy a file temporarily to a disk system that has better transfer access for the user. For this reason it is useful to have a single logical file name (LFN) associated with that site. For access by SRMs, the LFN is complemented with machine name and port at the site to produce a site URL (SURL). An example of such an SURL is, srm://sleepy.lbl.gov:4000/tmp/foo-123. Note that the protocol is "srm"; the machine name is "sleepy.lbl.gov" at port 4000; the directory name is "tmp"; and the file name is "foo-123." In general, using SURLs by SRMs is a way for a site to move files to any storage system they support while still having the same SURL visible externally through catalogs or other means.

2. **Transfer URL for accessing files.** Once the SRM is given the SURL through the function call to get a file, it decides which copy to provide the client. The file can be on a different machine and port than the SRM, and a protocol supported on that machine has to be provided for the client to use to transfer the file. Thus, the SRM returns a transfer URL (TURL). An example of a TURL for the SURL above might be gridftp://dm.lbl.gov:4010/home/level1/foo-123. Note that the transfer protocol provided is "gridftp"; the machine name and port are different, as well as the directory path. Usually, file names stay the same in the TURL, but returning a different file name is possible.

3. **Transfer protocol negotiation.** The protocol provided to the client must match the protocols the client has. For this reason, a request to get files has a parameter, where a client can specify an ordered list of preferred protocols. The SRM returns the highest possible protocol it supports. For example, the client can provide a protocols list: bbftp, gridftp, and ftp; and the SRM returns gridftp, since it does not have a server that supports bbftp.

4. **Putting files into SRMs.** In order to put files into SRMs, such as output files of a job, the SRM has to allocate space. The function call requesting to put a file specifies the SRUL of the file that will be put into the SRM, along with the file size. If the file size is unknown, the SRM

assumes a maximum default size. The SRM allocates space, given that the quota available to the user has enough space to hold this extra file and return a TURL the client can then put the file using the TURL transfer protocol into the allocated space. We note that an SRM may limit the number of TURLs it returns to the client because of space limitations or prevent the client from transferring too many files concurrently, thus overwhelming the storage system.

5. **Getting files from remote sites.** Since SURLs are used to identify precisely the SRM site that stores a particular file, it is possible in principle to request a file to be brought into a local SRM from a remote site, by simply providing it with the remote SURL. This requires that SRMs communicate with each other in a peer-to-peer fashion. A special copy function is provided that permits getting files from a source SRM to a target SRM in either *pull* or *push* mode.

   Not all SRM implementations support the copy function if the SRM is intended for use as a local SRM. In this case, the SRM only supports requests to get files from or to put files into the SRM. In dealing with such local SRMs, it is still possible to transfer files between sites. For example, the file transfer service (FTS)[37] developed at CERN for the LHC projects and described in Section 3.4 requests to get a file from the source SRM, gets back a source TURL, then requests to put a file from the target SRM, which allocates the space for it, and returns a target TURL. FTS then uses the two TURLs to perform a third-party transfer between the source and target disk caches. This mode of operation also gives FTS full control of the number of transfers requested, as well as monitoring their performance.

6. **Pinning files and releasing files.** When a file is requested from an SRM, there needs to be some guarantee that the file will stay in place until it is used. On the other hand, an SRM cannot afford to keep the file indefinitely or rely on the client to delete it when done. For example, a client can crash and never delete the requested file. For this reason SRM supports the concept of *pinning* a file for a specified lifetime. The length of the lifetime is negotiable, where it can be specified in a request, and the SRM can choose to return a shorter lifetime based on its local policies. The clock on the lifetime for a file starts at the time that the file is made available to the client. Thus, when requesting multiple files to be staged from an MSS, for example, each file may have a different start time on its lifetime.

   It is useful for the SRM to know when the client is finished accessing a file. In that case it can reuse the space by deleting the file if necessary. The function to let the SRM know that the client no longer needs the file is referred to as *releasing* the pin on the file. The reason for having a release concept rather than delete is that the SRM can in this case have the choice whether to keep the file longer (e.g., for another client) or delete it. That is, the SRM can make the decision about which files to delete when space is needed based on the history of requests to each file or based on the requests currently queued. For example, it is advantageous to keep files that are accessed often ("hot" files) longer in the disk cache. There are many such caching algorithms, the simplest but effective one being to remove the least recently used (LRU) files.

Pinning and releasing files is an essential function for supporting file streaming, described next.

7. **File streaming.** Suppose that a client running an analysis program requests 500 files, each 1 GB in size. Also, suppose that the analysis can start as soon as a subset of the files is available. Some or all of the files could be already in the SRM disk cache, or they have to be brought in from an MSS or remote sites. When a request is made to the SRM, it may allocate a quota that is smaller than the total size needed. For example, assume that the SRM allocated 100 GBs to this client. Obviously, this job cannot be accomplished if the client wishes to have all 500 files in the SRM cache before proceeding. For such tasks, the concept of file streaming is useful.

Basically, once the request is made, the SRM brings as many files as will fit into the user's quota space. If some files are already in the SRM's disk cache (for instance, they were brought in for another user previously), they are made available immediately. The client can check which files are there by issuing a status call, which returns TURLs for all the files currently available. The client can then get each file and release them when finished. For every file released, the SRM can bring in another file not yet accessed, thus streaming the files to the client. This is shown schematically in Figure 3.5, where requested files are staged from two near-line storage systems (could be local or remote) into the client's quota space.

Some projects choose not to support file streaming since they require all the files before analysis can start. However, even for such applications, file streaming can be used effectively by using the SRM disk cache as a temporary staging space for files on the way to the client's local disk.

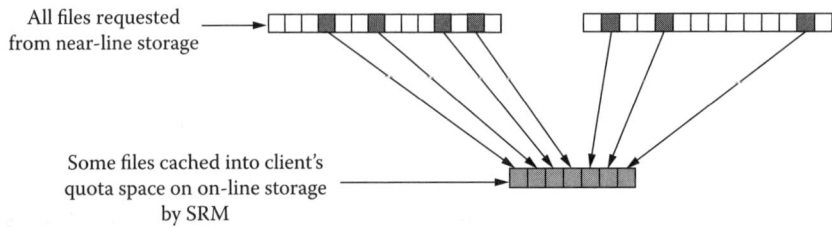

Figure 3.5 File streaming: stage, use, release, stage next.

We discuss such examples in some detail in Section 3.5.2 for systems in use by the ESG project and by the STAR HEP experiment.

8. **File sharing.** The term *file sharing* is used to convey the concept that the same physical copy of a file is shared by multiple users. Of course, this can be done only because these files cannot be modified, which is the typical use case that SRMs deal with. The value of file sharing is naturally saving storage resources, by avoiding the replication of the same file in the SRM cache, one for each user requesting it. However, supporting this feature is not straightforward because of the need to manage the lifetime of such a file.

    The lifetime of a file is associated with a particular request by a particular client. Suppose that a file was already brought to cache and was assigned a lifetime of an hour. Now, suppose that another client makes a request for the same file half an hour later, requesting a lifetime of another hour. This can be simply supported by extending the lifetime of the file. However, how would the SRM know when all lifetimes expired or the files' pins released so that the file can be disposed of? This requires that each lifetime is kept track of and marked as either *active, expired,* or *released.* A file can only be removed if all of its lifetimes either expired or were released.

    Another aspect is how to count a shared file against the quota of clients. Let us suppose that a file is already used by a client and a second client requests it. Should the space required for the file be counted against the second client's quota? In general, the answer is yes. However, such a consideration is a choice of the VO policies.

    Since it is complicated to manage file sharing, some implementations choose to avoid that altogether and replicate each file in the client's quota space. While this is wasteful of space, it only has a serious effect if files are shared heavily by clients. Since many applications focus on a subset of the files for their analysis (e.g., a group of scientists investigating jointly a certain phenomena), it is advisable to support file sharing.

9. **File lifetime expiration mode.** So far we have described files that are brought into a disk space on a temporary basis, thus having a lifetime associated with them. We refer to these files as *volatile* because they can be removed when the lifetime expires. If space is not extremely tight, volatile files are usually not removed immediately, thus having a high probability of being on the SRM cache if requested again often. Another use for volatile files is a case when an administrator for a group of researchers wants to make a set of files available to all for their analysis over a relatively long period. This is common in applications where data from experiments or simulations is shared by many researchers. In such a case, one can setup an SRM per region (e.g., southern Europe) and replicate files to that SRM for an extended period of time. A special function referred to as *bring online* permits long lifetimes (e.g., many months),

still allowing the SRM to garbage-collect the space at the end of that period in case the administrator neglects to do so. This is a very important capability to ensure that unnecessary files do not clog large shared disks.

When putting files into the SRM, it is often needed to ensure they are not lost, especially when it is the original copy of a file. For this reason, SRMs support a file type referred to as *permanent*, referring to files that can only be removed explicitly by the owner or by users given permission by the owner. While this is an obvious requirement, an SRM needs to ensure that such files are protected. For example, the file may reside in a storage space that is released by its owner (see space reservation, below). If such a space contains a permanent file, the SRM needs to take action such as archiving the file and/or notifying the owner before that space can be released.

Another interesting case is when files are "dumped" temporarily into a disk cache because of high transfer-rate requirements, with the intention of archiving them later and releasing them once archived. In such a case, files are given a lifetime long enough until the archiving completes. Normally, the files will be released after they are archived by the program responsible to archive them. However, in case of a mishap, it is necessary to protect these files as if they were permanent. We refer to such a file type as *durable* in the sense that a warning is issued by the SRM if their lifetime expires rather than removing the files. Concurrent with the warning, an SRM may take a default action, such as archiving these files so that the disk space can be released for other usage.

The three types of files, volatile, durable, and permanent, were found to be sufficient for all use cases encountered so far. In practice, some projects choose to have all files permanent in order to have full control of what is kept in the storage systems, and others always use volatile file types when data is replicated. So far, durable file types have not been used in practice.

10. **Asynchronous support of requests.** Multifile requests to get files from an SRM or to put files into an SRM are an important feature that allows clients to express a large request at once either by listing the set of files or by providing a directory name that contains all the files. Since such requests can take a long time to complete, it is not practical for them to be synchronous, that is, forcing the client to wait for their completion. Thus, all such requests need to be supported asynchronously. This means that some kind of a request token has to be returned when a request is made, and the client can at any time later use the request token to find out the status of the request.

Request tokens can be either supplied by the client or provided by the SRM. However, relying on client assignment of token names can cause errors, and therefore in the SRM specification it was decided that the SRM generates a request token that is guaranteed to be unique (usually based on the time of the request). Since such tokens can be lost by the client, SRMs also permit clients to optionally specify a user-generated request token in addition. Accordingly, there is an SRM function that allows getting back all system-generated tokens given a user-generated request token.

11. **Storage types.** The issue of how to characterize different storage systems in a uniform way was debated at length by SRM collaborators involved in its specification. The issue was how to have a meaningful but simple characterization. Eventually, two orthogonal concepts emerged: *access latency* and *retention quality* of the storage component. Access latency can assume two values, online and near-line, where online storage implies that a file residing on it is immediately available and near-line implies an access latency of having to bring the file online before it is available for access. Retention quality can assume three possible values representing a high, medium and low retention quality referred to as *custodial, output,* and *replica,* respectively. The interpretation of these quality types is unspecified, but it is expected that custodial quality will keep multiple copies (at least two), while replica quality can be an inexpensive disk. In case that replica quality storage fails, it is expected that the files on it were replicas that can be retrieved again.

The combination of these orthogonal properties can be used to interpret the type of storage used. For example, custodial–near-line usually implies a robotic tape but can also be a large archival disk. Similarly, replica–online usually implies an inexpensive disk, while custodial–online usually implies a RAID disk system. As will be discussed in the example in Section 3.4, the HEP community found that combinations of these are also valuable, for example, having the files both on disk and on tape at the same time.

12. **Space reservation.** Requesting space as needed from an SRM was a major addition in the specification of SRM v2.2. A requested space has a size and a lifetime. Both can be negotiated by the requester specifying minimum guaranteed size and maximum desired size. The SRM returns the minimum guarantee it can support and maximum "best-effort" size. Best effort implies that this amount of space is likely to be made available based on the current load of the system, but it may be reduced if the system's load increases. Note that in some organizations getting space is counted against a global allocation and thus it is not in the best interest of the client to request more space than it anticipates needing. Similar to lifetime of files, since spaces have a lifetime, they can be released if no longer needed.

When a storage space is reserved, a *space token* is assigned to it by the SRM. A user-defined name can also be assigned to refer to the space. When getting or putting files into an SRM, the space token can be specified, thus ensuring that clients use only space allocated to them.

Some organizations pre-assign spaces for certain usage, and the space tokens are made known to those permitted to use them.

In practice, allocating space to individual users requires a complicated allocation scheme where the amount of space used by each individual is controlled and monitored by the organization that owns the storage and compute resources, referred to as the virtual organization (VO). While SRMs can be expected to keep track and enforce space usage, they cannot be expected to implement various policies. Therefore, many organizations choose to manage space for groups of users, expecting each user in that group to behave responsibly and not "hog" the entire space. On the other side of the spectrum, SRMs can use quotas by default, which are specified in a *configuration file*. If all users accessing an SRM have the same privileges in terms of space quotas, this is sufficient. However, experience shows that it will be beneficial to assign quota privileges for individuals in storage space allocated to a group. Such a possibility is not available in the latest SRM v2.2 specification, as noted in Section 3.4.

13. **Directory management.** The initial version of the SRM specification assumed that it is not necessary to provide clients with the ability to organize their files and that the SRM can assign its own directory structure. Typically, a single flat directory was assigned for each user. However, as the number of users, groups, and projects grew, it became obvious that the support of a directory structure and functions similar to those supported by Unix as well as access control lists (ACLs) is needed. Thus, the usual functions of ls, mkdir, rmdir, mv, rm, and cp were added to the SRM specification. However, given the ability to have multiple spaces, the question was whether each space can have its own directory structure.

The conclusion to this question was that there should be a single directory for all spaces owned by a virtual organization. Furthermore, files, assignments to spaces is a property of the files, and therefore when a file is moved from one space to another, there is no effect on the directory structure; that is, the file stays in the same position in the directory. The important advantage of this choice is that files can be moved between spaces without effect on their path names.

A similar question arose about file types. Should all file types be separated into different directory structures? Here, too, the conclusion was that file types are the property of the file, and therefore a file can be changed, for example, from volatile to permanent without any effect on the directory structure and the file path name.

3.3.3 Conclusion

The description of the SRM functionality provided in this section makes it clear that the decisions on what to include in the standard specification are not simple and evolved over a period of seven years based on practical experience and requirements of multiple projects. Yet, it demonstrates that such a standardization effort to abstract the semantics of storage systems, to provide a uniform interface to heterogeneous storage systems, and to dynamically reserve and manage space is possible and valuable, especially as the storage hardware is changing over time. This effort makes it possible for large projects to select the functionality they wish to support, regardless of the underlying storage system. An example of such a large project and its use of SRMs is presented in the next section. Much of the discussion involves decisions that had to be made regarding the SRM functionality that the project as a whole will commit to support, given practical human resource limitations and deadlines.

3.4 Data Management in WLCG and EGEE
3.4.1 The WLCG Infrastructure

The Worldwide LHC Computing Grid (WLCG) is the largest grid infrastructure in operation today, comprising more than 250 sites spread over 45 countries on 5 continents. Its main mission is to provide resources for the storage and analysis of the data generated by the experiments at the large hadron collider (LHC) facility currently being commissioned at CERN, the European Laboratory for Particle Physics. There are four major experiment at the LHC: a large ion collector experiment (ALICE), a toroidal LHC apparatus (ATLAS), compact muon solenoid (CMS), and LHC beauty experiment (LHCb). Each experiment has its own computing model, but all rely on the WLCG for the necessary storage and computing resources. The WLCG itself comprises a federation of sufficiently compatible grids. The main contributors currently are the Enabling Grids for E-sciencE (EGEE)[11] project, the Open Science Grid (OSG),[12] and the Nordic Data Grid Facility (NDGF).[13] It is important to note, however, that each of the grids contributing to WLCG has been explicitly funded to provide infrastructures for e-science in general, and in particular for sciences other than particle physics, in contrast with the main mission of the WLCG. Other disciplines include biomedical research, computational chemistry, nuclear fusion, astronomy, geophysics, meteorology, and digital libraries. They usually are new to the ways of working in large international collaborations that have been normal in particle physics for decades. While many of the practices used in particle physics analysis may simply be copied, other disciplines also bring additional requirements, in particular with respect to security and privacy. Furthermore, most WLCG sites also participate in national or other international grids, which may yet pose other requirements on the services that some of the sites need to provide. Finally, the large WLCG sites have a history of providing computing and storage resources for pregrid scientific projects and typically have built up significant storage infrastructures that will also have to be used by grid projects. That is, data storage and retrieval on the grid will have to be compatible with preexisting basic infrastructures that will even differ significantly from site to site. In particular, CERN and other big computing centers each have their own MSSs, typically for many years. The different tape back-end systems include CASTOR,[4] Enstore,[5] and HPSS.[14] The corresponding HSM front-end systems in use are CASTOR[4] and dCache.[15] Disk-only storage is provided by dCache, DPM,[16] and StoRM[7] systems at this time. Some OSG sites use BeStMan,[6] which supports HPSS among other back ends.

3.4.1.1 The Tiers Model

The four experiments together are expected to produce between 10 and 20 PB of data per year. To aid in handling such very large quantities efficiently, the WLCG infrastructure has been divided into *tiers*. At the lowest level is Tier-0, CERN itself, where the experiments record their data and perform a first-pass analysis over it. One copy of this data is saved on tape in the computer center at CERN, while another copy is spread over 11 Tier-1 centers. These are large computer centers in Europe, the United States, Canada, and Taiwan, all connected to CERN through dedicated network links of at least 10 Gbps. Each Tier-1 center saves its fraction of shipped data on tape as well. Later the Tier-1 center typically reprocesses the data with better calibration parameters or improved software versions. The subsequent output is saved on tape and may need to be copied to one or more partner Tier-1 centers for better availability of the data for subsequent analyses. At the next level there are about 100 Tier-2 sites. A Tier-2 site normally comprises a CPU farm and disk storage of up to a few tens of TB. Most of the analysis is expected to be done at Tier-2 sites, which will download the necessary input data files from the Tier-1 sites. A Tier-1 site will have a cloud of Tier-2 sites around it, often in the same country or larger region, and dependent on support from their Tier-1 center. Some of the experiments foresee their Tier-2 sites to download large amounts of data from Tier-1 centers in other regions, though. A Tier-2 site is also used to produce simulated data, which it uploads to its Tier-1 center for safekeeping and further distribution as needed. A Tier-3 site typically amounts to a CPU farm at a university participating in one of the experiments. Tier-3 resources typically are used opportunistically. To complicate matters further, CERN also acts as a Tier-1 center, and any Tier-1 center can also act as a Tier-2 site.

3.4.1.2 The WLCG Data Management Services and Clients

The WLCG infrastructure makes available a set of data management services and interfaces that the LHC experiments can use in order to implement their data models.

The gLite[17] data management client tools allow a user to move data in and out of the grid, replicate files between storage elements,* interact with a file catalog, and more. High-level data management clients and services shield the user from the complexities of the storage services and catalog implementations as well as transport and access protocols.

The **file transfer service (FTS)**[37] allows for the scheduling of the transfer of data files between sites. The service is configured to allow for transfers only on predefined channels between peers configured at service startup. The FTS uses low-level services and tools to perform data transfers and related operations.

Another data management library worth mentioning is the **grid file access library (GFAL)**.[18] It interacts with grid file catalogs and storage services via the available control protocols. It allows applications to access files using abstractions such as the "logical file name" (LFN), a human-readable identifier of a file in the grid. Once presented with an LFN, the GFAL library contacts a grid file catalog (LFC, the LCG File Catalog) to retrieve a handle to the best replica available. Then, it negotiates with the corresponding storage service to determine which file access protocol will be used (POSIX, gsiftp, rfio, gsidcap, etc.).

Because of historical reasons and the untimely availability of general solutions, the WLCG experiments have developed their own data management frameworks to various degrees. For instance, a transfer service such as PhEDEx,[19] developed by the CMS collaboration, could have evolved to fulfill the role of the FTS, that is, scheduling not only within but also between concurrent experiments. Instead, PhEDEx has been modified to drive the common FTS from the CMS perspective. Another protocol for efficient file transfer and access has been developed by SLAC. This is the XROOTD[20] system that is being considered in particular for end-user analysis.[38]

3.4.2 High-Energy Physics Use Cases

Although there are different HEP experiments within the WLCG project, all of them follow a common way of organizing their basic distributed computing model. We first describe the general computing and data model that is applicable to all four experiments and outline experiment specific differences later whenever necessary.

---

*The term *storage element* is used to mean any type of grid-enabled storage system and the software managing it, including multiple storage components such as disks, disk arrays, and robotic tapes.

3.4.2.1 Constraints for Distributed Computing and Storage

All four experiments have the following items in common, which can be regarded as the main constraints for a distributed computing model:

Central data recording: Data coming from the experiment detectors (raw data) is recorded at CERN. Data is typically written once and never updated (i.e., **read-only data**).

Large data storage: Each experiment produces a few (5 to 10) petabytes of data each year that need to be stored permanently.

Data processing: Raw data needs to be processed in order to extract and summarize information that has relevance to physics. This processing of data is called *reconstruction* in HEP terminology and is typically very compute-intensive. The storage requirement for reconstructed data is smaller than for raw data but still in the order of many terabytes to a few petabytes per year.[21-24]

Distributed computing and storage centers: CERN is considered to be the main center to provide storage and processing power. However, each of the four experiments consists of a collaboration of many countries, almost all of which provide storage and computing capacity that is dedicated to one or more of the experiments. In this way, the overall computing power and storage capacity available to a single experiment are increased.

Distributed user community: A few hundred research institutes and universities participate in the LHC experiments, with physicists (the actual end users) distributed over the globe. Their common goal is to analyze physics data as if all of it were available locally, having transparent access and good response time also when accessing data remotely.

3.4.3 High-Level Physics Use Cases

In the following section we give an overview of the basic high-level use cases for the computing usage in HEP. These use cases are representative of the data model explained in the previous section.

3.4.3.1 Reconstruction

The raw data, whether real or simulated, must be reconstructed in order to provide physical quantities such as the identities, positions, and momenta of the particles of interest. The pattern recognition algorithms in the reconstruction program make use of calibration and alignment parameters to correct for any temporal changes in the response of the detectors and their electronics. This process is computationally very intensive and needs to be repeated a few times in order to accommodate improvements in the algorithms or in calibration and alignment parameters. Therefore, it cannot be executed entirely at CERN. Raw data is stored on tape at CERN and streamed to Tier-1 sites where the reconstruction program should start shortly on data that just arrived. For this use case, the storage requirements are the following:

- Specific data transfer servers with *wide area network (WAN) access* and adequately large buffers need to be in place in order to efficiently receive data coming from the Tier-0.
- *Discovery functions* should allow for the identification of the data services and buffers dedicated to the given experiments.
- Data transfer services should allow for *reliable* and *secure transfer* of big buffers of data. Such services should provide users with transfer scheduling and retry functionalities.
- Data transfer servers must be connected to the *tape storage* systems for persistent storage of the data.
- A proper *storage interface* to MSSs should be available in order to trigger and control store and stage operations in an implementation-independent way.
- Given the amount of data involved, it is desirable to avoid making multiple copies of the data. Therefore, the data needs to remain on disk for a time sufficient to reconstruct it, before it is deleted to make space for new data. The *pinning functionality* of SRMs allows for specifying a lifetime associated to the data stored in a given space.
- For a critical operation such as reconstruction of physics data, it is mandatory not to compete for resources with other experiments. Therefore, *dedicated resources* are normally required by the experiments.
- Furthermore, it is important that user activities do not interfere with production or import/export activities. Support is required for access control lists on spaces provided by the storage services, as well as mechanisms to block unwanted types of access to specific data buffers.

3.4.3.2 Mainstream Analysis

This use case can be considered as the standard, scheduled activity of a physics group in a certain university. The research group is interested in analyzing a certain dataset (typically consisting of many gigabytes or several terabytes of data) in a certain Tier-1 center that has free computing capacity. If the data is not available at that site, it needs to be transferred in a scheduled way and the operation might last for a few days. Once the data has arrived, computing-intensive physics analysis operations can be done on the specified data. For instance, the validation of reconstructed data is a process in which the validity of the used algorithms and parameters is assessed. This process implies access to 1%–2% of the total reconstructed data of an experiment.

It might imply running variations of the program several times on the same set of data. Once the process is finished, the result is stored on tape.

The implicit storage, data, and transfer requirements are as follows:

- Data needs to be accessible from the storage system, that is, MSSs and disk systems. The corresponding data servers need to provide the required performance.
- Data transfer tools need to be in place that have access to the source storage system and can transfer data to another storage system at a different site/tier. Since the physics activities and therefore also the data transfers are scheduled, the latter can be optimized: Bandwidth can be "reserved" by *prioritizing* the requests of a particular physics group and reducing the ones of other physics groups or individual users.
- Once data has arrived at the site, computing and storage *resources* must be dynamically or statically *reserved* for a particular user group.
- It should be possible to express *ownership of resources* and specify *authorization patterns*.
- In order to ensure resource sharing, *quotas* should essentially be enforced in a transparent way so that several groups within the experiment or even multiple experiments can concurrently use the resources at a site.
- Resource usage and *status* should be *monitored* and published so that busy resources are not selected for further computing and/or storage tasks.
- If the needed data is on tape, it must first be transferred to disk for online access. Therefore, transparent staging tools must be available.
- Specific *file access protocols* need to be supported by the storage facility, so that applications limited to using only those protocols can be executed.
- Once data is analyzed, the relevant output can be saved on tape if deemed important. Therefore, tools to archive the results on tape and register them on the grid are necessary.
- Physicists should be provided with the necessary tools to *manage space*, for instance in case the storage system does not remove unneeded files automatically.

Grid operators and/or site administrators that take care of the execution and monitoring of data transfers as well as the allocation of CPU power to the physics group can further support and optimize the actual execution of this use case scenario.

3.4.3.3 Calibration Study

During the run of the LHC, particles pass through detectors that have to be aligned and calibrated in order to allow for correct physics analysis. A physics group might work on the calibration study and detect problems with the calibration and alignment parameters. In such a case, some of the reconstruction algorithms need to be rerun, and new reconstructed data needs to be stored.

In this use case, there is a request for fast access to a substantial subset of the data and for a large amount of computing power at peak times. This may involve transferring raw data from tape to disk. Many tape drives can thus be busy in this task that typically has high priority. Once the set of calibration parameters proves to be accurate, it is stored in experiment-specific databases that are distributed to a few sites for reasons of performance and fault tolerance.

3.4.3.4 Chaotic Analysis

In contrast to the scheduled "mainstream analysis" of a particular physics group, here a single physicist working on a specific analysis might request access to a dataset that can be of any size, that is, it is not known a priori how much data would need to be made available locally or accessed through the WAN.

This use case is of particular importance for physicists, system administrators, operators, and developers, since it can create worst-case scenarios that stress the system. This use can also help detect scalability issues in many parts of the data access and storage system.

Because of this unpredictable behavior, it is very important to be able to control storage resource usage and access accurately in order to prevent problems. In particular, *quota and dynamic space reservation* become essential. Also important is the ability to control data and resource access through local policies and access control lists. For instance, the capability of staging files from tape to disk or to store results permanently on tape should be allowed only to users with certain roles and belonging to specific groups. Data processing managers within each experiment are allowed to check the resources available and ensure correct usage. They need to check for file ownership, correct placement, sizes, and so forth. They can delete files or move them to storage with appropriate quality of service whenever needed.

3.4.4 Storage Requirements

In this section we describe the current state and the continuous evolution of storage services available on the WLCG and EGEE infrastructures.

3.4.4.1 The Classic Storage Element and SRM v1.1

A grid-enabled storage facility is called a storage element (SE). Originally, an SE was nothing more than a GridFTP server in front of a set of disks, possibly backed by a tape system. Such a facility is called a classic SE. It was the first storage service implementation in the WLCG infrastructure. Many tens of classic SEs are still there today, but they are used by VOs other than the LHC experiments. Each supported VO has access to a part of the name space on the SE, typically corresponding to a file system dedicated to the VO. A shared file system with quotas would also work. A large MSS with a tape back-end may be configured such that files in certain subsets of the name space will be flushed to tape and recalled to disk as needed.

There are at least the following issues with the Classic SE:

- It usually is not easy to enlarge the amount of disk space available to a VO indefinitely. Various modern file systems can grow dynamically when new disks are made available through some logical volume manager, but a single file system may become an I/O bottleneck, even when the file system is built out of multiple machines in parallel (e.g., as a SAN or a cluster file system). Furthermore, commercial advanced file systems are expensive and may lead to vendor lock-in, while open-source implementations have lacked maturity (this is steadily improving, though). Instead, multiple file systems could be made available to a VO, mounted on different parts of the VO name space, but it usually is impossible to foresee in which parts more space will be needed. A site could grow its storage by setting up multiple GridFTP servers, all with their own file systems, but that may leave some of those servers idle while others are highly loaded. Therefore, the desire is for an SE to present itself under a single name, while making transparent use of multiple machines and their independent, standard file systems. This is one of the main reasons for developing the SRM concept.
- GridFTP servers lack advance space reservation: A client will have to try to find out which fraction of its data it actually can upload to a particular server, and look for another server to store the remainder.
- A GridFTP server fronting a tape system has no elegant means to signal that a file needs to be recalled from tape: The client will simply have to remain connected and wait while the recall has not finished. If it disconnects, the server might take that as an indication that the client is no longer interested in the file and that the recall should therefore be canceled. Furthermore, there is no elegant way to keep a file pinned on disk to prevent untimely cleanup by a garbage collector.
- A GridFTP server has no intrinsic means to replicate hot files for better availability. The host name of a GridFTP service could be a round-robin or load-balanced alias for a set of machines, but then each of them must have access to all the files. This could be implemented by some choice of shared file system, or by having the GridFTP server interact with a management service that will replicate hot files on the fly, making space by removing replicas of unpopular files as needed. Such functionality is naturally implemented by an SRM.

In the spring of 2004 the WLCG middleware releases started including SRM v1.1 client support in data management. The first SRM v1.1 service available on the WLCG infrastructure (to the CMS experiment) was a dCache instance at FNAL, the Tier-1 center where the dCache SRM is developed. The majority of the Tier-1 centers have since adopted dCache as MSS front-end. In the autumn of 2004 the CASTOR services at CERN and three Tier-1 centers also became accessible through SRM v1.1, while retaining a classic SE appearance in parallel for backward compatibility. In the spring of 2005, to assist in the configuration and operation of Tier-2 sites, the WLCG/EGEE middleware releases started including support for both dCache and Disk Pool Manager (DPM) installations. dCache has more options for advanced configurations (e.g., separation of read and write pools), but the DPM has been simpler to operate. In early 2008 CASTOR is in use at 7 WLCG sites, dCache at about 60, and the DPM at about 130.

Though the transition to SRM v1.1 has brought significant improvements to WLCG data management, it became clear that it still had important defects:

SRM v1.1 still lacks advance space reservation. It only allows for an implicit space reservation as the first part of a short-lived store operation. This does allow for the operation to be canceled cleanly when insufficient space happens to be available, though.

SRM v1.1 lacks an elegant prestaging functionality. When a file has to be recalled from tape, the client will either have to remain connected and wait, or it would have to resort to a server-specific protocol for having the file staged in advance.

There is no portable way to guarantee the removal of files that are no longer wanted. SRM v1.1 only has an advisory delete function, whose effects differ in different implementations. Client tools typically have to recognize the various implementations and invoke server-specific algorithms, contrary to the idea of a protocol standard.

SRM v1.1 lacks equivalents to basic file system operations for, for example, renaming files, removing directories, or changing permissions. Directories are created implicitly.

As for the Classic SE, files and directories owned by a VO have to be made writable for their whole VO by default, so that any member of the VO can write to the SE without further administrative operations or requiring a strict organization of the VO name space. Exceptions are made for the VO production managers, who are responsible for the vast majority of the data produced by a VO. Usually they ask for dedicated subsets of the VO name space, where only they can write. At the same time they can negotiate the desired quality of service, for example, dedicated disk pools.

3.4.4.2 The Storage Element Service

In the first quarter of 2005 the WLCG Baseline Services working group (BSWG)[25] was established in order to understand the experiment requirements for their data challenges. For each of the experiments a data challenge is a set of large-scale tests focused on verifying the readiness and functionality of its computing infrastructure. The BSWG report[25] includes an assessment of the main functionalities needed from storage services. It established that an SE is a logical entity that provides the following services and interfaces:

- An MSS that can be provided by either a pool of disk servers or more specialized high-performance disk-based hardware, or a disk cache front-end backed by a tape system.
- A storage interface to provide a common way to access the specific MSS, no matter what the implementation of the MSS is.
- A GridFTP service to provide data transfer in and out of the SE to and from the grid. This is the essential basic mechanism by which data is imported to and exported from the SE. The implementation of this service must scale to the bandwidth required. Normally, the GridFTP transfer will be invoked indirectly via the FTS or via the storage interface.
- Local POSIX-like input/output calls providing application access to the data on the SE.
- Authentication, authorization, and audit/accounting facilities. The SE should provide and respect ACLs for files and datasets, with access control based on the use of extended X.509 proxy certificates with a user distinguished name (DN) and attributes based on virtual organization membership service (VOMS) roles and groups. It is essential that an SE provide sufficient information to allow tracing of all activities for an agreed historical period, permitting audit on the activities. It should also provide information and statistics on the use of the storage resources, according to schema and policies.

A site may provide multiple SEs with different qualities of storage. For example, it may be considered convenient to provide an SE for data intended to remain available for extended periods and a separate SE for data that is transient—needed only for the lifetime of a job or set of jobs. Large sites with MSS-based SEs may also deploy disk-only SEs for such a purpose or for general use. Since most applications will not communicate with the storage system directly but will use higher-level applications such as ROOT,[26] it is clear that these applications must also be enabled to work with storage interfaces.

3.4.4.3 Beyond the WLCG Baseline Services Working Group

The BSWG required storage services to provide the features of SRM v1.1 along with a subset of SRM v2.1. By the end of 2005, however, it became clear that sufficiently compatible implementations would not be available before the spring of 2006, by which time it was foreseen to start WLCG Service Challenge 4, the last in a series of large-scale tests to assess the WLCG infrastructure readiness for handling LHC data taking and analysis. Therefore, in February 2006 an initiative was agreed to simplify the set of requirements.

A new working group was established, including storage system developers and managers, representatives from the experiments, and data-management middleware developers. At the end of May the first version of SRM v2.2 was agreed to. At the same time a WLCG-specific usage agreement (UA) spelled out that WLCG client middleware would only exercise a subset of the full functionality. This allowed the storage system developers to ignore features not required by the UA or to postpone their implementation.

The new set of requirements essentially was the following:

Only permanent files; that is, only the user can remove files.

Advance space reservation without streaming, initially only static, later also dynamic.

Quotas. Unfortunately not yet accepted as an SRM feature.

Permission functions with POSIX-like ACLs for directories and files. It must be possible to match permissions to the contents of the client's proxy credentials, that is, the distinguished name and/or a set of VOMS groups and roles.

It must be possible for privileged users, groups, and roles to have a better quality of service, for example, dedicated disk pools, higher priority.

Basic directory functions: mkdir; rmdir; rename (on the same SE); remove; list (up to a server-dependent maximum number of entries may be returned).

Data transfer control functions: stage-in and stage-out type functionality; pinning and unpinning; request status monitoring; request cancellation.

Paths relative to an implicit VO-specific base directory.

Paths should be orthogonal to quality of service (e.g., retention policy, access latency).

A method to discover the supported transfer protocols.

3.4.4.4 The Storage Classes

In the summer of 2006 the WLCG Storage Classes Working Group was established to understand the requirements of the LHC experiments in terms of *quality* of storage (storage classes) and how such requirements could be implemented in the various storage solutions available. For instance, this implies understanding how to assign disk pools for LAN or WAN access and trying to devise common configurations for VOs and recipes tailored per site.

The *storage class* determines the essential quality-of-service properties that a storage system needs to provide for given data.

The LHC experiments have asked for the availability of combinations of the following storage devices: tapes (or other reliable storage systems, always referred to as tape in what follows) and disks. A file residing on tape is said to be in Tape1.* A file residing on an experiment-managed disk is said to be in Disk1. Tape0 means that the file does not have a copy stored on a reliable storage system. Disk0 means that the disk where the copy of the file resides is managed by the system: If such a copy is not being used, the system can delete it.

The Storage Classes Working Group decided that only certain combinations (or storage classes) are needed for the time being, corresponding to specific choices for the Retention Policy and the Access Latency as defined by SRM v2.2:

Custodial–Near-line = Tape1Disk0 class

Custodial–Online = Tape1Disk1 class

Replica–Online = Tape0Disk1 class

Tape0Disk0 is not implemented. It is pure scratch space that could be emulated using one of the available classes and removing the data explicitly once done. However, it could be handy for LHC VOs to have such a type of space actually implemented eventually.

In the *custodial–near-line* storage class data is stored on some reliable secondary storage system (such as a robotic tape or DVD library). Access to data may imply significant latency. In WLCG this means that a copy of the file is on tape (Tape1). When a user accesses a file, the file is recalled in a cache that is managed by the system (Disk0). The file can be pinned for the time the application needs the file. However, the treatment of a pinned file on a system-managed disk is implementation dependent, some implementations choosing to honor pins until they expire or are released, and others removing unused online copies of files to make space for new requests.

In the *custodial–online* storage class, data is always available on disk. A copy of the data resides permanently on tape, DVD, or on a high-quality RAID system as well. The space owner (the virtual organization) manages the space available on disk. If no space is available in the disk area for a new file, the file creation operation fails. This storage class guarantees that a file is never removed by the system.

The *replica–online* storage class is implemented through the use of disk-based solutions not necessarily of high quality. The data resides on disk space managed by the virtual organization.

Through the storage system interface, it is possible to schedule storage class transitions for a list of files. Only the following transitions are allowed in WLCG:

---

*The term Tape1 was coined to mean that a single copy of the file is residing on tape. Accordingly, it is possible to have Tape2 (two copies), Tape3 (three copies), and so forth, and likewise for Disk1. However, it was decided to avoid the ability to specify multiple copies at the present time.

Tape1Disk1 → Tape1Disk0. On some systems this can be implemented as a metadata operation only, while other systems may require more operations to guarantee such a transition.

Tape1Disk0 → Tape1Disk1. This transition is implemented with some restrictions: The request will complete successfully but the files will remain on tape. The files will be actually recalled from tape to disk only after an explicit request is executed. This is done in order to avoid unnecessarily scheduling of a big set of files for staging and therefore to smooth operations, in particular for those MSSs that do not have a scheduler.

Tape0Disk1 ↔ Tape1DiskN. Such transitions are not supported at the start of LHC (if ever). For physics validation operations, since the amount of data to transfer to tape after the validation is not big (only 1%–2% of total data), a change from Tape0Disk1 to Tape1DiskN can be approximated by copying the files to another part of the name space, specifying Tape1DiskN as the new storage class, and then removing the original entries.

3.4.4.5 The Grid Storage Systems Deployment Working Group

In January 2007 the Grid Storage Systems Deployment (GSSD) working group[10,27–29] was established with the following main goals:

Testing of SRM v2.2 implementations for compliance and interoperability.

Establishing a migration plan from SRM v1.1 to SRM v2.2 so that the experiments can access the same data through the two protocols transparently.

Coordinating with sites, experiments, and developers the deployment of the various SRM v2.2 implementations and the corresponding storage classes.

Coordinating the definition and deployment of an information schema for the storage element to allow for the relevant aspects of SRM v2.2 services to be discoverable through the WLCG information system.

Coordinating the provision of the necessary information by the storage providers in order to monitor the status and usage of storage resources.

It took until the autumn of 2007 for CERN and the Tier-1 sites to start putting SRM v2.2 services into the WLCG production infrastructure, with known deficiencies to be corrected in later versions. Further improvements were implemented after significant operational experience had been gained during a final set of large-scale tests, the Common Computing Readiness Challenge, which took place in February and May 2008. The work of the GSSD has carried on throughout 2008 and into 2009.

3.4.5 Beyond WLCG: Data Management Use Cases in EGEE

The data management use cases of many other disciplines for which the EGEE infrastructure is intended have a lot in common with those of the LHC experiments, which have been the main driving force behind the development of the various data management services and client utilities available today. Other disciplines also pose additional requirements that matter less to the LHC experiments, at least for the time being:

Fully encrypted communication and storage for much-improved privacy. For example, in biomedical research it is vital that no data be accidentally exposed, whereas LHC expermenters do not mind if the vast majority of their data happens to be world-readable. The expermenters do not want their high-level analysis results to be exposed, but such results typically can be stored and processed locally without the need for grid services. On the other hand, encrypting and decrypting their huge volumes of low-level data on the fly probably would still be an unacceptable overhead for a long time. For the biomedical community, encrypted storage has been developed as a plug-in for the DPM storage element and the GFAL client suite. An additional necessary component is the Hydra distributed key service.[30]

Standard POSIX access to data served by storage elements. Most of the EGEE communities have legacy applications that expect to read and write files through standard POSIX I/O instead of GFAL,[17] ROOT.[26] XROOTD,[20] or server-dependent protocols like RFIO[4] and DCAP.[31] It was fairly straightforward for a classic SE to make its data available to a local batch farm via NFS, but this has not been possible for SRM services. An SRM should be able to replicate hot files (i.e., files that are accessed frequently) and direct clients to any disk server in a load-balanced way. It should also be able to clean up a replica without the risk that a client might still be reading it. These problems can be overcome, though. For example, the main objective for the StoRM implementation of SRM v2.2 is to allow secure POSIX access by legacy applications, typically through a cluster file system like GPFS and with just-in-time access control lists, so that a file is only accessible for a duration that a client has to negotiate with the SRM in advance. All of the SRM implementations in use on the EGEE infrastructure have expressed interest in developing NFSv4 interfaces, which would allow for standard POSIX access by clients without reducing vital server functionality. Finally, a transparent remote I/O library like Parrot (see Chapter 4) may be enhanced with plug-ins for some of the popular protocols that can be served. This will not help statically linked applications, though.

Availability of client utilities on many platforms. To optimize the use of available resources, the HEP experiments and laboratories have moved their computing infrastructures almost exclusively toward a few flavors of Linux distributions. Many other disciplines, however, find themselves with applications that will only run on platforms of other types. NFSv4 would help in this regard as well, but SRM clients and other data management utilities would still have to be ported to other popular platforms.

3.5 Examples of Using File Streaming in Real Applications

In the previous example, WLCG has chosen to have all files have a "permanent" type, and therefore file placement and removal are managed by administrators of VOs. In this section, we describe two example applications where file streaming is quite useful. In such applications, the streaming files are *volatile*; that is, they have a lifetime, and each file can be released by the client when it is done accessing it.

3.5.1 Robust File Replication in the STAR Experiment

Robust file replication of thousands of files is an extremely important task in data-intensive scientific applications. Moving the files by writing scripts is too tedious since the scientist needs to monitor for failures over long periods of time (hours, even days), and recover from such failures. Failures can occur in staging files from a MSS, or because of transient transfer failures over networks. This mundane, seemingly simple task is extremely time-consuming and prone to mistakes. We realized that SRMs are perfectly suited to perform such tasks automatically. The SRMs monitor the staging, transfer, and archiving of files, and recover from transient failures.

Figure 3.6 shows the sequence of actions for each file as a request of multiple files between two MSSs, such as NCAR-MSS and HPSS (a real use case between NCAR and LBNL). The request is made to the target SRM, and it issues a request for files to the source SRM. The source SRM stages files from its MSS into its disk cache and notifies the target SRM. The target SRM allocates space, notifies the source, and the transfer takes place. File streaming works as follows: After each file gets transferred, the space it occupies at the source cache is released, making space for additional files. Likewise, after every file gets archived, the space it occupies in the target cache is released, making space for additional files to be transferred. File streaming is essential for this operation, where files are pinned and released, since the number of files to be transferred can often be much larger than the space available in the disk cache, especially when the disk cache is shared by multiple users.

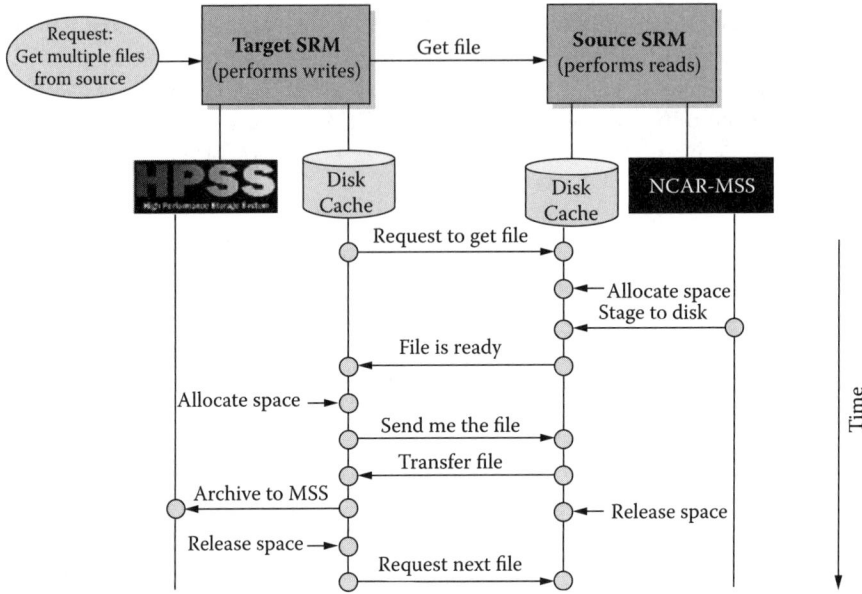

Figure 3.6 Sequence of operations between two cooperating SRMs to support robust file replication.

SRMs are also capable of performing the same operation of copying of an entire directory, by first getting the directory structure from the source system under SRM, establishing the same structure in the target system under SRM, and then starting the transfer process into the corresponding locations in the directories. This amounts to performing the Unix recursive remote copy command "rcp –r directory" between two remote sites for very large directories by using file streaming.

We note that the file staging and transfer can overlap in time for multiple files, and therefore there may be multiple files ready in the cache to be transferred, creating the opportunity to transfer multiple files concurrently. Another interesting property is that the entire process is self-regulating; that is, if the network is too slow, the quota in the source disk gets full, and it stops requesting files to be staged. Similarly, if the target disk gets full because the archiving is too slow, it stops asking the source to send files. Thus, there is no possibility for files to get lost. This setup has proven to work reliably even in the face of long maintenance periods (usually an hour or so) of the MSSs. In particular, this has been used for daily robust large-scale data replication for the STAR experiment[32] between the MSS at Brookhaven National Laboratory and Lawrence Berkeley National Laboratory since 2004. The volume needed to be transferred is about 10,000 files (about 1 TB) per week. This arrangement resulted in a 50X reduction in the error rates, from 1% to 0.02%. The same arrangement has been used periodically by the Earth System Grid (ESG),[33] as shown in Figure 3.6.

3.5.2 The Earth System Grid

The purpose of the ESG is to provide the worldwide community of climate scientists access to climate modelling data from multiple sources. This includes data from the Community Climate System Model (CCSM),[34] the Intergovernmental Panel on Climate Change (IPCC)[35] model, the Parallel Ocean Program (POP),[36] and others. In the current version of ESG, the data is stored in multiple distributed sites including NCAR and several Department of Energy Laboratories: LANL, LBNL, LLNL, and ORNL. As of the end of 2008, the ESG has about 9,000 registered users, over 200 TB of data that can be accessed by users, and so far supported over 66 TBs of data downloaded.

The ESG is built using a collection of tools for finding the desired data using metadata catalogs and search engines, for requesting aggregation of the data, and for moving subsets of the data to users. The current distributed organization is shown in Figure 3.7 where only the SRMs, the storage components (disks, MSS) and the GridFTP servers are shown. The SRMs call the GridFTP servers to perform the file transfers. The SRM at the portal has a large disk cache shared by all active clients. It allocates space for each user and brings in files from any of the other sites, either files residing on disks or files that SRMs bring into disks from MSSs if necessary. So far, file streaming has not been used in this version of ESG since the number of files allowed to be downloaded was limited. However, as the number of files requested grew, the disk cache became clogged and had to be enlarged. In anticipation of much larger use the next version of ESG, which will be a federated system with multiple portals and data nodes all over the world, will use file streaming in order to have more effective use of the shared disk caches and avoid clogging the disks.

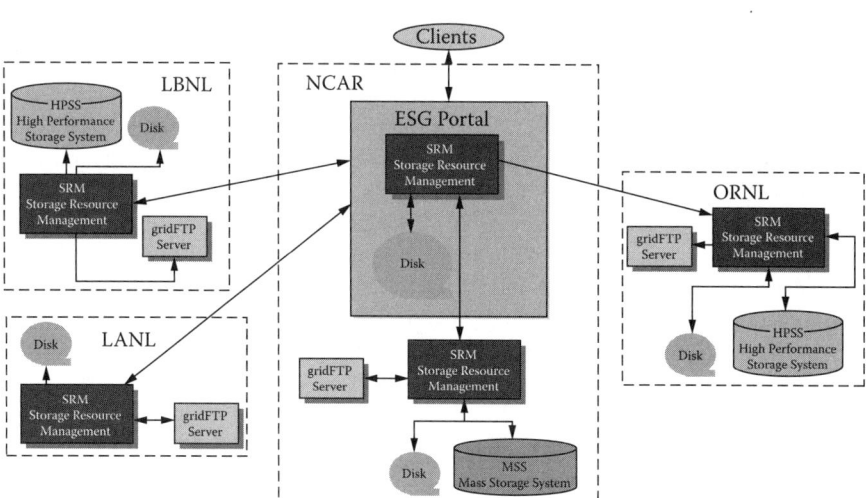

Figure 3.7 The Earth System Grid use of SRMs for file streaming to clients.

3.6 Conclusions and Future Work

Storage management is one of the most important enabling technologies for large-scale scientific investigations. Storage resource managers (SRMs) provide the technology needed to manage the rapidly growing distributed data volumes, as a result of faster and larger computational facilities. SRMs are software components whose function is to provide dynamic space allocation and file management on shared storage systems. They call on transport services to bring files into their spaces transparently and provide effective sharing of files. SRMs are based on a common specification that emerged over time and evolved into an international collaboration. This approach of an open specification that can be used by various institutions to adapt to their own storage systems has proven to be a remarkable success. SRMs are being used in production by multiple facilities to provide uniform interfaces to various storage systems at many sites in the United States, Canada, Europe, Asia, and elsewhere. They are being used in several application domains, quite extensively in high-energy and nuclear physics experiments, the Earth System Grid, and other application domains, including fusion modeling, biology, and medical applications.

Based on past experience we have identified several features that require further development and coordination. These include sophisticated aspects of resource monitoring that can be used for performance estimation, authorization enforcement in coordination with virtual organizations (VOs), and accounting tracking and reporting for the purpose of enforcing individual or group quotas in the use of SRM resources. Performance estimation is very important for the clients to decide whether to launch a request to a particular SRM, especially when that SRM is highly loaded. Such a task is nontrivial, in that an SRM has to estimate not only its own load (based on queue monitoring), but also take into account external performance estimations, such as the time to get files in or out of MSSs, or from another site over the network. Another item to be considered for support by SRMs is advance of space reservation. Currently, SRMs only respond to a space reservation request at the time the request is made. It is much harder to manage reservations for periods in the future. However, such provisioning capabilities have started to be supported by networks dedicated to research communities such as Energy Sciences Network (ESnet) and Internet2. Advance network provisioning and compute resource reservations will have to be coordinated with advance space reservation when supported by SRMs, in order to plan and perform complex coordinated tasks successfully. Finally, the issue of enforcing authorization is quite complex, and requires a foolproof, secure way of coordinating with a VO. In general, given that the SRM back-end storage resources are shared by multiple users, the quota assigned to a user (measured in gigabyte-minutes, for example) has to be centrally coordinated with a VO, or predetermined, in order for the SRM to enforce it. Currently, enforcement of space usage authorization is not uniformly supported by SRMs either. Adding such capabilities requires keeping track of usage by each user or group of users and providing a reporting mechanism to VOs. Security mechanisms have to be in place as well, in order to verify that a request is vetted by the VO.

Acknowledgments

The work in this chapter was partially supported by the Director, Office of Science, Office of Advanced Scientific Computing Research, of the U.S. Department of Energy under Contract No. DE-AC02-05CH11231. A large part of the work referred to in Section 3.4, Data Management in WLCG and EGEE, has been carried out in the WLCG and EGEE projects, supported by their respective funding agencies. EGEE is cofunded by the European Union under contract INFSO-RI-222667.

References

[1] http://www.sdsc.edu/srb/index.php/Main_Page. Accessed on July 20, 2009.

[2] J. Bent, V. Venkataramani, N. Leroy, A. Roy, J. Stanley, A. C. Arpaci-Dusseau, R. H. Arpaci-Dusseau, and M. Livny. NeST—A Grid Enabled Storage Appliance, in J. Weglarz, J. Nabrzyski, J. Schopf, and M. Stroinkski, editors, *Grid Resource Management*, chapter 22. Kluwer Academic Publishers, Norwall, MA, 2003.

[3] http://www.ogf.org/documents/GFD.129.pdf. Accessed on July 20, 2009.

[4] CASTOR: CERN Advanced STORage manager, http://castor.web.cern.ch/castor Accessed on July 20, 2009.

[5] http://www-ccf.fnal.gov/enstore/. Accessed on July 20, 2009.

[6] http://sdm.lbl.gov/bestman/. Accessed on July 20, 2009.

[7] http://storm.forge.cnaf.infn.it/. Accessed on July 20, 2009.

[8] A. Shoshani, A. Sim, J. Gu, Storage Resource Managers: Middleware Components for Grid Storage, Nineteenth IEEE Symposium on Mass Storage Systems, 2002.

[9] A. Shoshani, A. Sim, and J. Gu, Storage Resource Managers: Essential Components for the Grid, in Grid Resource Management: State of the Art and Future Trends, Edited by J. Nabrzyski, J. M. Schopf, J. Weglarz, Kluwer Academic Publishers, Norwall, MA, 2003.

[10] L. Abadie et al., Storage Resource Managers: Recent International Experience on Requirements and Multiple Co-Operating Implementations, Mass Storage Systems and Technologies, September 24–27, 2007, San Diego, California.

[11] http://public.eu-egee.org/. Accessed on July 20, 2009.

[12] http://www.opensciencegrid.org/. Accessed on July 20, 2009.

[13] http://www.nordugrid.org/. Accessed on July 20, 2009.

[14] http://www.hpss-collaboration.org/hpss/index.jsp. Accessed on July 20, 2009.

[15] http://www.dcache.org/. Accessed on July 20, 2009.

[16] http://www.gridpp.ac.uk/wiki/Disk_Pool_Manager. Accessed on July 20, 2009.

[17] http://glite.web.cern.ch/glite/. Accessed on July 20, 2009.

[18] J.-P. Baud, J. Casey, Evolution of LCG-2 Data Management, CHEP, La Jolla, California, March 2004.

[19] T. Barrass, D. Newbold, L. Tuura, The CMS PhEDEx System: A Novel Approach to Robust Grid Data Distribution, AHM 2005, September 19–22, 2005, Nottingham, UK.

[20] C. Boeheim, A. Hanushevsky, D. Leith, R. Melen, R. Mount, T. Pulliam, B. Weeks, *Scalla: Scalable Cluster Architecture for Low Latency Access Using xrootd and olbd Servers,* August 22, 2006, http://xrootd.slac.stanford.edu/. Accessed on July 20, 2009.

[21] Alice Collaboration, *Alice Technical Design Report of the Computing* CERN-LHCC-2005-018, ALICE, TDR-012, June 15, 2005: http://aliceinfo.cern.ch/static/Documents/TDR/Computing/All/alice_computing.pdf. Accessed on July 20, 2009.

[22] ATLAS Collaboration, *ATLAS Computing Technical Design Report* CERN-LHCC-2005-017, ATLAS, TDR-017, June 20, 2005: http://cern.ch/atlas-proj-computing-tdr/PDF/Computing-TDR-final-June20.pdf. Accessed on July 20, 2009.

[23] CMS Collaboration, *CMS Computing Technical Design Report* CERN-LHCC-2005-023, CMS, TDR-007 http://cdsweb.cern.ch/search?id=838359. Accessed on July 20, 2009.

[24] LHCb Collaboration *LHCb Computing Technical Design Report* CERN-LHCC-2005-019, LHCb, TDR-019, June 20, 2005: http://doc.cern.ch/archive/electronic/cern/preprints/lhcc/public/lhcc-2005-019.pdf. Accessed on July 20, 2009.

[25] The WLCG Baseline Service Working Group Report v1.0, 24 June 2005 http://lcg.web.cern.ch/LCG/peb/bs/BSReport-v1.0.pdf. Accessed on July 20, 2009.

[26] http://root.cern.ch/. Accessed on July 20, 2009.

[27] The WLCG SRM Development Group: https://twiki.cern.ch/twiki/bin/view/SRMDev/WebHome. Accessed on July 20, 2009.

[28] The WLCG Grid Storage System Deployment Working Group: https://twiki.cern.ch/twiki/bin/view/LCG/GSSD. Accessed on July 20, 2009.

[29] F. Donno et al., Storage Resource Manager version 2.2: Design, Implementation, and Testing Experience, International Conference in Computing in High Energy and Nuclear Physics, September 2–7, 2007, Victoria, B.C., Canada.

[30] S. Rafaeli, D. Hutchison, Hydra: A Decentralised Group Key Management, wetice, pp.62, Eleventh IEEE International Workshop on Enabling Technologies: Infrastructure for Collaborative Enterprises (WETICE'02), 2002.

[31] M. Ernst, P. Fuhrmann, T. Mkrtchyan, J. Bakken, I. Fisk, T. Perelmutov, D. Petravick, *Managed Data Storage and Data Access Services for Data Grids,* CHEP, La Jolla, California, March 2004.

[32] http://www.star.bnl.gov/. Accessed on July 20, 2009.

[33] http://www.earthsystemgrid.org/. Accessed on July 20, 2009.

[34] http://www.ccsm.ucar.edu/ Accessed on July 20, 2009.

[35] http://www.ipcc.ch/. Accessed on July 20, 2009.

[36] http://climate.lanl.gov/Models/POP/. Accessed on July 20, 2009.

[37] http://egee-jra1-dm.web.cern.ch/egee-jra1-dm/FTS. Accessed on July 20, 2009.

[38] http://xrootd.slac.stanford.edu. Accessed on July 20, 2009.

Part II

Data Transfer and Scheduling

Chapter 4

Coordination of Access to Large-Scale Datasets in Distributed Environments

Tevfik Kosar,[1] Andrei Hutanu,[1] Jon McLaren,[1] and Douglas Thain[2]

[1] *Louisiana State University, Baton Rouge, Louisiana*
[2] *University of Notre Dame, Notre Dame, Indiana*

Contents

4.1 Introduction 116
4.2 Background 118
4.3 Scheduling Data Movement 119
    4.3.1 Data Placement Job Types 120
    4.3.2 Job Scheduling Techniques 120
        4.3.2.1 Auxiliary Scheduling of Data Transfer Jobs 121
4.4 High-Performance Wide Area Data Transfers 125
    4.4.1 TCP Characteristics and Limitations 125
    4.4.2 Alternative Transport Protocols 127
    4.4.3 Current Status and Options 128
    4.4.4 Other Applications 130
    4.4.5 Challenges and Future Directions 131
4.5 Remote Input and Output 131
    4.5.1 Challenges of Remote I/O 133
    4.5.2 Case Study: Parrot and Chirp 134
    4.5.3 Open Problems in Remote I/O 136
4.6 Coscheduling Compute and Network Resources 136
    4.6.1 HARC: The Highly Available Resource Co-Allocator 137
    4.6.2 Architecture 137
    4.6.3 Reserving Network Resources Using HARC 139
    4.6.4 Experiments Using HARC 139
4.7 Conclusions 141
Acknowledgments 141
References 142

4.1 Introduction

Modern scientific applications and experiments become increasingly data intensive. Large experiments, such as high-energy physics simulations, genome mapping, and climate modeling generate data volumes reaching hundreds of terabytes.[41] Similarly, remote sensors and satellites are producing extremely large amounts of data for scientists.[19,82] In order to process these data, scientists are turning toward distributed resources owned by the collaborating parties to provide them the computing power and storage capacity needed to push their research forward. But the use of distributed resources imposes new challenges.[52] Even simply sharing and disseminating subsets of the data to the scientists' home institutions is difficult. The systems managing these resources must provide robust scheduling and allocation of storage and networking resources, as well as efficient management of data movement.

One benefit of distributed resources is that it allows institutions and organizations to gain access to resources needed for large-scale applications that they would not otherwise have. But in order to facilitate the sharing of compute, storage, and network resources between collaborating parties, middleware is needed for planning, scheduling, and management of the tasks as well as the resources. Existing research in this area has mainly focused on the management of compute tasks and resources, as they are widely considered to be the most expensive. As scientific applications become more data intensive, however, the management of storage resources and data movement between the storage and compute resources is becoming the main bottleneck. Many jobs executing in distributed environments fail or are inhibited by overloaded storage servers. These failures prevent scientists from making progress in their research.

According to the Strategic Plan for the U.S. Climate Change Science Program (CCSP), one of the main objectives of the future research programs should be "Enhancing the data management infrastructure," since "the users should be able to focus their attention on the information content of the data, rather than how to discover, access, and use it."[18] This statement by CCSP summarizes the goal of many cyberinfrastructure efforts initiated by DOE, NSF, and other federal agencies, as well as the research direction of several leading academic institutions.

Accessing and transferring widely distributed data can be extremely inefficient and can introduce unreliability. For instance, an application may suffer from insufficient storage space when staging in the input data, generating the output, and staging out the generated data to a remote storage. This can lead to trashing of the storage server and subsequent timeout due to too many concurrent read data transfers, ultimately causing server crashes due to an overload of write data transfers. Other third-party data transfers may stall indefinitely due to loss of acknowledgment. And even if transfer is performed efficiently, faulty hardware involved in staging and hosting can cause data corruption. Furthermore, remote access will suffer from unforeseeable contingencies such as performance degradation due to unplanned data transfers and intermittent network outages.

Efficient and reliable access to large-scale data sources and archiving destinations in a widely distributed computing environment brings new challenges:

Scheduling Data Movement. Traditional distributed computing systems closely couple data handling and computation. They consider data resources as second-class entities, and access to data as a side effect of computation. Data placement (i.e., access, retrieval, and/or movement of data) is either embedded in the computation and causes the computation to delay, or performed as simple scripts that do not have the privileges of a job. The insufficiency of the traditional systems and existing CPU-oriented schedulers in dealing with the complex data-handling problem has yielded a newly emerging era: the data-aware schedulers. One of the first examples of such schedulers is the Stork data placement scheduler that we have developed. Section 4.3 presents Stork and data-aware scheduling.

Efficient Data Transfers. Another important challenge when dealing with data transfers over wide area networks is efficiently utilizing the available network bandwidth. Data transfers over wide area, and in particular over high-capacity network links, make the performance limitations of the TCP protocol visible. In Section 4.4, we discuss these limitations and various alternatives. Much of the work related to wide area data transfer is focused on file or disk-to-disk transfer. We will also present other types of scenarios, how they are different, what the challenges are, and possible solutions.

Remote Access to Data. In some cases, transferring complete datasets to remote destinations for computation may be very inefficient. An alternate solution is performing remote I/O, where the files of interest stay in one place and the programs issue network operations to read or write small amounts of data that are of immediate interest. In this model, transfer protocols must be optimized for small operations, and the processing site may need no storage at all. In Section 4.5, we discuss advantages and challenges of remote I/O, and present the Parrot and Chirp technologies as a case study.

Coscheduling of Resources. Distributed applications often require guaranteed levels of storage space at the destination sites, as well as guaranteed bandwidth between compute nodes, or between compute nodes and a visualization resource. The development of a booking system for storage or network resources is not a complete solution, as the user is still left with the complexity of coordinating separate booking requests for multiple computational resources with their storage and network booking(s). We present a technique to coallocate computational and network resources in Section 4.6. This coscheduling technique can easily be extended to storage resources as well.

4.2 Background

In an effort to achieve reliable and efficient data placement, high-level data management tools such as the reliable file transfer service (RFT),[61] the lightweight data replicator (LDR),[51] and the data replication service (DRS)[20] were developed. The main motivation for these tools was to enable byte streams to be transferred in a reliable manner by handling possible failures like dropped connections, machine reboots, and temporary network outages automatically via retrying. Most of these tools are built on top of GridFTP,[2] which is a secure and reliable data transfer protocol especially developed for high-bandwidth wide area networks.

Beck et al. introduced logistical networking,[10] which performs global scheduling and optimization of data movement, storage, and computation based on a model that takes into account all the network's underlying physical resources. Systems such as the storage resource broker (SRB)[8] and the storage resource manager (SRM)[75] were developed to provide a uniform interface for connecting to heterogeneous data resources. SRB provides a single front-end that can access a variety of back-end storage systems. SRM is a standard interface specification that permits multiple implementations of the standard on top of storage systems. SRMs were discussed in detail in Chapter 3.

GFarm[65] provided a global parallel filesystem with online petascale storage. Their model specifically targets applications where data primarily consists of a set of records or objects that are analyzed independently. GFarm takes advantage of this access locality to achieve a scalable I/O bandwidth using a parallel file system integrated with process scheduling and file distribution. The Open-source Project for a Network Data Access Protocol (OPeNDAP) provides software which makes local multidimensional array data accessible to remote locations regardless of local storage format.[66] OPeNDAP is discussed in detail in Chapter 10.

OceanStore[54] aimed to build a global persistent data store that can scale to billions of users. The basic idea is that any server may create a local replica of any data object. These local replicas provide faster access and robustness to network partitions. Both GFarm and OceanStore require creating several replicas of the same data, but still they do not address the problem of scheduling the data movement when there is no replica close to the computation site.

Bent et al.[12] introduced a new distributed file system, the Batch-Aware Distributed File System (BADFS), and a modified data-driven batch scheduling system.[11] Their goal was to achieve data-driven batch scheduling by exporting explicit control of storage decisions from the distributed file system to the batch scheduler. Using some simple data-driven scheduling techniques, they have demonstrated that the new data-driven system can achieve orders of magnitude throughput improvements both over current distributed file systems such as the Andrew file system (AFS) as well as over traditional CPU-centric batch scheduling techniques which are using remote I/O.

One of the earliest examples of dedicated data schedulers is the Stork data scheduler.[53] Stork implements techniques specific to queuing, scheduling, and optimization of data placement jobs and provides a level of abstraction between the user applications and the underlying data transfer and storage resources. Stork introduced the concept that the data placement activities in a distributed computing environment need to be first-class entities just like computational jobs. Key features of Stork are presented in the next section.

4.3 Scheduling Data Movement

Stork is especially designed to understand the semantics and characteristics of data placement tasks, which can include data transfer, storage allocation and deallocation, data removal, metadata registration and unregistration, and replica location.

Stork uses the ClassAd[71] job description language to represent the data placement jobs. The ClassAd language provides a very flexible and extensible data model that can be used to represent arbitrary services and constraints. This flexibility allows Stork to specify job-level policies as well as global ones. Global policies apply to all jobs scheduled by the same Stork server. Users can override them by specifying job-level policies in job description ClassAds.

Stork can interact with higher-level planners and workflow managers. This allows the users to schedule both CPU resources and storage resources together. We have introduced a new workflow language capturing the data placement jobs in the workflow as well. The enhancements made to the workflow manager (i.e., DAGMan) allow it to differentiate between computational jobs and data placement jobs. The workflow manager can then submit computational jobs to a computational job scheduler, such as Condor or Condor-G, and the data placement jobs to Stork.

Stork also acts like an I/O control system (IOCS) between the user applications and the underlying protocols and data storage servers. It provides complete modularity and extensibility. The users can add support for their favorite storage system, data transport protocol, or middleware very easily. This is a crucial feature in a system designed to work in a heterogeneous distributed environment. The users or applications should not expect all storage systems to support the same interfaces to talk to each other. And we cannot expect all applications to understand all the different storage systems, protocols, and middleware. There needs to be a negotiating system between the applications and the data storage systems that can interact with all such systems easily and even translate different data transfer protocols to each other. Stork has been developed to be capable of this. The modularity of Stork allows users to easily insert plug-ins to support any storage system, protocol, or middleware.

Stork can control the number of concurrent requests coming to any storage system it has access to, and makes sure that neither that storage system nor the network link to that storage system get overloaded. It can also perform space allocation and deallocations to make sure that the required storage space is available on the corresponding storage system. The space allocations are supported by Stork as long as the underlying storage systems have support for it.

4.3.1 Data Placement Job Types

We categorize data placement jobs into seven types as follows:

transfer: This job type is for transferring a complete or partial file from one physical location to another one. This can include a get or put operation or a third-party transfer.

allocate: This job type is used for allocating storage space at the destination site, allocating network bandwidth, or establishing a light-path on the route from source to destination. Basically, it deals with all necessary resource allocations prerequired for the placement of the data.

release: This job type is used for releasing the corresponding resource which is allocated before.

remove: This job is used for physically removing a file from a remote or local storage server, tape, or disk.

locate: Given a logical file name, this job consults a metadata catalog service such as metadata catalog (MCAT)[8] or replica location service (RLS)[21] and returns the physical location of the file.

register: This type is used to register the file name to a metadata catalog service.

unregister: This job unregisters a file from a metadata catalog service.

The reason that we categorize the data placement jobs into different types is that all of these types can have different priorities and different optimization mechanisms.

4.3.2 Job Scheduling Techniques

We have applied some of the traditional job scheduling techniques common in computational job scheduling to the scheduling of data placement jobs:

First-Come, First-Served (FCFS) Scheduling: Regardless of the type of the data placement job and other criteria, the job that has entered into the queue of the data placement scheduler first is executed first. This technique, being the simplest one, does not perform any optimizations at all.

Shortest Job First (SJF) Scheduling: The data placement job which is expected to take the least amount of time to complete will be executed first. All data placement jobs except the transfer jobs have job completion time in the order of seconds, or minutes in the worst case. On the other hand, the execution time for the transfer jobs can vary from a couple of seconds to a couple of hours or even days. Accepting this policy would mean nontransfer jobs would be executed always before transfer jobs. This may cause big delays in executing the actual transfer jobs, which defeats the whole purpose of scheduling data placement.

Multilevel Queue Priority Scheduling: In this case, each type of data placement job is sent to separate queues. A priority is assigned to each job queue, and the jobs in the highest priority queue are executed first. To prevent starvation of the low-priority jobs, the traditional aging technique is applied. The hardest problem here is determining the priorities of each data placement job type.

Random Scheduling: A data placement job in the queue is randomly picked and executed without considering any criteria.

4.3.2.1 Auxiliary Scheduling of Data Transfer Jobs

The above techniques are applied to all data placement jobs regardless of the type. After this ordering, some job types require additional scheduling for further optimization. One such type is the data transfer jobs. The transfer jobs are the most resource consuming ones. They consume much more storage space, network bandwidth, and CPU cycles than any other data placement job. If not planned well, they can fill up all storage space, thrash and even crash servers, or congest all of the network links between the source and the destination.

Storage Space Management. One of the important resources that needs to be taken into consideration when making scheduling decisions is the available storage space at the destination. The ideal case would be the destination storage system supports space allocations, as in the case of NeST,[1] and before submitting a data transfer job, a space allocation job is submitted in the workflow. This way, it is assured that the destination storage system will have sufficient available space for this transfer.

Unfortunately, not all storage systems support space allocations. For such systems, the data placement scheduler needs to make the best effort in order to not overcommit the storage space. This is performed by keeping track of the size of the data transferred to and removed from each storage system that does not support space allocation. When ordering the transfer requests to that particular storage system, the remaining amount of available space, to the best of the scheduler's knowledge, is taken into consideration. This method does not assure availability of storage space during the transfer of a file, since there can be external effects, such as users who access the same storage system via

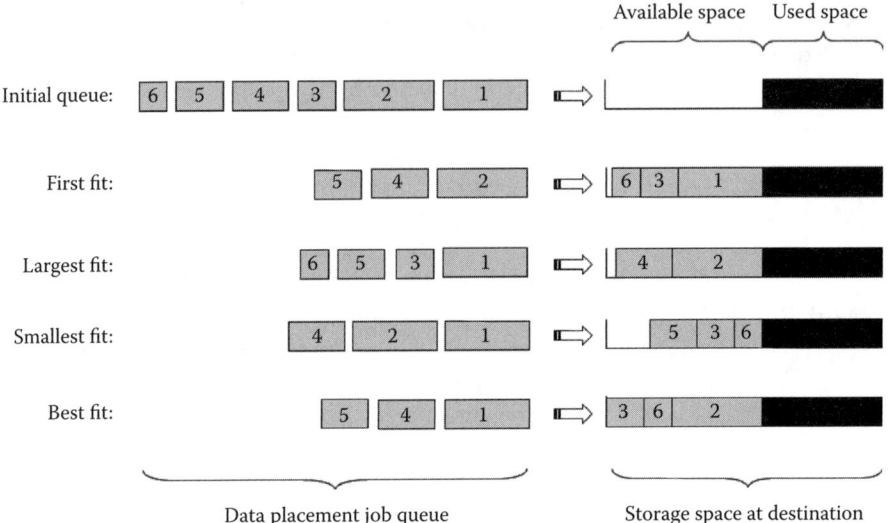

Figure 4.1 Storage space management: different techniques.

other interfaces without using the data placement scheduler. In this case, the data placement scheduler at least assures that it does not overcommit the available storage space, and it will manage the space efficiently if there are no external effects.

Figure 4.1 shows how the scheduler changes the order of the previously scheduled jobs to meet the space requirements at the destination storage system. In this example, four different techniques are used to determine in which order to execute the data transfer request without overcommitting the available storage space at the destination: *first fit*, *largest fit*, *smallest fit*, and *best fit*.

First Fit: In this technique, if the next transfer job in the queue is for data that will not fit in the available space, it is skipped for that scheduling cycle, and the next available transfer job with data size less than or equal to the available space is executed instead. It is important to point out that a complete reordering is not performed according to the space requirements. The initial scheduling order is preserved, but only requests that will not satisfy the storage space requirements are skipped, since they would fail anyway and also would prevent other jobs in the queue from being executed.

Largest Fit and Smallest Fit: These techniques reorder all of the transfer requests in the queue and then select and execute the transfer request for the file with the largest, or the smallest, file size. Both techniques have a higher complexity compared with the *first fit* technique, although they do not guarantee better utilization of the remote storage space.

Coordination of Access to Large-Scale Datasets 123

Best Fit: This technique involves a greedy algorithm that searches all possible combinations of the data transfer requests in the queue and finds the combination that utilizes the remote storage space best. Of course, it comes with a cost, which is a very high complexity and long search time. Especially in the cases where there are thousands of requests in the queue, this technique would perform very poorly.

Using a simple experiment setting, we will display how the built-in storage management capability of the data placement scheduler can help improve both overall performance and reliability of the system. In this experiment, we want to process 40 gigabytes of data, which consists of 60 files each between 500 megabytes and 1 gigabyte. First, the files need to be transferred from the remote storage site to the staging site near the compute pool where the processing will be done. Each file will be used as an input to a separate process, which means there will be 60 computational jobs followed by the 60 transfers. The staging site has only 10 gigabytes of storage capacity, which puts a limitation on the number of files that can be transferred and processed at any time.

A traditional scheduler would simply start all of the 60 transfers concurrently since it is not aware of the storage space limitations at the destination. After a while, each file would have around 150 megabytes transferred to the destination. But suddenly, the storage space at the destination would get filled, and all of the file transfers would fail. This would follow with the failure of all of the computational jobs dependent on these files.

On the other hand, Stork completes all transfers successfully by smartly managing the storage space at the staging site. At any time, no more than the available storage space is committed, and as soon as the processing of a file is completed, it is removed from the staging area to allow transfer of new files. The number of transfer jobs running concurrently at any time and the amount of storage space committed at the staging area during the experiment are shown in Figure 4.2 on the left side.

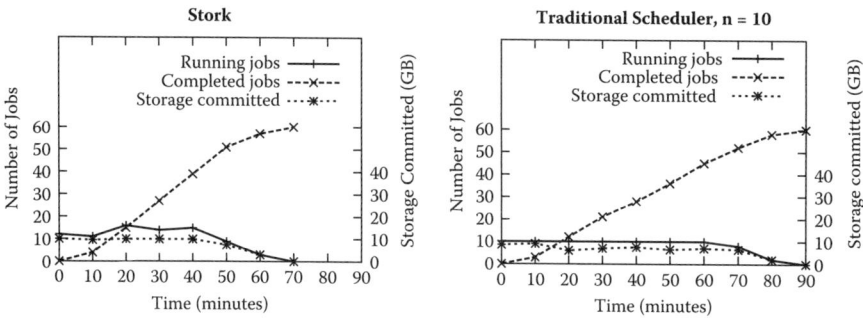

Figure 4.2 Storage space management: stork versus traditional scheduler.

In a traditional batch scheduler system, the user could intervene and manually set some virtual limits to the level of concurrency the scheduler can achieve during these transfers. For example, a safe concurrency limit would be the total amount of storage space available at the staging area divided by the size of the largest file that is in the request queue. This would assure the scheduler does not overcommit remote storage. Any concurrency level higher than this would have the risk of getting out of disk space anytime, and may cause failure of at least some of the jobs. The performance of the traditional scheduler with concurrency level set to 10 manually by the user in the same experiment is shown in Figure 4.2 on the right side.

Manually setting the concurrency level in a traditional batch scheduling system has three main disadvantages. First, it is not automatic; it requires user intervention and depends on the decision made by the user. Second, the set concurrency is constant and does not fully utilize the available storage unless the sizes of all the files in the request queue are equal. Finally, if the available storage increases or decreases during the transfers, the traditional scheduler cannot readjust the concurrency level in order to prevent overcommitment of the decreased storage space or fully utilize the increased storage space.

Storage Server Connection Management. Another important resource that needs to be managed carefully is the number of concurrent connections made to specific storage servers. Storage servers being thrashed or getting crashed due to too many concurrent file transfer connections has been a common problem in data-intensive distributed computing.

In our framework, the data storage resources are considered first-class citizens just like the computational resources. Similar to computational resources advertising themselves, their attributes, and their access policies, the data storage resources advertise themselves, their attributes, and their access policies as well. The advertisement sent by the storage resource includes the number of maximum concurrent connections it wants to take anytime. It can also include a detailed breakdown of how many connections will be accepted from which client, such as "maximum n GridFTP connections." and "maximum m HTTP connections."

This throttling is in addition to the global throttling performed by the scheduler. The scheduler will not execute more than, let us say, x amount of data placement requests at any time, but it will also not send more than y requests to server a, and more than z requests to server b, $y + z$ being less than or equal to x.

Other Scheduling Optimizations. In some cases, two different jobs request the transfer of the same file to the same destination. Obviously, all of these requests except one are redundant and wasting computational and network resources. The data placement scheduler catches such requests in its queue, performs only one of them, but returns success (or failure, depending on the return code) to all of such requests. We want to highlight that the redundant jobs are not canceled or simply removed from the queue. They still get honored, and the return value of the actual transfer is returned to them, but no redundant transfers are performed.

In some other cases, different requests are made to transfer different parts of the same file to the same destination. These requests are merged into one request, and only one transfer command is issued. But again, all of the requests get honored, and the appropriate return value is returned to all of them.

4.4 High-Performance Wide Area Data Transfers

An important set of challenges appears when dealing with data transfers over wide area networks. Perhaps the most visible issue in this context is that of the transport protocol. Data transfers over wide area, and in particular over high-capacity network links, make the performance limitations of the TCP (Transmission Control Protocol) visible. In this section, we will discuss these limitations and various alternatives in the following sections. Much of the work related to wide area data transfer is focused on file or disk-to-disk transfer. We will present other types of scenarios, how they are different, and what the challenges and applicable solutions are. This section concludes with a summary of remaining challenges and possible directions for future research.

4.4.1 TCP Characteristics and Limitations

Traditionally, applications using wide area networks have been designed to use the transmission control protocol (TCP) for reliable data communication.

TCP, defined in RFC (Requests for Comments) 793[*] but with numerous updates published since, provides byte stream semantics for end-to-end data transport. The two communicating entities create a TCP connection that can be used to send bytes in order and reliably (this defines byte stream semantics). Another important feature of TCP is that it implements congestion avoidance. Congestion appears when multiple concurrent streams go through the same link or network element (router) or when a stream traverses a low-capacity link. Congestion avoidance is necessary since if congestion occurs, the overall utility of the network would be reduced because of the capacity wasted with retransmissions and transmission of data that eventually is dropped. TCP's congestion avoidance mechanism is based on two algorithms: slow start and congestion avoidance. These algorithms utilize the notion of a *congestion window*, whose size is modified in response to packed

---
[*]http://www.faqs.org/rfcs/rfc768.html. Accessed July 16, 2009.

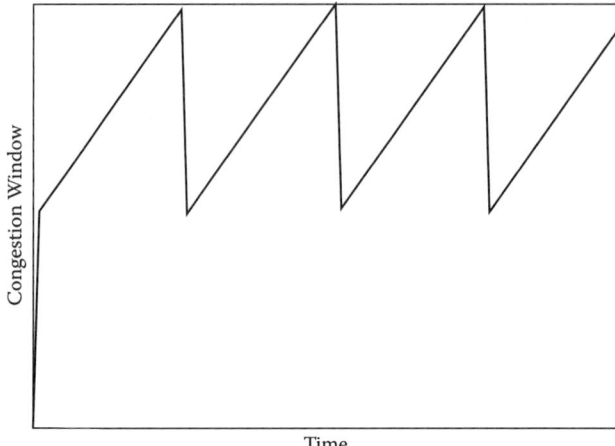

Figure 4.3 Theoretical TCP behavior on a link with no concurrent traffic.

acknowledgments and indication of packet loss. In TCP, packet loss is taken as an indication of congestion. These algorithms are described in more detail in RFC 2581.[*]

We can summarize that TCP works by increasing the congestion window with one segment each RTT (round-trip time for the connection) when no congestion is detected, and decreases the window to half when congestion is detected. This algorithm can be described as AIMD (additive increase multiplicative decrease). This behavior is illustrated in Figure 4.3.

A long history of utilization has shown that TCP is extremely successful in supporting the Internet traffic while avoiding a crash of the system. However, its characteristics create performance problems for situations where window sizes become very large, such as when using high-capacity wide area networks. First, the response to a congestion event (reducing the window to half) means that on a network link with no concurrent traffic, TCP will not be able to sustain utilization of the full link capacity. Second, the slow increase of the congestion window during the congestion avoidance stage means that on long-distance, high-capacity links, the time needed to increase the congestion window to reach the link capacity can be measured in hours. Also in practice, bandwidth utilization as a fraction of the total capacity will be lower on higher capacity networks since the probability of a single packet loss occurrence that is not caused by congestion is higher when the number of packets traversing the network increases, and the algorithm treats all packet losses as an indication of congestion.

---

[*]http://www.faqs.org/rfcs/rfc2581.html. Accessed July 16, 2009.

4.4.2 Alternative Transport Protocols

UDP, or the user datagram protocol,[68] is a thin transport protocol defined on top of IP that provides de-multiplexing of application messages between various applications running on the same host. In effect, UDP offers little more than IP, which is addressing, routing, and best-effort packet delivery. As such, UDP is perfectly suited for implementing various other transport protocols. SCTP, or the stream control transmission protocol, is the latest approved standard transport protocol.[78] Similar to TCP it provides reliable transmission and uses the same congestion control mechanism; however, the byte stream semantics have been replaced by message delivery. SCTP supports multiple delivery modes (ordered, partially ordered, and unordered).

As the number of applications that require high-performance wide area transfers has increased, so has the field of research in transport protocols. A survey on the subject[40] classifies newly emerging protocols in three categories: TCP variants, UDP based, and protocols requiring router support. A short list of some of these protocols and their distinctive design ideas regarding congestion control follows.

Scalable TCP[50] and HighSpeed TCP[28] use a similar congestion control detection and avoidance mechanism as standard TCP but use a modified response function to packet loss and acknowledgment. In H-TCP[57] the increase function a depends on the time elapsed since the last congestion event, and the decrease function b depends both on throughput and ratio between minimum and maximum delay experienced by a source.

TCP Vegas[15] introduced a congestion avoidance algorithm based on measuring time delays of individual packets. The congestion window is updated based on measured differences between packet round-trip transit times and ideal network round-trip time of the uncongested network. The same approach is used by a newer protocol, FAST TCP.[84] Compound TCP[48] combines these two approaches (delay based and loss based) into a single algorithm. Its congestion window has two components: a loss-based window, which is computed as in standard TCP, and a delay-based computed window added on top of it.

BI-TCP[88] introduces a new technique for determining the available window size called *binary search increase*. In response to acknowledgment BI-TCP increases its window to the midpoint between the current window size and the maximum known size if this increase does not exceed a certain threshold, or with the threshold value if it does (additive increase). When the window exceeds the maximum, the available bandwidth is probed using a symmetric response function. The response function was updated in CUBIC[72] from the binary search logarithmic to a strictly cubic function.

TCP Westwood and the updated version TCP Westwood+[63] use a technique called *adaptive decrease*, which in response to a congestion event tries to guess the available network bandwidth and sets the windows size to the guessed value. The response to acknowledgments remains that of standard TCP.

A different direction is represented by protocols that do not take a window-based approach and thus indirectly control the data transmission rate, but rather use a rate-based mechanism directly. These protocols use a fixed or variable data transmission rate and no explicit congestion window.

Reliable Blast UDP, or RBUDP,[39] is a rate-based protocol that performs reliable data transmission on top of UDP, and the data transmission rate is specified by the user. Its successor, LambdaStream,[87] automatically modifies the data transmission rate in case of congestion or when available bandwidth is detected. Congestion is detected when the data-receiving rate is smaller than the data-transmission rate, and detection is based on measuring inter-packet delays at both the sender and the receiver.

UDT[36] also uses a rate-based congestion control algorithm. On acknowledgments, each RTT rate is modified. The increase function depends on the estimated link bandwidth, which is probed periodically. On a negative feedback the rate is decreased by a constant factor. UDT also supports a messaging mode with optional ordering and reliability parameters.

The Group Transport Protocol, or GTP,[86] is a rate-based protocol design that uses receiver-based flow management. Its design considers the problem of multiple senders–single receiver flows explicitly. GTP uses flow statistics to estimate the capacity of each flow and allocates bandwidth to each according to the max–min fairness criteria.

The eXplicit Control Protocol (XCP)[49] is a router-supported transport protocol. Based on the observation that packet loss is not a reliable indication of congestion, XCP uses feedback from the network (routers) in order for the sender to correctly determine the degree of congestion in the network.

We should also note that with the advent of wide area optical network links, high-speed protocols for wide area transport are being researched and deployed outside the TCP/IP world. For example InfiniBand,* a high-speed interconnect protocol originally designed for local area networks, can be expanded to wide area networks using specialized hardware (Longbow by Obsidian Research†). Fibre Channel extension to wide area networks was also demonstrated by ADVA.‡ Fibre Channel is an interconnect technology originally designed for local area networks.§

4.4.3 Current Status and Options

Link utilization for dedicated or long-distance connections and fairness between concurrent transfers are two of the many possible metrics that can be used to evaluate a transport protocol. There is no clear consensus, but there

---

*http://www.infinibandta.org/specs/. Accessed July 16, 2009.
†http://www.obsidianresearch.com/. Accessed July 16, 2009.
‡http://www.advaoptical.com/. Accessed July 16, 2009.
§http://www.fibrechannel.org/technology/overview.html. Accessed July 16, 2009.

is work in progress in IETF* (Internet Engineering Task Force) on the list of metrics that should be used for evaluating a transport protocol. However, it is becoming clear that every transport protocol represents a trade-off point in the vast multidimensional space of evaluation metrics. There is no clear answer to the general question: *"What is the best protocol?"* as this answer depends on at least three important factors: What is the definition of *best* for the given scenario; what is the application type; and what are the deployment, network, and traffic constraints for the scenario?

Many evaluations and comparisons of transport protocols use the transfer of large data files over wide area networks as a motivating application scenario. File transfers need reliable transmission, are not highly sensitive to latency or throughput variations, are usually long-lived, use disk-to-disk transmission, and the data is organized as a linear sequence. The important measure for a file transfer application is the total time needed to transfer a file from the source to the destination.

One of the most utilized data transfer systems for grid and distributed applications today is GridFTP.[4,62] One approach that can be taken to overcome some of the limitations of TCP is to use parallel connections. GridFTP[3] and PSockets[76] are early adopters of this mechanism. On public or shared links, the improvement in data transfer rate may come at the cost of other streams reducing their transfer rate, which is not always acceptable. Also, on dedicated connections other protocols provide better performance. Recently, the Globus GridFTP implementation was modified to accept various data transport plug-ins. One of the currently available plug-ins uses the UDT transport protocol[16] mentioned previously.

One of the practical issues of high-speed protocols is that for some of them implementations are only available as kernel patches, and switching TCP variants in the kernel on a shared resource is in most situations undesirable. An alternative to kernel patches is the implementation of transport protocols in the user space, using UDP. This is the approach utilized for implementing transport protocols such as RBUDP or UDT. The UDT library also provides a framework for custom user-space implementations of various congestion-control algorithms.[35] The advantage of user-space protocol implementations is that they can be used in any application. The disadvantage is that they are more resource intensive; the CPU utilization on the machines is higher than that of kernel implementations for the same protocol. The packetization effort is also becoming more significant. The application or the transport protocol implementation must process each individual packet as it is sent or received from the network. If the packet size is small, the number of packets becomes too large for the application to be able to obtain high transmission rates. Many applications now require the network to support "jumbo" (or 9,000-byte) packets in order to obtain high transmission rates and to reduce CPU utilization.

---

*http://tools.ietf.org/html/draft-irtf-tmrg-metrics. Accessed July 16, 2009.

The increasing cost of protocol operations has been recognized by network controller providers who, as one possible solution, provide offload engines that move the processing from the CPU to the network controller.[22]

4.4.4 Other Applications

Although not traditionally seen as high-performance data transport applications, interactive applications using video transmission over long-distance networks have complex requirements regarding network transmission. Videoconferencing applications[42] require minimal latency in the video transmission path. Using uncompressed or low-rate compressed video are options that can be applied to videoconferencing. The data streams carrying uncompressed video can use in excess of 1 Gbps transmission rates and are highly sensitive to network latency and jitter. Fortunately, limited data loss is in some cases considered acceptable for video streams because for interactivity it is less costly to lose data than to retransmit it. Video transmission generally uses the standard Realtime Transmission Protocol* on top of plain UDP. Videoconferencing technology can be used for interactive remote visualization[45,47] as these two problems have similar challenges and requirements.

Another application is that of visualization of remote data. Often, scientists need to locally visualize datasets generated on remote supercomputers by scientific simulations. These datasets can be very large, and it is not desirable to transfer the entire data file locally for visualization. As a user is only interested in certain parts of the dataset at any given time, one alternative is to transfer only the section(s) of interest to the local machine. When the data is received, the visualization is updated and the user can move to another section of interest.[70] An important difference to file transfer is that the user may access different sections of the data object, depending on the type of data and analysis. The data transfer is not continuous. The user may analyze a certain section for a while, then move on to the next one. Data transfer is thus bursty, and a single data access may take as little as one or two seconds as the application needs to remain responsive. This is a significant difference to the file transfer scenario: for example, it is a major issue for remote visualization (but not necessary for file transfer) if the transport protocol takes several seconds to reach the maximum transfer rate. One solution that is applicable to dedicated network connections are transmission protocols with fixed data transmission rate. The data transmission rate needs to be computed based on network, compute, and storage resource availability.[17]

The destination of the transfer is the main memory of a local machine, and network utilization can be improved by using remote main memory to store (cached or prefetched) data of interest. We thus have disk-to-memory

---

*http://www.faqs.org/rfcs/rfc3550.html. Accessed July 16, 2009.

or memory-to-memory transfers. When disk is utilized as a data source (or destination), it is usually the disk that is the bottleneck in the system. For memory-to-memory transfers we see that the end hosts, and the efficiency of application and protocol implementations, are becoming the bottleneck. In addition to the main memory optimization, multiple remote resources (data servers) may be utilized in parallel to improve data transfer performance.

4.4.5 Challenges and Future Directions

With increasing availability of high-speed wide area network connections we expect the amount of interest in high-speed transfer protocols to increase. Although a significant number of transport protocols have been proposed, the availability of implementations for the end user is still very limited. We believe that the problem of finding efficient transport mechanism for applications will continue to be a difficult one in the near future. The need for developing new protocols and analyzing the existing ones will continue to increase as new requirements are formulated and new applications are emerging. Efficient implementations will be needed for these protocols to be adopted. A framework for evaluating transport protocols as well as frameworks for implementing new transport protocols will help with some of the issues. We believe that in the future different applications will use different transport protocols. Also a single application will use different protocols when executed in different network environments. Defining a set of application benchmarks and a set of representative network environments may help protocol designers in the future.

4.5 Remote Input and Output

An alternate form of access to distant data is remote I/O, where the files of interest stay in one place, and the programs issue network operations to read or write small amounts of data that are of immediate interest. In this model, transfer protocols must be optimized for small operations, and the processing site may need no storage at all. Figure 4.4 shows both of these modes of file access. There are several potential benefits for employing remote I/O:

- **Remote I/O simplifies usability.** It is not always easy for the end user to identify exactly what files he or she needs. A complex program or dataset can easily consist of hundreds of files and directories, of which not all are needed for any given task. In an interactive application, it may not even be possible to identify the needed files until runtime. When using remote I/O, the system fetches the files needed at runtime, and the user is freed from this burden.

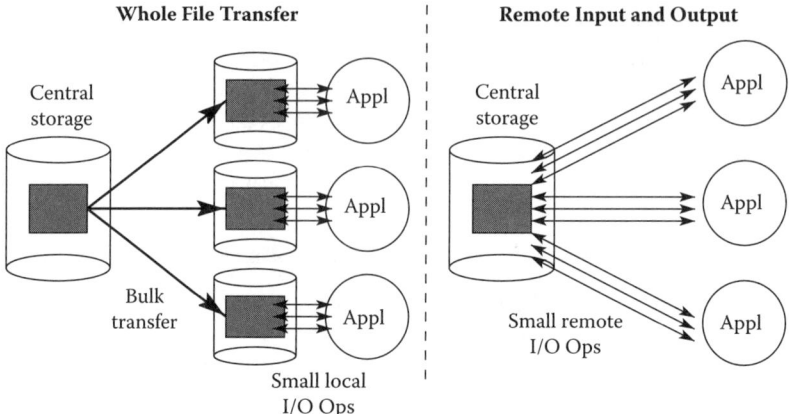

Figure 4.4 Whole file transfer versus remote input and output.

- **Remote I/O exploits selectivity.** When users explore very large datasets interactively, they may not need the entire contents of the repository. When using remote I/O, only the data that is actually needed for the current application is retrieved.
- **Remote I/O minimizes use of local storage.** The storage available where programs actually run—on user's desktops and in computing clusters—is not likely to have the capacity or performance of an institutional data storage system. By employing remote I/O, the local disk is removed from the system, improving performance, and increasing the possible execution sites for a given program.
- **Remote I/O minimizes initial response time.** When using file transfer for a large workload, no processing begins until an entire input file is transferred, and no output becomes available until some output file transfers complete. For long-running jobs, this can be a problem, because the user cannot even verify that the program is running correctly until the workload completes. When using remote I/O, outputs become immediately available to the end user for verification or further processing.

On the other hand, a remote I/O system may result in a large number of network round trips in order to service each small I/O request. If the remote I/O is performed over a high-latency wide area network, the result may be very low CPU utilization because the CPU is constantly waiting for a network operation to complete. In high-latency networks, it is more practical to perform whole file transfers. In practice, remote I/O is best used when the desired application is interactive, selectively uses data, or the execution nodes are storage constrained.

4.5.1 Challenges of Remote I/O

An ideal remote I/O system allows an unmodified application to access remote files in the exact same way as local files, differing perhaps only in performance and naming. That is, a user ought to be able to take any existing application and attach it to a remote I/O system without requiring any changes to the code or the local computing system.

In order to design a suitable remote I/O system for an application, we must address each of the following design challenges:

How should the application be attached to the remote I/O system? Distributed file systems such as the network file system (NFS)[74] and AFS[44] rely on kernel-level drivers and configuration in order to redirect applications to remote services. As a result, these traditional systems cannot be used in large-scale distributed systems where the end user is harnessing machines from multiple independent domains without any help from the local administrator. Instead, we must find a user-level mechanism for attaching the application to the remote I/O system.

One way to perform this attachment is by modifying libraries. If the application is written to use a middleware library like MPI-IO,[81] then that library can be modified to perform remote I/O as in RIO.[31] Another technique is to replace the standard C library. This can be done by recompiling the application, as in Condor,[58] or by preloading dynamic libraries as in Bypass[79] or XUnion.[83] Each of these techniques is effective on a subset of applications, but none applies to all possible binaries. For these reasons, we recommend the use of the ptrace interface, described below, for attaching applications to a remote I/O system.

What protocol should be used for remote I/O? A variety of data transfer protocols are in use on the Internet today. Unfortunately, few are directly suitable for carrying small I/O operations over the network. HTTP[26] is designed for moving entire files, and has no facilities for querying or managing directories. FTP[69] and variants such as GridFTP[5] provide some directory capabilities, but are also focused on large-scale data movement, requiring a new TCP connection for every open file. For the typical application that opens hundreds or thousands of files to access executables, libraries, and data files, one TCP connection for each file results in poor performance and possibly resource exhaustion. NFS[74] and its many variants[6,9,27,32,43] would appear to be more well suited, but the NFS protocol is difficult to disentangle from a kernel-level implementation, due to the use of persistent numeric file handles to identify files. For these reasons, we recommend that remote I/O requires a custom network protocol that corresponds more closely to the Unix I/O interface.

What security mechanisms are needed for remote I/O? A traditional Unix file system has very limited security mechanisms, in which users are identified by integers, and a total of nine mode bits are used to specify access controls. A remote I/O system used in the context of a larger distributed system must have more sophisticated kinds of access control. Multiple users form virtual organizations[30] that wish to share data securely across the wide area. Some users require elaborate encryption and authorization mechanisms, while others are content to identify users by simple host names. A suitable security mechanism for remote I/O must be flexible enough to accommodate all of these scenarios without overly burdening the user that has simple needs. For these reasons, we recommend the use of multiple authentication techniques with full text subject names and access control lists.

4.5.2 Case Study: Parrot and Chirp

As a case study of a complete system for remote I/O, we present a discussion of the tools Parrot and Chirp. Figure 4.5 shows how Parrot and Chirp work together to provide seamless access to remote storage using the Unix I/O interface. These tools have been used to provide remote I/O services for a variety of applications in bioinformatics,[14] biometrics,[64] high-energy physics,[80] and molecular dynamics.[85]

Parrot is a general-purpose tool for attaching applications to remote I/O systems. Parrot does not require changes to applications or special privileges to install, so it is well suited for use in large distributed systems such as computing grids. Parrot works by trapping an application's system calls through the debugging interface. System calls unrelated to I/O (such as sbrk) are allowed to execute unmodified. System calls related to I/O (such as open, read, and stat) are handled by Parrot itself, and the results returned to the application. Parrot is not limited to single process programs: It can run complex multiprocess scripts and interpreted languages such as Perl and Java. An interactive session using Parrot can be started by simply invoking:

% parrot tcsh

Parrot can be thought of as a user-level operating system. It contains multiple drivers that implement access to remote services. Each driver is represented

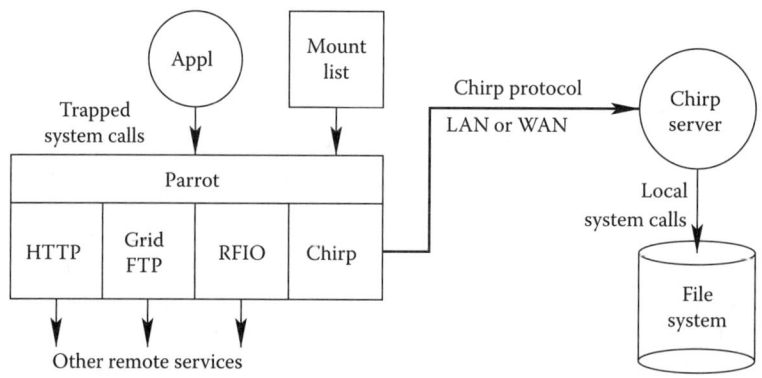

Figure 4.5 Using Parrot and Chirp for remote I/O.

in the root of the file namespace visible to applications run within Parrot. For example, the HTTP driver makes HTTP servers visible under files named like /http/server.somewhere.edu/mydata. Drivers are provided for a variety of protocols such as GridFTP,[5] Nest[13] and RFIO.[7] However, for the reasons described above, none of these protocols is perfectly suited for performing remote I/O.

To address this problem, we created the Chirp protocol and server in order to provide the precise Unix I/O semantics required by Unix applications. A Chirp server is a user-level process that can be deployed by any user without special privileges. It exports a local file system to remote users via a protocol that closely resembles the Unix I/O interface, containing operations such as open, read, stat, and so forth. Like the other protocols, data on a Chirp server can be accessed under the path name /chirp/server.somewhere.edu/mydata.

Each Chirp server periodically makes itself known to a global catalog server by sending a UDP packet listing its name, address, and other vital information. Information about all known Chirp servers can be obtained by querying the catalog server via a Web browser, or by executing ls /chirp within Parrot.

Chirp provides a flexible security model that is suited to the cooperative nature of large-scale distributed computing. The user may choose from a variety of mechanisms, depending on the degree of security required.

Upon connecting to a Chirp server, the client may authenticate with Kerberos,[77] the Globus Grid Security Infrastructure,[29] or by simple hostnames. Depending on what method is chosen, a given user might be known by any of the following subject names:

```
kerberos:smith@somewhere.edu
globus:/O=Somewhere/CN=Smith
hostname:client.somewhere.edu
```

Each directory in a Chirp server is protected by an access control list that lists acceptable subjects and access rights. Clearly, this access control list cannot be stored in the standard nine bits reserved for access control in Unix. Instead, each directory has a hidden file .__acl, which can be manipulated with the commands parrot_getacl and parrot_setacl. For example, the following access control list grants read access to any client at the campus computing center, any client holding Globus credentials issued by the university, and all other rights to a single Kerberos user:

```
hostname:*.hpc.somewhere.edu    RL
globus:/O=Somewhere/*           RL
kerberos:smith@somewhere.edu    RWLDA
```

Using Parrot and Chirp together, a user can run any sort of program on one machine and use it to access data on any other machine on the Internet as if it were in the local file system, while protected by strong security mechanisms.

The system can easily be deployed into an existing computational grid. The user must simply adjust their submission scripts to execute, for example, `parrot myjob.exe` instead of just `myjob.exe`.

To redirect access to remote files, the user may either adjust the program arguments manually, or provide Parrot with a *mount list*, which is a text file that redirects file names to other locations. For example, the following mount list would cause an application to load its files from Chirp server `alpha`, libraries from `beta`, and use temporary space on `gamma`. In this way, a program or script with hard-coded path names can be employed without modification.

```
/bin    /chirp/alpha.somewhere.edu/bin
/lib    /chirp/beta.somewhere.edu/lib
/tmp    /chirp/gamma.somewhere.edu/tmp
```

4.5.3 Open Problems in Remote I/O

There are some interesting avenues of future research in remote I/O:

Combining Remote I/O and Directed Transfer. As noted above, remote I/O is most useful when an application has highly selective behavior on a large dataset. However, there are many cases where the needs of an application may be practically known. For example, it may be clear that a job will need certain executables and libraries, but the exact input data is unknown. In these cases, it would be advantageous to perform an efficient bulk transfer of the known needs, and then rely upon remote I/O for the unknown needs at runtime. This is related to the idea of push-caching.[37]

Exploiting Common Access Patterns. Many programs perform complex but predictable patterns of access across a file system. For example, a search for an executable or a library requires a predictable search through a known list of directories. An `ls -l` requires a directory listing followed by a metadata fetch on each file. Although each of these cases can be individually optimized, a more general solution would allow the client to express a complex set of operations to be forwarded to the server and executed there, in the spirit of active storage.[73]

4.6 Coscheduling Compute and Network Resources

The increasing availability of high-bandwidth, low-latency optical networks promises to enable the use of distributed scientific applications as a day-to-day activity, rather than simply for demonstration purposes. However, in order to enable this transition, it must also become simple for users to reserve all the resources the applications require.

The reservation of computational resources can be achieved on many supercomputers using advance reservation, now available in most commercial and research schedulers. However, distributed applications often require guaranteed levels of bandwidth between compute resources. At the network level there are switches and routers that support the bandwidth allocation over network links, and/or the configuration of dedicated end-to-end lightpaths. These low-level capabilities are sufficient to support the development of prototype middleware solutions that satisfy the requirements of these applications.

However, the development of a booking system for network resources is not a complete solution, as the user is still dealing with the complexity of coordinating separate booking requests for multiple computational resources with their network booking(s). Even if there is a single system available that can reserve all the required compute resources, such as Moab, or GUR (which can reserve heterogeneous compute resources), this does not address the need to coordinate the scheduling of compute and network resources. A co-allocation system that can deal with multiple types of resources is required.

4.6.1 HARC: The Highly Available Resource Co-Allocator

HARC, the Highly Available Resource Co-allocator,[38,60] is an open-source software infrastructure for reserving resources for use by distributed applications. From the client's perspective, the multiple resources are reserved in a single atomic step, so that all of the resources are obtained if possible; if not, no resources are reserved. We call this all-or-nothing reservation of multiple resources *co-allocation* or *co-scheduling*. Our definition is slightly more general than that commonly found in the literature, in that we do not mandate that all the reservations have the same start and end times. Rather, HARC allows a different time window to be specified for each resource being reserved; the case where all of these windows are the same is just a special case.

Here, we will briefly sketch the architecture of HARC, showing how the co-allocation process works, and explaining how the system achieves its high availability. We will then show how HARC can be used to co-schedule Compute and Network resources, and also show how HARC can be extended to cover more types of resources. Finally, some results of our experiences using HARC to reserve network resources, as part of both special demonstrations and on a more production-like basis, are presented.

4.6.2 Architecture

To provide the required atomic behavior, HARC treats the co-allocation process as a *transaction*, and a *transaction commit* protocol is used to reserve the multiple resources. When designing HARC, we wanted to provide a co-allocation *service*, but without introducing a single point-of-failure into the architecture. For these reasons, HARC was designed around the *Paxos Commit* protocol[34] by Gray and Lamport. Paxos Commit replaces the single

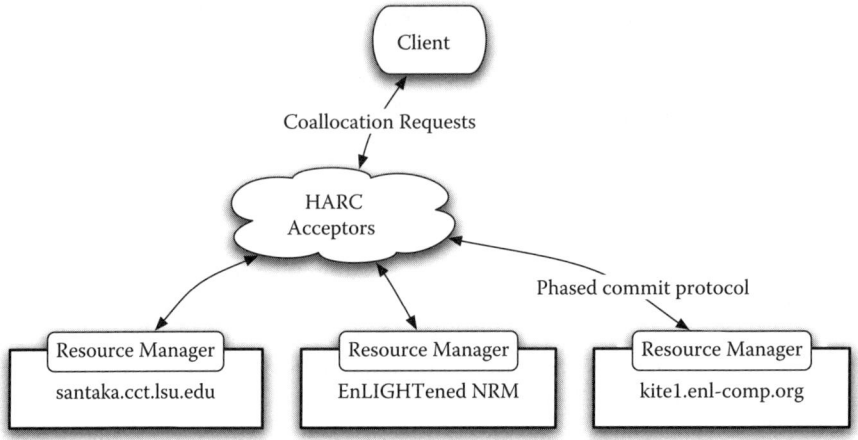

Figure 4.6 HARC architecture. NRM = network resource manager.

Transaction Manager process that you find in the classic *two-phase commit* protocol[56]* with multiple, replicated processes known as *Acceptors*.† The overall system makes progress as long as a majority of Acceptors remain operational. By deploying sufficient Acceptors, it is possible to construct systems that have a very long mean time to failure, even if the mean time to failure of each individual Acceptor is quite short.‡

The HARC Architecture is shown in Figure 4.6. To reserve resources using HARC,

1. the client makes a request, for example, from the command line, via the Client API,
2. the request goes to the HARC Acceptors, which manage the coallocation process, and
3. the Acceptors talk to individual resource managers, which make the individual reservations by talking to the underlying scheduler for their resource.

HARC was designed so that it could be extended by the Grid community to handle new types of resources, without needing modifications to the existing code. Specifically, the required steps for extending HARC to handle a new resource type are designing the description of the new resource type (XML);

---

*Or see: http://en.wikipedia.org/wiki/Two-phase-commit_protocol. Accessed July 16, 2009.

†This terminology comes from the Paxos Consensus algorithm,[55] upon which Paxos Commit is based.

‡Let us conservatively assume that the mean time to failure (MTTF) of a single Acceptor is 1 week, and that following a failure, it takes an average time of 4 hours to bring the Acceptor back online. Then, a deployment with seven Acceptors would have an MTTF of over 10 years[60] [Sec. 2.3].

adding client code that provides a method for encoding requests as XML; and adding a Resource Manager module which decodes requests and talks to the underlying scheduler for the resource, to make reservations. The scheduler needs to support reservations.

4.6.3 Reserving Network Resources Using HARC

One of the main goals of the EnLIGHTened Computing Project (2005–2007)[23] was to enable coordinated use of compute and network resources. This required network resources such as those in the EnLIGHTened testbed (shown in Figure 4.7 (a)) to be managed using middleware; previously this task required sending emails to network administrators. To enable this, a HARC network resource manager (NRM) component was developed.[59]

In the case of compute resources, there are a number of schedulers that HARC can communicate with (e.g., Moab, LoadLeveler), but there is no equivalent product for optical networks. Currently, there are a number of prototype network schedulers (e.g., References 25 and 68) however these are often site-specific, so we wrote our own simple scheduler for the EnLIGHTened network.

The software on the optical elements (Layer 1 optical network switches from Calient) in the testbed supports generalized multi-protocol label switching (GMPLS), and connections across the testbed can be initiated by sending a command to a switch at either end of the connection. A dedicated light path can be set up between any two entities at the edge of the cloud shown in Figure 4.7 (b). These are either routers (UR2, UR3) or compute nodes (RA1, VC1, CH1).

The NRM accepts requests for network connections on a first-come, first-served basis. Requests specify the start and end points of the connection (using the three-letter acronyms shown in the figure). The following command line example requests eight cores on machines at Louisiana State University and MCNC as well as a dedicated light path connecting them:

```
$ harc-reserve -c santaka.cct.lsu.edu/8 \
               -c kite1.enlightenedcomputing.org/8 \
               -n EnLIGHTened/BT2-RA1 -s 12:00 -d 1:00
```

4.6.4 Experiments Using HARC

HARC has been successfully used to co-schedule compute and network resources in a number of large-scale demonstrations, most notably in the high-profile EnLIGHTened/G-lambda experiments at Global Lambda Grid Workshop (GLIF) 2006 and Supercomputing Conference (SC)'06, where compute resources across the United States and Japan were co-scheduled together with end-to-end optical network connections.[25]* The HARC NRM has also been

---

*Also see http://www.gridtoday.com/grid/884756.html. Accessed July 16, 2009.

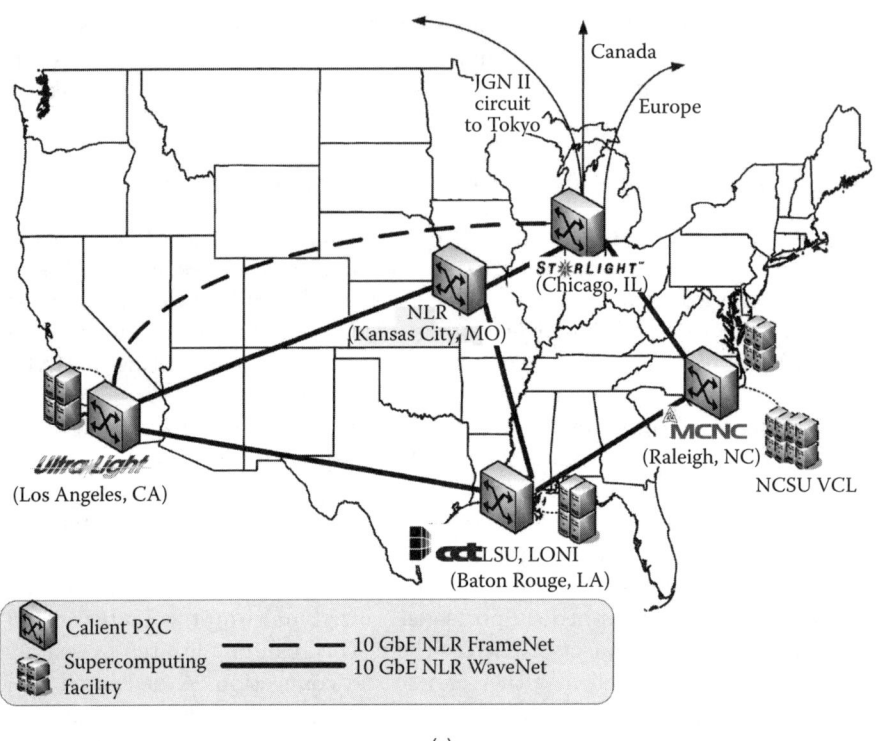
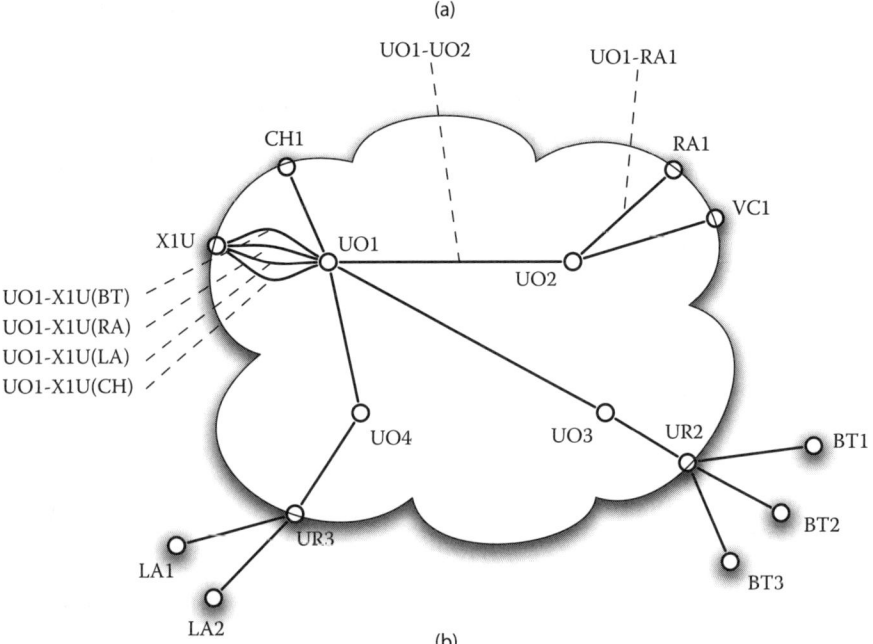

Figure 4.7 The EnLIGHTened testbed: (a) geographically; (b) as seen by HARC NRM.

used on a more regular basis to schedule a subset of the optical network connections used to broadcast lectures from Louisiana State University as part of an education experiment carried out in 2007.[46*]

HARC Compute Resource Managers are now deployed on several machines in the TeraGrid,[†] LONI,[‡] and UK NGS[§] infrastructures. This growing, supported infrastructure was used in demos by the GENIUS project.[33] at SC'07

Although the early results are encouraging, if the advance scheduling of light paths is to become a production activity, then the network scheduling service(s) need to be properly integrated with the other control/management plane software to ensure that these activities do not interfere with the pre-scheduled light paths (and vice versa).

4.7 Conclusions

The increase in the demand for large-scale data processing has necessitated collaboration and sharing of data collections among the world's leading education, research, and industrial institutions and use of distributed resources owned by collaborating parties. Efficient and reliable access to large-scale data sources and archiving destinations in a widely distributed computing environment brings new challenges such as scheduling data movement and placement, efficient use of wide area network links, remote access to partial data, and co-scheduling of data storage, network, and computational resources. In this chapter, we tried to address the challenges and possible solutions in each of these areas with specific examples and case studies. We believe that these examples are only some of the initial steps taken toward the solution of these problems, and there is still a lot to be done to break the barriers in front of the domain scientists to easily and efficiently access and use distributed datasets.

Acknowledgments

The work presented in this chapter is in part sponsored by the National Science Foundation under award numbers CNS-0846052 (Career-Kosar), CNS-0643229 (Career-Thain), CNS-0619843 (PetaShare), EPS-0701491

---

[*]This was the first distance-learning course ever offered in uncompressed high-definition video. Participating locations included other sites in Louisiana, Masaryk University in the Czech Republic, University of Arkansas, and North Carolina State University.
[†]http://www.teragrid.org/. Accessed July 16, 2009.
[‡]http://www.loni.org/. Accessed July 16, 2009.
[§]http://www.ngs.ac.uk/. Accessed July 16, 2009.

(Cyber-Tools), CNS-0509465 (Enlightened), by the SURA SCOOP Program (ONR N00014-04-1-0721, NOAA NA04NOS4730254) and by the Board of Regents, State of Louisiana, under contract numbers DOE/LEQSF (2004-07), NSF/LEQSF (2007-10)-CyberRII-01, and LEQSF(2007-12)-ENH-PKSFI-PRS-03.

References

[1] NeST: Network Storage Technology. http://www.cs.wisc.edu/condor/nest/.

[2] B. Allcock, J. Bester, J. Bresnahan, A. Chervenak, I. Foster, C. Kesselman, S. Meder, V. Nefedova, D. Quesnel, and S. Tuecke. Secure, efficient data transport and replica management for high-performance data-intensive computing. In *Proceedings of IEEE Mass Storage Conference*, April 2001.

[3] B. Allcock, J. Bester, J. Bresnahan, A. L. Chervenak, I. Foster, C. Kesselman, S. Meder, V. Nefedova, D. Quesnel, and S. Tuecke. Data management and transfer in high-performance computational grid environments. *Parallel Comput.*, 28(5):749–771, 2002.

[4] W. Allcock, J. Bester, J. Bresnahan, S. Meder, P. Plaszczak, and S. Tuecke. GridFTP: Protocol extensions to FTP for the Grid. *GWD-R (Recommendation)*, April 2003.

[5] W. Allcock, A. Chervenak, I. Foster, C. Kesselman, and S. Tuecke. Protocols and services for distributed data-intensive science. In *Proceedings of Advanced Computing and Analysis Techniques in Physics Research*, pp. 161–163, 2000.

[6] P. Andrews, P. Kovatch, and C. Jordan. Massive high-performance global file systems for grid computing. In *Supercomputing*, Seattle, WA, November 2005 pp. 53–67.

[7] O. Barring, J. Baud, and J. Durand. CASTOR project status. In *Proceedings of Computing in High Energy Physics*, Padua, Italy, 2000.

[8] C. Baru, R. Moore. A. Rajasekar, and M. Wan. The SDSC Storage Resource Broker. In *Proceedings of CASCON*, Toronto, Canada, 1998.

[9] A. Batsakis and R. Burns. NFS-CD: Write-Enabled Cooperative Caching in NFS. IEEE Transactions on Parallel & Distributed Systems, March 2008 (vol. 19 no. 3), pp. 323–333.

[10] M. Beck, T. Moore, J. Blank, and M. Swany. Logistical networking. In *Active Middleware Services*, S. Hariri, C. Lee, and C. Raghavendra, editors. Kluwer Academic Publishers, 2000.

[11] J. Bent. Data-driven batch scheduling. Ph.D. Dissertation, University of Wisconsin-Madison, 2005.

[12] J. Bent, D. Thain, A. Arpaci-Dusseau, and R. Arpaci-Dusseau. Explicit control in a batch-aware distributed file system. In *Proceedings of the First USENIX/ACM Conference on Networked Systems Design and Implementation*, 2004.

[13] J. Bent, V. Venkataramani, N. LeRoy, A. Roy, J. Stanley, A. Arpaci-Dusseau, R. Arpaci-Dusseau, and M. Livny. Flexibility, manageability, and performance in a grid storage appliance. In *IEEE Symposium on High Performance Distributed Computing*, Edinburgh, Scotland, July 2002.

[14] C. Blanchet, R. Mollon, D. Thain, and G. Deleage. Grid deployment of legacy bioinformatics applications with transparent data access. In *IEEE Conference on Grid Computing*, September 2006.

[15] L. S. Brakmo, S. W. O'Malley, and L. L. Peterson. TCP Vegas: New techniques for congestion detection and avoidance. In *SIGCOMM '94: Proceedings of the Conference on Communications Architectures, Protocols and Applications*, pp. 24–35, New York, NY, 1994. ACM.

[16] J. Bresnahan, M. Link, G. Khanna, Z. Imani, R. Kettimuthu, and I. Foster. Globus gridftp: What's new in 2007 (invited paper). In *Proceedings of the First International Conference on Networks for Grid Applications*, 2007.

[17] C. Toole Jr. and A. Hutanu. Network flow-based resource brokering and optimization techniques for distributed data streaming over optical networks. In *MG '08: Proceedings of the 15th ACM Mardi Gras Conference*, pp. 1–8, New York, NY, 2008. ACM.

[18] CCSP. Strategic plan for the US climate change science program. CCSP Report, 2003.

[19] E. Ceyhan and T. Kosar. Large scale data management in sensor networking applications. In *Proceedings of Secure Cyberspace Workshop*, Shreveport, Louisiana, November 2007.

[20] A. Chervenak, R. Schuler, C. Kesselman, S. Koranda, and B. Moe. Wide area data replication for scientific collaborations. In *Proceedings of the 6th IEEE/ACM International Workshop on Grid Computing*, November 2005.

[21] L. Chervenak, N. Palavalli, S. Bharathi, C. Kesselman, and R. Schwartzkopf. Performance and scalability of a Replica Location Service. In *Proceedings of the International Symposium on High Performance Distributed Computing Conference (HPDC-13)*, Honolulu, Hawaii, June 2004.

[22] A. Currid. TCP offload to the rescue. *ACM Queue*, 2(3), May 2004.

[23] EnLIGHTened Computing: Highly-dynamic Applications Driving Adaptive Grid Resources. http://www.enlightenedcomputing.org.

[24] A. Takefusaa et al. G-lambda: Coordination of a grid scheduler and lambda path service over GMPLS. *Future Generation Computing Systems*, 22:868–875, 2006.

[25] S. R. Thorpe et al. G-lambda and EnLIGHTened: Wrapped in middleware, co-allocating compute and network resources across Japan and the U.S. In *Proceedings of Gridnets 2007: First International Conference on Networks for Grid Applications (Lyon, France, October 2007)*.

[26] R. Fielding, J. Gettys, J. Mogul, H. Frystyk, L. Masinter, P. Leach, and T. Berners-Lee. Hypertext transfer protocol (HTTP). Internet Engineering Task Force Request for Comments (RFC) 2616, June 1999.

[27] R. Figueiredo, N. Kapadia, and J. Fortes. The PUNCH virtual file system: Seamless access to decentralized storage services in a computational grid. In *IEEE High Performance Distributed Computing*, San Francisco, CA, August 2001.

[28] S. Floyd. Highspeed TCP for large congestion windows. Internet draft, work in progress ftp://ftp.rfc-editor.org/in-notes/rfc3649.txt, December 2003.

[29] I. Foster, C. Kesselman, G. Tsudik, and S. Tuecke. A security architecture for computational grids. In *ACM Conference on Computer and Communications Security*, pp. 83–92, San Francisco, CA, November 1998.

[30] I. Foster, C. Kesselman, and S. Tuecke. The anatomy of the grid: Enabling scalable virtual organizations. *Lecture Notes in Computer Science*, 2150, 2001.

[31] I. Foster, D. Kohr, R. Krishnaiyer, and J. Mogill. Remote I/O: Fast access to distant storage. In *Workshop on I/O in Parallel and Distributed Systems (IOPADS)*, 1997.

[32] A. Ganguly, A. Agrawal, P. O. Boykin, and R. J. Figueiredo. WOW: Self organizing wide area overlay networks of workstations. *Journal of Grid Computing*, 5(2), June 2007.

[33] GENIUS: Grid Enabled Neurosurgical Imaging Using Simulation. http://wiki.realitygrid.org/wiki/GENIUS.

[34] J. Gray and L. Lamport. Consensus on transaction commit. Technical Report MSR-TR-2003-96, Microsoft Research, January 2004. http://research.microsoft.com/research/pubs/view.aspx?tr_id=701.

[35] Y. Gu and R. Grossman. Supporting configurable congestion control in data transport services. In *Supercomputing 2005*, November 2005.

[36] Y. Gu and R. L. Grossman. Udt: Udp-based data transfer for high-speed wide area networks. *Comput. Networks*, 51(7):1777–1799, 2007.

[37] J. Gwertzman and M. Seltzer. The case for geographical push-caching. In *Hot Topics in Operating Systems*, 1995 pp. 51–55.

[38] HARC: The Highly-Available Resource Co-allocator. http://www.cct.lsu.edu/HARC.php. Accessed on July 20, 2009.

[39] E. He, J. Leigh, O. Yu, and T. A. DeFanti. Reliable blast UDP: Predictable high performance bulk data transfer. In *CLUSTER '02: Proceedings of the IEEE International Conference on Cluster Computing*, pp. 317–324, Washington, D.C. USA, 2002. IEEE Computer Society.

[40] E. He, P. Vicat-Blanc, and M. Welzl. A survey of transport protocols other than standard TCP. *Global Grid Forum* document GFD55, November 2005.

[41] T. Hey and A. Terefethen. The data deluge: An e-science perspective. In *Grid Computing — Making the Global Infrastructure a Reality*, Chapter 36, pp. 809–824. Wiley and Sons, 2003.

[42] P. Holub, L. Matyska, M. Liška, L. Hejtmánek, J. Denemark, T. Rebok, A. Hutanu, R. Paruchuri, J. Radil, and E. Hladká. High-definition multimedia for multiparty low-latency interactive communication. *Future Generation Computer Systems*, 22(8):856–861, 2006.

[43] P. Honeyman, W. A. Adamson, and S McKee. GridNFS: Global storage for global collaboration. In *Local to Global Data Interoperability-Ehallenges and Technologies*, 2005. (IEEE Cat. No. 05EX1096).

[44] J. Howard, M. Kazar, S. Menees, D. Nichols, M. Satyanarayanan, R. Sidebotham, and M. West. Scale and performance in a distributed file system. *ACM Transactions on Computer Systems*, 6(1):51–81, February 1988.

[45] A. Hutanu, G. Allen, S. D. Beck, P. Holub, H. Kaiser, A. Kulshrestha, M. Liška, J. MacLaren, L. Matyska, R. Paruchuri, S. Prohaska, E. Seidel, B. Ullmer, and S. Venkataraman. Distributed and collaborative visualization of large datasets using high-speed networks. *Future Generation Computer Systems*, 22(8):1004–1010, 2006.

[46] A. Hutanu, M. Liška, P. Holub, R. Paruchuri, D. Eiland, S. R. Thorpe, and Y. Xin. Uncompressed hd video for collaborative teaching — an experiment. In *Proceedings of CollaborateCom*, 2007.

[47] B. Jeong, L. Renambot, R. Jagodic, R. Singh, J. Aguilera, A. Johnson, and J. Leigh. High-performance dynamic graphics streaming for scalable adaptive graphics environment. In *SC '06: Proceedings of the 2006 ACM/IEEE conference on Supercomputing*, p. 108, New York, NY, 2006. ACM.

[48] Q. Zhang, K. Tan, J. Song, and M. Sridharan. Compound TCP: A scalable and TCP-friendly congestion control for high-speed networks. In *Fourth International Workshop on Protocols for Fast Long-Distance Networks*, 2006.

[49] D. Katabi, M. Handley, and C. Rohrs. Congestion control for high bandwidth-delay product networks. In *SIGCOMM '02: Proceedings of the 2002 conference on applications, technologies, architectures, and protocols for computer communications*, pp. 89–102, New York, NY, 2002. ACM.

[50] T. Kelly. Scalable TCP: Improving performance in highspeed wide area networks. *SIGCOMM Comput. Commun. Rev.*, 33(2):83–91, 2003.

[51] S. Koranda and M. Moe. Lightweight data replicator. http://www.ligo.caltech.edu/docs/G/G030623-00/G030623-00.pdf.

[52] T. Kosar. A new paradigm in data intensive computing: Stork and the data-aware schedulers. In *Proceedings of Challenges of Large Applications in Distributed Environments (CLADE 2006) Workshop*, Paris, France, June 2006.

[53] T. Kosar and M. Livny. Stork: Making data placement a first class citizen in the Grid. In *Proceedings of the 24th Int. Conference on Distributed Computing Systems (ICDCS 2004)*, Tokyo, Japan, March 2004.

[54] J. Kubiatowicz, D. Bindel, Y. Chen, S. Czerwinski, P. Eaton, D. Geels, R. Gummadi, S. Rhea, H. Weatherspoon, W. Weimer, C. Wells, and B. Zhao. Oceanstore: An architecture for global-scale persistent storage. In *Proceedings of the Ninth International Conference on Architectural Support for Programming Languages and Operating Systems (ASPLOS 2000)*, 2000 pp. 190–201.

[55] L. Lamport. Paxos made simple. In ACM SIGACT news distributed computing column 5. *SIGACT News*, 32(4):18–25, 2001.

[56] B. W. Lampson and H. E. Sturgis. Crash recovery in a distributed data storage system. Technical Report (Unpublished), Xerox Palo Alto Research Center, 1976, 1979. http://research.microsoft.com/Lampson/21-CrashRecovery/Acrobat.pdf.

[57] D. Leith and R. Shorten. H-TCP: TCP for high-speed and long-distance networks. In *Second International Workshop on Protocols for Fast Long-Distance Networks*, 2004.

[58] M. Litzkow, M. Livny, and M. Mutka. Condor — a hunter of idle workstations. In *Eighth International Conference of Distributed Computing Systems*, June 1988.

[59] J. MacLaren. Co-allocation of compute and network resources using HARC. In *Proceedings of "Lighting the Blue Touchpaper for UK*

e-Science: *Closing Conference of the ESLEA Project.* PoS(ESLEA) 016, 2007. http://pos.sissa.it/archive/conferences/041/016/ESLEA_016.pdf.

[60] J. MacLaren. HARC: The Highly-Available Resource Co-allocator. In Robert Meersman and Zahir Tari, editors, *OTM Conferences 2007, Part II: Proceedings of GADA'07*, volume 4804 of *Lecture Notes in Computer Science*, pp. 1385–1402. Springer Verlag, 2007.

[61] R. K. Madduri, C. S. Hood, and B. Allcock. Reliable file transfer in grid environments. In *Proceedings of 27th Annual IEEE Conference on Local Computer Networks*, November 2002 pp. 737–738.

[62] I. Mandrichenko, W. Allcock, and T. Perelmutov. Gridftp v2 protocol description. *GWD-R (Recommendation)*, May 2005.

[63] S. Mascolo, L. A. Grieco, R. Ferorelli, P. Camarda, and G. Piscitelli. Performance evaluation of Westwood+ TCP congestion control. *Perform. Eval.*, 55(1-2):93-111, 2004.

[64] C. Moretti, J. Bulosan, D. Thain, and P. Flynn. All-pairs: An abstraction for data intensive cloud computing. In *International Parallel and Distributed Processing Symposium*, 2008.

[65] Y. Morita, H. Sato, Y. Watase, O. Tatebe, S. Segiguchi, S. Matsuoka, N. Soda, and A. DellAcqua. Building a high performance parallel file system using grid datafarm and root I/O. In *Proceedings of the 2003 Computing in High Energy and Nuclear Physics (CHEP03)*, 2003.

[66] OPeNDAP. Open-source project for a network data access protocol. http://www.opendap.org/.

[67] C. Palansuriya, M. Büchli, K. Kavoussanakis, A. Patil, C. Tziouvaras, A. Trew, A. Simpson, and R. Baxter. End-to-end bandwidth allocation and reservation for grid applications. In *Proceedings of BROAD-NETS 2006*. http://www.x-cd.com/BroadNets06CD/pdfs/87.pdf, October 2006.

[68] J. Postel. User datagram protocol, some 786, August 1988.

[69] J. Postel and J. Reynolds. File transfer protocol (FTP). Internet Engineering Task Force Request for Comments 959, October 1985.

[70] S. Prohaska, A. Hutanu, R. Kahler, and H. Hege. Interactive exploration of large remote micro-CT scans. In *VIS '04: Proceedings of the Conference on Visualization '04*, pp. 345–352, Washington, DC, 2004. IEEE Computer Society.

[71] R. Raman, M. Livny, and M. Solomon. Matchmaking: Distributed resource management for high throughput computing. In *Proceedings of the Seventh IEEE International Symposium on High Performance Distributed Computing (HPDC7)*, Chicago, Illinois, July 1998, pp. 1–7.

[72] I. Rhee and L. Xu. Cubic: A new TCP-friendly high-speed TCP variant. In *Third International Workshop on Protocols for Fast Long-Distance Networks*, February 2005.

[73] E. Riedel, G. A. Gibson, and C. Faloutsos. Active storage for large scale data mining and multimedia. In *Very Large Databases (VLDB)*, 1998, 62–73.

[74] R. Sandberg, D. Goldberg, S. Kleiman, D. Walsh, and B. Lyon. Design and implementation of the Sun network file system. In *USENIX Summer Technical Conference*, pp. 119–130, 1985.

[75] A. Shoshani, A. Sim, and J. Gu. Storage resource managers: Essential components for data grids. In *Grid Resource Management: State of the Art and Future Trends*, Edited by Jarek Nabrzyski, Jennifer M. Schopf, Jan Weglarz, Kluwer Academic Publishers, Norwell, Massachusetts, 2003, pp. 329–348.

[76] H. Sivakumar, S. Bailey, and R. L. Grossman. Psockets: The case for application-level network striping for data intensive applications using high speed wide area networks. In *Supercomputing '00: Proceedings of the 2000 ACM/IEEE Conference on Supercomputing (CDROM)*, p. 37, Washington, D.C., 2000. IEEE Computer Society.

[77] J.G. Steiner, C. Neuman, and J. I. Schiller. Kerberos: An authentication service for open network systems. In *Proceedings of the USENIX Winter Technical Conference*, pp. 191–200, 1988.

[78] R. Stewart, Q. Xie, K. Morneault, C. Sharp, H. Schwarzbauer, T. Taylor, I. Rytina, M. Kalla, L. Zhang, and V. Paxson. Stream control transmission protocol Internet Engineering Task Force Request for Comment 2960, 2000.

[79] D. Thain and M. Livny. Multiple bypass: Interposition agents for distributed computing. *Journal of Cluster Computing*, 4:39–47, 2001.

[80] D. Thain, C. Moretti, and I. Sfiligoi. Transparently distributing CDF software with parrot. In *Proceedings of Computing in High Energy Physics*, Elsevier Science Publishers, Amsterdam, The Netherlands, 2006.

[81] R. Thakur, W. Gropp, and E. Lusk. Data sieving and collective I/O in ROMIO. In *Symposium on the Frontiers of Massively Parallel Computation*, 1999.

[82] S. Tummala and T. Kosar. Data management challenges in coastal applications. *Journal of Coastal Research*, Coastal Education & Research Foundation, Inc. West Palm Beach Florida, 2007.

[83] E. Walker. A distributed file system for a wide-area high performance computing infrastructure. In *USENIX Workshop on Real Large Distributed Systems*, Seattle, WA, November 2006.

[84] D. X. Wei, C. Jin, S. H. Low, and S. Hegde. Fast TCP: motivation, architecture, algorithms, performance. *IEEE/ACM Trans. Netw.*, 14(6):1246–1259, 2006.

[85] J. Wozniak, P. Brenner, D. Thain, A. Striegel, and J. Izaguirre. Generosity and gluttony in GEMS: Grid enabled molecular simulations. In *IEEE High Performance Distributed Computing*, July 2005.

[86] R. X. Wu and A. A. Chien. GTP: Group transport protocol for lambda-grids. In *Cluster Computing and the Grid*, pp. 228–238, April 2004.

[87] C. Xiong, J. Leigh, E. He, V. Vishwanath, T. Murata, L. Renambot, and T. A. DeFanti. Lambdastream — a data transport protocol for streaming network-intensive applications over photonic networks. In *Third International Workshop on Protocols for Fast Long-Distance Networks*, February 2005.

[88] L. Xu, K. Harfoush, and I. Rhee. Binary increase congestion control for fast long distance networks. In *INFOCOM 2004*, volume 4, pp. 2514–2524, March 2004.

Chapter 5

High Throughput Data Movement

Scott Klasky,[1] Hasan Abbasi,[2] Viraj Bhat,[3] Ciprian Docan,[3] Steve Hodson,[1] Chen Jin,[1] Jay Lofstead,[2] Manish Parashar,[3] Karsten Schwan,[2] and Matthew Wolf[2]

[1] *Oak Ridge National Laboratory*
[2] *Georgia Institute of Technology*
[3] *Rutgers, The State University of New Jersey*

Contents

5.1 Introduction 151
5.2 High-Performance Data Capture 155
    5.2.1 Asynchronous Capture of Typed Data 155
    5.2.2 DataTaps and DataTap Servers 159
    5.2.3 High-Speed Asynchronous Data Extraction Using DART ... 166
    5.2.4 In-Transit Services 168
        5.2.4.1 Structured Data Transport: EVPath 168
        5.2.4.2 Data Workspaces and Augmentation of Storage Services 169
        5.2.4.3 Autonomic Data Movement Services Using IQ-Paths 170
5.3 Autonomic Services for Wide-Area and In-Transit Data 171
    5.3.1 An Infrastructure for Autonomic Data Streaming 172
    5.3.2 QoS Management at In-Transit Nodes 175
5.4 Conclusions 176
References 177

5.1 Introduction

In this chapter, we look at technology changes affecting scientists who run data-intensive simulations, particularly concerning the ways in which these computations are run and how the data they produce is analyzed. As computer systems and technology evolve, and as usage policy of supercomputers often permits very long runs, simulations are starting to run for over 24 hours and produce unprecedented amounts of data. Previously, data produced by supercomputer applications was simply stored as files for subsequent analysis, sometimes days or weeks later. However, as the amount of the data becomes very large and/or the rates at which data is produced or consumed by supercomputers become very high, this approach no longer works, and high-throughput data movement techniques are needed.

Consequently, science-driven analytics over the next 20 years must support high-throughput data movement methods that shield scientists from machine-level details, such as the throughput achieved by a file system or the network bandwidth available to move data from the supercomputer site to remote machines on which the data is analyzed or visualized. Toward this end, we advocate a new computing environment in which scientists can ask, "What if I increase the pressure by a factor of 10?" and have the analytics software run the appropriate methods to examine the effects of such a change without any further work by the scientist. Since the simulations in which we are interested run for long periods of time, we can imagine scientists doing in-situ visualization during the lifetime of the run. The outcome of this approach is a paradigm shift in which potentially plentiful computational resources (e.g., multicore and accelerator technologies) are used to replace scarce I/O (Input/Output) capabilities by, for instance, introducing high-performance I/O with visualization, without introducing into the simulation code additional visualization routines.

Such "analytic I/O" efficiently moves data from the compute nodes to the nodes where analysis and visualization is performed and/or to other nodes where data is written to disk. Furthermore, the locations where analytics are performed are flexible, with simple filtering or data reduction actions able to run on compute nodes, data routing or reorganization performed on I/O nodes, and more generally, with metadata generation (i.e., the generation of information about data) performed where appropriate to match end-user requirements. For instance, analytics may require that certain data be identified and tagged on I/O nodes while it is being moved, so that it can be routed to analysis or visualization machines. At the same time, for performance and scalability, other data may be moved to disk in its raw form, to be reorganized later into file organizations desired by end users. In all such cases, however, high-throughput data movement is inexorably tied to data analysis, annotation, and cataloging, thereby extracting the information required by end users from the raw data.

In order to illustrate the high-throughput data requirements associated with data-intensive computing, we describe next in some detail an example of a real, large-scale fusion simulation. Fusion simulations are conducted in order to model and understand the behavior of particles and electromagnetic waves in tokomaks, which are devices designed to generate electricity from controlled nuclear fusion that involves the confining and heating of a gaseous plasma by means of an electric current and magnetic field. There are a few small devices already in operation, such as DIII-D[1] and NSTX,[2] and a large device in progress, ITER,[3] being built in southern France.

The example described next, and the driver for the research described in this chapter, is the gyrokinetic toroidal code (GTC)[4] fusion simulation that scientists ran on the 250+ Tflop computer at Oak Ridge National Laboratory (ORNL) during the first quarter of 2008. GTC is a state-of-the-art global fusion code that has been optimized to achieve high efficiency on a single computing node and nearly perfect scalability on massively parallel computers. It uses the particle-in-cell (PIC) technique to model the behavior of particles and electromagnetic waves in a toroidal plasma in which ions and electrons are confined by intense magnetic fields. One of the goals of GTC simulations is to resolve the critical question of whether or not scaling in large tokamaks will impact ignition for ITER.

In order to understand these effects and validate the simulations against experiments, the scientists will need to record enormous amounts of data. The particle data in the PIC simulations is five-dimensional, containing three spatial dimensions and two velocity dimensions. The best estimates are that the essential information can be 55 GB of data written out every 60 seconds. However, since each simulation takes 1.5 days, and produces roughly 150 TB of data (including extra information not included in our previous calculation), it is obvious that there will not be enough disk space for the next simulation scheduled on the supercomputer unless the data is archived on the high-performance storage system, HPSS, while the simulation is running.[5] Moving the data to HPSS, running at 300 MB/s still requires staging simulations, one per week. This means that runs will first need to move the data from the supercomputer over to a large disk. From this disk, the data can then move over to HPSS, at the rate of 300 MB/s.

Finally, since human and system errors can occur, it is critical that scientists monitor the simulation during its execution. While running on a system with 100,000 processors, every wasted hour results in 100,000 wasted CPU hours. Obviously we need to closely monitor simulations in order to conserve the precious resources on the supercomputer, and the time of the application scientist after a long simulation. The general analysis that one would do during a simulation can include taking multidimensional FFTs (fast fourier transforms) and looking at correlation functions over a specified time range, as well as simple statistics. Adding these routines directly to the simulation not only complicates the code, but it is also difficult to make all of the extra routines scale as part of the simulation. To summarize, effectively running the large simulations to enable cutting-edge science, such as the GTC fusion simulations described above, requires that the large volumes of data generated must be (a) moved from the compute nodes to disk, (b) moved from disk to tape, (c) analyzed during the movement, and finally (d) visualized, all while the simulation is running. Workflow management tools can be used very effectively for this purpose, as described in some detail in Chapter 13.

In the future, codes like GTC, which models the behavior of the plasma in the center of the device, will be coupled with other codes, such as X-point gyrokinetic guiding center (XGC1),[6] which models the edge of the plasma.

The early version of this code, called XGC0,[7] is already producing very informative results that fusion experimentalists are beginning to use to validate against experiments such as DIII-D and NSTX. This requires loose coupling of the kinetic code, XGC0,[7] with GTC and other simulation codes. It is critical that we monitor the XGC0 simulation results and generate simple images that can be selected and displayed while the simulation is running. Further, this coupling is tight, that is, with strict space and time constraints, and the data movement technologies must be able to support such a coupling of these codes while minimizing programming effort. Automating the end-to-end process of configuring, executing, and monitoring of such coupled-code simulations, using high-level programming interfaces and high-throughput data movement is necessary to enable scientists to concentrate on their science and not worry about all of the technologies underneath.

Clearly a paradigm shift must occur for researchers to dynamically and effectively find the needle in the haystack of data and perform complex code coupling. Enabling technologies must make it simple to monitor and couple codes and to move data from one location to another. They must empower scientists to ask "what if" questions and have the software and hardware infrastructure capable of answering these questions in a timely fashion. Furthermore, effective data management is not just becoming important—it is becoming absolutely essential as we move beyond current systems into the age of exascale computing. We can already see the impact of such a shift in other domains; for example, the Google desktop has revolutionized desktop computing by allowing users to find information that might have otherwise gone undetected. These types of technologies are now moving into leadership-class computing and must be made to work on the largest analysis machines. High-throughput end-to-end data movement is an essential part of the solution as we move toward exascale computing. In the remainder of the chapter, we present several efforts toward providing high-throughput data movement to support these goals.

The rest of this chapter will focus on the techniques that the authors have developed over the last few years for high-performance, high-throughput data movement and processing. We begin the next section with a discussion of the Adaptable IO System (ADIOS), and show how this can be extremely valuable to application scientists and lends itself to both synchronous and asynchronous data movement. Next, we describe the Georgia Tech DataTap method underlying ADIOS, which supports high-performance data movement. This is followed with a description of the Rutgers DART (decoupled and asynchronous remote data transfers) method, which is another method that uses remote direct memory access (RDMA) for high-throughput asynchronous data transport and has been effectively used by applications codes including XGC1 and GTC. Finally, we describe mechanisms, such as autonomic management techniques and in-transit data manipulation methods, to support complex operations over the LAN and WAN.

5.2 High-Performance Data Capture

A key prerequisite to high-throughput data movement is the ability to capture data from high-performance codes with low overheads such that data movement actions do not unnecessarily perturb or slow down the application execution. More succinctly, data capture must be flexible in the overheads and perturbation acceptable to end-user applications. This section first describes the ADIOS API and design philosophy and then describes two specific examples of data capture mechanisms, the performance attained by them, and the overheads implied by their use.

5.2.1 Asynchronous Capture of Typed Data

Even with as few as about 10,000 cores, substantial performance degradation has been seen due to inappropriately performed I/O. Key issues include I/O systems difficulties in dealing with large numbers of writers into the same file system, poor usage of I/O formats causing metadata-based contention effects in I/O subsystems, and synchronous I/O actions unable to exploit communication/computation overlap. For example, when a simulation attempts to open, and then write one file per processor, the first step is to contact the metadata service of the parallel file system, issuing tens of thousands of requests at once. This greatly impacts the speed of I/O. Furthermore, scientific data is generally written out in large bursts. Using synchronous I/O techniques makes the raw speed to write this data the limiting factor. Therefore, if a simulation demands that the I/O rate take less than 5 percent of the calculation cost, then the file system must be able to write out, for example, 10 TB of data every 3600 seconds (generated in burst mode at a rate of 56 GB/sec). Using asynchronous techniques instead would only require a sustained 2.8 GB/sec write in the same case.

A first step to addressing these problems is to devise I/O interfaces for high-performance codes that can exploit modern I/O techniques while providing levels of support to end users that do not require them to have intimate knowledge of underlying machine architectures, I/O, and communication system configurations.

The Adaptable I/O System, ADIOS, is a componentization of the I/O layer. It provides the application scientist with easy-to-use APIs, which are almost as simple as standard FORTRAN write statements. ADIOS separates the metadata "pollution" away from the API, and allows the application scientist to specify the variables in their output in terms of groups. For example, let us suppose that a user has a variable, zion, which is associated with the ion particles of the plasma. The variable has the units of m/s (meters/second), and has the long name of ion parameters. Conventionally, all of this metadata must be written in the code, which involves placing these statements inside the Fortran/C code. In the ADIOS framework, the application scientist creates an XML file that contains this information, along with the specification of the method for each group, such as message passing interface input/output (MPI-IO), or portable operating system interface for unix (POSIX). The method declarations can be switched at runtime and allow the scientist to change from POSIX I/O, to MPI-IO,[8] to asynchronous methods such as the DataTap services[9] and the DART system[10] described below. By allowing the scientist to separate out the I/O implementation from the API, users are allowed to keep their code the same and only change the underlying I/O method when they run on different computers. Another advantage of specifying the information in this manner is that the scientist can just maintain one write statement for all of the variables in a group, thus simplifying their programs. This system also allows the user to move away from individual write statements, and as a result, the system can buffer the data and consequently write large blocks of data, which works best in parallel file systems. A small example of an XML file is as follows.

< **ioconfig** >

  <**datatype** name="restart">

    <scalar name="mi" path="/param" type="integer"/>

    <dataset name="zion" type="real" dimensions="n,1:4,2,mi"/>

    <data−attribute name="units" path="/param" value="m/s"/>

    <data−attribute name="long{_}name" path="/param" value="ion parameters"/>

< /**datatype** >

<**method** priority="1" method="DATATAP" iterations="1" type="diagnosis srv=ewok001.ccs.ornl.gov</**method**>

< /**ioconfig** >

Most importantly, however, ADIOS can provide such methods with rich metadata about the data being moved, thereby enabling the new paradigms for high-throughput data movement. These new paradigms include: (1) compact binary data transmission using structure information about the data (e.g., for efficient interpretation of data layout and access to and manipulation of select data fields); (2) the ability to operate on data as it is being moved (e.g., for online data filtering or data routing); and (3) the ability to use appropriate underlying *transport mechanisms* (e.g., such as switching from MPI-I/O to POSIX, to netCDF, to HDF-5). Furthermore, we envision building a code-coupling framework extending the ADIOS APIs that will allow scientists to try different mathematical algorithms by simply changing the metadata. Beyond providing information about the structure of the data, ADIOS also has built-in support for collecting and forwarding to the I/O subsystem key performance information, enabling dynamic feedback for scheduling storage-related I/O, the external configuration of data collection and storage/processing mechanisms, and value-added, additional in-flight and offline/near-line processing of I/O data—for example, specifying that the data be reduced in size (using a program that is available where the data is) before the data is written to the disk.

ADIOS encodes data in a compact, tagged, binary format for transport. This can either be written directly to storage or parsed for repackaging in another format, such as HDF-5 or netCDF. The format consists of a series of size-marked elements, each with a set of tags-values pairs to describe the element and its data. For example, an array is represented by a tag for a name, a tag for a data path for HDF-5 or similar purposes, and a value tag. The value tag contains rank of the array, the dimensional magnitude of each rank, the data type, and the block of bytes that represent the data. In the previous example where we showed the XML data markup, the large array written during the restarts for GTC is zion. Zion has rank=4, the dimensions are (n × 4 × 2 × mi) on each process, and the block of bytes will be 4*n*4*2*mi bytes. The remainder of this section demonstrates the utility of ADIOS for driving future work in high-throughput data movement, by using it with two different asynchronous data capture and transport mechanisms: the Rutgers DART system and the Georgia Tech LIVE DataTap system. Before doing so, we first summarize some of the basic elements of ADIOS.

ADIOS exploits modern Web technologies by using an external XML configuration file to describe all of the data collections used in the code. The file describes for each element of the collection the data types, element paths (similar to HDF-5 paths), and dynamic and static array sizes. If the data represents a mesh, information about the mesh, as well as the global bounds of an array, and the ghost regions* used in the MPI programming is encoded in the XML structure. For each data collection, it describes the transport mechanism selection and parameters as well as pacing information for timing the data transmissions. With this information, the ADIOS I/O implementation can then control when, how, and how much data is written at a time, thereby affording efficient overlapping with computation phases of scientific codes and proper pacing to optimize the write performance of the storage system.

A basic attribute of ADIOS is that it de-links the direct connection between the scientific code and the manipulation of storage, which makes it possible to add components that manipulate the I/O data outside the realm of the supercomputer's compute nodes. For example, we can inject filters that generate calls to visualization APIs like Visit,[11] route the data to potentially

---

*Ghost regions are regions that overlap adjacent grid cells.

multiple destinations in multiple formats, and apply data-aware compression techniques.

Several key high-performance computing (HPC) applications' driving capacity computing for petascale machines have been converted to using ADIOS, with early developments of ADIOS based on two key HPC applications: GTC (a fusion modeling code) and Chimera (an astrophysics supernova code).[12] Prior to its use of ADIOS, GTC employed a mixture of MPI-IO, HDF-5, netCDF, and straight Fortran I/O; and Chimera used straight Fortran I/O routines for writing binary files. Both of these codes provided different I/O requirements that drove the development of the API. Specifically, in GTC, there are seven different data formats, corresponding to various restart, diagnostic, and analysis values. Some of the data storage format requirements, such as combining some data types together in the same file, represent a good exercise of the capabilities of the ADIOS API. Chimera exercises the ADIOS API in three different ways. First, it contains approximately 475 different scalars and arrays for a single restart format. Second, this data ideally needs to be stored in both an efficient binary format and a readable text format. Third, the large number of elements encouraged the development of an experimental data reading extension for the ADIOS API that follows similar API semantics as used for writing, and leverages the writing infrastructure as much as possible. A simple code fragment showing how ADIOS is used is presented below:

```
call adios{\_}init ('config.xml')

...

! do main loop

call adios{\_}begin{\_}calculation ()

! do non-communication work

call adios{\_}end{\_}calculation ()

...

! perform restart write

...

! do communication work

call adios{\_}end{\_}iteration ()

! end loop
```

...

```
call adios{\_}finalize ()
```

Adios_init () initiates parsing of the configuration file generating all of the internal data type information, configures the mechanisms described above, and potentially sets up the buffer. Buffer creation can be delayed until a subsequent call to adios_allocate_buffer if it should be based on a percentage of free memory or other allocation-time-sensitive considerations.

Adios_begin_calculuation () and adios_end_calculation () provide the "ticker" mechanism for asynchronous I/O, providing the asynchronous I/O mechanism with information about the compute phases, so that the I/O can be performed at times when the application is not engaged in communications. The subsequent "end" call indicates that the code wishes to perform communication, ratcheting back any I/O use of bandwidth.

Adios_end_iteration () is a pacing function designed to give feedback to asynchronous I/O to gauge what progress must be made with data transmission in order to keep up with the code. For example, if a checkpoint/restart[*] is written every 100 iterations, the XML file may indicate an iteration count that is less than 100, to evacuate the data in order to accommodate possible storage congestion or other issues, such as a high demand on the shared network.

Adios_finalize () indicates the code is about to shut down and any asynchronous operations need to complete. It will block until all of the data has been drained from the compute node.

We describe next two asynchronous I/O mechanisms underlying ADIOS. Note that these mechanisms target current day supercomputers, which are typically composed of login nodes, I/O nodes, and compute nodes.

5.2.2 DataTaps and DataTap Servers

DataTap addresses the following performance issues for high-throughput data movement:

- Scaling to large data volumes and large numbers of I/O clients given limited I/O resources
- Avoiding excessive CPU and memory overheads on the compute nodes
- Balancing bandwidth utilization across the system
- Offering additional I/O functionality to the end users, including on-demand data annotation and filtering

---

[*]When running simulations with many time steps, it is customary to write out checkpoint/restart data following a number of time steps, in case the computation needs to backtrack, thus avoiding repeating the computation from the start.

In order to attain these goals, it is necessary to move structured rather than unstructured data, meaning, as expressed above in describing the ADIOS API, efficiency in data movement is inexorably tied to knowledge about the type and structure of the data being moved. This is because such knowledge makes it possible to manipulate data during movement, including routing it to appropriate sites, reorganizing it for storage or display, filtering it, or otherwise transforming it to suit current end-user needs. Next we describe the efficient, asynchronous data capture and transport mechanisms that underlie such functionality:

- **DataTaps**—flexible mechanisms for extracting data from or injecting data into HPC computations; efficiency is gained by making it easy to vary I/O overheads and costs in terms of buffer usage and CPU cycles spent on I/O and by controlling I/O volumes and frequency. DataTaps move data from compute nodes to DataTap servers residing on I/O nodes.

- **Structured data**—structure information about the data being captured, transported, manipulated, and stored enables annotation or modification both synchronously and asynchronously with data movement.

- **I/O graphs**—explicitly represent an application's I/O tasks as configurable overlay[*] topologies of the nodes and links used for moving and operating on data, and enable systemwide I/O resource management. I/O graphs start with the lightweight DataTaps on computational nodes; traverse arbitrary additional task nodes on the petascale machine (including compute and I/O nodes as desired); and "end" on storage, analysis, or data visualization engines. Developers use I/O graphs to flexibly and dynamically partition I/O tasks and concurrently execute them across petascale machines and the ancillary engines supporting their use.

The simple I/O graphs shown in Figure 5.1 span compute to I/O nodes. This I/O graph first filters particles to only include interesting data—say, within some bounding boxes or for some plasma species. The filtering I/O node then forwards the particle data to other I/O nodes, which in turn forward particle information to in situ visualization clients (which may be remotely accessed), and to storage services that store the particle information two different ways—one in which the particles are stored based on the bounding box they fall in, and one in which the particles are stored based on the timestep and compute node in which the information was generated.

---

[*]Overlay networks are virtual networks of nodes on top of another physical network. For the I/O graphs, data moves between nodes in the I/O graph overlay via logical (virtual) links, whereas in reality it may traverse one or more physical links between the nodes in the underlying physical network.

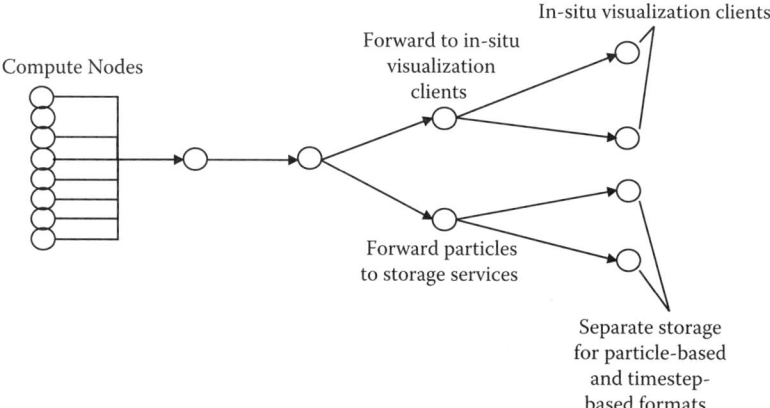

Figure 5.1 I/O graph example.

- **Scheduling techniques** dynamically manage DataTap and I/O graph execution, taking into account the I/O costs imposed on petascale applications.

- **Experimental results** attained with DataTaps and I/O graphs demonstrate several important attributes of I/O systems that benefit petascale machines. First, asynchronous I/O makes it possible to carry out I/O actions while massively parallel processor (MPP) computations are ongoing. This computation–I/O overlap improves throughput substantially, compared with the synchronous methods used by current file systems. Second, when performing I/O asynchronously, we demonstrated that it can scale without perturbing the applications running on compute nodes. For instance, sustained high-bandwidth data extraction (over 900 MB/s) has been achieved on the Cray XT4 without undue application perturbation and with moderate buffering requirements.[9]

DataTaps Implementation A DataTap is a request-read service designed to address the difference between the available memory on typical MPP compute partitions and that on I/O and service nodes. We assume the existence of a large number of compute nodes producing data—DataTap clients—and a smaller number of I/O nodes receiving the data—DataTap servers. The DataTap client issues a data-available request to the DataTap server, encodes the data for transmission, and registers this buffer with the transport for remote read. For very large data sizes, the cost of encoding data can be significant, but it will be dwarfed by the actual cost of the data transfer.[13–15] On receipt of the request, the DataTap server issues a read call. The DataTap server feeds an I/O graph, which can replicate the functionality of writing the output to a file, or it can be used to perform "in-flight" data transformations.

The design and implementation of DataTap servers deal with several performance issues and constraints present on modern MPP machines. First, due to the limited amount of memory available on the DataTap server, the server only issues a read call if there is memory available to complete it. Second, since buffer space used by asynchronous I/O on compute nodes is limited, the server issues multiple read calls each time it operates. Third, the next generation of DataTap servers will install controls on the speed and timing of reading data from DataTap buffers. The goal is to prevent perturbation caused when I/O actions are performed simultaneously with internal communications of application code (e.g., MPI collectives). Additional constraints result from in-transit actions performed by I/O graphs; these are evaluated in our ongoing and future work.

The current implementation of DataTap leverages existing protocols (i.e., Cray Portals and InfiniBand RDMA). Since the abstraction presented to the programmer is inherently asynchronous and data driven, data movement can take advantage of data object optimizations like message aggregation, data filtering, or other types of in-transit data manipulations, such as data validation. In contrast, the successful paradigm of MPI-IO, particularly when coupled with a parallel file system, heavily leverages the file nature of the data target and utilizes the transport infrastructure as efficiently as possible within that model. That inherently means the underlying file system concepts of consistency, global naming, and access patterns will be enforced at higher levels as well. By adopting a model that allows for the embedding of computations within the transport overlay, it is possible to delay execution of or entirely eliminate those elements of the file object that the application does not immediately require. If a particular algorithm does not require consistency (as is true of some highly fault-tolerant algorithms), then it is not necessary to enforce it from the application perspective. Similarly, if there is an application-specific concept of consistency (such as validating a checkpoint file before allowing it to overwrite the previous checkpoint file), then that could be enforced, as well as all of the more application-driven specifications mentioned earlier.

DataTaps leverage extensive prior work with high-performance data movement, including (1) efficient representations of meta-information about data structure and layout using a Portable Binary I/O format (PBIO[16]); which enables (2) high performance and "structure-aware" manipulations on data in flight, carried out by dynamically deployed binary codes and using higher level tools with which such manipulations can be specified, termed XChange.[17]; (3) a dynamic overlay (i.e., the I/O graph) optimized for efficient data movement, where data fast path actions are strongly separated from the control actions necessary to build, configure, and maintain the overlay[*,18]; and (4) a lightweight object storage facility (LWFS [19]) that provides flexible, high-performance data storage while preserving access controls on data.

---

[*]Such control actions are referred to as the control layer for the overlay network.

Because the DataTap API is not common to GTC or other current MPP applications, we use the ADIOS system to make DataTap (and structured stream) integration easier. By employing this API, a simple change in an entry in the XML file causes GTC, for example, to use synchronous MPI-IO, POSIX, our asynchronous DataTap servers, parallel-netCDF, HDF-5, NULL (no I/O performed), or other transports. Further, each data grouping, such as a restart versis diagnostic output, can use different transports, at no loss in performance compared with the direct use of methods like MPI-IO. The outcome is that integration details for downstream processing are removed from MPP codes, thereby permitting the user to enable or disable integration without the need for recompilation or relinking. A key property of structured streams preserved by ADIOS is the description of the structure of data to be moved in addition to extents or sizes. This makes it possible to describe semantically meaningful actions on data in ADIOS, such as chunking it for more efficient transport, filtering it to remove uninteresting data for analysis or display,[20] and similar actions.

We describe next the use of DataTap with the GTC code, as an example. Once GTC had been modified to use ADIOS, it was configured to use the DataTap as an output transport. The DataTap uses some of the compute node memory, storage that would otherwise be available to GTC. This method allows the application to proceed with computation as the data is moved to the DataTap server. Once at the DataTap server, the data is forwarded into the I/O graph.

The GTC simulations use several postprocessing tasks. These include the visualization of the simulated plasma toroid with respect to certain parameters. We describe next how the visualization data is constructed using DataTap and the I/O graph. The visualization that has proven useful is a display of the electrostatic potential at collections of points in a cross-section of the simulated toroid, called poloidal planes. The poloidal plane is described by a grid of points, each of which has a scalar value—the electrostatic potential at that grid vertex. To construct an image, this grid can be plotted in two or three dimensions, with appropriate color values assigned to represent the range of potential values. The visualization can be constructed after the simulation is run by coordinating information across several output files. Using the I/O graph components, we can recover this information with minimal impact on the application and, equally importantly, while the application is running. This permits end users to rapidly inspect simulation results while the MPP code is executing.

The DataTap is comprised of two separate components, the server and the client. The DataTap server operates on the I/O or service nodes, while the DataTap client is an I/O method provided to GTC through the ADIOS API. Because the number of compute nodes is so much greater than the number of service nodes, there is a corresponding mismatch between the number of DataTap clients and servers. To take advantage of asynchronicity, the DataTap client only issues a transfer request to the DataTap server instead of sending the entire data packet to the server. Also, to enable asynchronous communication the data is buffered before the data transfer request is issued. We use PBIO[21] to marshal the data into a buffer reserved for DataTap usage. The use of the buffer consumes some of the memory available to GTC but allows the application to proceed without waiting for I/O. The application only blocks for I/O while waiting for a previous I/O request to complete.

Once the DataTap server receives the request, it is queued up locally for future processing. The queuing of the request is necessary due to the large imbalance in the total size of the data to be transferred and the amount of memory available on the service node. For each request the DataTap server issues an RDMA read request to the originating compute node.

To maximize the bandwidth usage for the application, the DataTap server issues multiple RDMA read requests concurrently. The number of requests is predicated on the available memory at the service nodes and the size of the data being transferred. Also to minimize the perturbation caused by asynchronous I/O, the DataTap server uses a scheduling mechanism so as not to issue read requests when the application is actively using the network fabric. Once the data buffer is transferred over, the DataTap server sends the buffer to the I/O graph for further processing.

DataTap Evaluation To evaluate the efficiency and performance of the DataTap we look at the bandwidth observed at the DataTap server (at the I/O node). In Figure 5.2 we evaluate the scalability of our two DataTap implementations by looking at the maximum bandwidth achieved during data transfers. The InfiniBand DataTap (on a Linux Cluster) suffers a performance degradation due to the lack of a reliable datagram transport in our current hardware. However, this performance penalty only affects the first iteration of the data transfer, where connection initiation is performed. Subsequent transfers use cached connection information for improved performance. For smaller data sizes the Cray XT3 is significantly faster than the InfiniBand DataTap. The InfiniBand DataTap offers higher maximum bandwidth due to more optimized memory handling on the InfiniBand DataTap; we are currently addressing this for the Cray XT3 version.

In GTC's default I/O pattern, the dominant cost is from each processor's writing out the local array of particles into a separate file. This corresponds to writing out something close to 10% of the memory footprint of the code, with the write frequency chosen so as to keep the average overhead of I/O within a reasonable percentage of total execution time. As part of the standard process of accumulating and interpreting this data, these individual files are then aggregated and parsed into time series, spatially bounded regions, and so forth, depending on downstream needs.

To demonstrate the utility of structured streams in an application environment, we evaluated GTC on a Cray XT3 development cluster at ORNL with two different input set sizes. For each, we compared GTCs runtime for three different I/O configurations: no data output, data output to a per-mpi-process Lustre file, and data output using a DataTap (Table 5.1). We observed

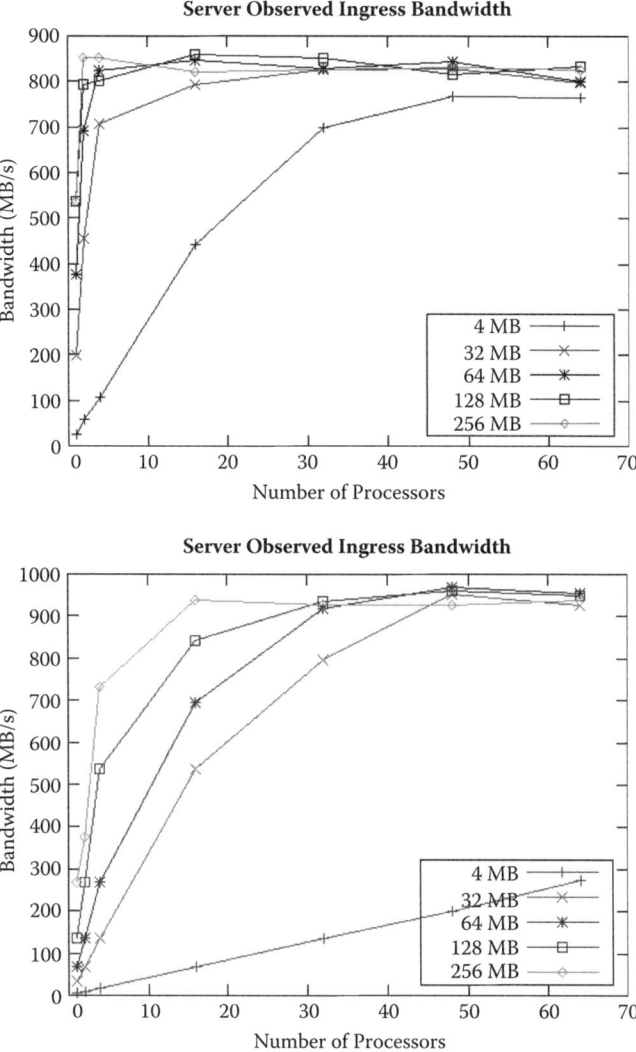

Figure 5.2 DataTap performance, on the left from the Cray XT3, and on the right from Infiniband.

a significant reduction in the overhead caused by the data output as the input set size increases, from about 9% on Lustre to about 3% using DataTap.[*]

The structured stream is configured with a simple I/O graph: DataTaps are placed in each of the GTC processes, feeding out asynchronously to an I/O node. From the I/O node, each message is forwarded to a graph node

---

[*]We define the I/O overhead as (time with I/O—total time with no I/O)/total time with no I/O.

TABLE 5.1 Comparison of GTC run times on the ORNL Cray XT3 development machine for two input sizes using different data output mechanisms

Run Parameters	Time for 100 iterations (582,410 ions)	Time for 100 iterations (1,164,820 ions)
No Output	213	422
Lustre	232	461
DataTap	220	435

where the data is partitioned into different bounding boxes. Once the data is received by the DataTap server, we filter the data based on the bounding box and then transfer the data for visualization. Copies of both the whole data and the multiple small partitioned datasets are then forwarded on to the storage nodes. Since GTC has the potential of generating PBs of data, we find it necessary to filter/reduce the total amount of data. The time taken to perform the bounding box computation is 2.29s and the time to transfer the filtered data is 0.037s. In the second implementation we transfer the data first and run the bounding box filter after the data transfer. The time taken for the bounding box filter is the same (2.29s) but the time taken to transfer the data increases to 0.297s. The key is not the particular values for the two cases but rather the relationship between them, which shows the relative advantages and disadvantages. In the first implementation the total time taken to transfer the data and run the bounding box filter is lower, but the computation is performed on the DataTap server. This increases the server's request service latency. For the second implementation, the computation is performed on a remote node and the impact on the DataTap is reduced. The value of this approach is that it allows an end user to compose a utility function that takes into account the cost in time **at a particular location**. Since most centers charge only for time on the big machines, oftentimes the maximum utility will show that filtering should be done on the remote nodes. If the transmission time to the remote site was to increase and slow down the computation more than the filtering time, higher utility would come from filtering the data before moving it. Thus, it is important that the I/O system be flexible enough to allow the user to switch between these two cases.

5.2.3 High-Speed Asynchronous Data Extraction Using DART

As motivated previously, scientific applications require a scalable and robust substrate for managing the large amounts of data generated and for asynchronously extracting and transporting them between interacting components. DART (decoupled and asynchronous remote transfers)[10] is an alternate design strategy to DataTap described above, and it is an efficient data transfer substrate that effectively addresses the requirements described above. Unlike DataTap, which attempts to develop an overall data management framework, DART is a thin software layer built on RDMA technology to enable fast, low-overhead, and asynchronous access to data from a running simulation, and support high-throughput, low-latency data transfers. The design and prototype implementation of DART using the Portals RDMA library on the Cray XT3/XT4 at ORNL are described next. DART has been integrated with the applications simulating fusion plasma in a Tokamak, described above, and is another key component of ADIOS.

The primary goal of DART is to efficiently manage and transfer large amounts of data from applications running on the compute nodes of an HPC system to the service nodes and remote locations, to support remote application monitoring, data analysis, coupling, and archiving. To achieve these goals, DART is designed so that the service nodes asynchronously extract data from the memory of the compute nodes, and so we offload expensive data I/O and streaming operations from the compute nodes to these service nodes. DART architecture contains three key components as shown in Figure 5.3: (1) a thin client layer (DARTClient), which runs on the compute nodes of an HPC system and is integrated with the application; (2) a streaming server (DARTSServer), which runs independently on the service nodes and is responsible for data extraction and transport; and (3) a receiver (DARTReceiver), which runs on remote nodes and receives and processes data streamed by DARTSServer.

A performance evaluation using the GTC simulation demonstrated that DART can effectively use RDMA technologies to offload expensive I/O operations to service nodes with very small overheads on the simulation itself, allowing a more efficient utilization of the compute elements, and enabling efficient online data monitoring and analysis on remote clusters.

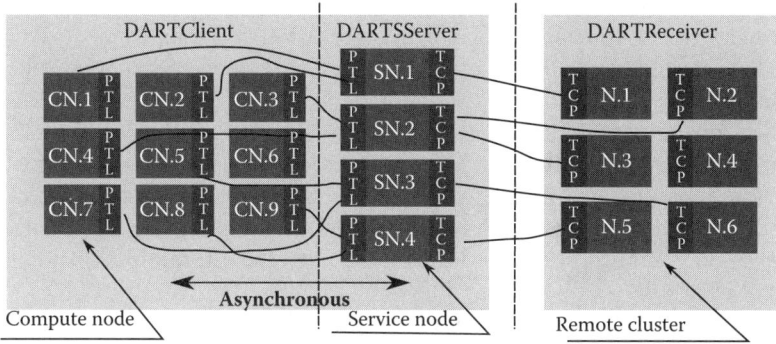

Figure 5.3 Architectural overview of DART.

5.2.4 In-Transit Services

In addition to the data movement and low-level data capture interfaces described above, applications that require high-throughput adaptive I/O must also depend on robust transport and specialization services. Such specialization services are required to perform "in-transit" data inspection and manipulation, including filtering, aggregation, or other types of processing actions that tune the data output to the current user- or application-specific requirements. The use of specialization services jointly with the basic data movement results in attainment of adaptive I/O services, needed to address the dynamism in application inputs and outputs, computational and communication loads and operating conditions, and end-user interests. We focus here on techniques for the autonomic tuning of these transports to provide the user-level quality of information and specification of utility that next-generation application data flows require. An example of this is a scientist who is particularly interested in one type of interatomic bond during a molecular dynamics simulation and who is, under bandwidth constraints, willing to rely on a specialized transport service that filters out simulation output not related to atoms involved in such bonds, or that gives those data outputs higher priority compared with other outputs. The detection of the bandwidth limitation and the selection of the appropriate specialization action (i.e., filtering or change in priority) should happen autonomically, without additional intervention of the simulation user.

5.2.4.1 Structured Data Transport: EVPath

After data events have been captured through the DataTap implementation of the ADIOS interface, an event processing architecture is provided in support of high-performance data streaming in networks with internal processing capacity. EVPath, the newest incarnation of a publish/subscribe infrastructure developed over many years,[22,23] is designed to allow for easy implementation of overlay networks with active data processing, routing, and management at all points within the overlay. In addition, EVPath allows the use of a higher-level control substrate to enable global overlay creation and management. Domain-specific control layers allow the management of the overlay to best utilize the underlying physical resources and provide for overlays that best address the application needs. For instance, the IFLOW management layer described in Reference 14 is best suited for large-scale, wide-area, streaming applications, whereas another version of a control layer is more suitable for a massively parallel processor (MPP) such as the Cray system, where management components on compute nodes have limited ability for interaction with external control entities.

The basic building block in EVPath is a *stone*. An overlay *path* is comprised of a number of connected stones. A stone is a lightweight entity that roughly corresponds to processing points in a dataflow diagram. Stones can perform different types of data filtering and data transformation, as well as transmission of data between processes over network links.

EVPath is designed to support a flexible and dynamic computational environment where stones might be created on remote nodes and possibly relocate during the course of the computation. In order to support such an environment, we use a sandboxed version of C, coupled with a dynamic code-generation facility to allow native binary transformation functions to be deployed anywhere in the system at runtime.[24] The interface allows for the specification of data gateways (pass/no-pass) and data transformations (sum aggregation trees), and calls out to more specialized code (for example, invocation of a signed, shared library for performing FFTs). From these elements, the application user can specify in much greater detail how the interaction between the output of the running code and the data stored for later use should look.

5.2.4.2 Data Workspaces and Augmentation of Storage Services

As a concrete example of the user-driven interfaces that can be provided for application scientists, it is useful to consider the concept of data workspaces. In a data workspace, users are provided with an execution model (i.e., a semitransparent way of creating and submitting batch MPI jobs), along with a way for specifying the data control networks for how this data should move and be interpreted while in transit from the computing resource to the storage. Note that this concept interacts cleanly with the concept of a workflow—it is a part of a rich transport specification that then feeds the manipulation of the data once it has reached disk.

As an example of this concept, a team at Georgia Institute of Technology has built a data workspace for molecular dynamics applications that can make synchronous tests of the quality of the data and use that to modify the priority and even the desirability of moving that data into the next stage of its workflow pipeline.[17] Specifically, this workspace example modifies a storage service (ADIOS) that the molecular dynamics program invokes. As an example scenario, consider an application scientist who runs the parallel data output through an aggregation tree so that there is a single unified dataset (rather than a set of partially overlapping atomic descriptors), and then undergoes data quality and timeliness evaluation. Raw atomic coordinate data is compared to a previous graph of nearest neighbors through the evaluation of a central symmetry function to determine if any dislocations (seed of crack formation) have occurred in the simulated dataset. The frequency of the data storage is then changed, in this particular case, dependent on whether the data is from before, during, or after the formation of a crack, since the data during the crack formation itself is of the highest scientific value.

Similarly, in Reference 17, data quality can be adapted based on a requirement for timeliness of data delivery—if a particular piece of data is too large to be delivered within the deadline, user-defined functions can be chosen autonomously to change data quality so as to satisfy the delivery timeline.

In the case of an in-line visualization annotation, one could consider deploying a host of visualization-related functions—changing color depth, changing frame rate, changing resolution, visualization-specific compression techniques, and so forth. Based on the user-specified priorities (color is unimportant, but frame rate is crucial), the in-transit manipulation of the extracted data allows for a much higher fidelity interaction for the application scientist. As the adaptation of the data stream becomes more complex, it leads naturally to discussion of full-fledged autonomic control of the network and computer platforms, which is the topic of the next sections.

5.2.4.3 Autonomic Data Movement Services Using IQ-Paths

Among data-driven high-performance applications, such as data mining and remote visualization, the ability to provide quality of service (QoS) guarantees is a common characteristic. However, due to most networking infrastructure being a shared resource, there is a need for middleware to assist end-user applications in best utilizing available network resources.

An IQ-Path is a novel mechanism that enhances and complements existing adaptive data streaming techniques. First, IQ-Paths dynamically measure[25,26] and also predict the available bandwidth profiles on the network links. Second, they extend such online monitoring and prediction to the multilink paths in the overlay networks used by modern applications and middleware. Third, they offer automated methods for moving data traffic across overlay paths, including splitting a data stream across multiple paths and dynamically differentiating the volume and type of data traffic on each path. Finally IQ-Paths use statistical methods to capture the noisy nature of available network bandwidth, allowing a better mapping to the underlying best-effort network infrastructure.

The overlay implemented by IQ-Paths has multiple layers of abstraction. First, its *middleware underlay*—a middleware extension of the network underlay proposed in Reference 27—implements the execution layer for overlay services. The underlay is comprised of processes running on the machines available to IQ-paths, connected by logical links and/or via intermediate processes acting as router nodes. Second, underlay nodes continually assess the qualities of their logical links as well as the available resources of the machines on which they reside. Figure 5.4 illustrates an overlay node part of an IQ-Path. The routing and scheduling of application data is performed with consideration of path information generated by the monitoring entities. The service guarantees provided to applications are based on such dynamic resource measurements, runtime admission control, resource mapping, and a self-regulating packet routing and scheduling algorithm. This algorithm, termed PGOS (predictive guarantee overlay scheduling), provides probabilistic guarantees for the available bandwidth, packet loss rate, and round-trip time (RTT) attainable across the best-effort network links in the underlay. More information on IQ-Paths can be found in Reference 28.

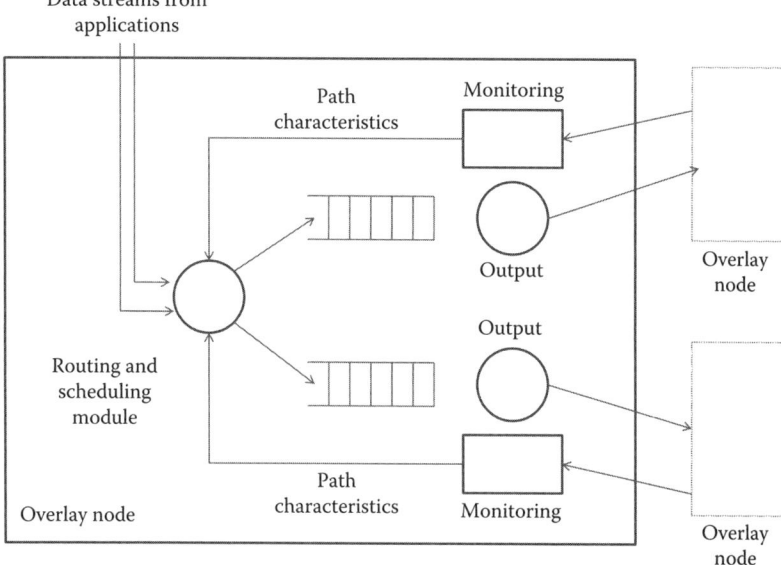

Figure 5.4 Data movement services using IQ-Path.

IQ-Paths and PGOS provide the following services:

- **Probabilistic and "violation bound" guarantees.** Using the PGOS algorithm, service guarantees can be provided using network behavior prediction. PGOS can ensure that applications receive the bandwidths they require with high levels of accuracy (e.g., an application receives its required bandwidth 99% of the time), and that the occurrence of any violations, such as missed packet deadlines, is bound (e.g., only 0.1% of packets miss their deadline).
- **Reduced jitter.** Buffering requirements are minimized by reducing jitter in time-sensitive applications.
- **Differentiated streaming services.** Higher priority streams receive better service when network approaches maximum utilization.
- **Full bandwidth utilization.** Even with guarantees, the available and utilized bandwidths are not sacrificed.

5.3 Autonomic Services for Wide-Area and In-Transit Data

Complementary of the low-overhead asynchronous data extraction capabilities provided by ADIOS and its underlying mechanisms (i.e., DataTap and DART), wide-area streaming aims at efficiently and robustly transporting the extracted data from live simulations to remote services. In the previous sections we talked about services that worked on the local area network, and in this section we discuss services that must work over the wide area network. For example, in the context of the DOE SciDAC CPES fusion simulation project,[29] a typical workflow consists of coupled simulation codes—the edge turbulence particle-in-cell (PIC) code (GTC) and the microscopic MHD code (M3D)—running simultaneously on thousands of processors at various supercomputing centers. The data produced by these simulations must be streamed to remote sites and transformed along the way, for online simulation monitoring and control, simulation coupling, data analysis and visualization, online validation, and archiving. Wide-area data streaming and in-transit processing for such a workflow must satisfy the following constraints: (1) Enable high-throughput, low-latency data transfer to support near real-time access to the data. (2) Minimize related overhead on the executing simulation. Since the simulation is long running and executes in batch for days, the overhead due to data streaming on the simulation should be less than 10% of the simulation execution time. (3) Adapt to network conditions to maintain desired quality of service (QoS). The network is a shared resource and the usage patterns vary constantly. (4) Handle network failures while eliminating data loss. Network failures can lead to buffer overflows, and data has to be written to local disks to avoid loss. However, this increases overhead on the simulation and the data is not available for real-time remote analysis and visualization. (5) Effectively manage in-transit processing while satisfying the above requirements. This is particularly challenging due to the heterogeneous capabilities and dynamic capacities of the in-transit processing nodes.

5.3.1 An Infrastructure for Autonomic Data Streaming

The data streaming service described in this section is constructed using the Accord programming infrastructure,[30–32] which provides the core models and mechanisms for realizing self-managing Grid services. These include autonomic management using rules as well as model-based online control. Accord extends the service-based Grid programming paradigm to relax static (defined at the time of instantiation) application requirements and system/application behaviors and allows them to be dynamically specified using high-level rules. Further, it enables the behaviors of services and applications to be sensitive to the dynamic state of the system and the changing requirements of the application, and to adapt to these changes at runtime. This is achieved by extending Grid services to include the specifications of policies (in the form of high-level rules) and mechanisms for self-management, and providing a decentralized runtime infrastructure for consistently and efficiently enforcing these policies to enable autonomic self-managing functional interaction and composition behaviors based on current requirements, state, and execution context.

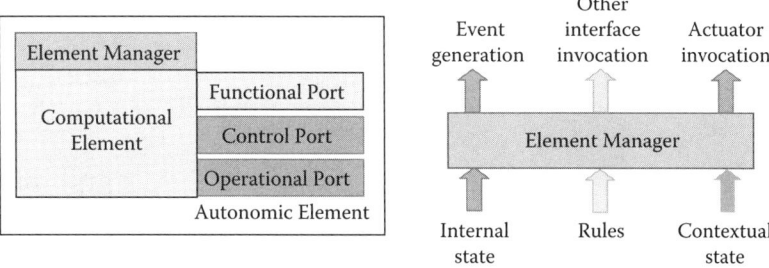

Figure 5.5 An autonomic service in Accord.

An autonomic service extends a Grid service (such as the in-transit services described above) with a control port, as shown in Figure 5.5. The control augments the functional and operational ports that typically define computational elements, and supports external monitoring and steering. An autonomic service also encapsulates a service manager, shown in Figure 5.5 on the right, which monitors and controls the runtime behaviors of the managed service according to changing requirements and state of applications as well as their execution environment based on user-defined rules. As shown in the figure, the manager uses the local state of the element as well as its context along with user-defined rules to generate adaptations as well as management events. The control port (Figure 5.5, left) consists of sensors that enable the state of the service to be queried, and actuators that enable the behaviors of the service to be modified. Rules are simple if-condition-then-action statements described using XML and include service adaptation and service interaction rules. Accord is part of Project AutoMate,[32] which provides the required middleware services.

The element (service) managers within the Accord programming system are augmented with online controllers[33,34] as shown in Figure 5.6. The figure shows

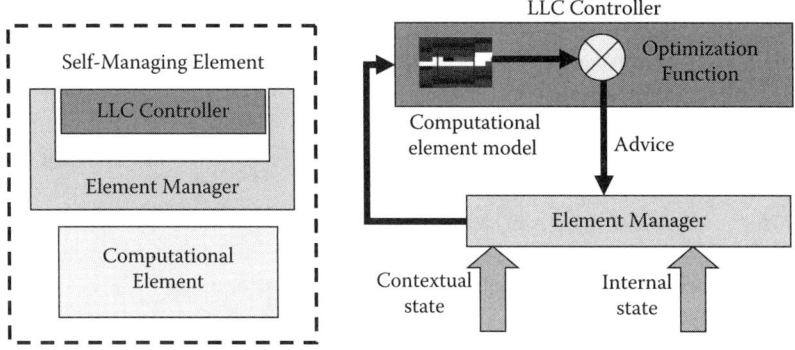

Figure 5.6 Self-management using model-based control in Accord.

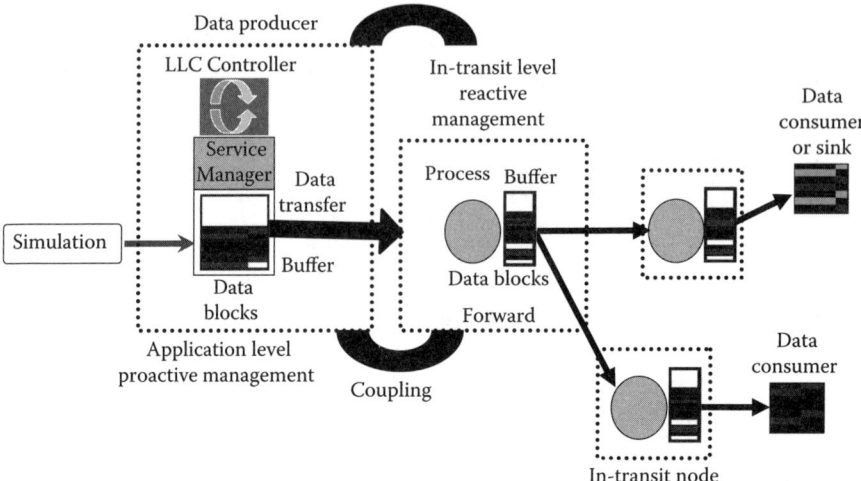

Figure 5.7 Overview of the two-level cooperative QoS management strategy.

the complementary relationship of an element manager and the limited look-ahead controller (LLC) within an autonomic element. Each manager monitors the state of its underlying elements and their execution context, collects and reports runtime information, and enforces the adaptation actions decided by the controller. These managers thus augment human-defined rules, which may be error-prone and incomplete, with mathematically sound models, optimization techniques, and runtime information. Specifically, the controllers decide when and how to adapt the application behavior, and the managers focus on enforcing these adaptations in a consistent and efficient manner.

We use the Accord programming system described above to address end-to-end QoS management and control at two levels shown in Figure 5.7. The first level in this figure is at the end points using adaptive buffer management mechanisms and proactive QoS management strategies based on online control and user-defined policies.[5,33,34] The second level shown in the figure is at the in-transit processing nodes, which are resources in the data path between the source and the destination, using reactive runtime management strategies and adaptive buffer management mechanisms.[35,36] These two levels of management operate cooperatively to address overall application constraints and QoS requirements.

QoS management at application end-points The QoS management strategy at the application end-points combines model-based LLCs and policy-based managers with adaptive multithreaded buffer management.[37] The application-level data streaming service consists of a service manager and an LLC controller. The QoS manager monitors state and execution context, collects and reports runtime information, and enforces adaptation actions determined by its controller. Specifically, the controller decides when and how to adapt the application behavior, and the QoS manager focuses on enforcing these adaptations in a consistent and efficient manner. The effectiveness of this strategy was experimentally demonstrated in Reference 5 which showed that it reduced overheads on the simulation (less than 5%) as well as buffer overflow and data loss.

5.3.2 QoS Management at In-Transit Nodes

In-transit data processing is achieved using a dynamic overlay of available nodes (workstations or small to medium clusters, etc.) with heterogeneous capabilities and loads—note that these nodes may be shared across multiple applications flows. The goal of in-transit processing is to opportunistically process as much data as possible before the data reaches the sink. The in-transit data processing service at each node performs three tasks, namely, processing, buffering and forwarding, and the processing depends on the capacity and capability of the node and the amount of processing that is still required for a data block at hand. The basic idea is that the in-transit data processing service at each node completes at least its share of the processing (which may be predetermined or dynamically computed) and can perform additional processing if the network is too congested for forwarding. Key aspects of the in-transit QoS management include: (1) adaptive buffering and data streaming that dynamically adjusts buffer input and buffer drainage rates, (2) adaptive run-time management in response to network congestions by dynamically monitoring the utility and tradeoffs of local computation versus data transmission, and (3) signal the application end-points about network state to achieve cooperative end-to-end self-management—that is, the in-transit management reacts to local services while the application end-point management responds more intelligently by adjusting its controller parameters to alleviate these congestions.

Experiments conducted using the cooperative end-to-end self-managing data streaming using the GTC fusion application[5,35] have shown that adaptive processing by the in-transit data processing service during congestions decreases the average percent idle time per data block from 25% to 1%. Furthermore, coupling end-point and in-transit level management during congestion reduces percent average buffer occupancy at in-transit nodes from 80% to 60.8%. Higher buffer occupancies at the in-transit lead to failures and result in in-transit data being dropped, and can impact the QoS of applications at the sink. Finally, end-to-end cooperative management decreases the amount of data lost due to congestions at intermediate in-transit nodes, increasing the QoS at the sink. For example, if the average processing time per data block (1 block is 1 MB) is 1.6 sec at the sink, cooperative management saves about 168 sec (approx. 3 minutes) of processing time at the sink.

5.4 Conclusions

As the complexity and scale of current scientific and engineering applications grow, managing and transporting the large amounts of data they generate is quickly becoming a significant bottleneck. The increasing application runtimes and the high cost of high-performance computing resources make online data extraction and analysis a key requirement in addition to traditional data I/O and archiving. To be effective, online data extraction and transfer should impose minimal additional synchronization requirements, should have minimal impact on the computational performance, maintain overall QoS, and ensure that no data is lost.

A key challenge that must be overcome is getting the large amounts of data being generated by these applications off the compute nodes at runtime and over to service nodes or another system for code coupling, online monitoring, analysis, or archiving. To be effective, such an online data extraction and transfer service must (1) have minimal impact on the execution of the simulations in terms of performance overhead or synchronization requirements, (2) satisfy stringent application/user space, time, and QoS constraints, and (3) ensure that no data is lost. On most expensive HPC resources, the large numbers of compute nodes are typically serviced by a smaller number of service nodes where they can offload expensive I/O operations. As the result, the I/O substrate should be able to asynchronously transfer data from compute nodes to a service node with minimal delay and overhead on the simulation. Technologies such as RDMA allow fast memory access into the address space of an application without interrupting the computational process, and provide a mechanism that can support these requirements.

In this chapter we described the ADIOS I/O system and underlying mechanisms, which represent a paradigm shift in which I/O in high-performance scientific application is formulated, specified, and executed. In this new paradigm, the construction of the writes and reads within the application code is decoupled from the specification of how that I/O should occur at runtime. This allows the end user substantial additional flexibility in making use of the latest in high-throughput and asynchronous I/O methods without rewriting (or even relinking) their code. The underlying mechanisms include low-level interfaces which enable lightweight data capture, asynchronous data movement, and specialized adaptive transport services for MPP and wide-area systems. Our experiences with a number of fusion and other codes have demonstrated the effectiveness, efficiency, and flexibility of the ADIOS approach and the accompanying technologies such as DataTaps, I/O graphs, DART, and the autonomic data management, transport, and processing services. These services use metadata that effect I/O operations and access of parallel file systems. Other aspects of metadata are discussed in Chapter 12.

References

[1] D3D, D3D https://fusion.gat.com/global/DIII-D, 2008. Accessed July 13, 2009.

[2] M. Ono et al., Exploration of sperical torus physics in the NSTX device, *Nuclear Fusion*, vol. 40, pp. 557–561, 2000.

[3] ITER, ITER http://www.iter.org, 2008.

[4] Z. Lin, S. Ethier, T. S. Hahm, and W. M. Tang, Size scaling of turbulent transport in magnetically confined plasmas, *Phys. Rev. Letters*, vol. 88, 2002.

[5] V. Bhat, M. Parashar, and N. Kandasamy, Autonomic data streaming for high performance scientific applications, in *Autonomic Computing: Concepts, Infrastructure and Applications*, M. Parashar and S. Hariri, Eds.: CRC Press, 2006, pp. 413–433.

[6] C. S. Chang, S. Ku, M. Adams, F. Hinton, D. Keyes, S. Klasky, W. Lee, Z. Lin, and S. Parker, Gyrokinetic particle simulation of neoclassical transport in the pedestal/scrape-off region of a tokamak plasma, *Institute of Physics Publishing Journal of Physics: Conference Series*, pp. 87–91, 2006.

[7] S. Ku, H. Baek, and C. S. Chang, Property of an X-point generated velocity-space hole in a diverted tokamak plasma edge, *Physics of Plasmas*, Vol. 11, 5626–5633, 2004.

[8] R. Thakur, W. Gropp, and E. Lusk, On implementing MPI-IO portably and with high performance, in *Proceedings of the 6th* Workshop on I/O in Parallel and Distributed Systems, 1999, pp. 23–32.

[9] M. W. Hasan Abbasi, and K. Schwan, LIVE data workspace: A flexible, dynamic and extensible platform for petascale applicatons, in *IEEE Cluster 2007*, 2007.

[10] C. Docan, M. Parashar, and S. Klasky, High speed asynchronous data transfers on the Cray XT3, *Cray User Group Conference*, May 3rd, 2007.

[11] VisIt Visualization Tool: https://wci.llnl.gov/codes/visit/.

[12] O. Messer, S. Bruenn, et al., Petascale supernove simulation with CHIMERA, *Journal of Physics Conference Series*, vol. 78, 2007, pp. 1–5.

[13] G. Eisenhauer, F. Bustamante, and K. Schwan, A middleware toolkit for client-initiated service specialization, in *PODC Middleware Symposium*, 2000.

[14] F. E. Bustamante, G. Eisenhauer, K. Schwan, and P. Widener, Efficient wire formats for high performance computing, in *Proceedings of the 2000 ACM/IEEE Conference on Supercomputing (CDROM)*, p. 39, 2000.

[15] G. Eisenhauer, F. E. Bustamante, and K. Schwan, Event services for high performance computing, in *Proceedings of the Ninth IEEE International Symposium on High Performance Distributed Computing (HPDC'00)*, p. 113, 2000.

[16] G. Eisenhauer, PBIO http://www.cc.gatech.edu/systems/projects/PBIO. Accessed July 13, 2009.

[17] M. Wolf, H. Abbasi, B. Collins, D. Spain, and K. Schwan, Service augmentation for high end interactive data services, in *IEEE Cluster Computing Conference (Cluster '05)*, Boston, MA, 2005.

[18] V. Kumar, B. F. Cooper, Z. Cai, G. Eisenhauer, and K. Schwan, Resource-aware distributed stream management using dynamic overlays, in *the 25th IEEE International Conference on Distributed Computing Systems (ICDCS-2005)*, Columbus, OH, 2005.

[19] P. W. Ron Oldfield, A. Maccabe, Lee Ward, and T. Kordenbrock, Efficient data-movement for lightweight I/O, in *The 2006 IEEE International Conference on Cluster Computing (Cluster 2006)*, Barcelona, 2006.

[20] M. Wolf, Z. Cai, W. Huang, and K. Schwan, SmartPointers: Personalized scientific data portals in your hand, in *Proceedings of the IEEE/ACM SC2002 Conference*, 2002, p. 20.

[21] G. Eisenhauer, Portable self-describing binary data streams, 1994.

[22] G. Eisenhauer, The ECho Event Delivery System (GIT-CC-99-08), in *Georgia Tech College of Computing Technical Reports* (ftp://ftp.cc.gatech.edu/pub/coc/tech_reports), 1999.

[23] G. Eisenhauer, F. Bustamante, and K. Schwan, Native data representation: An efficient wire format for high-performance distributed computing, *IEEE Transactions on Parallel and Distributed Systems*, vol. 13, pp. 1234–1246, Dec 2002.

[24] B. Plale, G. Eisenhauer, L. K. Daley, P. Widener, and K. Schwan, Fast heterogeneous binary data interchange for event-based monitoring, in *International Conference on Parallel and Distributed Computing Systems (PDCS2000)*, 2000.

[25] M. Jain and C. Dovrolis, End-to-end available bandwidth: Measurement methodology, dynamics, and relation with TCP throughput, *Proceedings of the 2002 SIGCOMM conference*, vol. 32, pp. 295–308, 2002.

[26] M. Jain and C. Dovrolis, Pathload: A measurement tool for end-to-end available bandwidth, *Passive and Active Measurements*, vol. 11, pp. 14–25, 2002.

[27] N. Akihiro, P. Larry, and B. Andy, A routing underlay for overlay networks, in *Proceedings of the 2003 Conference on Applications, Technologies, Architectures, and Protocols for Computer Communications*, Karlsruhe, Germany: ACM, 2003.

[28] Z. Cai, V. Kumar, and K. Schwan, IQ-Paths: predictably high performance data streams across dynamic network overlays, *J. Grid Computing*, Vol. 5, 2007, pp. 129–150.

[29] S. Klasky, M. Beck, V. Bhat, E. Feibush, B. Ludascher, M. Parashar, A. Shoshani, D. Silver, and M. Vouk, Data management on the fusion computational pipeline, *Journal of Physics: Conference Series*, vol. 16, pp. 510–520, 2005.

[30] H. Liu, V. Bhat, M. Parashar, and S. Klasky, An autonomic service architecture for self-managing grid applications, in *6th International Workshop on Grid Computing (Grid 2005)*, Seattle, WA, USA, 2005, pp. 132–139.

[31] H. Liu and M. Parashar, Accord: A programming framework for autonomic applications, *IEEE Transactions on Systems, Man and Cybernetics, Special Issue on Engineering Autonomic Systems*, vol. 36, pp. 341–352, 2006.

[32] M. Parashar, H. Liu, Z. Li, V. Matossian, C. Schmidt, G. Zhang, and S. Hariri, AutoMate: Enabling autonomic applications on the grid, *Cluster Computing* vol. 9, pp. 161–174, April 2006.

[33] V. Bhat, M. Parashar, H. Liu, M. Khandekar, N. Kandasamy, and S. Abdelwahed, Enabling self-managing applications using model-based online control strategies, in *3rd IEEE International Conference on Autonomic Computing*, Dublin, Ireland, 2006, pp. 15–24.

[34] V. Bhat, M. Parashar, M. Khandekar, N. Kandasamy, and S. Klasky, A self-managing wide-area data streaming service using model-based online control, in *7th IEEE International Conference on Grid Computing (Grid 2006)*, Barcelona, Spain, 2006, pp. 176–183.

[35] V. Bhat, M. Parashar, and S. Klasky, Experiments with in-transit processing for data intensive grid workflows, in *8th IEEE International Conference on Grid Computing (Grid 2007)*, 2007, pp. 193–200.

[36] V. Bhat, M. Parashar, H. Liu, M. Khandekar, N. Kandasamy, S. Klasky, and S. Abdelwahed, A self-managing wide-area data streaming service, *Cluster Computing: The Journal of Networks, Software Tools, and Applications*, vol. 10, pp. 365–383, December 2007.

[37] V. Bhat, S. Klasky, S. Atchley, M. Beck, D. McCune, and M. Parashar, High performance threaded data streaming for large scale simulations, in *5th IEEE/ACM International Workshop on Grid Computing (Grid 2004)*, Pittsburgh, PA, 2004, pp. 243–250.

Part III

Specialized Retrieval Techniques and Database Systems

Chapter 6

Accelerating Queries on Very Large Datasets

Ekow Otoo and Kesheng Wu

Lawrence Berkeley National Laboratory

Contents

6.1	Introduction	184
6.2	Characteristics of Scientific Data	185
6.3	A Taxonomy of Index Methods	189
6.4	Single-Attribute Index Schemes	193
	6.4.1 Sequential Scan Access	193
	6.4.2 Tree-Structured Indexing	194
	6.4.3 Hashing Schemes	196
6.5	Multidimensional Index Schemes	196
	6.5.1 Multidimensional Tree-Structured Indexing Methods	200
	6.5.1.1 K-D-Tree and LSD-Tree	200
	6.5.1.2 R-Tree and Its Variants	200
	6.5.2 Multidimensional Direct Access Methods	203
	6.5.3 Hybrid Indexing Methods	203
	6.5.3.1 Quaternary/Hierarchical Triangular Mesh	205
	6.5.4 Simple Scanned Access of Multidimensional Datasets	207
6.6	Bitmap Index	208
	6.6.1 The Basic Bitmap Index	209
	6.6.2 Compression	210
	6.6.3 Bitmap Encoding	213
	6.6.4 Binning	215
	6.6.5 Implementations	217
6.7	Data Organization and Parallelization	218
	6.7.1 Data Processing Systems	219
	6.7.1.1 Column-Based Systems	219
	6.7.1.2 Special-Purpose Data Analysis Systems	219
	6.7.1.3 MapReduce Parallel Analysis System	220
	6.7.1.4 Custom Data Processing Hardware	221

6.7.2 Indexes Still Useful 221
6.7.3 Using Indexes to Make Smart Iterators 224
6.8 Summary and Future Trends 226
Acknowledgment 227
References 227

6.1 Introduction

One of the primary goals of a data management system (DBMS) is to retrieve the records under its control upon user requests. In the SQL language, such retrievals are typically formed as queries. Answering these queries efficiently is a key design objective of a data management system. To achieve this goal, one needs to consider many issues including data organization, available methods for accessing the data, user interface, effective query execution planning, and the overall system design. In this chapter, we primarily focus on the aspects that have the most direct influence on the efficiency of query processing, which primarily include three of them: the data organization, access methods, and query execution planning.[25,39,47,67,79,99]

Usually a query can be answered in different ways, for example, the tables and columns involved may be retrieved in different orders or through different access methods.[23,44,81] However, these choices are built on top of a set of good access methods and data organizations. Therefore, we choose to concentrate more on the issues of data organizations and access methods. Furthermore, many common types of queries on scientific data do not require complex execution plans, as we explain in the next section. Thus, optimizing the execution plan is less important than the other two issues. Furthermore, much of the scientific data is not under the control of a DBMS system, but is under the control of some stand-alone systems or emerging scientific DBMSs. A discussion on the core data access methods and data organizations may influence the design and implementation of such systems.

On the issue of data organization, a fundamental principle of the database research is the separation of logical data organization from the physical data organization. Since most DBMSs are based on software that does not have direct control of the physical organization on secondary storage systems, we primarily concentrate on logical organization in this chapter. One common metaphor of logical data organization is the relational data model, consisting of tables with rows and columns. Sometimes, a row is also called a tuple, a data record, or a data object; and a column is also known as an attribute of a record, or a variable in a dataset.

There are two basic strategies of partitioning a table: the row-oriented organization that places all columns of a row together, and the column-oriented organization that places all rows of a column together. The row-oriented organization is also called the horizontal data organization, while the column-oriented organization is also known as the vertical data organization. There are many variations based on these two basic organizations. For example, a large table is often horizontally split into partitions, where each partition is then further organized horizontally or vertically. Since the organization of a partition typically has more impact on query processing, our discussion will center around how the partitions are organized. The data organization of a system is typically fixed; therefore, to discuss data organization we cannot avoid touching on different systems even though they have been discussed elsewhere already. Most notably, Chapter 7 has extensive information about systems with vertical data organizations.

This chapter primarily focuses on access methods and mostly on indexing techniques to speed up data accesses in query processing. Because these methods can be implemented in software and have great potential of improving query performance, there have been extensive research activities on this subject. To motivate our discussion, we review key characteristics of scientific data and queries in the next section. In Section 6.3, we present a taxonomy of index methods. In the following two sections, we review some well-known index methods, with Section 6.4 on single-column indexing and Section 6.5 on multidimensional indexing. Given that scientific data are often high-dimensional data, we present a type of index that has been demonstrated to work well with this type of data. This type of index is the bitmap index; we devote Section 6.6 to discussing the recent advances on the bitmap index. In Section 6.7 we revisit the data organization issue by examining a number of emerging data processing systems with unusual data organizations. All these systems do not yet use any indexing methods. We present a small test to demonstrate that even such systems could benefit from an efficient indexing method.

6.2 Characteristics of Scientific Data

Scientific databases are massive datasets accumulated through scientific experiments, observations, and computations. New and improved instrumentations now not only provide better data precision but also capture data at a much faster rate, resulting in large volumes of data. Ever-increasing computing power is leading to ever-larger and more realistic computation simulations, which also produce large volumes of data. Analysis of these massive datasets by domain scientists often involves finding some specific data items that have some characteristics of particular interest. Unlike the traditional information management system (IMS), such as management of bank records in the 1970s and 1980s where the database consisted of a few megabytes of records that have a small number of attributes, scientific databases typically consist of terabytes of data (or billions of records) that have hundreds of attributes. Scientific databases are generally organized as datasets. Often these datasets are not under the management of traditional DBMS systems, but merely appear as a collection of files under a certain directory structure or following certain naming conventions. Usually, the files follow a format or schema agreed among the domain scientists.

An example of such a scientific dataset with hundreds of attributes is the data from the High Energy Physics (HEP) STAR experiments,[87] that maintains billions of data items (referred to as *events*) on over hundred attributes. Most of the data files are in a format called ROOT.[18,70] To search for a subset of the billions of events that satisfy some conditions based on a small number of attributes requires special data-handling techniques beyond traditional database systems. We address specifically some of the techniques for efficiently searching through massively large scientific datasets in this chapter.

The need for efficient search and subset extraction from very large datasets is motivated by the requirements of numerous applications in both scientific domains and statistical analysis. Here are some such application domains:

- high-energy physics and nuclear data generations from experiments and simulations
- remotely sensed or in situ observations in the earth and space sciences, (e.g., data observations used in climate models)
- seismic sounding of the earth for petroleum geophysics (or similar signal processing endeavors in acoustics/oceanography)
- radio astronomy, nuclear magnetic resonance, synthetic aperture radar, and so forth
- large-scale supercomputer-based models in computational fluid dynamics (e.g., aerospace, meteorology, geophysics, astrophysics), quantum physics, chemistry, and so forth
- medical (tomographic) imaging (e.g., CAT, PET, MRI)
- computational chemistry
- bioinformatic, bioengineering, and genetic sequence mapping
- intelligence gathering, fraud detection, and security monitoring
- geographic mapping and cartography
- census, financial, and other statistical data

Some of these applications are discussed in References 36, 92, and 95. Compared with the traditional databases managed by commercial DBMSs, one immediate distinguishing property of scientific datasets is that there is almost never any simultaneous read and write access to the same set of data records. Most scientific datasets are *read-only* or *append-only*. Therefore, there is a potential to significantly relax the ACID* properties observed by a typical

---

*Atomicity, consistency, isolation and durability

DBMS system. This may give rise to different types of data access methods and different ways of organizing them as well.

Consider a typical database in astrophysics. The archived data include observational parameters such as the detector, the type of observation, coordinates, astronomical object, exposure time, and so forth. Besides the use of data-mining techniques to identify features, users need to perform queries based on physical parameters such as magnitude of brightness, redshift, spectral indexes, morphological type of galaxies, photometric properties, and so forth, to easily discover the object types contained in the archive. The search usually can be expressed as constraints on some of these properties, and the objects satisfying the conditions are retrieved and sent downstream to other processing steps such as statistics gathering and visualization.

The datasets from most scientific domains (with the possible exception of bioinformatics and genome data), can be mostly characterized as time-varying arrays. Each element of the array often corresponds to some attribute of the points or cells in two- or three-dimensional space. Examples of such attributes are temperature, pressure, wind velocity, moisture, cloud cover, and so on in a climate model. Datasets encountered in scientific data management can be characterized along three principle dimensions:

Size: This the number of data records maintained in the database. Scientific datasets are typically very large and grow over time to be terabytes or petabytes. This translates to millions or billions of data records. The data may span hundreds to thousands of disk storage units and often are archived on robotic tapes.

Dimensionality: The number of searchable attributes of the datasets may be quite large. Often, a data record can have a large number of attributes, and scientists may want to conduct searches based on dozens or hundreds of attributes. For example, a record of a high-energy collision in the STAR experiment[87] is about 5 MB in size, and the physicists involved in the experiment have decided to make 200 or so high-level attributes searchable.[101]

Time: This concerns the rate at which the data content evolves over time. Often, scientific datasets are constrained to be *append-only* as opposed to frequent random insertions and deletions as typically encountered in commercial databases.

Traditional DBMSs such as ORACLE, Sybase, and Objectivity have not had much success in scientific data management. These have had only limited applications. For example a traditional relational DBMSs, MySQL, is used to manage the metadata, while the principal datasets are managed by domain-specific DBMSs such as ROOT.[18,70] It has been argued by Gray et al.[35] that managing the metadata with a nonprocedural data manipulation language combined with *data indexing* is essential when analyzing scientific datasets.

Index schemes that efficiently process queries on scientific datasets are only effective if they are built within the framework of the underlying physical data organization understood by the computational processing model. One example of a highly successful index method is the bitmap index method,[4,21,48,102,103] which is elaborated upon in some detail in Section 6.6. To understand why traditional DBMSs and their accompanying index methods such as B-tree, hashing, R-Trees, and so forth, have been less effective in managing scientific datasets, we examine some of the characteristics of these applications.

Data Organizational Framework: Many of the existing scientific datasets are stored in custom-formatted files and may come with their own analysis systems. ROOT is a very successful example of such a system.[18,70] Much of astrophysics data are stored in FITS format,[41] and many other scientific datasets are stored in NetCDF format[61] and HDF format.[42] Most of these formats including FITS, NetCDF, and HDF are designed to store arrays, which can be thought of as a vertical data organization. However, ROOT organizes data as objects and is essentially row-oriented.

High-Performance Computing (HPC): Data analysis and computational science applications, for example, Climate Modeling, have application codes that run on high-performance computing environments that involve hundreds or thousands of processors. Often these parallel application codes utilize a library of data structures for hierarchical structured grids where the grid points are associated with a list of attribute values. Examples of such applications include finite element, finite difference, and adaptive mesh refinement (AMR) method. To efficiently output the data from the application programs, the data records are often organized in the same way as they are computed. The analysis programs have to reorganize them into a coherent, logical view, which adds some unique challenges for data access methods.

Data-Intensive I/O: Often, highly parallel computations in HPC also perform data-intensive data inputs and outputs. A natural approach to meet the I/O throughput requirements in HPC is the use of parallel I/O and parallel file systems. To meet the I/O bandwidth requirements in HPC, the parallel counterparts of data formats such as NetCDF and HDF/HDF5 are applied to provide consistent partitioning of the dataset into chunks that are then striped over disks of a parallel file system. While such partitioning is efficient during computations that produce the data, the same partitioning is usually inefficient for later data analysis. Reorganization of the data or an index structure is required to improve the efficiency of the data analysis operations.

None-Transactional ACID Properties: Most scientific applications do not access data for analysis while concurrently updating the same data records. The new data records are usually added to the data in large chunks. This allows the data management system to treat access control in a much more optimistic manner than is possible with traditional DBMS systems. This feature will be particularly important as data management systems evolve to take advantage of multicore architectures and clusters of such multicore computers, where concurrent accesses to data is a necessity.

More discussion about the differences between scientific and commercial DBMSs is presented in Section 7.6 in the context of SciDB.

6.3 A Taxonomy of Index Methods

An *access method* defines a data organization, the data structures, and the algorithms for accessing individual data items that satisfy some query criteria. For example, given N records, each with k attributes, one very simple access method is that of a sequential scan. The records are stored in N consecutive locations, and for any query the entire set of records is examined one after the other. For each record, the query condition is evaluated; and if the condition is satisfied, the record is reported as a hit of the query. The data organization for such a sequential scan is called the *heap*. A general strategy to accelerate this process is to augment the heap with an *index scheme*.

An index scheme is the data structure and its associated algorithms that improve the data accesses such as insertions, deletions, retrievals, and query processing. The usage and preference of an index scheme for accessing a dataset is highly dependent on a number of factors including the following:

Dataset Size: One factor is whether the data can be contained entirely in memory or not. Since our focus is on massively large scientific datasets, we will assume the latter with some consideration for main memory indexes when necessary.

Data Organization: The datasets may be organized into fixed-size *data blocks* (also referred to as *data chunks* or *buckets* at times). A data block is typically defined as a multiple of the physical page size of disk storage. A data organization may be defined to allow for future insertions and deletions without impacting the speed of accessing data by the index scheme. On the other hand, the data may be organized and constrained to be *read-only*, *append-only*, or both. Another influencing data organization factor is whether the records are of fixed length or variable length. Of particular interest in scientific datasets are those datasets that are mapped into very large k-dimensional arrays. To partition the array into manageable units for transferring between memory and disk storage, fixed-size subarrays called *chunks* are used. Examples of such data organization methods are NetCDF,[61] HDF5[42] and FITS.[41]

Index Type: A subset of the attributes of a record that can uniquely identify a record of the dataset is referred to as the *primary key*. An index constructed using the *primary key* attributes is called the *primary index*, otherwise it is a *secondary index*. An index may also be classified as either *clustered* or *nonclustered* according to whether records whose primary keys are closely similar to each other are also stored in close proximity to each other. A metric of similarity of two keys is defined by the collating sequence order of the index keys.

Class of Queries: The adoption of a particular index scheme is also highly dependent on the types of queries that the dataset is subsequently subjected to. Let the records of a dataset have k attributes $A = \{A_1, A_2, \ldots A_k\}$, such that a record $r_i = \langle a_1, a_2, \ldots a_k \rangle$, where $a_i \in A_i$. Typical classes of queries include the following:

Exact Match: Given k values $\langle v_1, v_2, \ldots v_k \rangle$ an exact-match query asks for the retrieval of a record $r_i = \langle a_1, a_2, \ldots a_k \rangle$ such that $a_i = v_i, 1 \leq i \leq k$.

Partial Match: Let $A' \subseteq A$ with $a'_j \in A'_j$ and $a_j \in A_j$, respectively, for $j \leq |A'|$, where $|A'| = k' \leq k$. Given values $\{v_1, v_2, \ldots v_{k'}\}$ for the k' attributes of A', a partial-match query asks for the retrieval of all records whose attribute values match the corresponding specified value, that is, $a'_j = v_j$, for $a'_j \in A'_j, 1 \leq j \leq k'$. The exact-match query is a special class of a partial-match query.

Orthogonal Range: For categorical attributes we assume that an ordering of the attribute values is induced by the collating sequence order of the values, and for numeric attribute values the ordering is induced by their respective values. Then, given k closed intervals of values $\langle [l_1, h_1], [l_2, h_2], \ldots [l_k, h_k] \rangle$, an orthogonal-range query asks for the retrieval of all records $r_i = \langle a_1, a_2, \ldots a_k \rangle$ such that $l_i \leq a_i \leq h_i, 1 \leq i \leq k$.

Partial Orthogonal Range: This is similar to orthogonal-range query but only a subset of the attributes is mentioned in the query. The orthogonal-range query is a special case of a partial-orthogonal-match query. Since partial-orthogonal-range query subsumes the preceding classes, we will use the measure of the efficiency of processing partial-orthogonal-range queries and the measure of the efficiency of processing partial-match queries as the comparative measures of the efficiencies of the various indexing schemes to be addressed. If the columns involved in a partial-orthogonal-range query vary from one query to the next, such queries are also known as *ad hoc range queries*. In the above definition, the condition on each column is of the form $l_j \leq a_j \leq h_j$. Since it specifies two boundaries of the query range, we say it is a two-sided range. If the query range only involves one boundary, for example, $l_j \leq a_j, a_j \leq h_j, l_j < a_j$, and $a_j > h_j$, it is a one-sided range.

Other Query Processing Considerations: There are numerous query processing conditions that influence the design and choice of an index scheme besides the preceding ones. For example, the orthogonal range query is very straightforward to specify by a set of closed intervals. The range coverage may be circular, spherical, or some arbitrary polygonal shape. There is a considerable body of literature that covers these query classes.[79]

Attribute Types: The data type of the attribute keys plays a significant role in the selection of an index scheme used in an access method. For example, when the data type is a long alphabetic string or bit-strings, the selection of an index scheme may be different from that when the data type of the index is an integer value.

Index Size Constraint: The general use of an index structure is to support fast access of data items from stored datasets. One desirable feature of an index is that its size be relatively small compared with the base data. An index may contain relatively small data structures that can fit entirely in memory during a running session of the application that uses the index. The memory resident information is then used to derive the relative positions or block addresses of the desired records. Examples of these indexing schemes include inverted indexes,[47] bitmap indexing,[21,103] and direct access index (or *hashing*[47,79]). However, when the index structure is sufficiently large, this will require it to be stored on a disk storage system and then page-in the relevant buckets of the index into memory on demand in a manner similar to the use of B-Tree index and its variants.[24]

Developing an efficient access method in scientific applications is one of the first steps in implementing an efficient system for a data-intensive application. The question that immediately arises then is what measures constitute a good metric for evaluating an efficient index scheme. Let an index scheme G be built to access N data items. For a query Q, the most important metrics are the query response time and the memory requirement—how long and how much memory does it take to answer the query. Often, the query response time is dominated by I/O operations, and the number of bytes (or disk sectors) read and written is sometimes used as a proxy for the query response time.

Additional metrics for measuring an indexing scheme include *recall* denoted by ρ, *precision* denoted by π, and *storage utilization* denoted by $\mu(G)$. Let $r(Q)$ denote number of correct results retrieved for a query, and let $R(Q)$ denote the actual number of results returned by a query in using G. Let $c(Q)$ denote the actual number of correct results desired where $c(Q) \leq r(Q) \leq R(Q)$. The *precision* is defined as the ratio of desired correct results to the number of correct results, that is, $\pi = c(Q)/r(Q)$; and the *recall* is the ratio of the number of correct results to the number of retrieved results, that is, $\rho = r(Q)/R(Q)$. Let $S(G)$ denote the actual storage used by an index scheme for

N data items where an average index record size is b. The storage utilization is defined as $\mu(G) = bN/S(G)$.

Another metric, sometimes considered, is the time to update the index. This is sometimes considered as two independent metrics: the *insertion* time and the *deletion* time. In scientific data management, deletions are hardly of concern.

There is a large body of literature on index structures. It is one of the highly researched subjects in computer science. Extensive coverage of the subject is given by References 25, 39, 47, 79, 99. Together, these books cover most of the indexing structures used in datasets and databases of various application domains. Index schemes may be classified into various classes. Figure 6.1 shows a taxonomy used to guide our review of known indexing methods.

Each class is partitioned into three subclasses based on whether accessing the data is by *direct access* method, *tree-structured* index, or some combination of tree-structured index and a direct access method, which we term as a *hybrid index*. Each of these subclasses can be subsequently grouped according to whether the search key is a *single attribute* or a combination of *multiple attributes* of the data item. Such combined *multiattribute* index schemes are also referred to as *multidimensional* index schemes.

Since the classification of hybrid indexes is fuzzy, we will discuss them under direct access methods in this chapter. Further, to restrict the classification space to a relatively small size, we will not distinguish between single-attribute

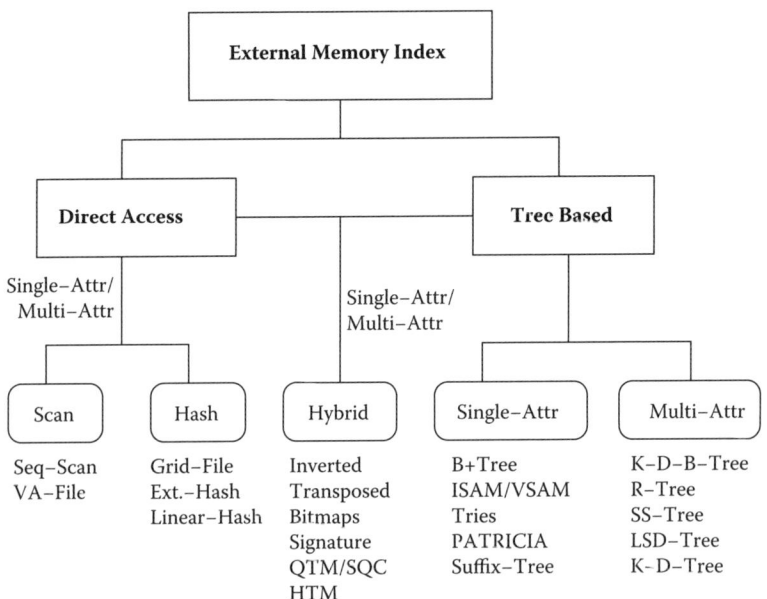

Figure 6.1 A taxonomy of indexed access methods.

and multiattribute indexing when we discuss both the direct access and the hybrid index methods. There is a further level of classification that distinguishes index schemes into those that are more suitable for static data and those that more are suitable for dynamic data. Most large-scale scientific datasets are append-only and as such are not typically subjected to concurrent reads and writes in a manner that requires models of transactional accessing. As such, we do not consider index methods under further classification of dynamic versus static methods.

In this chapter, we focus primarily on disk-based datasets that are stored on external memory and are paged in and out of memory during processing using *memory buffers* or *cache pools*. Consequently, we address access methods, denoted as the highlighted ovals of the leaf nodes of our classification hierarchy shown in Figure 6.1, where we also show some examples of the indexing methods that have been used in practice. In the next sections we describe methods of accessing massively large datasets by (1) simple scans, (2) hashing, (3) single-attribute indexes, (4) multiattribute indexes, and (5) hybrid schemes that are comprised of one or more combinations of hashing and scanning, hashing and tree indexes, or hashing, tree index, and scanning. Note also that methods that generate a single key by combining the multiple attribute values of the data, either by concatenation or bit interleaving, and then using the generated key in a single-attribute index scheme, are considered under the hybrid methods.

6.4 Single-Attribute Index Schemes

6.4.1 Sequential Scan Access

A natural approach to searching unordered datasets without an index is by a simple sequential scan. This gives an $O(N)$ search time for datasets of N items. Although seemingly naive and simplistic, it has the advantage that insertions are fast since new data items are simply appended at the end of the existing file. If order is maintained on either a single or combined attributes, searches can be performed in $O(\log N)$ time, using a binary search. Of course, this requires a preprocessing cost of sorting the items. As new records are added, one needs to ensure that the sorted order is preserved.

For very large high-dimensional datasets, where searches are allowed on any combination of attributes, it has been shown that such simple sequential scans of processing queries can be just as effective, if not more efficient, as building and maintaining multidimensional indexes,[14,82,97] particularly when nearest neighbor queries are predominant. With parallel computing and parallel file systems, a sequential scan can achieve near-linear speed-up with a balanced data partition. We revisit the case of sequential scans for high-dimensional datasets in Section 6.5.4.

6.4.2 Tree-Structured Indexing

Most applications require accessing data both sequentially and randomly. The tree-structured indexing methods facilitate both of them. These indexing schemes are sometimes referred to as multilevel indexing schemes. A file organization scheme, representative of a tree structured index, is the *indexed sequential access method* (ISAM). The early commercial design and implementation of the indexed sequential access method was by IBM over 50 years ago. The ISAM index method is a *static* index method. Many versions of tree-index schemes that allow *dynamic* updating of the tree structure were introduced later, including the well-known B-tree indexes and their many variations. In order to illustrate the main concept of tree-structures indexes, we discuss next the ISAM index using the example in Figure 6.2.

The index keys in the table shown in Figure 6.2 are first grouped into first-level nodes, or blocks, of size b records per node. The idea is illustrated in Figure 6.3, which represents the ISAM organization of the blocked indexes shown in Figure 6.2. In Figure 6.3, $b = 2$. The lowest index keys (alternatively the largest index key) of a node, each paired with its node address, are formed into records that are organized as the next higher (or second) level index of the first-level nodes. The process is recursively repeated until a single node (i.e., the root node) of size b can contain all the index records that point to the lower-level nodes.

Searching for a record begins at the root node and proceeds to the lowest-level node, that is, the leaf node, where a pointer to the record can be found. Within each node, a sequential or binary search is used to determine the next lower level node to access. The search time for a random record is

Rec No	Y	X	Label
1	y0	x6	A
2	y2	x6	B
3	y4	x3	C
4	y5	x3	D
5	y2	x1	E
6	y4	x1	F
7	y1	x7	G
8	y2	x5	H
9	y1	x5	I
10	y6	x4	J
11	y7	x2	K
12	y6	x7	L
13	y5	x2	M
14	y3	x0	N
15	y4	x5	N

X-Vals	Rec No
x6	1
x6	2
x3	3
x3	4
x1	5
x1	6
x7	7
x5	8
x5	9
x4	10
x2	11
x7	12
x2	13
x0	14
x5	15

X-Index List	
x0 → {14}	
x1 → {5, 6}	B_0
x2 → {11, 13}	
x3 → {3, 4}	B_1
x4 → {10}	
x5 → {8, 9, 15}	B_2
x6 → {1, 2}	
x7 → {7, 12}	B_3

Figure 6.2 An example of a single-attribute index.

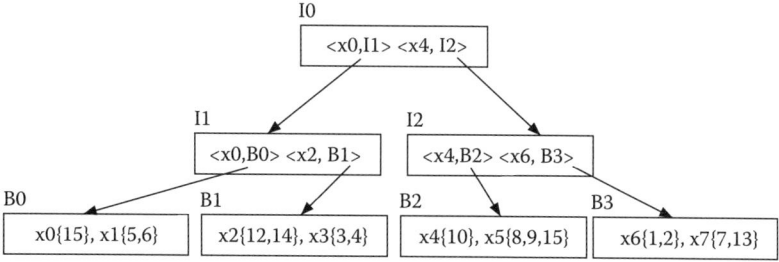

Figure 6.3 Illustrative tree-structured index by ISAM organization. The data blocks are labeled I0, ..., I2, B0, ..., B3.

$O(\log_b(N/b))$ given a file of N data items. Note that we are concerned with indexing very large datasets so that in practice $b >> 2$. In this scheme the ISAM index was mapped directly to the layout of data on a disk storage where the root level index searches give the proper cylinder number of the record. The first track of each cylinder gives the track number, and this corresponds to the second level of indexing for searches. At the lowest level, this corresponds to the locations of records within a track. A sequential scan at this level is used to locate the records. The ISAM index method is a *static* index method. Subsequent insertions require the records to be managed as overflow records that are periodically merged by reorganizing the entire ISAM index.

The *indexed sequential organization* illustrates the structural characteristic of tree-structured index schemes. To circumvent the static limitations of the ISAM, the *B*-Tree indexing scheme was developed. Detailed coverage of the *B*-Tree and its variants such as B^+-Tree, B^*-Tree, and Prefix-*B*-Tree is given in References 9, 24, 25, and 47. The *B*-Tree is a dynamic, height-balanced index scheme that grows and shrinks by recursively splitting and merging nodes from the lowest level of the index tree up to the root node. The VSAM file organization[55] is a B^+-Tree that is mapped directly to the layout of data on disk storage.

Tree-structured index schemes typically maintain the fixed-sized keys, such as integers or fixed-length character strings, in the index nodes. When the keys are variable-length strings, then rather than storing the entire keys, only sufficient strings of leading characters that form separator keys are stored. This is the basic idea of the prefix-*B*-Tree. An alternative index method for long alphabetic strings is the use of a *trie*.[39,47,79,99] Suppose the keys are formed from a domain of alphabet set Σ with cardinality $|\Sigma|$. Each node of the trie-index at level i is comprised of all occurrences of the distinct i^{th} characters of keys with the same $i-1$ prefix string. A node in a trie index structure has size of at most $|\Sigma|$ entries, where each entry is pair of a character and a pointer to the lower level node.

A special kind of trie, called the suffix tree,[37,39,43,47] can be used to index all suffixes in a text in order to carry out fast full or partial text searches. A basic trie implementation has the disadvantage of having a single path that is not space efficient. A more space-efficient implementation of a trie index is the *PATRICIA* index.[47,58] PATRICIA stands for practical algorithm to retrieve information coded in alphanumeric. Tries are the fundamental index schemes for string-oriented databases, for example, very large text databases used in information-retrieval problems, genome sequence databases, and bioinformatics. Tries can be perceived as generic index methods with variants such as *Suffix-Trees*, *String B-Trees*, and *Burst Tries*.[31,37,39,43,47,99]

6.4.3 Hashing Schemes

Hashing methods are mechanisms for accessing records by the address of a record. The address is computed using a key of the record, usually the *primary key* or some unique combination of attribute values of the record. The method involves a set of n disk blocks, also termed *buckets*, with index values numbered from 0 to $n-1$. A bucket is a unit of storage containing one or more records. For each record $\langle k_i, r_i \rangle$ whose key is k_i and entire record is r_i, a hash function $H()$ is used to compute the bucket address I where the record is stored, that is, $I = H(k_i)$. The result I of computing the hash function forms the index into the array of buckets or data blocks that hold the records. A hashing scheme can either be *static* or *dynamic*.

In static hashing, the number of buckets is preallocated. Records with different search-key values may map to the same bucket, in which case we say that a *collision* occurs. To locate a record in a bucket with multiple keys, the entire bucket has to be searched sequentially. Occasionally a bucket may have more records that hash into it than it can contain. When this occurs it is handled by invoking some overflow handling methods. Typical overflow handling methods include separate chaining, rehashing, coalesced chaining, and so forth. To minimize the occurrence of overflow, a hash function has to be chosen to distribute the keys uniformly over the allocated buckets. Hashing methods on single attributes are discussed extensively in the literature.[25,47,79]

An alternative to static hashing is dynamic hashing. Dynamic hashing uses a dynamically changing function that allows the addressed space, that is, the number of allocated buckets, to grow and shrink with insertions and deletions of the records, respectively. It embeds the handling of overflow records dynamically into its primary address space to avoid explicit management of overflow buckets. Various techniques for dynamic hashing have been proposed. Notable among these are dynamic hashing,[51] linear hashing,[54] and extendible hashing.[30,68]

6.5 Multidimensional Index Schemes

Multidimensional indexing arises from the need for efficient query processing in numerous application domains where objects are generally characterized by *feature vectors*. They arise naturally from representation and querying of

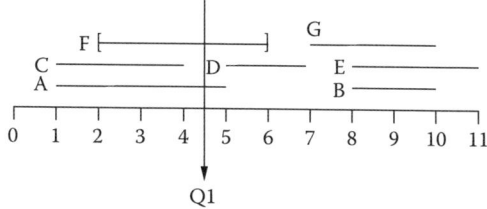

Figure 6.4 A collection of X-axis parallel line segments.

geometric objects such as points, lines, and polyhedra in computational geometry, graphics, multimedia, and spatial databases. Other applications that have seen considerable research in multidimensional data structures include data mining of scientific databases and geographic information systems. Objects characterized by k-dimensional feature vectors are typically represented either by their spatial extents in the appropriate metric space or mapped as points in k-dimensional metric space that defines an appropriate metric measure of relationship between any two points.[15,33,74,79] Depending on the types of queries on the objects, different data structures that recognize the spatial extents of the objects can be utilized. A more general approach, however, is still by mapping of these objects as points, and then partitioning either the points or the embedding space. To illustrate these general approaches of representing objects that can be characterized by feature vectors, we consider the representation of line segments as in Figure 6.4 that are subjected to stabbing line and intersection queries.

Two well-known memory resident data structures for representing such line segments for efficient processing of the stabbing line queries are the *interval tree* and *segment tree*.[74,79] They also have corresponding disk based counterparts.[96] Since each line segment is characterized by a vector $\langle lx, rx \rangle$ of its left and right x-values, these can be mapped as points in two-dimensional space as shown in Figure 6.5. The stabbing line query $Q1 = 4.5$ translates

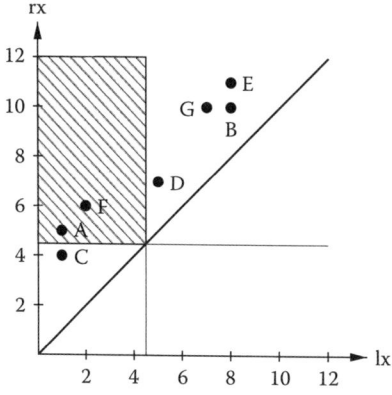

Figure 6.5 Shaded region representing a stabbing query.

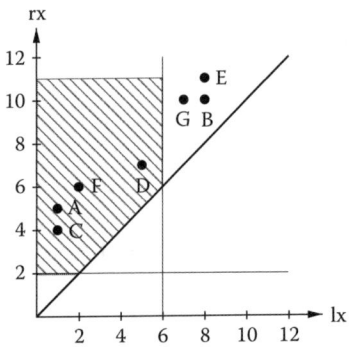

Figure 6.6 Shaded region representing an intersection query.

to a range query of finding all line segments whose $lx \leq 4.5$ and $rx \geq 4.5$ as depicted in Figure 6.5. Similarly, a query that asks for all lines that overlap line $F = \langle 2,6 \rangle$ also translates to a range query of finding all lines whose $lx \leq 6$ and $rx \geq 2$. The shaded region of Figure 6.6 is the area where the points of the response to query lie. In the illustrative example given, all lines are mapped onto points that lie above the 45-degree line. It is possible to choose a mapping that distributes the points uniformly in the embedding space. The basic idea is easily extended to objects of arbitrary shapes and dimensions such as rectangles, circles, spheres, and so forth. For example, a collection of rectangles may be represented either explicitly as rectangles with the appropriate index structure, such as R-trees[94] layered over it to process queries on them, or mapped as points in a four-dimensional data space. In the subsequent discussion, we focus mostly on datasets that are k-dimensional and are perceived as points in k-dimensional space.

A sample of a two-dimensional dataset is given in the table shown in Figure 6.7. The first three columns give the Y-values, the X-values, and the labels, respectively. The mapping of the dataset as points in a two-dimensional space is shown in Figure 6.8 and is based on the Y- and X-values. The indexing mechanism used to access the data is highly dependent on the physical organization of the datasets. In high-dimensional, very large, scientific datasets, the datasets are typically vertically partitioned and stored by columns. Such physical storage of datasets is amenable to efficient access by bitmaps and compressed bitmap indexes.[4,21,38,48,102,103] Another popular physical representation of datasets, particularly when the datasets are perceived as points in a k-dimensional data space, is by tessellating the space into rectangular regions and then representing each region as a *chunk* of data so that all the points in a region are clustered in the same *chunk*. A chunk typically corresponds to the physical page size of disk storage but may span more than one page. The different methods of tessellating and indexing the corresponding chunks

Accelerating Queries on Very Large Datasets 199

Y	X	Label	Linear Quad-Code	Linear Binary Gray-Code
y0	x6	A	110	010001 ⇒ 30
y2	x6	B	130	011011 ⇒ 18
y4	x3	C	211	101100 ⇒ 55
y5	x3	D	213	101110 ⇒ 52
y2	x1	E	021	001011 ⇒ 13
y4	x1	F	201	101001 ⇒ 49
y1	x7	G	113	010010 ⇒ 28
y2	x5	H	121	011111 ⇒ 21
y1	x5	I	103	010111 ⇒ 26
y6	x4	J	320	110110 ⇒ 36
y7	x2	K	232	100101 ⇒ 57
y6	x7	L	331	110010 ⇒ 35
y5	x2	M	212	101111 ⇒ 53
y3	x0	N	023	001000 ⇒ 15
y4	x5	N	301	111101 ⇒ 41

Figure 6.7 A sample multiattribute data.

give rise to the numerous multidimensional indexing methods described in the literature. For example, the tessellation of the data space may be

- overlapping or non-overlapping,
- flat or hierarchical, and
- based on equalizing the points in each region as much as possible or based on generating equal regions of spatial extents.

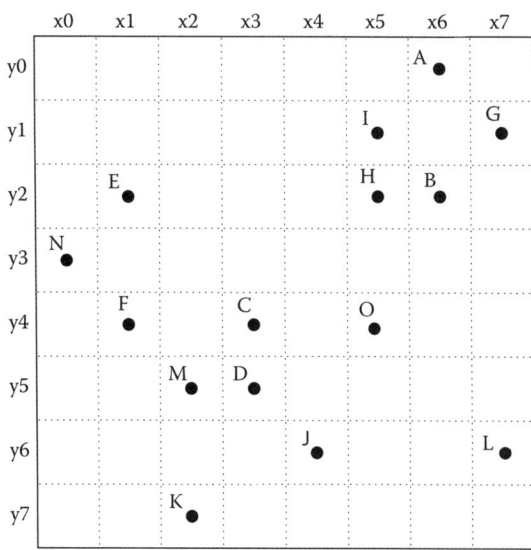

Figure 6.8 Multi-attribute data in 2D space.

These chunking methods give rise to typical formats for *array files*. Examples of *array files* include the file formats known as NetCDF (network common data format)[61] and HDF4/HDF5 (hierarchical data format).[42] We briefly explore some of the widely used indexing methods that are based on tessellation of the data space.

6.5.1 Multidimensional Tree-Structured Indexing Methods

From the general approach of designing multidimensional index schemes, we examine now some special cases that lead to the class of tree-structured multidimensional indexing.

6.5.1.1 K-D-Tree and LSD-Tree

The *K-D-Tree*[11,25,33] was designed as a memory-resident data structure for efficient processing of range, partial-match, and partial-range queries. Given the k-dimensional data space of a collection of points, the K-D-Tree is derived by tessellating the space with (k-1)-dimensional hyperplanes that are parallel to all but the axis being split. The split plane occurs at the position that splits the number of points as evenly as possible within the subspace being split. The choice of which axis to split is done cyclically. The subspace tessellation proceeds recursively until the number of points in a region forms the desired chunk size of the data space. Figure 6.9a illustrates the idea for a 2D-space tessellation that is represented explicitly as a tree structure in Figure 6.9b, for a chunk size of 2. Such a K-D-Tree is referred to as a *region* K-D-Tree as opposed to a *point* K-D-Tree. Rather than maintaining the leaf nodes, that is, the data chunks, in memory, these can be maintained on disks while the internal nodes of comparator values remain in memory. This variant of the K-D-Tree is referred to as the LSD-Tree (or local split-decision tree) and is more suitable for large disk-resident datasets. The idea is that when insertion causes a data chunk to exceed its capacity, a decision can be made to split the data chunk and to insert the split value as an internal node with pointers to the newly created leaves of data chunks. The memory-resident tree nodes can always be maintained persistently by offloading onto disk after a session and reloading before a session.

The complexity of building a K-D-Tree of N points takes $O(N/B \log_2(N/B))$ for a chunk size of B. Insertion and deletion of a new point into a K-D-Tree takes $O(\log(N/B))$ time. One key feature of the K-D-Tree is that partial-match and partial-range queries involving s of k dimensions take $O((N/B)^{1-s/k} + r)$ time to answer, where r is the number of the reported points, and k the dimension of the K-D-Tree.

6.5.1.2 R-Tree and Its Variants

The R-Tree, first proposed by Guttman,[40] is an adaptation of the B^+-Tree to efficiently index objects contained in k-dimensional bounding boxes. The

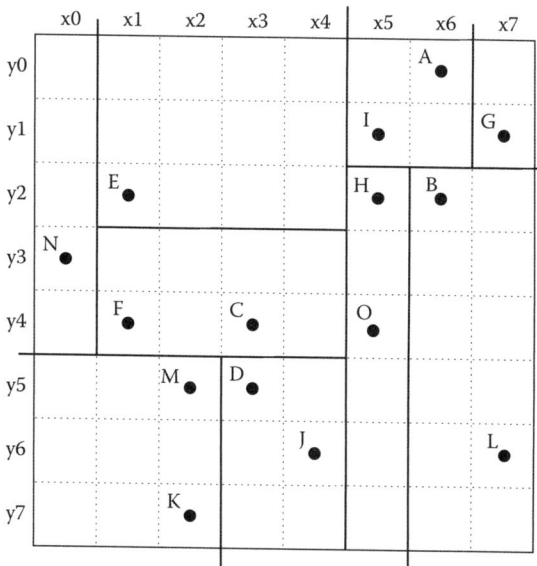

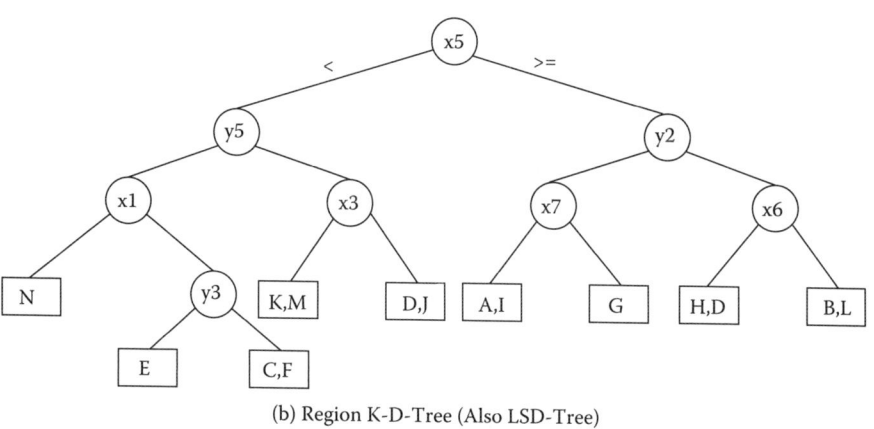

Figure 6.9 K-D-Tree indexing of 2D dataset.

R-Tree is a height-balanced tree consisting of internal and leaf nodes. The entries in the leaf nodes are pairs of values of the form $\langle RID, R \rangle$, where RID is the row identifier and R is a vector of values that defines the minimum rectilinear bounding box (MBR) enclosing the object. Note that the object may simply be a point. An internal node is a collection of entries of the form $\langle ptr, R \rangle$, where ptr is a pointer to a lower-level node of the R-Tree, and R is a minimum bounding box that encloses all MBRs of the lower-level node.

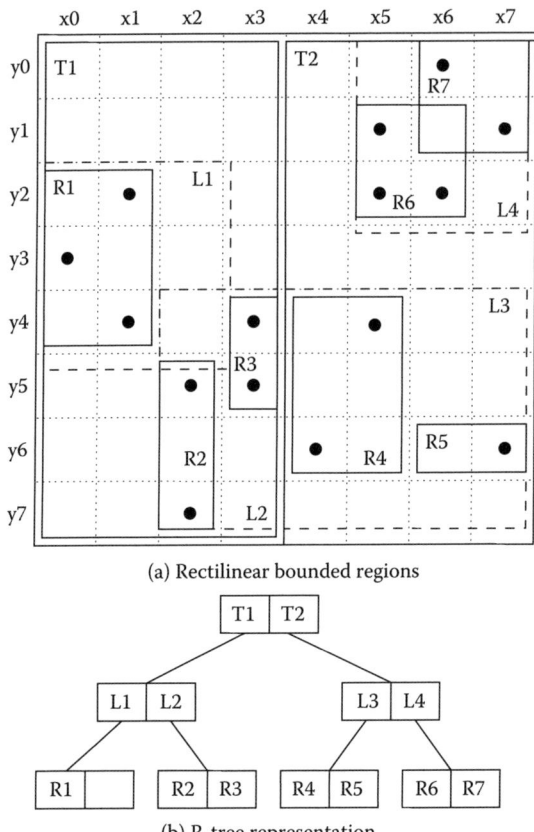

Figure 6.10 R-Tree indexing of bounded rectilinear 2D dataset.

Figure 6.10a shows a collection of points grouped into rectangular boxes of at most three points per box and indexed by an R-Tree.

Like the B^+-Tree, an R-Tree specifies the maximum number of entries B, that can be contained in a node and satisfies the following properties:

1. A leaf node contains between $B/2$ and B entries unless it is the root node.
2. For an entry $\langle RID, R \rangle$ in a leaf node, R is the minimum rectilinear bounding box of the object RID.
3. An internal node contains between $B/2$ and B entries unless it is the root node.
4. For an entry $\langle ptr, R \rangle$ in an internal node, R is the minimum rectilinear bounding box that encloses the MBRs in the node pointed to by ptr.
5. A root node can contain at least two children unless it is a leaf node.
6. All leaf nodes appear at the same level.

The R-Tree representation of the rectangular regions of Figure 6.10a is shown in Figure 6.10b. Since the introduction of the R-Tree, several variants have been introduced. The R-Tree portal[94] gives implementation codes and papers of the different variants. It has had numerous applications in spatial databases, geographic information systems (GIS), very large scale integration (VLSI) design and applications that depend on nearest-neighbor searching in low multidimensional space. The R-Tree and its variants use rectilinear bounding boxes. The use of other geometric shapes as bounding boxes, such as circles/spheres, has led to the development of similar index schemes such as the SR-Tree.[46]

6.5.2 Multidimensional Direct Access Methods

The conceptualized optimal multidimensional access method is one that can be correctly defined as a *dynamic order-preserving multidimensional extendible hashing method*. The idea is that the location where a record is stored is derived by a simple computation; the utilized data space of the dataset will grow and shrink with insertions and deletions of data items, and accessing records in consecutive key order should be just as efficient as a simple sequential scan of the data items given the first key value. Such a storage scheme is impossible to realize. However, numerous close approximations to it have been realized. Notable among these is the *Grid-File*.[62] Other related multidimensional storage schemes are the optimal partial-match retrieval method,[2] the multidimensional extendible hashing,[69] and the BANG-file.[32] Other similar methods are also presented in References 33 and 79.

Consider the mapping of the dataset of Figure 6.7, as points in a two dimensional space shown in Figure 6.8. The mapping is based on the Y- and X-values. A two-dimensional Grid-File partitions the space rectilinearly into a first-level structure of a grid directory. The Y-values define a Y-axis that is split into I_Y segments. Similarly the X-values define an X-axis that is split into I_X segments. The grid directory is comprised of an array of $I_Y \times I_X$ elements. An element of the grid directory array stores the bucket address of the data buckets where the data item is actually stored. Given a key value $\langle y, x \rangle$, the y-value is mapped onto a y-coordinate value i_y, and the x-value is mapped onto an x-coordinate value i_x. A lookup of the grid-directory array entry $\langle i_y, i_x \rangle$ gives the bucket address where the record can be found or inserted. Since grid-directory and data buckets are disk resident, accessing an element using the grid directory requires at most two disk accesses. The basic idea, illustrated with a two-dimensional key space, can easily be generalized to arbitrary number of dimensions. For a dataset of n-buckets, a partial-match retrieval that specifies *s* out of *k*-dimensional values can be performed in $O(n^{1-s/k} + r)$ time, where *r* is the number of records in the response set.

6.5.3 Hybrid Indexing Methods

The term *hybrid index* refers to index structures that are composed of two or more access structures such as hashing (or direct access method), tree-based

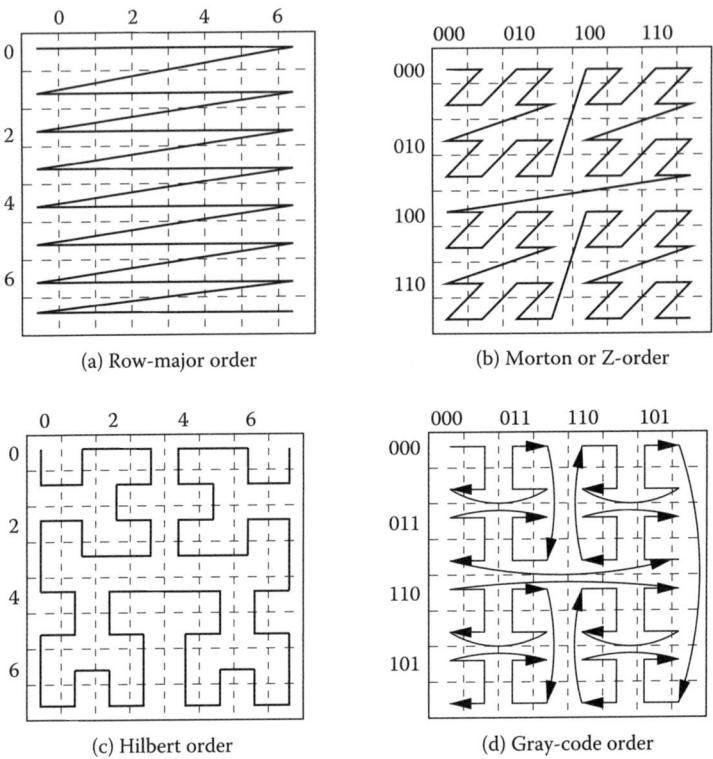

Figure 6.11 Examples of two-dimensional space curves.

indexes, and simple sequential scans. In the preceding section on *Grid-File*, we saw that in the two-dimensional grid-directory mapping, the key value of the form $\langle y,x \rangle$ is first translated into a $\langle i_y,i_x \rangle$ coordinate index that is used as an index of an array entry. The use of the $\langle i_y,i_x \rangle$ coordinate index is actually a mapping onto a one-dimensional array, which in a two-dimensional array happens to be either a row-major or a column-major addressing method. In general, given a k-dimensional index $K = \langle i_0,i_1 \ldots i_{k-1} \rangle$, one can generate a one-dimensional mapping denoted as I_K, of the k-dimensional key, and then use this in any of the tree-based index schemes described in the preceding sections. There are different methods by which one-dimensional mapping can be formed. These are formally referred to as *space-filling curves*.[78] Figure 6.11 gives some typical space-filling curves generated from a two-dimensional index and mapped onto one-dimensional codes.

The most popular of the space-filling curves that have been used in numerous applications is the Morton or Z-order mapping. It is generated simply by k-cyclic interlacing of the bits of the binary representation of the k-dimensional index of the data-space. Its simplicity, and the fact that it is consistent with the *linear quad-tree* encoding of the space, makes it one of the most widely used encoding methods in applications.[34,52,71,79] In Chapter 9, Z-order mapping is used for hierarchical indexing of multi-resolution data. Alternative methods to the Z-order encoding are the Hilbert order encoding[57] and the Gray-Code encoding.[79] In the table shown in Figure 6.7, column 4 shows the linear quad-code (or *Z-order* code), generated from an 8×8 grid partitioning of the data space. Note that linear quad-code is a base k string of digits formed by taking k-bits of the Z-order codes. The Gray-Code encodings of the points are shown in column 5. Figures 6.12a and 6.12b depict the spatial representations and code labels of points in a two-dimensional data space for the Z-order and Gray-Code encoding, respectively.

6.5.3.1 Quaternary/Hierarchical Triangular Mesh

While most database applications deal with planar and hyper-rectilinear regions, some large scientific applications deal with datasets that lie on spherical surfaces. Examples of these are datasets from climate models, GIS, and astronomy. The approach to indexing regions on spherical surfaces is similar to the method of linear quad-code or Morton-sequence order encoding of planar regions. The basic idea is to approximate the sphere by a base *platonic solid* such as a tetrahedron, hexahedron (or cube), octahedron, icosahedron, and so forth. If we consider, say, the octahedron, a spherical surface is approximated at level 0 by eight planar triangular surfaces. By bisecting the midpoints of each edge and pushing the midpoints along a ray from the center of the sphere that passes through the midpoint to the surface, 4×8 triangular surfaces are generated at level 1. Using such a recursive edge bisection procedure, the solid formed from the patches of triangular planes approximates closer and closer to the sphere. The process is depicted in Figure 6.13. Two base platonic solids that have been frequently used in such approximation schemes of the sphere are the octahedron and the icosahedron shown in Figure 6.14. Consider the use of an inscribed octahedron as the base solid. Indexing a point on a spherical surface, at any level of the tessellation, amounts then to indexing its projection on the triangular patch at the level where the point lies. If we now pair the triangular upper and lower triangular patches to form four quadrants, then the higher-level tessellation of each quadrant is similar to the higher-level tessellation of planar regions that can now be encoded using any of the space-filling curve encoding schemes. In Figure 6.14a we illustrate such a possible encoding with the Z-order encoding. Variations of the technique just described, according to whether the base-inscribing platonic solid is either an octahedron, cube, or an icosahedron, and the manner of labeling the triangular regions, form the basis of the various techniques known as the Quaternary Triangular Mesh (QTM), Hierarchical Triangular Mesh, Semi-Quadcode (SQC), and so forth.[7,8,53,79,93]

206 Scientific Data Management

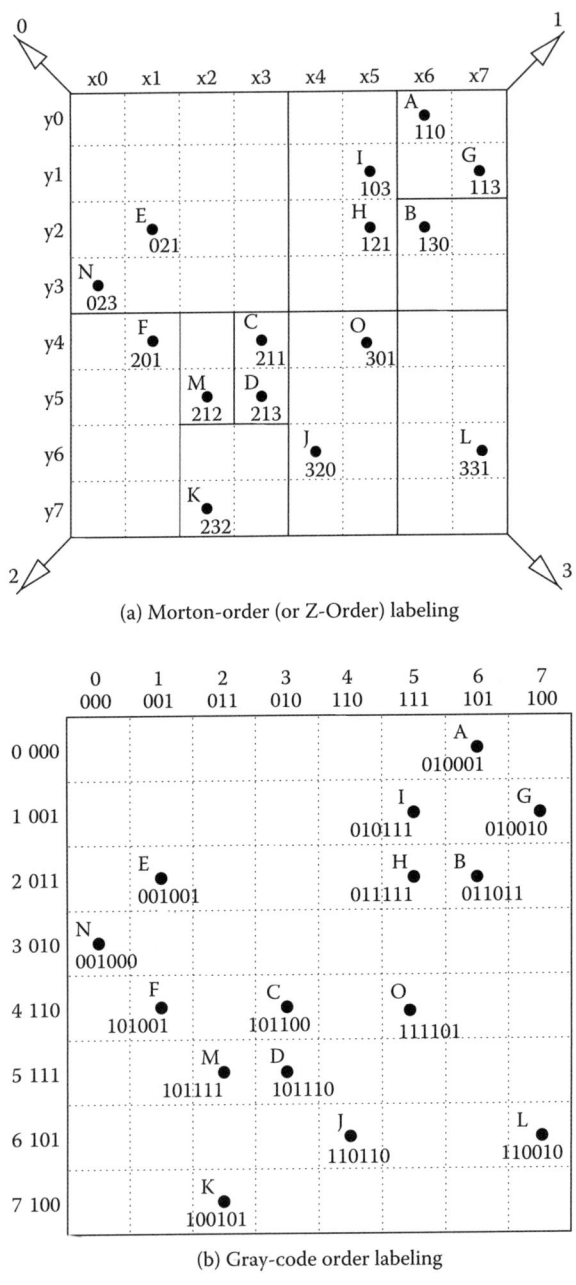

Figure 6.12 Hybrid indexing by space-filling curves of 2D dataset.

Accelerating Queries on Very Large Datasets

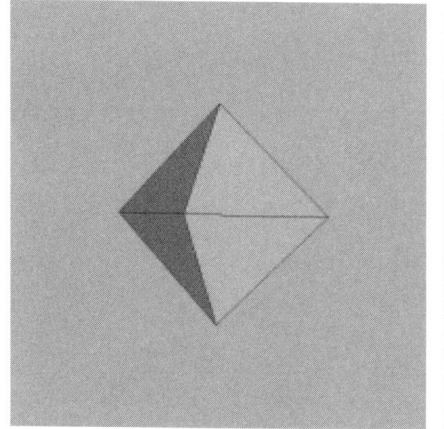
(a) Inscribed base octahedron

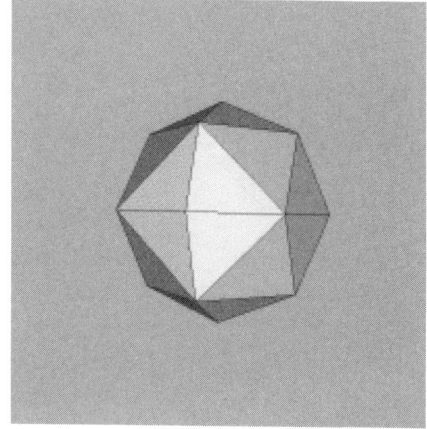
(b) Level 1 hierarchical tessellation

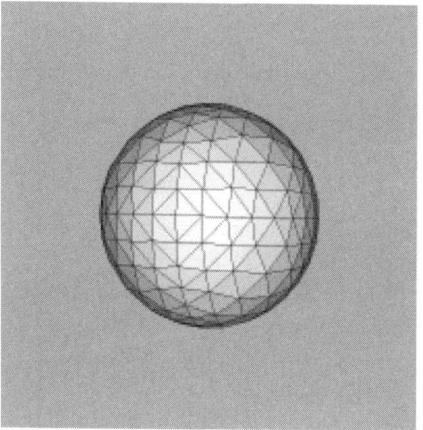
(c) Level 3 hierarchical tessellation

Figure 6.13 Spherical indexing with HTM (see http://www.sdss.jhu.edu/htm).

6.5.4 Simple Scanned Access of Multidimensional Datasets

Most of the multidimensional access methods described in preceding sections are only suitable for low-dimensional datasets of the order of $k \leq 8$. For datasets with higher dimensionality the *curse of dimensionality* sets in. The term *curse of dimensionality* was coined by Bellman to describe the problem caused by the exponential increase in volume associated with adding dimensions to a metric space.[10] This has two main implications on an indexing method. As the number of dimensions of data increases, the index size increases as well. Such indexes are usually only effective for exact-match queries and range queries where all indexed dimensions are used; if only a few of

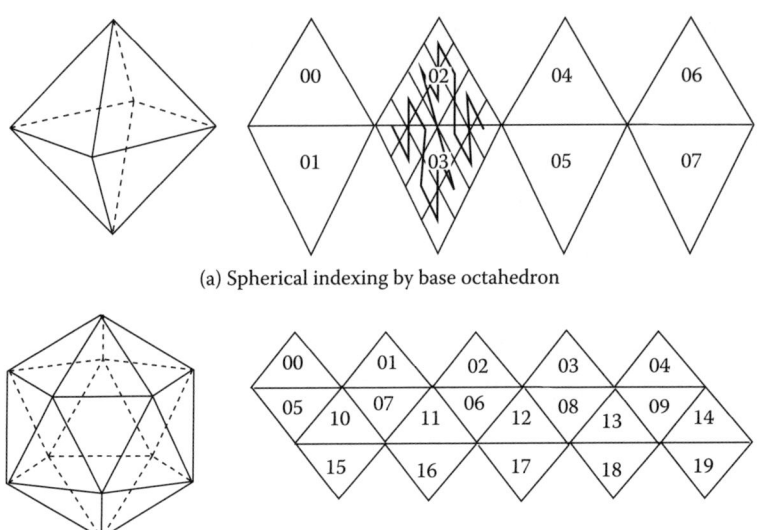

(a) Spherical indexing by base octahedron

(b) Spherical indexing by base icosahedron

Figure 6.14 Spherical indexing with quaternary/hierarchical triangular mesh.

the indexed dimensions are used, the effectiveness of the index deteriorates dramatically.

Many index methods have been proposed to address the curse of dimensionality, including the well-known X-Tree[13] and pyramid tree.[12] These methods typically address the index size issue, but fail to address the performance issue on partial-range queries. For both the X-Tree and the pyramid tree, where a k-dimensional data structure is partitioned into n buckets, the complexity of processing a partial-range query that specifies s out of k dimensions, is $O(n^{1-s/k} + r)$. If $s = d$, which is the case for an exact-match query or a full range query, $O(1)$ buckets are accessed. If k is large and $s = 1$, nearly all the pages are accessed. This prompted the question of whether a simple sequential scan is satisfactory for high-dimensional datasets.[14,82] In the case of similarity searches in high-dimensional space, one effective method is the use of sequential scan but with the attribute values mapped onto fixed length-strings of about 64~128 bits. The method is termed the VA-File approach.[97] Other, more efficient, methods are the bitmap indexes described in the next section.

6.6 Bitmap Index

In this section, we separately review the current work on bitmap indexes because they are more effective in accelerating query processing on large scientific datasets than other techniques reviewed earlier. The bitmap indexes

Accelerating Queries on Very Large Datasets 209

RID	A	bitmap index			
		=0	=1	=2	=3
1	0	1	0	0	0
2	1	0	1	0	0
3	2	0	0	1	0
4	2	0	0	1	0
5	3	0	0	0	1
6	3	0	0	0	1
7	1	0	1	0	0
8	3	0	0	0	1
		b_1	b_2	b_3	b_4

Figure 6.15 An illustration of the basic bitmap index for a column **A** that can only take on four distinct values from 0 to 3. Note: RID = row identifiers.

are generally efficient for answering queries. In fact, certain compressed bitmap indexes are known to have the theoretically optimal computational complexity.[103] They are relatively compact compared with common implementations of B-Trees, and they scale well for high-dimensional data and multidimensional queries. Because they do not require the data records to be in any particular order, they can easily take on data with any organization to improve the overall data processing task beyond the querying step.

In this section, we explain the key concept of bitmap index and review three categories of techniques for improving bitmap indexes: compression, encoding, and binning. We end this section with two case studies on using a particular implementation of bitmap indexes called *FastBit*.*

6.6.1 The Basic Bitmap Index

Figure 6.15 shows a logical view of the basic bitmap index. Conceptually, this index contains the same information as a B-tree[24,65]. The key difference is that a B-tree would store a list of row identifiers (RIDs) for each distinct value of column **A**, whereas a bitmap index represents the same information as sequences of bits, which we call *bitmaps*. In this basic bitmap index, each bitmap corresponds to a particular value of the column. A bit with value 1 indicates that a particular row has the value represented by the bitmap. What is required here is a durable mapping from RIDs to positions in the bitmaps.[63]

The mapping used by the first commercial implementation of bitmap index[65] is as follows. Let m denote the maximum number of rows that can fit on a page; assign m bits for each page in all bitmaps. The first record in a page is represented by the first bit assigned for the page, the second record by the second bit, and so on. If a page contains less than m records, then the unused bitmaps in the bitmap are left as 0. An additional existence bitmap may be

---
*FastBit software is available from https://codeforge.lbl.gov/projects/fastbit/.

used to indicate whether a bit position in a bitmap is used or not. Such an existence bitmap may also be used to indicate whether a particular record has been deleted without recreating the whole bitmap index. This mapping mechanism is robust to changes and can be applied to all bitmap indexes of a table.

In most scientific applications, data records are stored in densely packed arrays[35,49,59]; therefore, a more straightforward mapping between the RIDs and positions in bitmaps can be used. Furthermore, most scientific data contain only fixed-sized data values, such as integers and floating-point values, and are stored in multidimensional arrays. In these cases, the array index is a durable mapping between bit positions and data records. Usually such RIDs are not stored explicitly.

The bitmap indexes are particularly useful for query-intensive applications, such as data warehousing and on-line analytical processing (OLAP).[66,112] One of the key reasons is that queries can be answered with bitwise logical operations on the bitmaps. In the example shown in Figure 6.15, a query "$A < 2$" can be answered by performing bitwise OR on b_1 and b_2 ($b_1 \mid b_2$). Since most computer hardware support such bitwise logical operations efficiently, the queries can be answered efficiently in general. Another key reason is that answers from different bitmap indexes can be easily combined. This is because the answers from each bitmap index are a bitmap, and combining the different answers simply requires additional bitwise logical operations. Because combining answers from different indexes efficiently is such an important consideration, a number of DBMS that do not support bitmap indexes, such as PostgreSQL and MS SQL server, even convert intermediate solutions to bitmaps to combine them more effectively.

Because results from different indexes can be efficiently combined, a bitmap index is built for one column only, and composite bitmap indexes for multiple columns are rarely used. This simplifies the decisions on what indexes to build because one does not need to consider composite indexes. This also simplifies the query optimization because there are fewer indexes to consider.

The biggest weakness of the basic bitmap index is that its size grows linearly with the number of distinct values of the column being indexed. Next, we review three sets of strategies to control the index sizes and improve the query response time, namely, compression, encoding, and binning.

6.6.2 Compression

Each individual bitmap in a bitmap index can be compressed with a data compression method.[60] Any lossless compression may be used. However, the specialized bitmap compression methods typically offer faster bitwise logical operations and consequently faster query response time.[3,45] The most widely used bitmap compression method is byte-aligned bitmap code (BBC).[4,5] More recently, another bitmap compression method called word-aligned hybrid (WAH) code was shown to perform bitwise logical operations more than 10 times faster than BBC.[107,109]

WAH gains its speed partly from its simplicity. For long sequences of 0s or 1s, it uses run-length encoding to represent them, and for relatively short sequences of mixed 0s and 1s, it represents the bits literally. Hence, it is a hybrid of two methods. Another key feature that enables it to achieve performance is that the compressed data are word aligned. More specifically, WAH-compressed data contains two types of words: *literal* words and *fill* words.

A *literal* word contains one bit to indicate its type and uses the remaining bits to store the bitmap literally. On a 32-bit word system, it may use the most significant bit to indicate the type of the word, and use the remaining 31 bits to store the bitmap. A fill word similarly needs 1 bit to indicate its type. It uses another bit to indicate whether the bits are all 0s or all 1s, and the remaining bits are used to store the number of bits in a bitmap it represents. The number of bits represented by a WAH fill word is always a multiple of the number of bits stored in a literal word. Therefore, the length of a fill is stored as this multiple instead of the actual number of bits. For example, a fill of 62 bits will be recorded as being of length 2 because it is two times the number of bits that can be stored in a literal word (31). This explicitly enforces the word-alignment requirement and allows one to easily figure out how many literal words a fill word corresponds to during a bitwise logical operation. Another important property is that it allows one to store any fill in a single fill word as long as the number of bits in a bitmap can be stored in a word. This is an important property that simplifies the theoretical analysis of WAH compression.[103] An illustration of WAH compression on a 32-bit machine is shown in Figure 6.16a.

Figure 6.16b shows some examples to illustrate the effects of compression on overall query response time. In this case the commercial DBMS implementation of compressed bitmap index (marked as "DBMS bitmap index") uses BBC compression, while "FastBit index" uses WAH compression. The query response times reported are average time values over thousands of ad hoc range queries that produce the same number of hits. Over the whole range of different numbers of hits, the WAH compressed indexes answer queries about 14 times faster than the commercial bitmap indexes.

In addition to being efficient in timing measurements, WAH-compressed basic bitmap index is also theoretically optimal. In the worst case, the query response time is a linear function of the number of hits according to our analysis in References 102 and 103. A few of the best B-tree variants have the same theoretical optimality as the WAH-compressed bitmap index.[24] However, bitmap indexes are much faster in answering queries that return more than a handful of hits as illustrated in Figure 6.16. Since the basic bitmap index contains the same information as a typical B-tree, it is possible to switch between bitmaps and RID lists to always use the more compact representation as suggested in the literature.[65] This is an alternative form of compression that was found to perform quite well in a comparison with WAH-compressed indexes.[63]

The bitmap compression methods are designed to reduce the index sizes, and they are quite effective at this. Discussions on how each compression

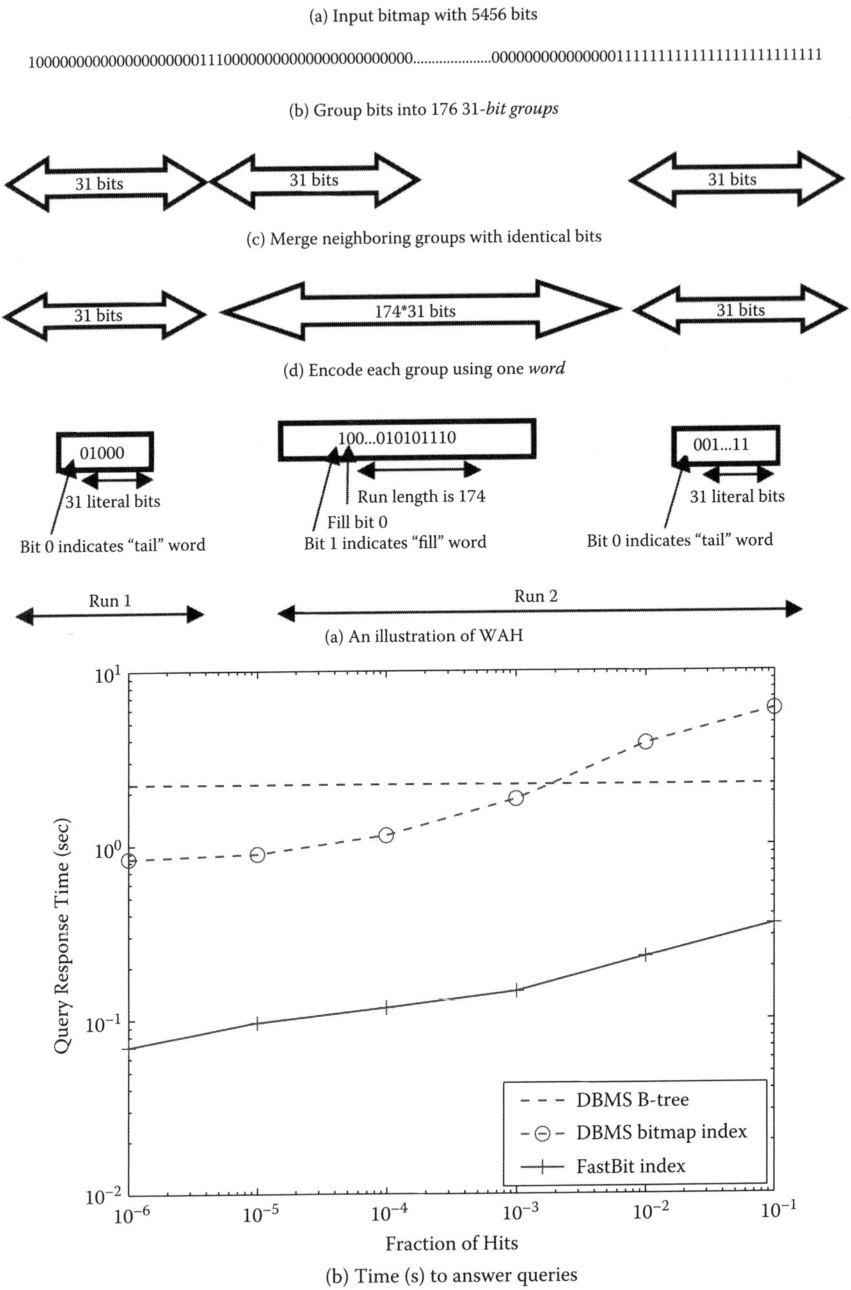

(a) Input bitmap with 5456 bits

1000000000000000000001110000000000000000000000...........0000000000000001111111111111111111111111

(b) Group bits into 176 31-*bit groups*

⟵ 31 bits ⟶ ⟵ 31 bits ⟶ ... ⟵ 31 bits ⟶

(c) Merge neighboring groups with identical bits

⟵ 31 bits ⟶ ⟵ 174*31 bits ⟶ ⟵ 31 bits ⟶

(d) Encode each group using one *word*

| 01000 | 100...010101110 | 001...11 |

31 literal bits / Run length is 174, Fill bit 0 / 31 literal bits
Bit 0 indicates "tail" word / Bit 1 indicates "fill" word / Bit 0 indicates "tail" word

⟵ Run 1 ⟶ ⟵ Run 2 ⟶

(a) An illustration of WAH

(b) Time (s) to answer queries

Figure 6.16 The effects of compression on query response time. The faster WAH compression used in FastBit reduces the query response time by an order of magnitude. (Illustration adapted from Stockinger, 16, and 2006. *Bitmap indicies for data warehouses.* Idea Group, Inc. used with permission from TGI Global.[89])

method controls the index size are prominent in many research articles. Since there is plenty of information on the index sizes, we have chosen to concentrate on the query response time. Interested readers can obtain more information on discussions of the index sizes in References 45 and 109.

6.6.3 Bitmap Encoding

Bitmap encoding techniques can be thought of as ways of manipulating the bitmaps produced by the basic bitmap index to either reduce the number of bitmaps in an index or reduce the number of bitmaps needed to answer a query. For example, to answer a range query of the form "**A** < 3" in the example given in Figure 6.15, one needs to OR the three bitmaps b_1, b_2, and b_3. If most of the queries involve only one-sided range conditions as in this example, then it is possible to store C bitmaps that correspond to $\mathbf{A} \leq a_i$ for each of the C distinct values of **A**. We call C the *column cardinality* of **A**. Such a bitmap index would have the same number of bitmaps as the basic bitmap index, but can answer all one-sided range queries by reading one bitmap. This is the *range encoding* proposed by Chan and Ioannidis.[21] The same authors also proposed another encoding method called the *interval encoding* that uses about half as many bitmaps as the basic bitmap index, but answers any range queries with only two bitmaps.[22] The encoding used in the basic bitmap index is commonly referred to as the *equality encoding*. Altogether, there are three basic bitmap encoding methods: equality, range, and interval encodings.

The three basic bitmap encodings can be composed together to form two types of composite encoding schemes: multicomponent encoding[21,22] and multilevel encoding.[86,108] The central idea of multicomponent encoding is to break the key value corresponding to a bitmap into multiple components in the same way an integer number is broken into multiple digits in a decimal representation. In general, each "digit" may use a different basis size. For an integer attribute with values between 0 and 62, we can use two components of basis sizes 7 and 9, and index each component separately. If the equality encoding is used for both components, then we use 7 bitmaps for one component and 9 bitmaps for the other. Altogether we use 16 bitmaps, instead of 63 had the equality encoding been used directly on the key values. It is easy to see that using more components can reduce the number of bitmaps needed, which may reduce the index size. To carry this to the extreme, we can make all base sizes 2 and use the maximum number of components. This particular multicomponent encoding can be optimized to be the binary encoding,[100] which is also known as the bit-sliced index.[64,66] This encoding produces the minimum number of bitmaps, and the corresponding index size is the smallest without compression. A number of authors have explored different strategies of using this type of encoding.[20,112]

To answer a query using a multicomponent index, all components are typically needed, therefore, the average query response time may increase with the number of components. It is unclear how many components would offer the best performance. A theoretical analysis concluded that two components offer the best space-time trade-off.[21] However, practitioners have stayed away from two-component encodings; existing commercial implementations either use one-component equality encoding (the basic bitmap index) or the binary encoding. This discrepancy between the theoretical analysis and the current best practice is because the analysis has failed to account for compression, which is a necessary part of a practical bitmap index implementation.

A multilevel index is composed of a hierarchy of nested bins on a column. Since a level in such an index is a complete index on its own, a query may be answered with one or a combination of different levels of a multilevel index. Therefore, this type of composite index offers a different type of trade-off than the multicomponent index.[86,108] We will give more detailed information about the multilevel indexes in the next subsection after we explain the basic concept of binning.

Because of the simplicity of WAH compression, it is possible to thoroughly analyze the performance of WAH-compressed indexes.[110] This analysis confirms the merit of the basic equality encoding and the binary encoding. Among the multilevel encodings, the new analysis reveals that two levels are best for a variety of parameter choices. More specifically, it identifies three two-level encoded indexes that have the same theoretical optimality as the WAH-compressed basic index, but can answer queries faster on average. Figure 6.17

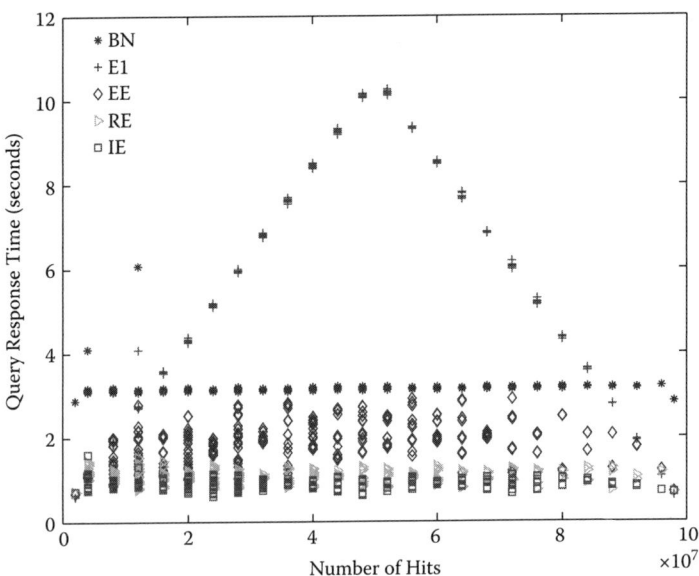

Figure 6.17 (See color insert following page 224.) The query response time of five different bitmap encoding methods with WAH compression (BN: binary encoding, E1: the basic one-component equality encoding, EE: two-level equality-equality encoding, RE: two-level range-equality encoding, IE: two-level interval-equality encoding).

shows some timing measurements to support the analysis. In this case, we see that two-level encodings (equality-equality encoding EE, range-equality encoding RE, and interval-equality encoding IE) can be as much as ten times faster than the basic bitmap index (marked E1). On the average, the two-level encoded indexes are about three to five times faster than both the basic bitmap index and the binary encoded index (BN).[110]

6.6.4 Binning

Scientific data often contains floating-point values with extremely high column cardinality. For example, the temperature and pressure in a combustion simulation can take on a large range of possible values, and each value rarely repeats in a dataset. The basic bitmap index will generate many millions of bitmaps in a typical dataset. Such indexes are typically large and slow to work, even with the best compression and bitmap encoding. We observed that such precise indexing is often unnecessary since the applications do not usually demand full precision. For example, a typical query involving pressure is of the form "pressure $> 2 \times 10^7$ Pascal." In this case, the constant in the query expression has only one significant digit. Often, such constants have no more than a few significant digits. One may take advantage of this observation and significantly reduce the number of bitmaps used in a bitmap index.

In general, the technique of grouping many values together is called *binning*.[48,83,88,105] The values placed in a bin are not necessarily consecutive values.[48] However, the most common forms of binning always place values from a consecutive range into a bin. For example, if the valid pressure values are in the range between 0 and 10^9, we may place values between 0 and 1 in the first bin, values between 1 and 10 in the second bin, values between 10 and 100 in the third bin, and so on. This particular form of binning is commonly known as logarithmic binning. To produce a binned index that will answer all range conditions using one-digit query boundaries, we can place all values that round to the same one-digit number into a bin.*

A simple way to divide all pressure values between 0 and 10^9 into 100 bins would be to place all values between $i \times 10^7$ and $(i+1) \times 10^7$ in bin i. We call them *equal-width bins*. Since each equal-width bin may contain a different number of records, the corresponding bitmaps will have varying sizes, and the amount of work associated with them will be different too. One way to reduce this variance is to make sure that each bin has the same number of records. We call such bins *equal-weight bins*. To produce such equal-weight bins, we need to first find the number of occurrences for each value. Computing such detailed histograms may take a long time and a large amount of memory, because there may be a large number of distinct values. We can sample the data to produce

---

*A caveat: We actually split all values that round to a low-precision number $\bar{x}$ into two bins: one for those that round up to $\bar{x}$ and one for those that round down to $\bar{x}$.

an approximate histogram, or produce a set of fine-grain, equal-width bins and coalesce the fine bins into the desired number of nearly equal-weight bins.

The multilevel bitmap indexes are composed of multiple bitmap indexes, each one corresponding to a different granularity of binning. To make it easier to reuse information from different levels of binning, we ensure that the bin boundaries from coarser levels are a subset of those used for the next finer level of bins. This generates a hierarchy of bins. To minimize the average query processing cost, the multilevel bitmap indexes mentioned in the previous subsection always use equal-weight bins for the coarse levels. These indexes all use two levels of bins, with the fine level having one bitmap for each distinct value. We consider such indexes to be *precise* indexes because they can answer any queries with the bitmaps, without accessing the base data.

Even though binning can reduce the number of bitmaps and improve the query response time in many cases, for some queries, however, we have to go back to the base data to answer the queries accurately. For example, if we have 100 equal-width bins for pressure between 0 and 10^9, then the query condition "pressure > 2.5×10^7" can be resolved with the index only. We know bins 0 and 1 contain records that satisfy the query condition, and bins 3 and onward contain records that do not satisfy the condition, but we are not sure which records in bin 2 satisfy the condition. We need to examine the actual values of all records in bin 2 to decide. In this case, we say that bin 2 is the *boundary bin* of the query and call the records in bin 2 *candidates* of the query. The process of examining the raw data to resolve the query accurately is called *candidate checking*. When a candidate check is needed, it often dominates the total query response time. There are a number of different approaches to minimize the impact of candidate checks. One approach is to reorder the expression being evaluated to minimize the overall cost of candidate checks.[88] Another approach is to place the bin boundaries to minimize the cost of evaluating a fixed set of queries.[48,75–77]

Both approaches mentioned above do not actually reduce the cost of a candidate-checking operation. More recently, a new approach was proposed to do just that.[111] It does so by providing a clustered copy named *order-preserving bin-based clustering* (OrBiC) of the base data. Since the values of all records in a bin are organized contiguously, the time needed for a candidate-checking operation is minimized. In tests, this approach was shown to significantly outperform the unbinned indexes. Figure 6.18 shows some performance numbers to illustrate the key advantages of the new approach. In Figure 6.18a, we see that the binned index with OrBiC outperforms the one without OrBiC for all query conditions tested. In Figure 6.18b, we see how the advantage of OrBiC varies with the number of bins. Clearly, we see that the advantage of OrBiC is significantly affected by the number of bins used. The analysis provided by the authors can predict the optimal number of bins for simple types of data,[111] but additional work is needed to determine the number of bins for more realistic data.

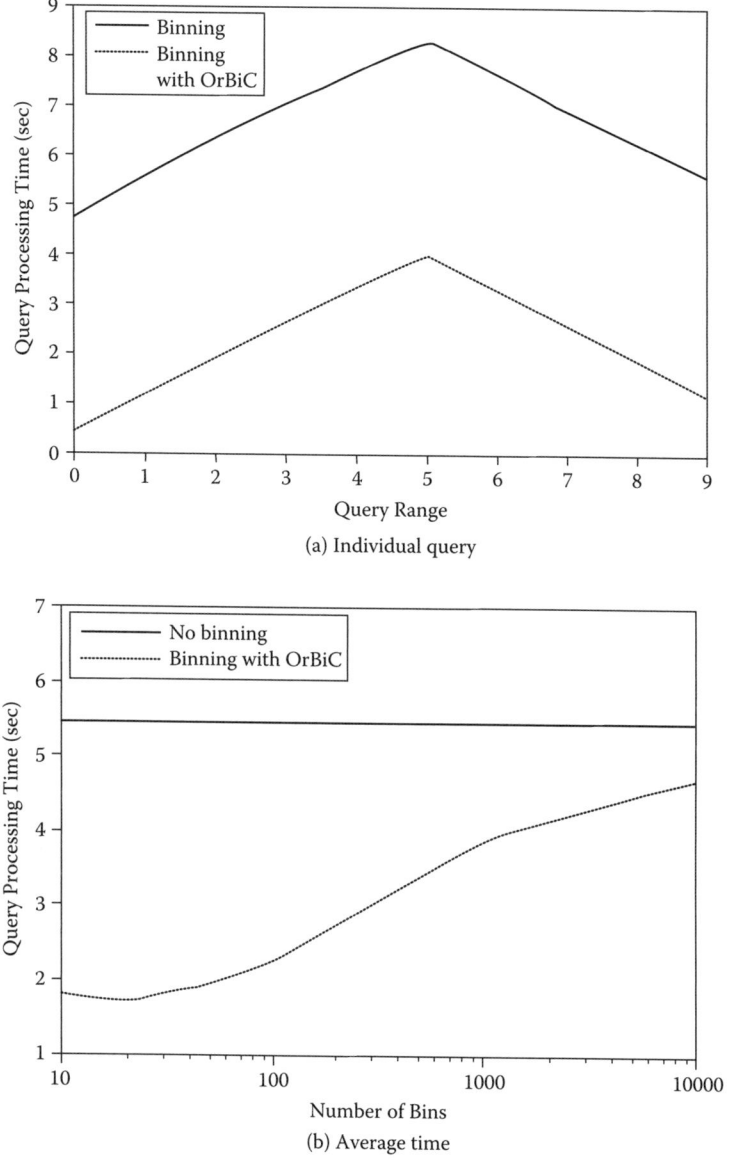

Figure 6.18 Time needed to process range queries.

6.6.5 Implementations

The first commercial implementation of a bitmap index was in Model 204 from the early 1980s,[65] and it is still available as a commercial product from Computer Corporation of America. A number of popular DBMS products have since implemented variants of bitmap index. For example, ORACLE has a BBC compressed bitmap index, IBM DB2 has the Encoded Vector Index, IBM Informix products have two versions of bitmap indexes (one for low-cardinality data and one for high-cardinality data), and Sybase IQ data warehousing products have two versions of bitmap indexes as well. These bitmap index implementations are either based on the basic bitmap index or the bit-sliced index, which are the two best choices among all multicomponent bitmap indexes.[110]

There are a number of research prototypes with numerous bitmap indexes.[63,106] In particular, FastBit is freely available for anyone to use and extend. We next briefly describe some of the key features of the FastBit software.

FastBit is distributed as C++ source code and can be easily integrated into a data processing system. On its own, it behaves as a minimalistic data warehousing system with column-oriented data organization. Its strongest feature is a comprehensive set of bitmap indexing functions that include innovative techniques in all three categories discussed above. For compression, FastBit offers WAH as well as the option to uncompress some bitmaps. For encoding, FastBit implements all four theoretically optimal compressed bitmap indexes in addition to a slew of bitmap encodings proposed in the research literature. For binning, it offers the unique low-precision binning as well as a large set of common binning options such as equal-width, equal-weight, and log-scale binning. Because of the extensive indexing options available, it is a good tool for conducting research in indexing. In 2007, two PhD theses involving FastBit software were successfully completed, which demonstrated the usefulness of FastBit as a research tool.[72,85] FastBit has also been successfully used in a drug screening software, TrixX-BMI, and was shown to speed up virtual screening by 12 times on average in one case and hundreds of times in another.[80] The chapter on visualization, Chapter 9, describes another application of using FastBit for network traffic analysis. Later in Section 6.7.3 we will briefly describe another application of using FastBit in analysis of high-energy physics data.

6.7 Data Organization and Parallelization

In this section, we briefly review a number of data management systems to discuss the different aspects of data organization and their impact on query performance. Since many of the systems are parallel systems, we also touch on the issue of parallelization. Most of the systems reviewed here do not have extensive indexing support. We also present a small test comparing one of these systems against FastBit to demonstrate that indexing could improve the query performance. Finally, we discuss the Grid Collector as an example of a smart iterator that combines indexing methods with parallel data processing to significantly speed up large-scale data analysis.

6.7.1 Data Processing Systems

To access data efficiently, the underlying data must be organized in a suitable manner, since the speed of query processing depends on the data organization. In most cases, the data organization of a data processing system is inextricably linked to the system design. Therefore we cannot easily separate the data organization issue from the systems that support them. Next, we review a few example systems to see how their data organization affects the query processing speed. Since most of the preceding discussion applies to the traditional DBMS systems, we will not discuss them any further.

6.7.1.1 Column-Based Systems

The column-based systems are extensively discussed in Chapter 7. Here, we will only mention some names and give a brief argument on their effectiveness.

There are a number of commercial database systems that organize their data in column-oriented fashion, for example, Sybase IQ, Vertica, and Kx Systems.[98] Among them, Kx Systems can be regarded as an array database because it treats an array as a first-class citizen like an integer number. There are a number of research systems that use vertical data organization as well, for example, C-Store,[90,91] MonetDB,[16,17] and FastBit. One common feature of all these systems is that they logically organize values of a column together. This offers a number of advantages. For example, a typical query only involves a small number of columns; the column-oriented data organization allows the system to only access the columns involved, which minimizes the I/O time. In addition, since the values in a column are of the same type, it is easier to determine the location of each value and avoid accessing irrelevant rows. The values in a column are more likely to be the same as values from different columns in row-oriented data organization, which makes it more effective to apply compression on data.[1]

6.7.1.2 Special-Purpose Data Analysis Systems

Most of the scientific data formats such as FITS, NetCDF, and HDF5 come with their own data access and analysis libraries, and can be considered as special-purpose data analysis systems. By far the most developed of such systems is ROOT.[18,19,70] ROOT is a data management system developed by physicists originally for high-energy physics data. It currently manages many petabytes of data around the world, more than many of the well-known commercial DBMS products. ROOT uses an object-oriented metaphor for its data: a unit of data is called an object or an event (of high-energy collision), which corresponds to a row in a relational table. The records are grouped into files, and the primary access method to records in a file is to iterate through them with an iterator. Once an event is available to the user, all of its attributes are available. This is essentially the row-oriented data access. In recent versions of ROOT, it is possible to split some attributes of an event to store them separately. This provides a means to allow for column-oriented data access.

ROOT provides an extensive set of data analysis frameworks, which makes analyses of high-energy physics data convenient and interactive. Its interpreted C++ environment also offers the possibility of infinitely complex analysis that some users desire. Since each ROOT file can be processed independently, the ROOT system also offers huge potential for parallel processing on a cluster of commodity computers. This is a nice feature that enabled the cash-strapped physicists to effectively process petabytes of data before anyone else could. The ROOT system is now being extensively used by many scientific applications, and has even gained some fans in the commercial world. More information about ROOT can be found at http://root.cern.ch/.

6.7.1.3 MapReduce Parallel Analysis System

The MapReduce parallel data analysis model has gained considerable attention recently.[27,28,50] Under this model, a user only needs to write a `map` function and a `reduce` function in order to make use of a large cluster of computers. This ease of use is particularly attractive because many other parallel data analysis systems require much more programming effort. This approach has been demonstrated to be effective in a number of commercial settings.

There are a number of different implementations of the MapReduce system following the same design principle. In particular, there is an open-source implementation from the Apache Hadoop project that is available for anyone to use. To use this system, one needs to place the data on a parallel file system supported by the MapReduce run-time system. The run-time system manages the distribution of the work onto different processors, selecting the appropriate data files for each processor, and passing the data records from the file to the `map` and `reduce` functions. The run-time system also manages the coordination among the parallel tasks, collects the final results, and recovers any errors.

The MapReduce system treats all data records as key/value pairs. The primary mechanism offer in this model is an iterator (identified by a key). Recall that the ROOT system also provides a similar iterator for data access. Another similarity is that both ROOT and MapReduce can operate on large distributed data. The key difference between ROOT and MapReduce is that the existing MapReduce systems rely on underlying parallel file systems for managing and distributing the data, while the ROOT system uses a set of daemons to deliver the files to the parallel jobs. In a MapReduce system, the content of the data is opaque to the run-time system and the user has to explicitly extract the necessary information for processing. In the ROOT system, an event has a known definition and accessing the attributes of an event therefore requires less work.

The data access mechanism provided by a MapReduce system can be considered as row-oriented because all values associated with a key are read into memory when the iterator points to the key/value pair. On structured data, for a typical query that requires only a small number of attributes, a MapReduce system is likely to deliver poorer performance than a parallel column-oriented system such as MonetDB, C-Store, or Vertica. The MapReduce system is proven effective for unstructured data.

6.7.1.4 Custom Data Processing Hardware

The speed of accessing secondary storage in the past few decades practically remains unchanged compared with the increases in the speed of main memory and CPU. For this reason, the primary bottleneck for efficient data processing is often the disk. There have been a number of commercial efforts to build data processing systems using custom hardware to more efficiently answer queries. Here we very briefly discuss two such systems: Netezza[26] and Teradata.[6,29]

Netezza attempts to improve query processing speed by having smart disk controllers that can filter data records as they are read off the physical media. In a Netezza server, there is a front-end system that accepts the usual SQL commands, so the user can continue to use the existing SQL code developed for other DBMS systems. Inside the server, an SQL query is processed on a number of different snippet processing units (SPUs), where each SPU has its own disk and processing logic. The results from different SPUs are gathered by the front-end host and presented to the user. In general, the idea of offloading some data processing to the disk controllers to make an *active storage* system could benefit many different applications.[56,73]

The most unique feature of Teradata's warehousing system is the BYNET interconnect that connects the main data access modules called AMPs (access module processors). The design of BYNET allows bandwidth among the AMPs to scale linearly with the number of AMPs (up to 4096 processors). It also is fault tolerant and performs automatic load balancing. The early versions of AMPs are similar to Netezza's "smart disk controllers;" however, the current version of AMPs are software entities that utilize commodity disk systems.

To the user, both Netezza and Teradata behave as a typical DBMS system, which is a convenience feature for the user. On disks, both systems appear to follow the traditional DBMS systems, that is, storing their data in the row-oriented organization. Potentially, using the column-oriented organization may improve their performances. Teradata has hash and B-Tree indexes, while Netezza does not use any index method.

6.7.2 Indexes Still Useful

Many of the specialized data management systems mentioned above do not employ any index method. When the analysis task calls for all or a large fraction, say one-tenth, of the records in a dataset, then having an index may not accelerate the overall data processing. However, there are plenty of cases where indexes can dramatically speed up the query processing. For example, for interactive exploration, one might select a few million records from a dataset with billions of records. Going through the whole dataset to find a few million is likely to take longer than using an effective index. Another example is in query estimation. Often, before users commit to evaluate subsets of data records, they may want to know the size of the result and the possible processing time. This task is usually better accomplished with indexes. Additionally, indexes may provide a faster way to gather some statistics. To illustrate the usefulness of an index in accomplishing such tasks, we next give a specific example of answering count queries from the Set Query Benchmark.

The Set Query Benchmark is a benchmark designed to evaluate the performance of OLAP-type applications.[63,67] The test data contains 12 columns of uniform random numbers plus a sequence number that serves as a row identifier. Our test uses queries of the form "Select count(*) From Bench Where ...," which we call count queries.[63] We use a test dataset with 100 million rows instead of the original specification of a million rows. In the test dataset, we also adjust the column cardinality of the two random columns to be 25 million and 50 million instead of 250,000 and 500,000 as the benchmark specification indicated. We note that such uniform random values with extremely high column cardinalities are the worst type of data for the compressed bitmap indexes. The table in Figure 6.19 contains the names of these count queries, Q1, Q2A, Q2B, and so on. Each of these queries has a number of different instances that involve different columns or different query conditions. Figure 6.19a shows the total time to complete all instances of a query, and Figure 6.19b shows the query response time for each instance of Q1.

The query response times are gathered from two systems: one with a compressed bitmap index and the other with compressed columns but without any index. The bitmap index is from FastBit and the indexless system is a commercial product that reorders and compresses the base data. This indexless system is advertised as the fastest data processing system. During our testing, we consulted with the vendor to obtain the best ordering and compression options available in early 2007. The two systems are run on the same computer with a 2.8 GHz Intel Pentium CPU and a small hardware RAID that is capable of supporting about 60 MB/s throughput.

The time results in Figure 6.19 clearly indicate that indexes are useful for these queries. Overall, one can answer all the queries about 11 times faster using bitmap indexes than using the compressed base data. Figure 6.19b shows some performance details that help to explain the observed differences. The horizontal axis in Figure 6.19b is the column cardinality. The best reordering strategy that minimizes the overall query response time is to order the lowest cardinality column first, then the next lowest cardinality column, and so on. More specifically, the column with cardinality 2 (the lowest cardinality) is completely sorted; when the lowest cardinality column values are the same, the rows are ordered according to the column with cardinality 4; when the first two columns have the same value, the rows are ordered according to the next

Accelerating Queries on Very Large Datasets

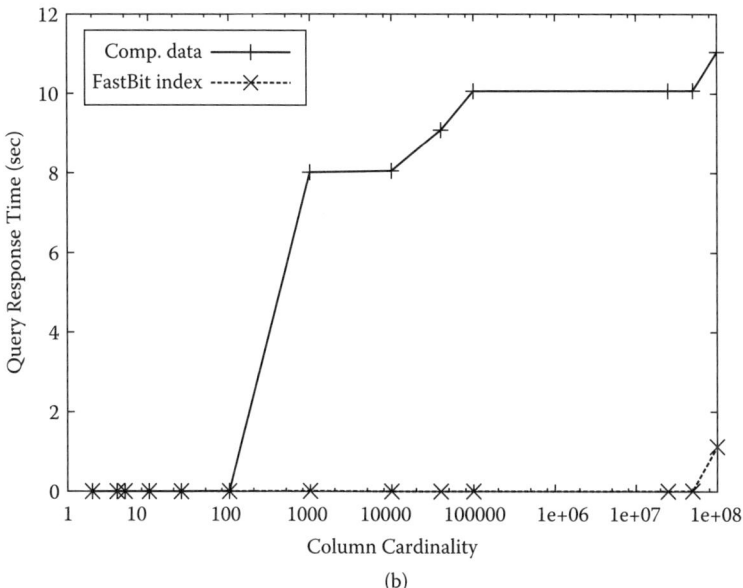

Figure 6.19 Time needed to answer count queries from the Set Query Benchmark with 100 million records: (a) total query response time; (b) time to answer-Q1 queries.

lowest cardinality column. This process continues on all columns including the column of sequence numbers. After this reordering, the first few columns are very nearly sorted and can be compressed very well. Their compressed sizes are much smaller than the original data, and Q1 queries (Select count(*) From Bench Where :col = 2, where :col is the name of a column in the test dataset) can be answered very quickly. In these cases, the time required by two test systems are about the same. However, when higher cardinality columns are involved in the queries, the indexless system requires much more time. Overall, the total time required to answer all 13 instances of Q1 is about 1.2 seconds with the FastBit indexes, but about 66 seconds with the compressed base data. The indexless system works well on queries Q4A0 and Q4B0 because these queries only involve low-cardinality columns where the compressed columns are very small.

Note that the bitmap indexes were built on the data that were reordered in the same way that the indexless system did. Had we not reordered the data and built the bitmap indexes on the data in the original order, the differences would be smaller because the reordering also helps reduce the sizes of bitmap indexes. Nevertheless, even without reordering the data, the overall performance of compressed bitmap index is about three times better than the indexless system. In short, the indexes are very useful in some applications; indexless approaches are unlikely to completely replace systems with indexes.

6.7.3 Using Indexes to Make Smart Iterators

The previous set of tests demonstrates that one can use indexes to count the number of results of queries. Next, we present another example where indexes are used to make "smart iterators" to speed up data analysis. More specifically, we present a case where FastBit is used to implement an efficient event iterator for a distributed data analysis of a large collection of high-energy physics data from STAR.

The Event Iterator was implemented for a plug-in to the STAR analysis framework called Grid Collector.[101,104] STAR is a high-energy physics experiment that collects billions of events on collisions of high-energy nuclei. The records about collisions are organized into files with a few hundred events each. Most analysis tasks in STAR go through a relatively small subset of the events (from a few thousand events to a few million events out of billions) to gather statistics about various attributes of the collisions. The basic method for analyzing the volumes of data collected by the STAR experiment is to specify a set of files and then iterate through the collision events in the files. Typically, a user program filters the events based on a set of high-level summary attributes called *tags* and computes statistics on a subset of desired events. The Grid Collector allows the user to specify the selection criteria as conditions on the tags and directly deliver the selected events to the analysis programs. Effectively, the Grid Collector replaces the existing simple iterator with a smart iterator that understands the conditions on the tags and extracts events satisfying the specified conditions for analysis. This removes the need for users to manage the files explicitly, reduces the amount of data read from the disks, and speeds up the overall analysis process as shown in Figure 6.20.

In Figure 6.20a, we show a block diagram of the key components of the Grid Collector. The main Grid Collector component can be viewed as an event catalog that contains all tags of every collision event collected by the STAR experiment. This component runs as a server and is also responsible for extracting the tag values, building bitmap indexes, resolving conditions on tags to identify relevant data files, locating these files, and coordinating with various storage resource managers[84] (see Chapter 3 for details) to retrieve the files when necessary. All of these tasks together deliver the selected events to the analysis code.

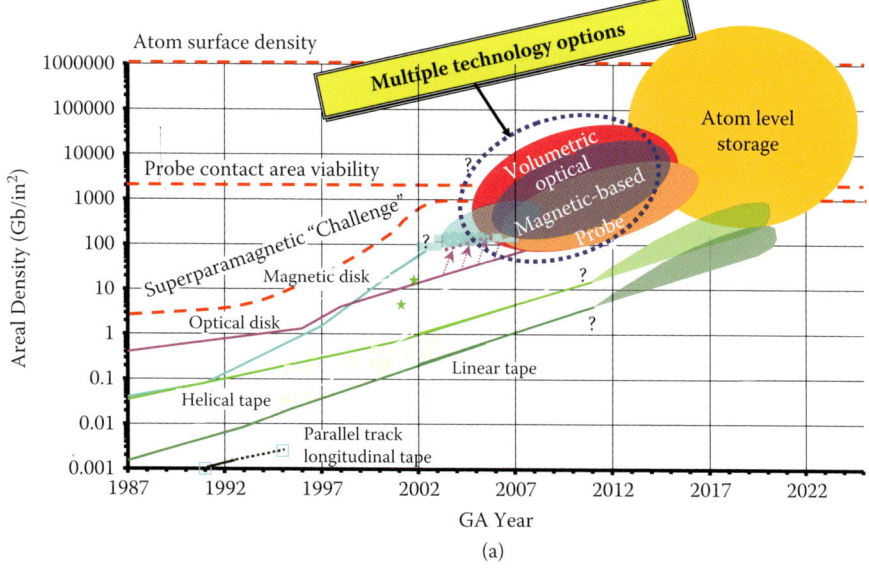

(a)

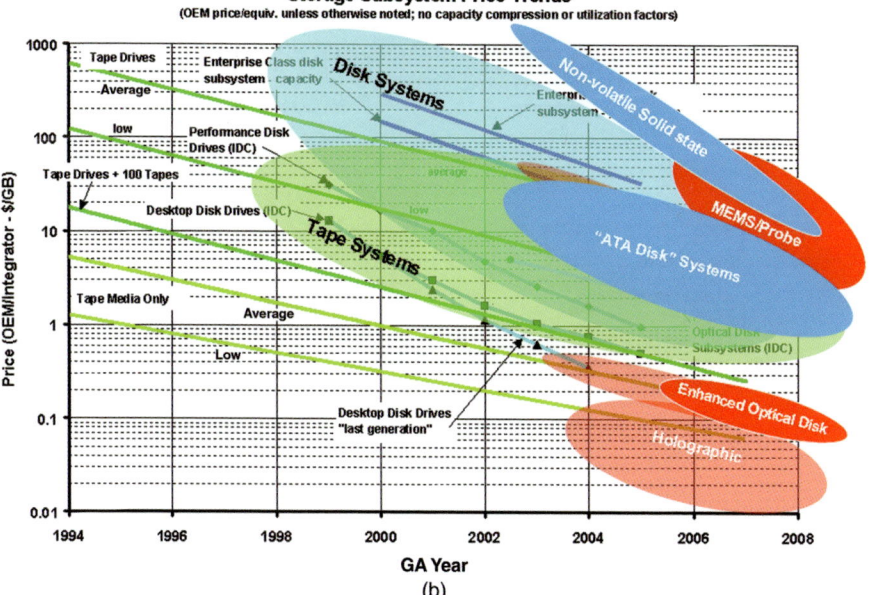

(b)

COLOR FIGURE 1.3 This figure summarizes the evolution of area density and price for storage devices (Courtesy of Sun Microsystems). Item (a) summarizes the trends in the areal density for different classes of storage media. Graph (b) shows the cost per bit for different classes of storage media.

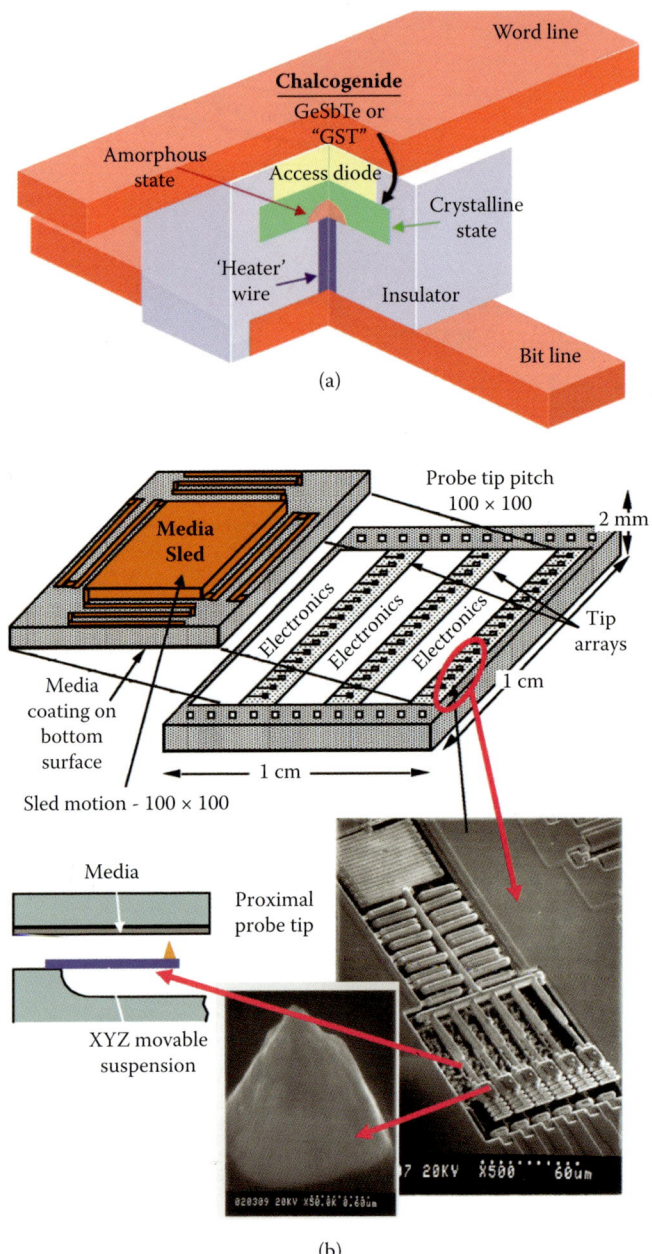

COLOR FIGURE 1.5 Part (a) shows a schematic of a single cell of a phase-change memory cell that uses electrically induced heating to induce phase changes in the embedded Chalcogenide (GeSbTe) material. Part (b) shows the architecture of the Micro-Electro-Mechanical system (MEMS)-based Millipede storage system that employs small STM heads to directly manipulate a polymer medium for storage.

COLOR FIGURE 2.3 Panasas storage clusters are built of bladeserver shelves containing object-serving StorageBlades; metadata-managing DirectorBlades; batteries; and redundant power, cooling, and networking.

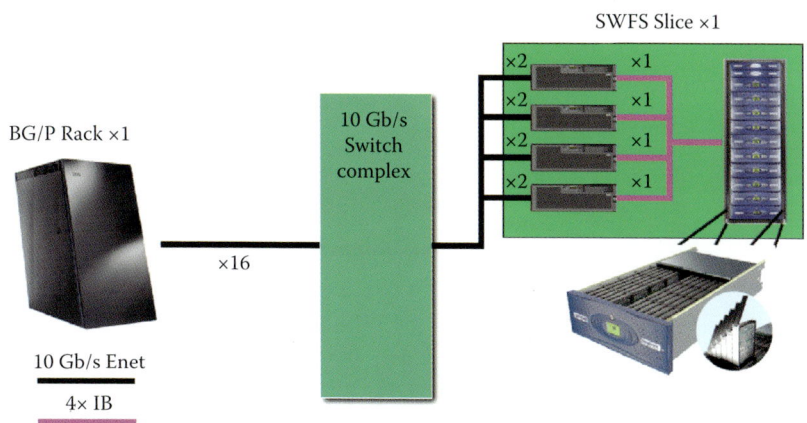

COLOR FIGURE 2.6 Each rack of Blue Gene/P hardware in the ALCF system is connected to storage via 16 10-gigabit-per-second links. On the storage side, servers are grouped into units of 4 servers attached to a single rack of disks via InfiniBand. There are 17 of these storage "slices" providing storage services to the ALCF system.

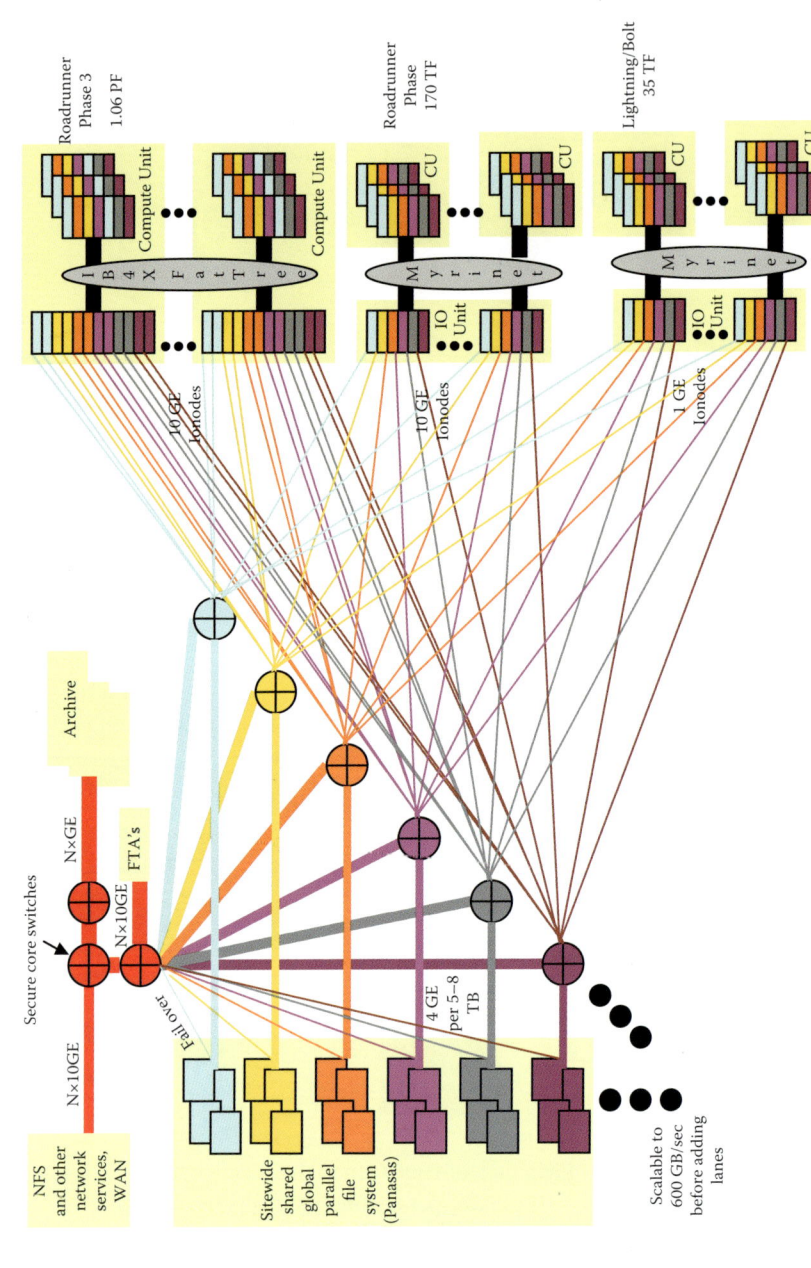

COLOR FIGURE 2.8 Los Alamos secure petascale infrastructure diagram with Roadrunner.

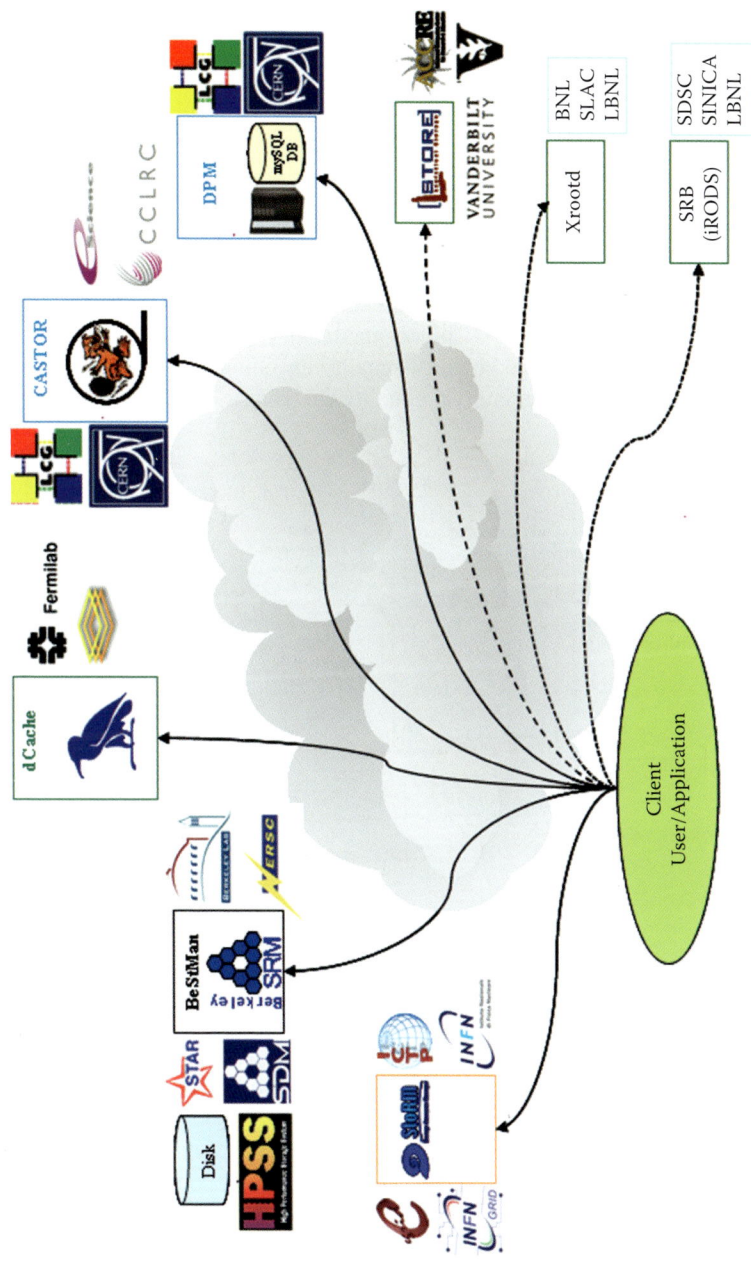

COLOR FIGURE 3.4 Interoperability of SRM v2.2 implementations.

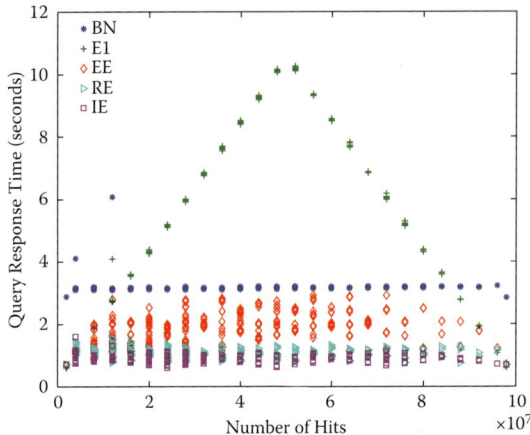

COLOR FIGURE 6.17 The query response time of five different bitmap encoding methods with WAH compression (BN: binary encoding, E1: the basic one-component equality encoding, EE: two-level equality-equality encoding, RE: two-level range-equality encoding, IE: two-level interval-equality encoding).

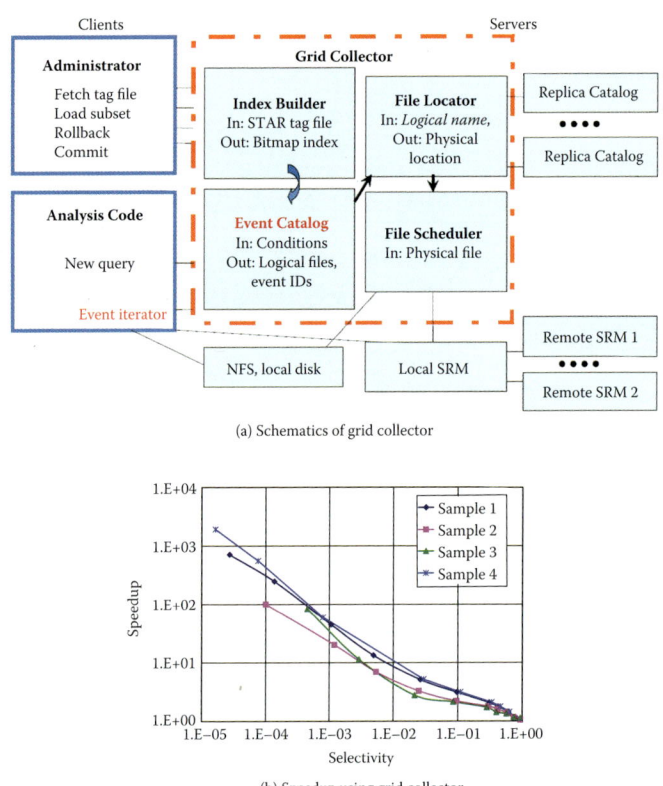

COLOR FIGURE 6.20 Using grid collector can significantly speed up analysis of STAR data. (a) schematics of grid collector, (b) speed-up using grid collector. Wu, K., J. Gu, J. Lauret, A.M. Poskanzer, A. Shoshani, and W.-M. Zhang. 2005. Facilitating efficient selective access from datagrids. *International Super computer Conference 2005*, Heidelberg, Germany.[101,104]

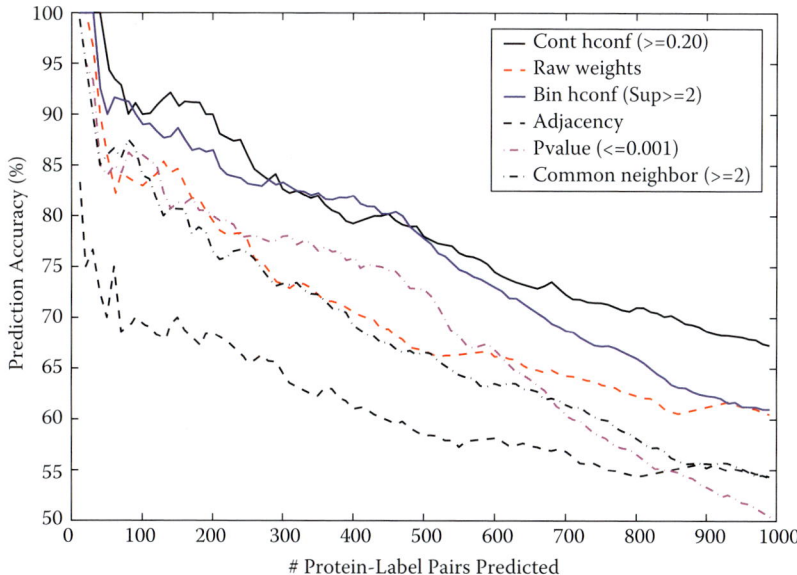
(a) Accuracy of top 1000 predictions on the combined network

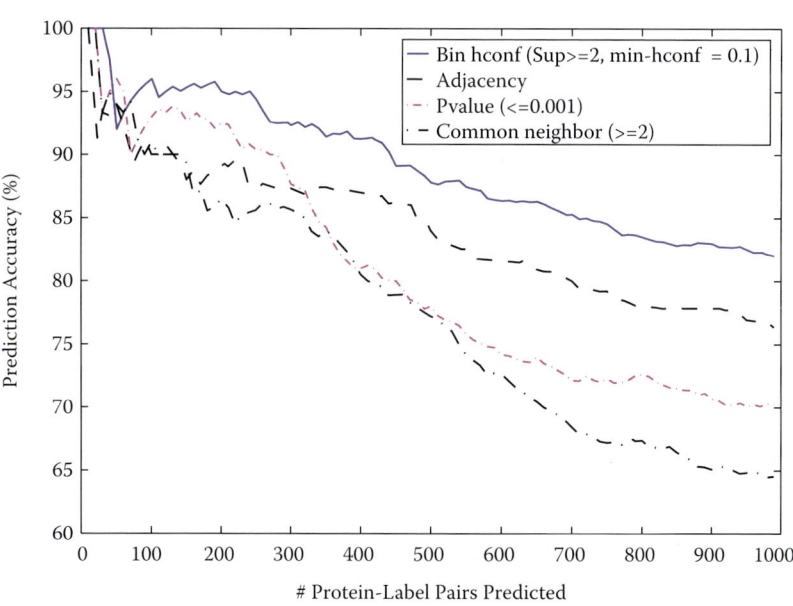
(b) Accuracy of top 1000 predictions on the DIPCore network

COLOR FIGURE 8.4 Comparison of performance of various transformed networks and the input networks.

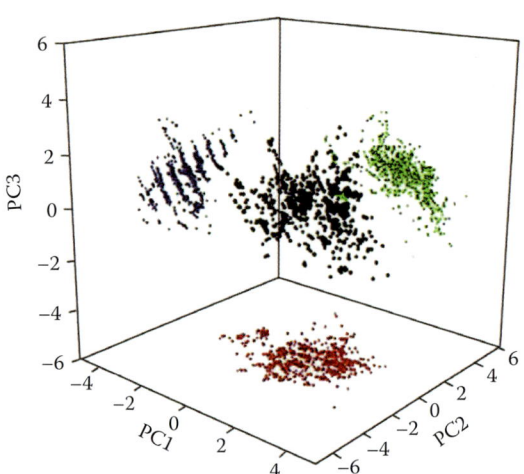

COLOR FIGURE 8.5 A three-dimensional PCA plot of a multivariate database of high-temperature superconductors. The initial dataset has 600 compounds with 8 attributes each. The projections in the PC1, PC2, and PC3 space are shown.

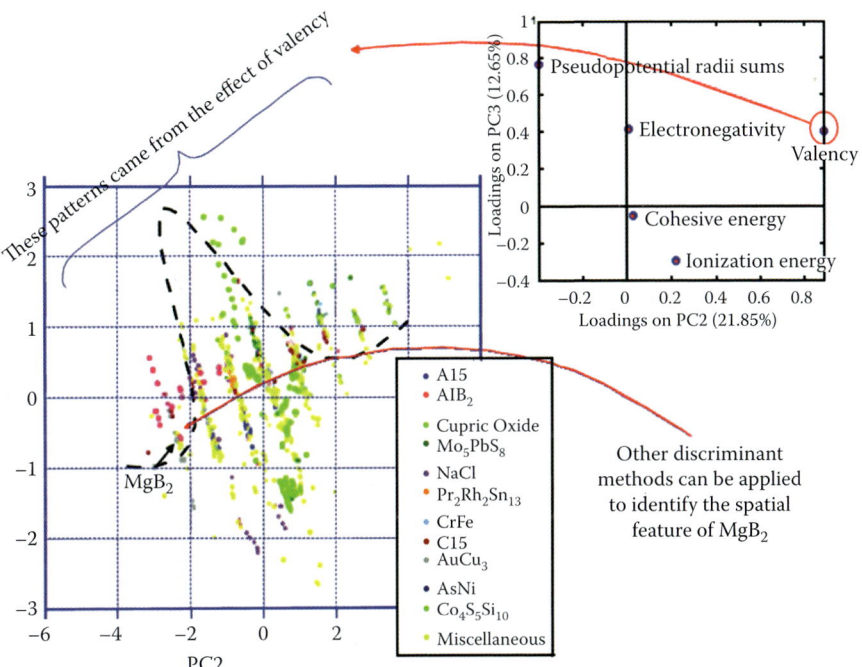

COLOR FIGURE 8.6 Interpreting results by comparing the scoring plot on the left and the loading plot on the right. The dashed curve shows the trend in transition temperature across the linear clusters. The peak corresponds to those with the highest recorded temperatures. By comparison to the earlier scoring plot, and using a different visualization scheme, we can now capture a more complete perspective of trends in this multivariate dataset.

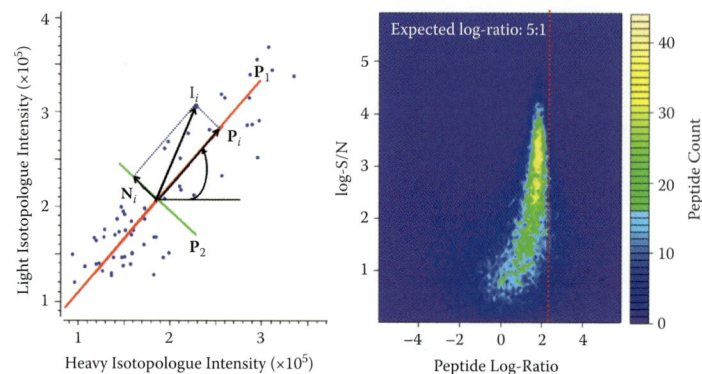

COLOR FIGURE 8.13 Peptide abundance ratio and signal-to-noise ratio estimation via PCA (left). Signal-to-noise ratio is inversely correlated with the variability and bias of peptide abundance ratio (right).

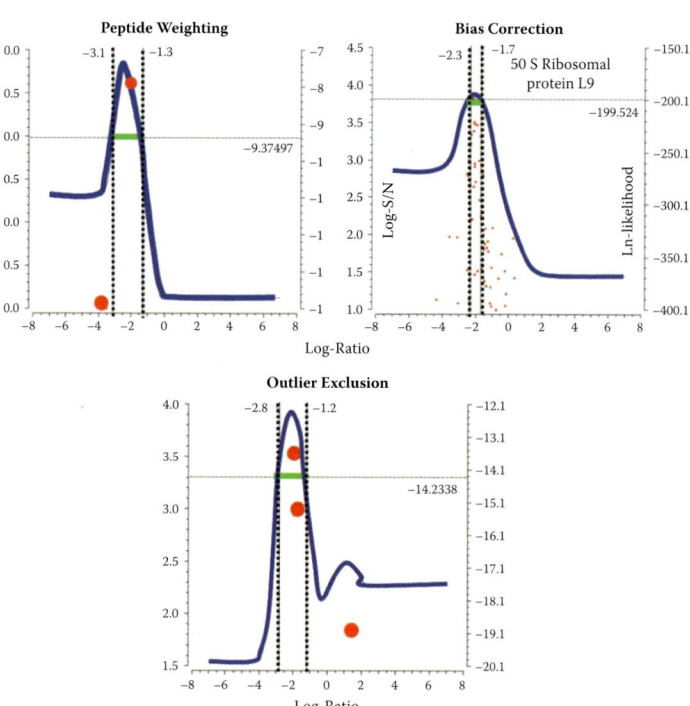

COLOR FIGURE 8.14 Protein abundance ratio and confidence interval estimation via profile likelihood algorithm.

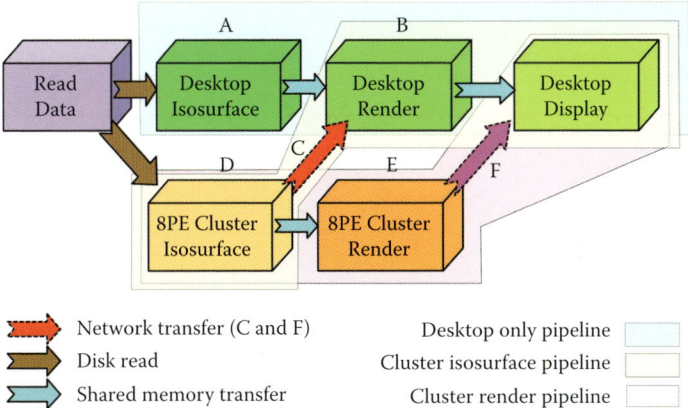

COLOR FIGURE 9.1 This image shows three different partitionings of a simple visualization pipeline consisting of four stages: data I/O, computing an isosurface, rendering isosurface triangles, and image display. In one partitioning, all operations happen on a single desktop machine (A-B, blue background). In another, data is loaded onto an eight-processor cluster for isosurface processing, and the resulting isosurface triangles are sent to the desktop for rendering (D-C-B, yellow background). In the third, data is loaded onto the cluster for isosurface processing and rendering, and the resulting image is sent to the desktop for display (D-E-F, magenta background).

COLOR FIGURE 9.2 Shown is a 36-domain dataset. The domains have thick black lines and are colored red or green. Mesh lines for the elements are also shown. To create the data-set sliced by the transparent gray plane, only the red domains need to be processed. The green domains can be eliminated before ever being read in.

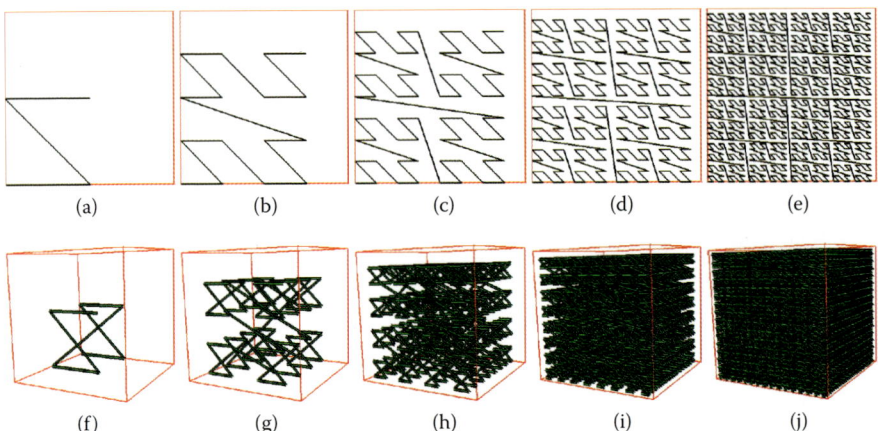

COLOR FIGURE 9.4 (a–e) The first five levels of resolution of the 2D Lebesgue's space-filling curve. (f–j) The first five levels of resolution of the 3D Lebesgue's space filling curve.

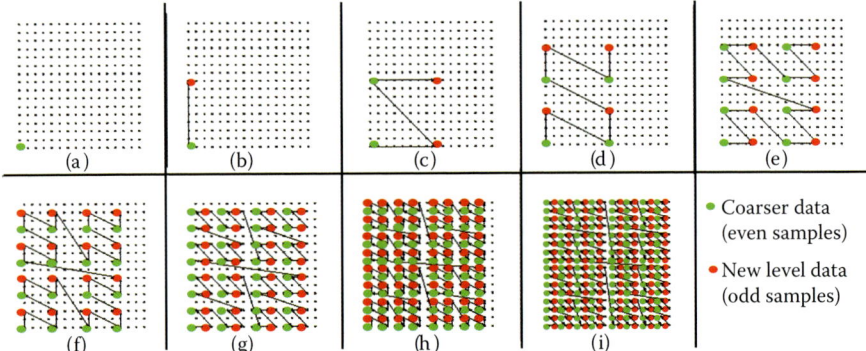

- Coarser data (even samples)
- New level data (odd samples)

COLOR FIGURE 9.5 The nine levels of resolution of the binary tree hierarchy defined by the 2D space-filling curve applied on 16 × 16 rectilinear grid. The coarsest level of resolution (a) is a single point. The number of points that belong to the curve at any level of resolution (b) to (i) is double the number of points of the previous level.

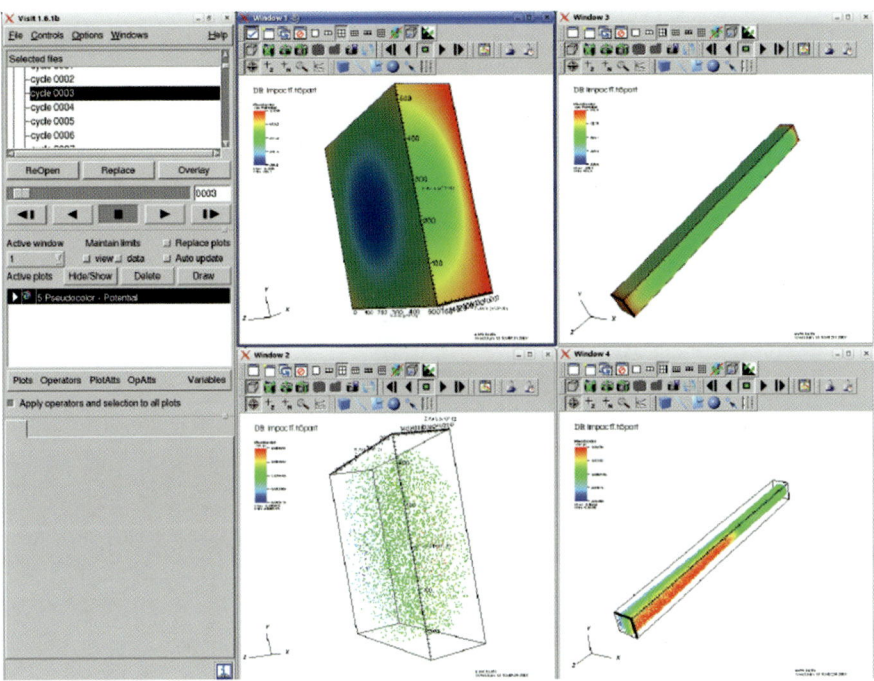

COLOR FIGURE 9.10 H5Part readers are included with visualization and data analysis tools in use by the accelerator modeling community. This image shows H5Part data loaded into VisIt for comparative analysis of multiple variables at different simulation timesteps.

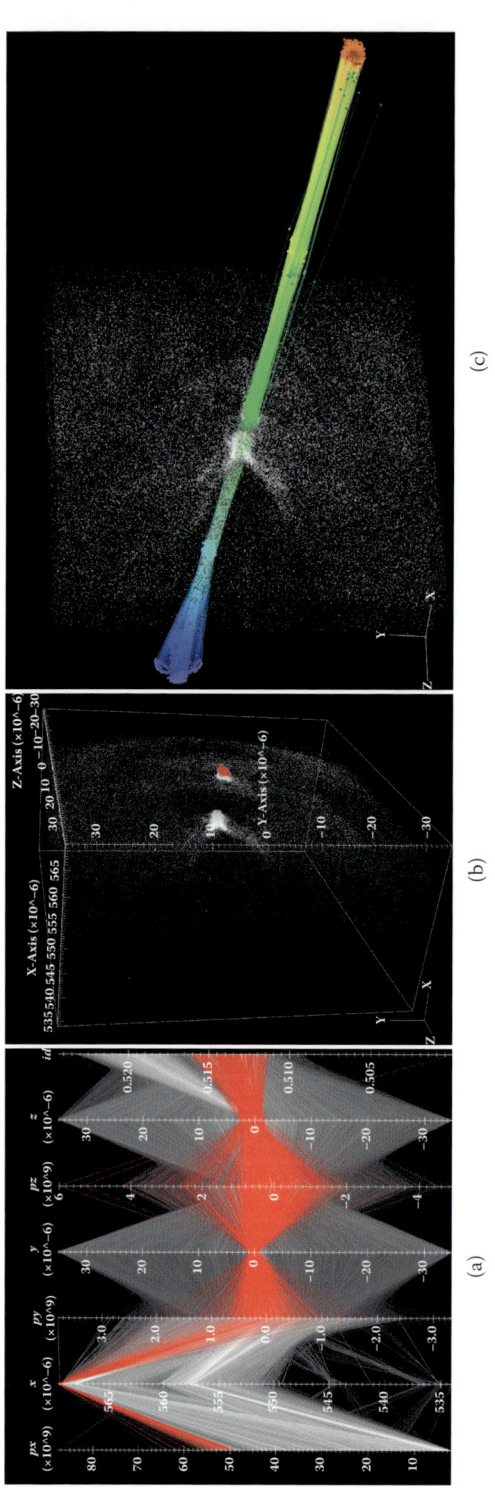

COLOR FIGURE 9.12 (a) Parallel coordinates of timestep $t = 12$ of the 3D dataset. Context view (gray) shows particles selected with $px > 2 * 10^9$. The focus view (red) consists of particles selected with $px > 4.856 * 10^{10}$ && $x > 5.649 * 10^{-4}$, which indicates particles forming a compact beam in the first wake period following the laser pulse. (b) Pseudocolor plot of the context and focus particles. (c) Traces of the beam. We selected particles at timestep $t = 12$, then traced the particles back in time to timestep $t = 9$ when most of the selected particles entered the simulation window. We also trace the particles forward in time to timestep $t = 14$. In this image, we use color to indicate px. In addition to the traces and the position of the particles, we also show the context particles at timestep $t = 12$ in gray to illustrate where the original selection was performed. We can see that the selected particles are constantly accelerated over time (increase in px) since their colors range from blue (relatively low levels of px) to red (relatively high levels of px) as they move along x over time.

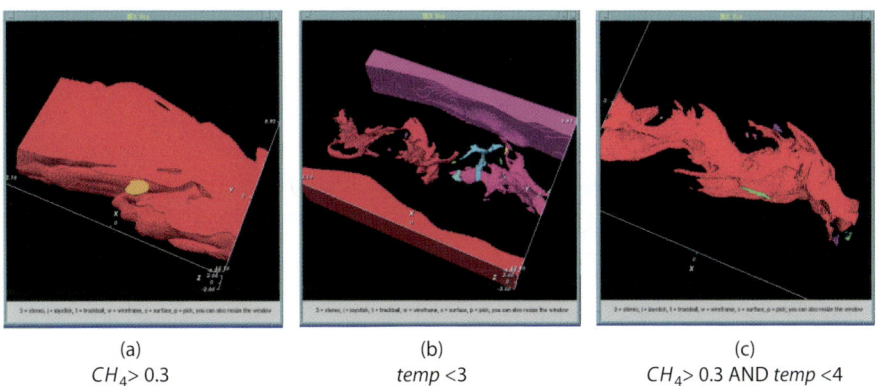

(a) $CH_4 > 0.3$
(b) $temp < 3$
(c) $CH_4 > 0.3$ AND $temp < 4$

COLOR FIGURE 9.13 A visualization of flames in a high-fidelity simulation of methane-air jet. The images show the cells in a 3D block-structured dataset that were returned by three different queries.

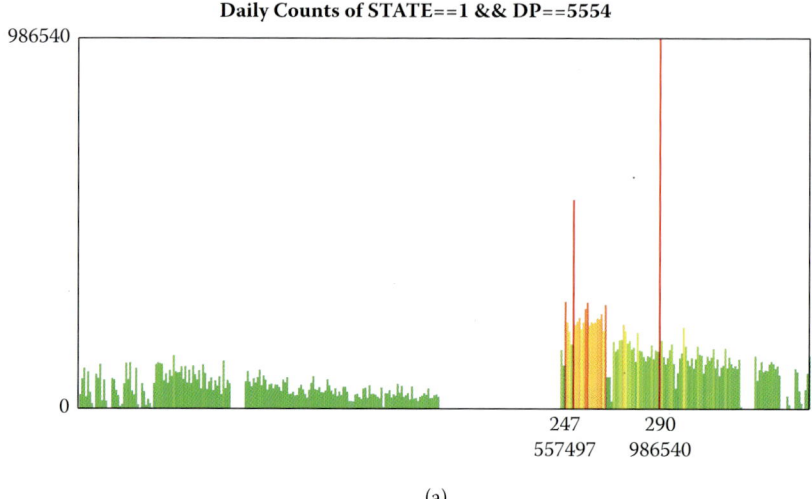

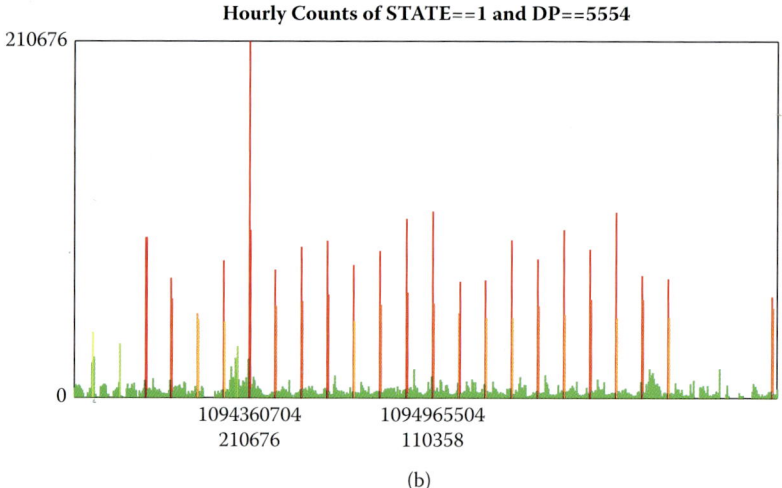

COLOR FIGURE 9.14 (a) Histogram of suspicious activity levels over a one-year period at one-day temporal resolution. (b) Suspicious activity levels over a four week period at one-hour temporal resolution. (c) Suspicious activity levels over a one-day period at one-hour temporal resolution. Forensic network traffic analysis is conducted by examining histograms of suspicious traffic activity at varying temporal resolution. These examples go from coarse, per-day resolution over a one-year time window down to per-minute resolution over a five-day window and show a regular pattern of systematic network attacks that occur with temporal regularity.

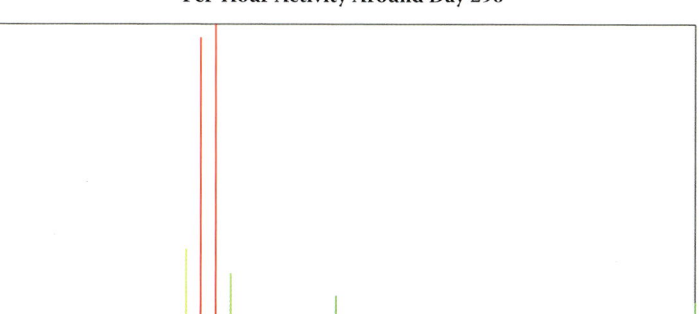

(c)

COLOR FIGURE 9.14   (*Continued*)

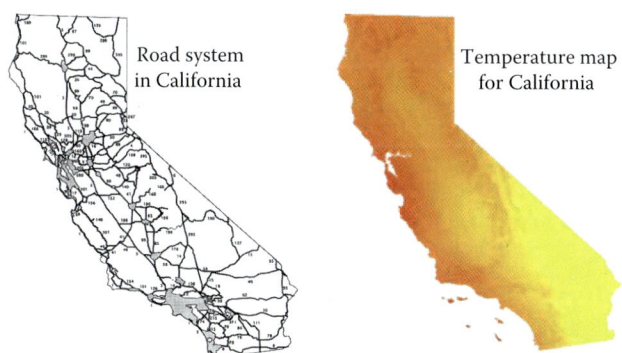

COLOR FIGURE 10.1   Examples of object-based (left) and field-based (right) geospatial data representation.

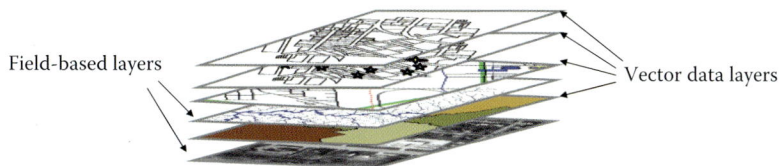

COLOR FIGURE 10.2   Illustration of an overlay of themes in a GIS. Georeferenced and aligned layers include both vector data and field-based data.

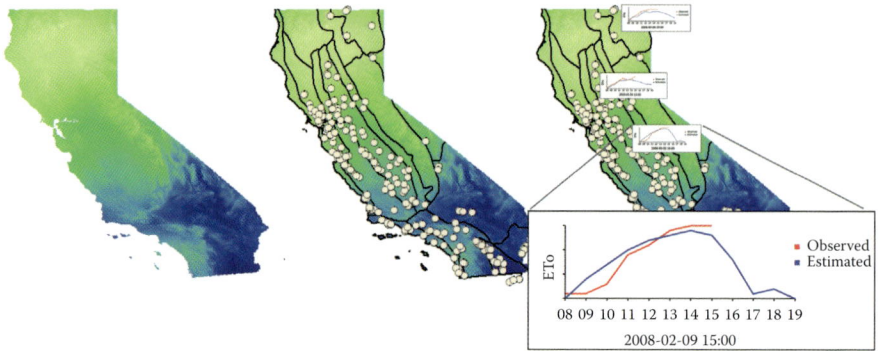

COLOR FIGURE 10.7 Main stages in the integration of sensor and model-generated data streams. Left: Estimated ETo map for the current hour; Center: Eco-regions and station locations overlayed; Right: Real-time charts for selected locations including the model prediction for a several-hour period and the actual observed values until current time.

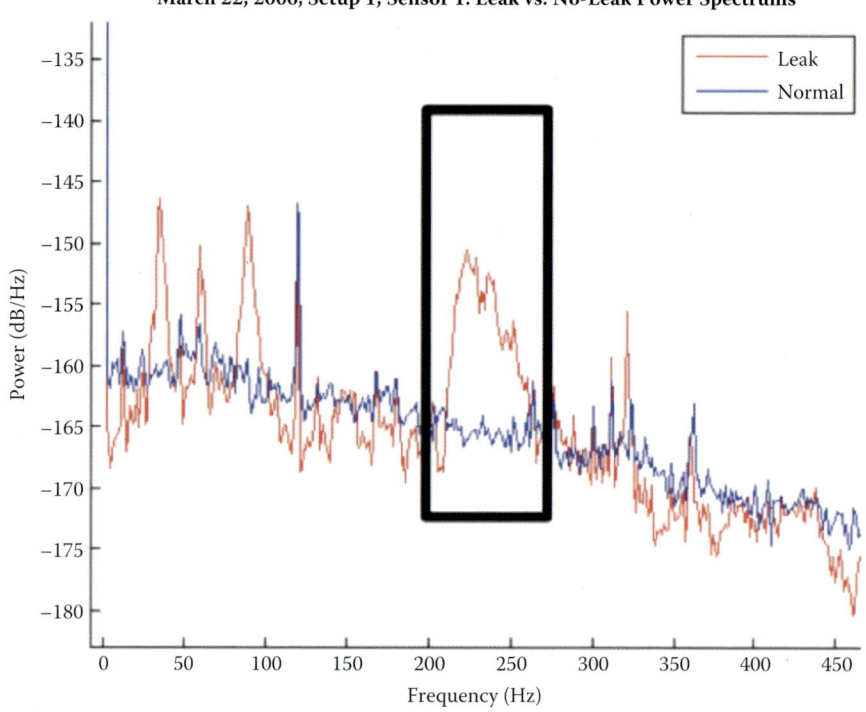

COLOR FIGURE 11.1 A leak shows up as additional energy in characteristic frequency bands.

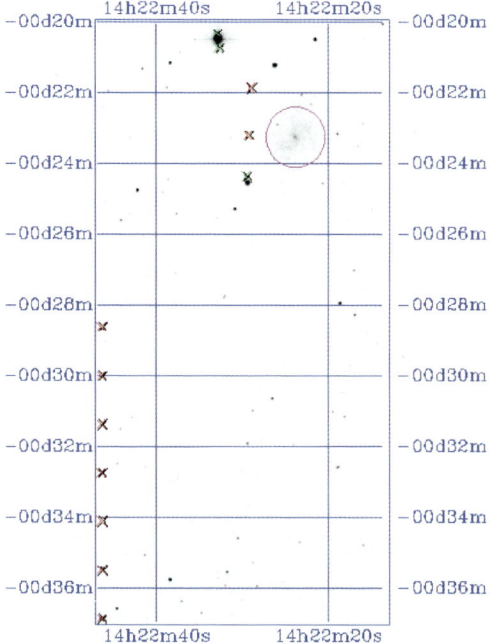

```
ORDATE  = '000503 '       / Observation Ref Date (yymmdd)
DAYNUM  = '1160 '         / Observation Day Num
FN_PRFX = 'j1160059'      / .rdo and .par filename prefix
TYPE    = 'sci '          / Scan type: dar flt sci cal tst
SCANNO  =            59   / Scan Number
SCANDIR = 'n '            / Scan Dire ction: n, s, -
COMMENT                   (OV)
STRIP_ID=        301788 / Strip ID (OV)
POSITNID= 's001422 '      / Position ID (OV)
ORIGIN  = '2MASS '        / 2MASS Survey Camera
CTYPE1  = 'RA---SIN'      / Orthographic Projection
CTYPE2  = 'DEC--SIN'      / Orthographic Projection
CRPIX1  =         256.5 / Axis 1 Reference Pixel
CRPIX2  =         512.5 / Axis 2 Reference Pixel
CRVAL1  =    215.6251831 / RA  at Frame Center, J2000 (deg)
CRVAL2  =    -0.4748106667 / Dec at Frame Center, J2000 (deg)
CROTA2  =    1.900065243E-05 / Image Twist +AXIS2 W of N, J2000 (deg)
CDELT1  =    -0.0002777777845 / Axis 1 Pixel Size (degs)
CDELT2  =    0.0002777777845 / Axis 2 Pixel Size (degs)
USXREF  =         -256.5 / U-scan X at Grid (0,0)
USYREF  =         19556. / U-scan Y at Grid (0,0)
```

COLOR FIGURE 12.2 Example of image metadata in astronomy. On the top is an image of the galaxy NGC 5584, shown at the center of the purple circle. The image was measured as part of the Two Micron All Sky Survey (2MASS). The crosses locate the positions of artifacts in the image. At bottom is a sample of the metadata describing the image, written in the form of keyword=value pairs, in compliance with the definition of the Flexible Image Transport System (FITS) in universal use in astronomy.

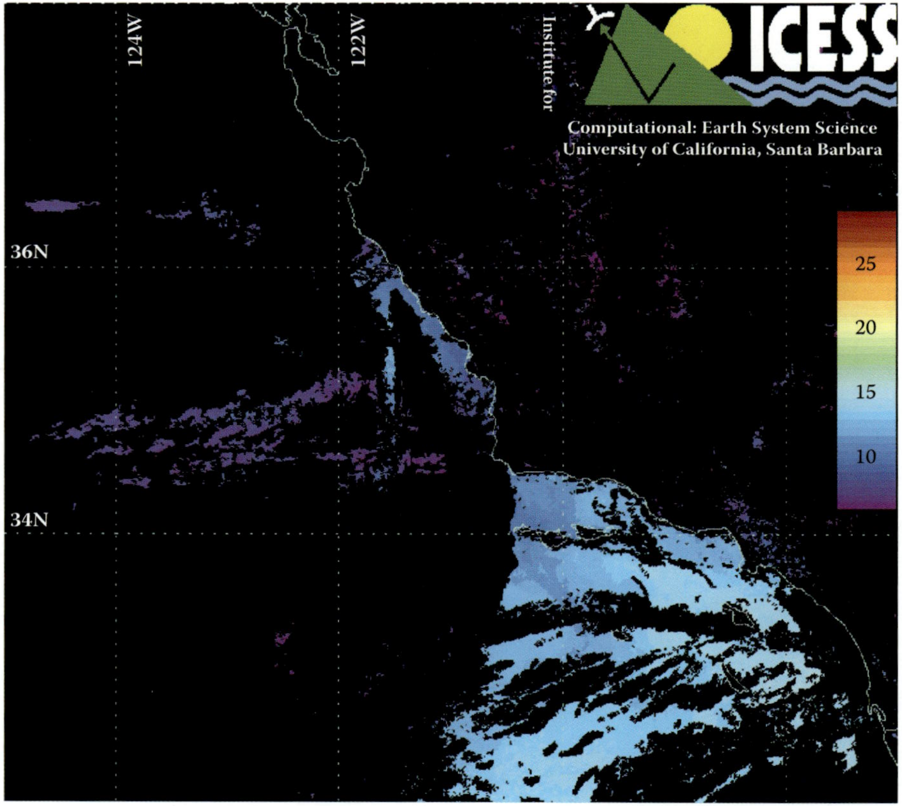

COLOR FIGURE 12.4 Map of sea surface temperature taken February 18, 2009, 10:42 p.m. GMT.

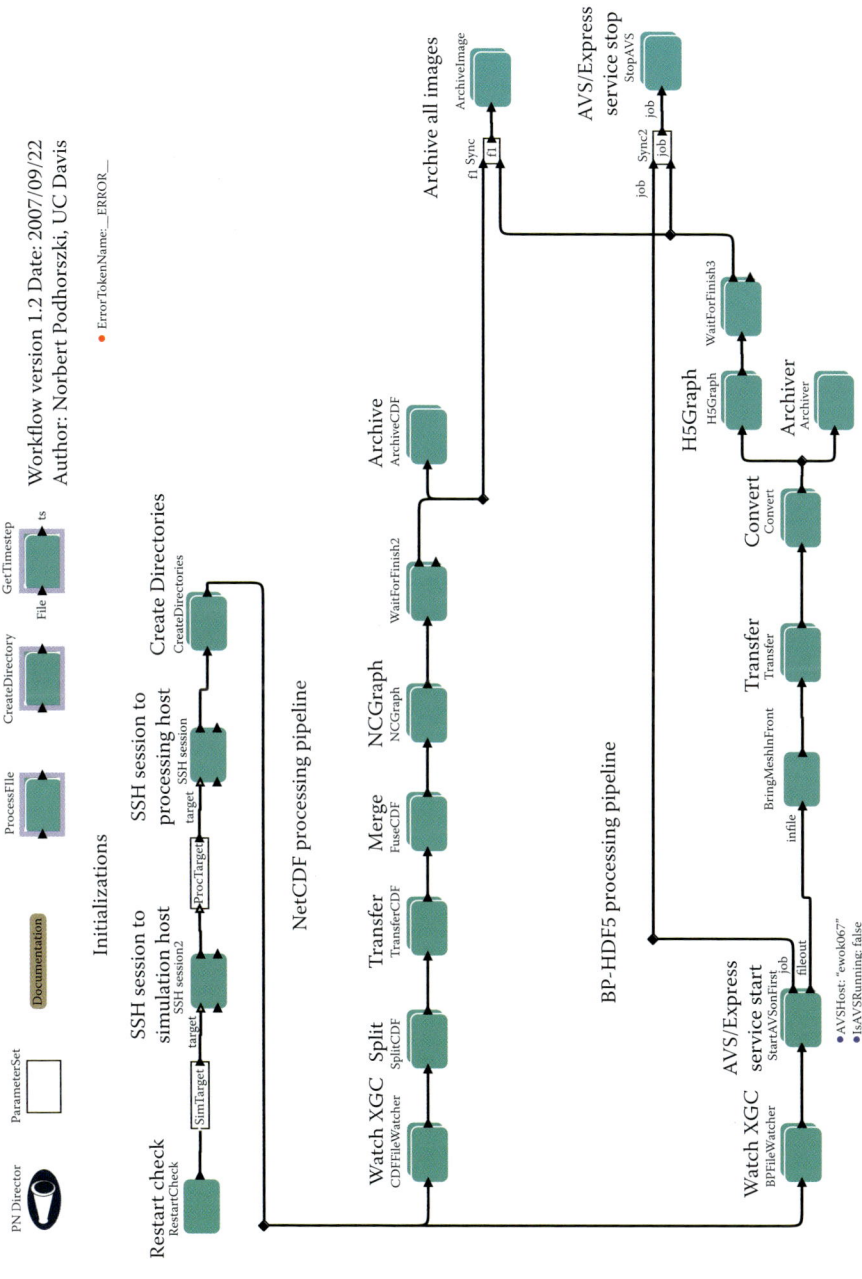

COLOR FIGURE 13.1 CPES fusion simulation monitoring workflow (in Kepler).

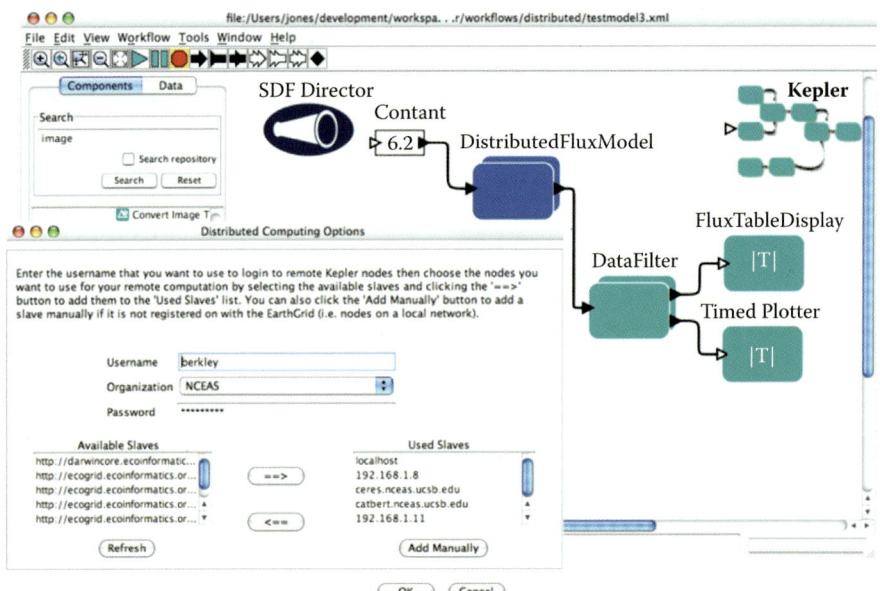

COLOR FIGURE 13.2 Kepler supports execution of workflows on remote peer nodes and remote clusters. Users indicate which portions of a workflow should be remotely executed by grouping them in a distributed composite component (shown in blue in the workflow). The user selects from a list of available remote nodes for execution (see dialog), and Kepler calculates a schedule and stages each data token before execution on one of the set of selected remote nodes.

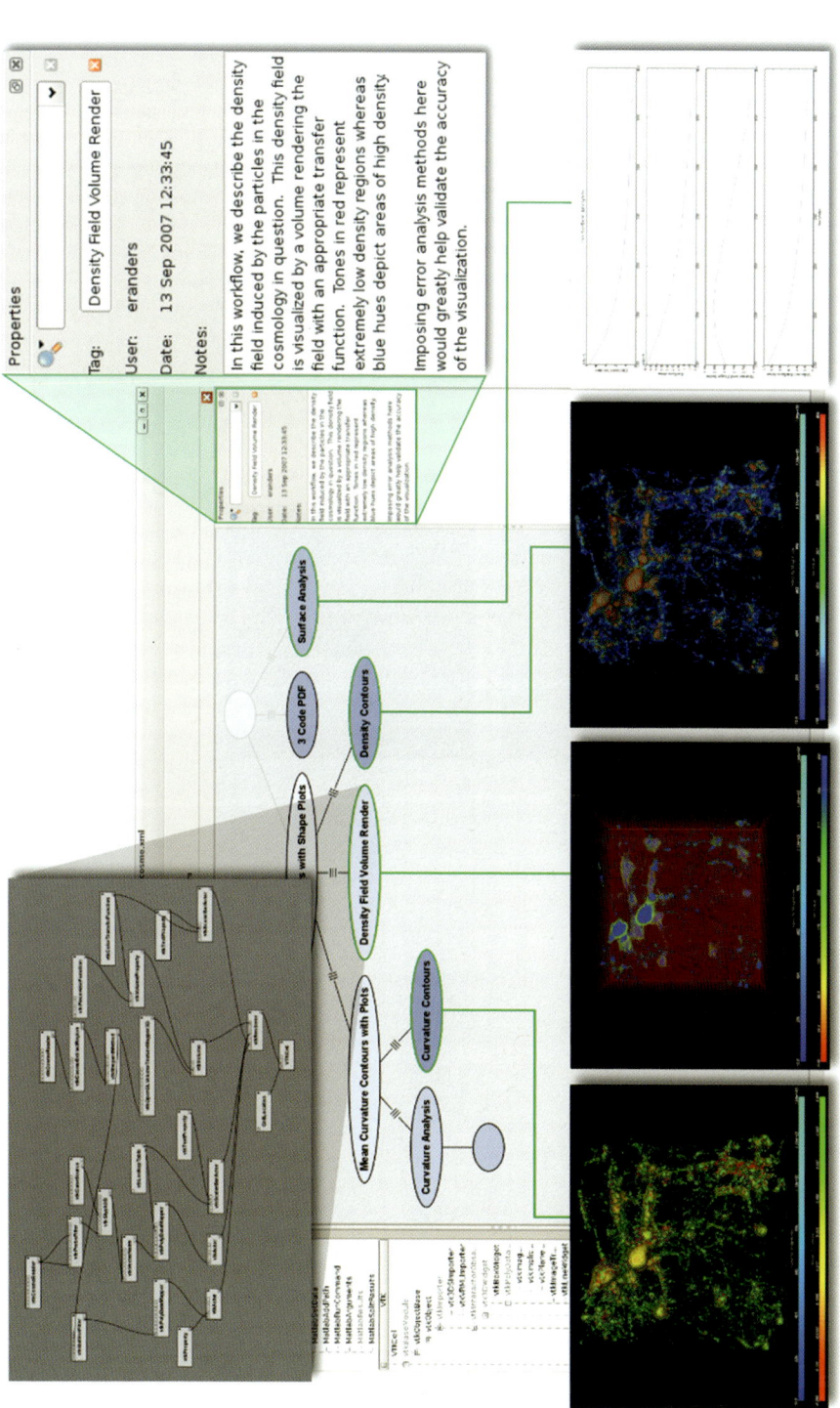

COLOR FIGURE 13.4 In VisTrails, workflow evolution provenance is displayed as a history tree, each node representing a workflow that generates a visualization. This tree allows a user to return to previous workflow versions and to be reminded of the actions that led to a particular result. Additional metadata stored with each version includes free-text notes and the user who created it.

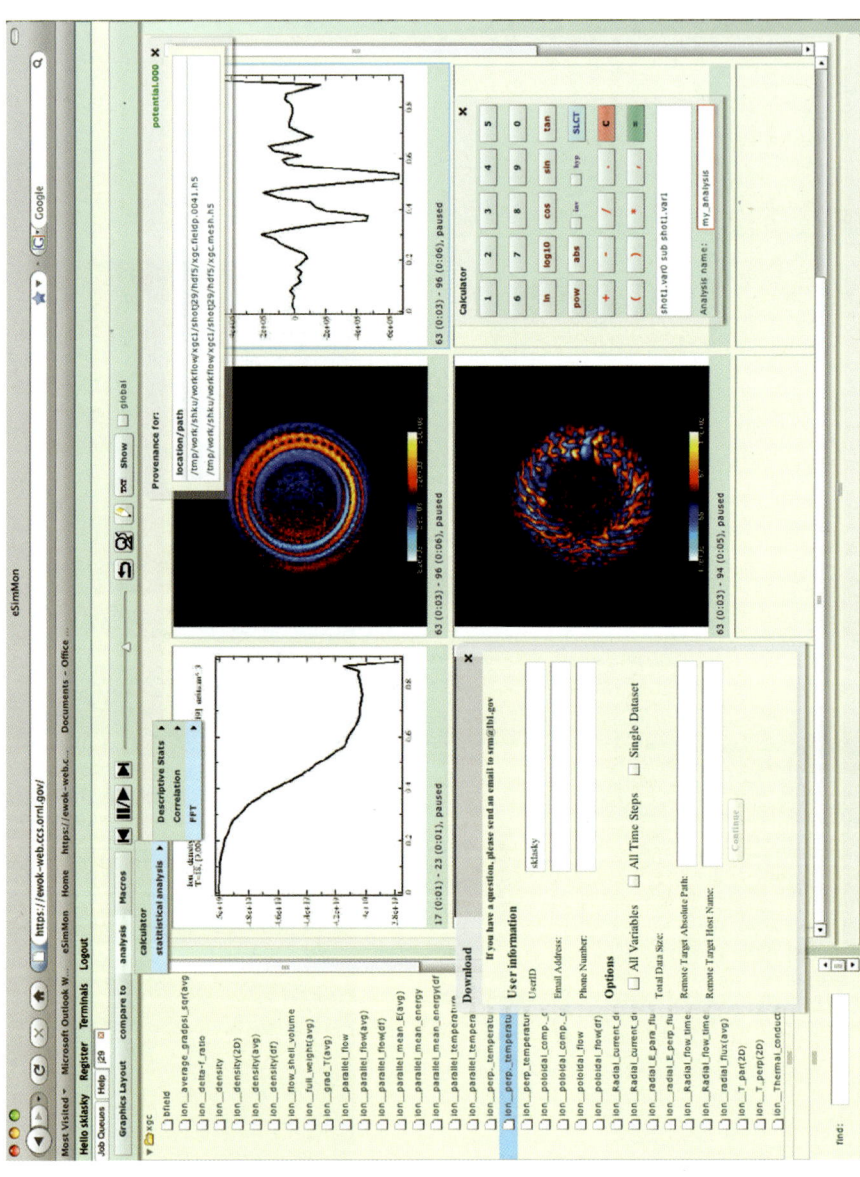

COLOR FIGURE 13.6 SDM dashboard with on-the-fly visualization.

Accelerating Queries on Very Large Datasets 225

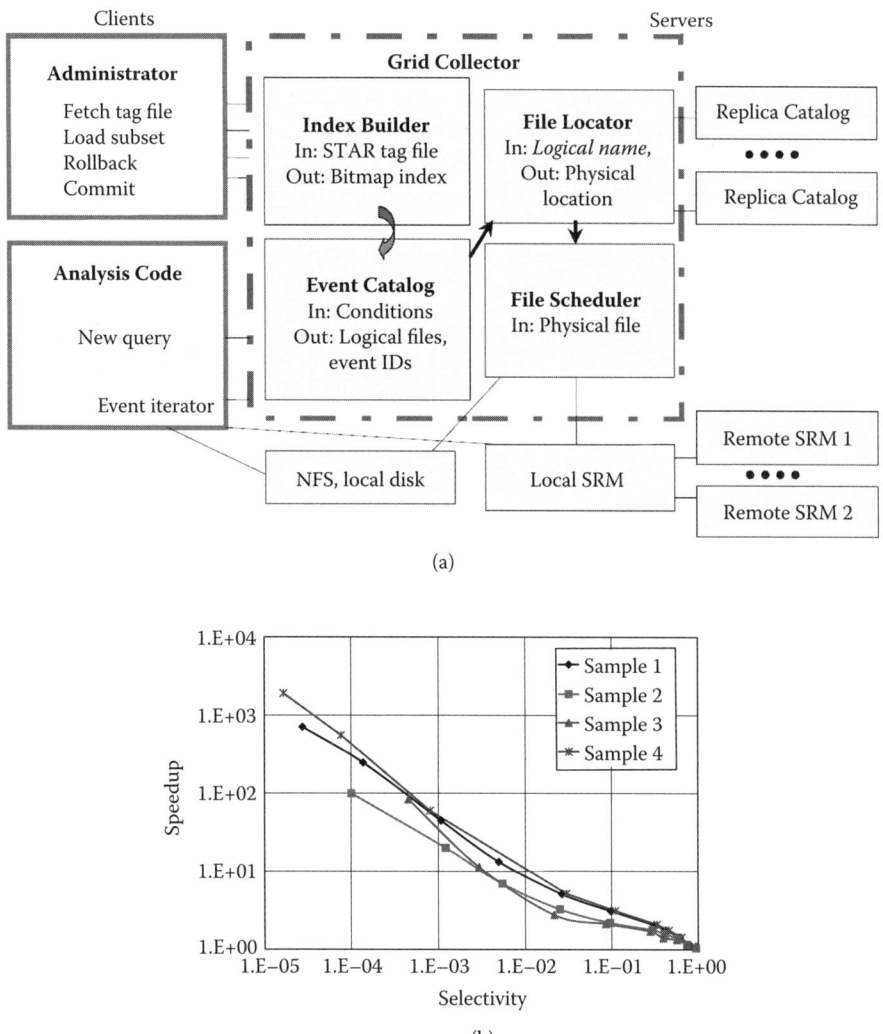

Figure 6.20 (See color insert following page 224.) Using grid collector can significantly speed up analysis of STAR data. (a) schematics of grid collector, (b) speed-up using grid collector. Wu, K, J. Gu, J. Lauret, A.M. Poskanzer, A. Shoshani, and W.-M. Zhang. 2005. Facilitating efficient selective access from datagrids. *International Super computer Conference 2005*, Heidelberg, Germany.[101,104]

In Figure 6.20b, we show the observed speed-up values versus the selectivity of the analysis task. In this case, the selectivity is defined as the fraction of events in the data files that are selected for an analysis task. The Grid Collector speeds up analysis tasks primarily by reducing the amount of disk pages accessed. Because the selected events typically are randomly scattered in the data files, the data files are compressed in blocks,[18] and the analysis jobs often involve a significant amount of computation; therefore, the speed-up is not the inverse of selectivity. However, as the selectivity decreases, the average speed-up value increases. When one out of 1,000 events is selected, the speed-up values are observed to be between 20 and 50. Even if 1 in 10 events is used in an analysis job, the observed speed-up is more than 2. STAR has hundreds of users at their various analysis facilities, and improving these facilities' overall throughput by a factor of 2 is a great benefit to the whole community.

As mentioned before, other parallel data management systems such as Hadoop currently iterate through all data records as well. A smart iterator similar to that of Grid Collector could benefit such a system as well.

6.8 Summary and Future Trends

In this chapter, we discussed two basic issues for accelerating queries on large scientific datasets, namely indexing and data organization. Since the data organizations are typically tied to an individual data management system, we also briefly touched on a number of different systems with distinct data organization schemes.

The bulk of this chapter discusses different types of index methods, most of which are better suited for secondary storage. Applications that use scientific data don't require simultaneous read and write accesses of the same dataset. This allows the data and indexes to be packed more tightly than in transactional applications. Furthermore, the indexes can be designed to focus more on query processing speed and less on updating of individual records. In general, scientific data tend to have a large number of searchable attributes and require indexes on every searchable attribute, whereas a database for a banking application may require only one index for the primary key.

After reviewing many of the well-known multidimensional indexes, we concluded that the bitmap indexes are the most appropriate indexing schemes for scientific data. We reviewed some recent advances in bitmap index research and discussed their uses in two examples. These examples use an open-source bitmap index software called FastBit. The first example demonstrated the usefulness of indexes by measuring the time needed to answer a set of queries from the Set Query Benchmark. We saw that FastBit outperforms the best available indexless system by an order of magnitude. This demonstrates that there are situations where the use of an index significantly improves performance of an application. The second example demonstrated the use of FastBit indexes to implement a smart iterator for a distributed data analysis framework. Since an iterator is a convenient way to access large datasets on parallel systems, the example demonstrated an effective way of using indexes for parallel data analysis.

As datasets grow in size, all data analyses are likely to be performed on parallel computers. Off-loading some data processing tasks to the disk controller (as the Netezza system does) or other custom hardware could be an effective strategy to improve the efficiency of query processing. However, advanced indexing techniques will continue to be an indispensable tool for analyzing massive datasets.

Acknowledgment

This work was supported by the Director, Office of Advanced Scientific Computing Research, Office of Science, of the U.S. Department of Energy, under Contract No. DE-AC02-05CH11231.

References

[1] D. Abadi, S. R. Madden, and M. C. Ferreira. Integrating compression and execution in column-oriented database systems. In *SIGMOD*. ACM, 2006.

[2] A. V. Aho and J. D. Ullman. Optimal partial-match retrieval when fields are independently specified. *ACM Trans. Database Syst.*, 4(2):168–179, 1979.

[3] S. Amer-Yahia and T. Johnson. Optimizing queries on compressed bitmaps. In *International Conference on Very Large Data Bases*, Cairo, Egypt, September 2000. Morgan Kaufmann.

[4] G. Antoshenkov. Byte-aligned bitmap compression. In *Data Compression Conference (DCC)*, March 28, 1995.

[5] G. Antoshenkov and M. Ziauddin. Query processing and optimization in ORACLE RDB. *VLDB Journal*, 5:229–237, 1996.

[6] C. Ballinger and R. Fryer. Born to be parallel: Why parallel origins give Teradata an enduring performance edge. *IEEE Data Eng. Bull.*, 20(2):3–12, 1997. An updated version of this paper is available at http://www.teradata.com/library/pdf/eb3053.pdf.

[7] P. Barrett. Application of linear quadtree to astronomical database. *Astronomical Data Analysis Software and Systems IV*, 77:1–4, 1995.

[8] J. Bartholdi and P. Goldsman. Continuous indexing of hierarchical subdivisions of the globe. *Int. J. Geographical Information Science*, 15(6):489–522, 2001.

[9] R. Bayer and E. McCreight. Organization and maintenance of large ordered indexes. *Acta Informatica*, 1:173–189, 1972.

[10] R. Bellman. *Adaptive control processes: A guided tour*. Princeton: Princeton University Press, 1961.

[11] J. L. Bentley. Mutidimensional binary search tree in database applications. *IEEE Trans. Soft. Eng.*, SE-5(4):333–337, 1979.

[12] S. Berchtold, C. Bohn, and H-P. Kriegel. The pyramid technique: Towards breaking the curse of dimensionality. In *SIGMOD*, pp. 142–153, 1998.

[13] S. Berchtold, D. Keim, and H-P. Kriegel. The X-tree: An index structure for high-dimensional data. In *VLDB*, pp. 28–39, 1996.

[14] K. S. Beyer, J. Goldstein, R. Ramakrishnan, and U. Shaft. When is "nearest neighbor" meaningful? In *ICDT*, volume 1540 of *Lecture Notes in Computer Science*, pp. 217–235. Springer, 1999.

[15] C. Böhm, S. Berchtold, and H.-P. Kriegel. Multidimensional index structures in relational databases. *Journal of Intelligent Info Syst.*, 15(1):322–331, 2000.

[16] P. A. Boncz, S. Manegold, and M. L. Kersten. Database architecture optimized for the new bottleneck: Memory access. In *The VLDB Journal*, pp. 54–65, 1999.

[17] P. A. Boncz, W. Quak, and M. L. Kersten. Monet and its geographic extensions: A novel approach to high performance GIS processing. In *EDBT'96*, volume 1057 of *Lecture Notes in Computer Science*, pp. 147–166. Springer, 1996.

[18] R. Brun and F. Rademakers. ROOT: An object oriented data analysis framework. *Nuclear instruments & methods in physics research, Section A*, 289 (1–2):81–86, 1997.

[19] J. J. Bunn and H. B Newman. Data intensive grids for high energy physics, 2003. Available at http://pcbunn.cithep.caltech.edu/Grids/GridBook.htm.

[20] A. Chadha, A. Gupta, P. Goel, V. Harinarayan, and B. R. Iyer. Encoded-vector indices for decision support and warehousing, 1998. US Patent 5,706,495.

[21] C.-Y. Chan and Y. E. Ioannidis. Bitmap index design and evaluation. *SIGMOD Rec.*, 27(2):355–366, 1998.

[22] C. Y. Chan and Y. E. Ioannidis. An Efficient Bitmap Encoding Scheme for Selection Queries. In *SIGMOD*, Philadelphia, PA, June 1999. ACM Press.

[23] S. Chaudhuri. An overview of query optimization in relational systems. In *PODS'98*, 1998.

[24] D. Comer. The ubiquitous B-tree. *ACM Comput. Surveys*, 11(2):121–137, 1979.

[25] C. Cormen, T. Stein, C. Leiserson, and R. Rivest. *Introduction to Algorithms*. MIT Press, Cambridge, Mass., 2nd edition, 2001.

[26] G. S. Davidson, K. W. Boyack, R. A. Zacharski, S. C. Helmreich, and J. R. Cowie. Data-centric computing with the Netezza architecture. Technical Report SAND2006-3640, Sandia National Laboratories, 2006.

[27] J. Dean and S. Ghemawat. MapReduce: Simplified data processing on large clusters. In *OSDI'04*, 2004.

[28] J. Dean and S. Ghemawat. MapReduce: Simplified data processing on large clusters. *Commun. ACM*, 51(1):107–113, 2008.

[29] D. J. DeWitt, M. Smith, and H. Boral. A single-user performance evaluation of the Teradata database machine. In *Proceedings of the 2nd International Workshop on High Performance Transaction Systems*, pp. 244–276, London, UK, 1989. Springer-Verlag.

[30] R. Fagin, J. Nievergelt, N. Pippenger, and H. R. Strong. Extendible hashing—a fast access method for dynamic files. *ACM Trans. on Database Syst.*, 4(3):315–344, 1979.

[31] P. Ferragina and R. Grossi. The string B-tree: A new data structure for string search in external memory and its applications. *J. ACM*, 46(2):236–280, 1999.

[32] M. Freeston. The bang file: A new kind of grid file. In *SIGMOD*, pp. 260–269, 1987.

[33] V. Gaede and O. Günther. Multidimensional access methods. *ACM Computing Surveys*, 30(2):170–231, 1998.

[34] I. Gargantini. An effective way to represent quad-trees. *Comm. ACM*, 25(12):905–910, 1982.

[35] J. Gray, D. T. Liu, N. Nieto-Santisteban, A. Szalay, D. J. DeWitt, and G. Heber. Scientific data management in the coming decade. *SIGMOD Rec.*, 34(4):34–41, 2005.

[36] J. Gray, D. T. Liu, M. Nieto-Santisteban, A. Szalay, D. J. DeWitt, and G. Heber. Scientific data management in the coming decade. *SIGMOD Rec.*, 34(4):34–41, 2005.

[37] R. Grossi and J. S. Vitter. Compressed suffix arrays and suffix trees with applications to text indexing and string matching. *SIAM J. Comput.*, 35(2):378–407, 2005.

[38] A. Gupta, K. C. Davis, and J. Grommon-Litton. Performance comparison of property map and bitmap indexing. In *DOLAP'02*, pp. 65–71. ACM, 2002.

[39] D. Gusfield. *Algorithms on strings, trees and sequences: Computer science and computational biology.* Cambridge University Press, Cambridge, MA, 1997.

[40] A. Guttman. R-trees: A dynamic index structure for spatial searching. In *SIGMOD*, pp. 47–57. ACM, 1984.

[41] R. J. Hanisch, A. Farris, E. W. Greisen, W. D. Pence, B. M. Schlesinger, P. J. Teuben, R. W. Thompson, and A. Warmork III. Definition of the flexible image transport system (FITS). *Astronomy and Astrophysics. Supp. Ser.*, 376:359–380, 2001.

[42] HDF5 home page. http://www.hdfgroup.org/HDF5.

[43] S. Heinz, J. Zobel, and H. E. Williams. Burst tries: A fast, efficient data structure for string keys. *ACM Trans. Inf. Syst.*, 20(2):192–223, 2002.

[44] M. Jarke and J. Koch. Query optimization in database systems. *ACM Computing Surveys*, 16(2):111–152, 1984.

[45] T. Johnson. Performance measurements of compressed bitmap indices. In *VLDB*, Edinburgh, Scotland, September 1999. Morgan Kaufmann.

[46] N. Katayama and S. Satoh. The SR-tree: An index structure for high-dimensional nearest neighbor queries. In *SIGMOD'97*, pp. 369–380. ACM, 1997.

[47] D. Knuth. *The art of computer programming: Sorting and searching*, volume 3. Addison-Wesley, Reading, MA, 2nd edition, 1973.

[48] N. Koudas. Space efficient bitmap indexing. In *CIKM'00*, pp. 194–201. ACM, 2000.

[49] Z. Lacroix and T. Critchlow, editors. *Bioinformatics: Managing scientific data.* San Francisco: Morgan Kaufmann, 2003.

[50] R. Lämmel. Google's mapreduce programming model—revisited. *Science of Computer Programming*, 70(1):1–30, 2008.

[51] P. Larson. Dynamic hashing. *BIT*, 18:184–201, 1978.

[52] J. K. Lawder and P. J. H. King. Using space-filling curves for multidimensional indexing. In *BNCOD*, volume 1832 of *Lecture Notes in Computer Science*. Springer, 2000.

[53] M. Lee and H. Samet. Navigating through triangle meshes implemented as linear quadtrees. *ACM Transactions on Graphics*, 19(2):79–121, 2000.

[54] W. Litwin. Linear hashing: A new tool for table and file addressing. In *VLDB'80*, pp. 212–223, 1980.

[55] D. Lovelace, R. Ayyar, A. Sala, and V. Sokal. VSAM demystified. Technical Report Redbook Series SG246105, IBM, 2001.

[56] X. Ma and A. L. N. Reddy. MVSS: An active storage architecture. *IEEE Transactions on Parallel and Distributed Systems*, PDS-14(10):993–1005, October 2003.

[57] B. Moon, H. V. Jagadish, and C. Faloutsos. Analysis of the clustering properties of the Hilbert space-filing curve. *IEEE Trans. on Knowledge and Data Eng.*, 13(1), 2001.

[58] D. R. Morrison. Patricia: Practical algorithm to retrieve information coded in alphanumeric. *J. ACM*, 15(4):514–534, 1968.

[59] R. Musick and T. Critchlow. Practical lessons in supporting large-scale computational science. *SIGMOD Rec.*, 28(4):49–57, 1999.

[60] M. Nelson and J. L. Gailly. *The Data Compression Book*. M&T Books, New York, NY, 2nd edition, 1995.

[61] NetCDF (network common data form) home page. http://www.unidata.ucar.edu/software/netcdf/

[62] J. Nievergelt, H. Hinterberger, and K. C. Sevcik. The grid file: An adaptable symmetric multikey file structure. *ACM Trans. on Database Syst.*, 9(1):38–71, 1984.

[63] E. O'Neil, P. O'Neil, and K. Wu. Bitmap index design choices and their performance implications. In *IDEAS 2007*, pp. 72–84, 2007.

[64] P. O'Neil and D. Quass. Improved query performance with variant indexes. In *SIGMOD*, Tucson, AZ, May 1997. ACM Press.

[65] P. O'Neil. Model 204 architecture and performance. In *Second International Workshop in High Performance Transaction Systems*. Springer Verlag, 1987.

[66] P. O'Neil. Informix indexing support for data warehouses. *Database Programming and Design*, 10(2):38–43, February 1997.

[67] P. O'Neil and E. O'Neil. *Database: Principles, programming, and performance*. San Francisco: Morgan Kaugmann, 2nd edition, 2000.

[68] E. J. Otoo. Linearizing the directory growth of extendible hashing. In *ICDE*. IEEE, 1988.

[69] E. J. Otoo. A mapping function for the directory of a multidimensional extendible hashing. In *VLDB*, pp. 493–506, 1984.

[70] F. Rademaker and R. Brun. ROOT: An object-oriented data analysis framework. *Linux Journal*, July 1998. ROOT software is available from http://root.cern.ch/.

[71] F. Ramsak, V. Markl, R. Fenk, M. Zirkel, K. Elhardt, and R. Bayer. Integrating the UB-tree into a database system kernel. In *VLDB'2000*, 2000.

[72] F. R. Reiss. *Data triage*. PhD thesis, UC Berkeley, June 2007.

[73] E. Riedel, G. Gibson, and C. Faloutsos. Active storage for large-scale data mining and multimedia. In *VLDB'98*, pp. 62–73, New York, NY, August 1998. Morgan Kaufmann Publishers Inc.

[74] P. Rigaux, M. Scholl, and A. Voisard. *Spatial databases, with applications to GIS*. Morgan Kaufmann, San Francisco, 2002.

[75] D. Rotem, K. Stockinger, and K. Wu. Minimizing I/O costs of multidimensional queries with bitmap indices. In *SSDBM, Vienna, Austria, July 2006*. IEEE Computer Society Press, 2005.

[76] D. Rotem, K. Stockinger, and K. Wu. Optimizing candidate check costs for bitmap indices. In *CIKM, Bremen, Germany, November 2005*. ACM Press, 2005.

[77] D. Rotem, K. Stockinger, and K. Wu. Optimizing I/O costs of multidimensional queries using bitmap indices. In *DEXA, Copenhagen, Denmark, August 2005*. Springer Verlag, 2005.

[78] H. Sagan. *Space-filling curves*. Springer-Verlag, New York, 1994.

[79] H. Samet. *Foundations of multidimensional and metric data structures*. Morgan Kaufmann, San Francisco, CA, 2006.

[80] J. Schlosser and M. Rarey. TrixX-BMI: Fast virtual screening using compressed bitmap index technology for efficient prefiltering of compound libraries. ACS Fall 2007, Boston, MA, 2007.

[81] T. K. Sellis. Multiple-query optimization. *ACM Transactions on Database Systems*, 13(1):23–52, 1988.

[82] U. Shaft and R. Ramakrishnan. Theory of nearest neighbors indexability. *ACM Trans. Database Syst.*, 31(3):814–838, 2006.

[83] A. Shoshani, L. M. Bernardo, H. Nordberg, D. Rotem, and A. Sim. Multidimensional indexing and query coordination for tertiary storage management. In *SSDBM*, pp. 214–225, 1999.

[84] A. Sim and A. Shoshani. The storage resource manager interface specification, 2008. http://www.ogf.org/documents/GFD.129.pdf (also in: http://sdm.lbl. gov/srm-wg/doc/GFD.129-OGF-GSM-SRM-v2.2-080523.pdf).

[85] R. R. Sinha. *Indexing Scientific Data*. PhD thesis, UIUC, 2007.

[86] R. R. Sinha and M. Winslett. Multi-resolution bitmap indexes for scientific data. *ACM Trans. Database Syst.*, 32(3):16, 2007.

[87] STAR: Solenoidal tracker at RHIC (STAR) experiment.

[88] K. Stockinger, K. Wu, and A. Shoshani. Evaluation strategies for bitmap indices with binning. In *DEXA*, Zaragoza, Spain, September 2004. Springer-Verlag.

[89] K. Stockinger and K. Wu. *Bitmap indices for data warehouses*, Chapter VII, pp. 179–202. Idea Group, Inc., 2006.

[90] M. Stonebraker, C. Bear, U. Çetintemel, M. Cherniack, T. Ge, N. Hachem, S. Harizopoulos, J. Lifter, J. Rogers, and S. B. Zdonik. One size fits all? Part 2: Benchmarking studies. In *CIDR*, pp. 173–184. www.crdrdb.org, 2007.

[91] S. Stonebraker, D. J. J. Abadi, A. Batkin, X. Chen, M. Cherniack, M. Ferreira, E. Lau, A. Lin, S. Madden, E. O'Neil, P. O'Neil, A. Rasin, N. Tran, and S. Zdonik. C-store: A column-oriented DBMS. In *VLDB'05*, pp. 553–564. VLDB Endowment, 2005.

[92] A. Szalay, P. Kunszt, A. Thakar, J. Gray, and D. Slutz. Designing and mining multi-terabyte astronomy archives: The Sloan digital sky survey. In *SIGMOD*, Dallas, Texas, USA, May 2000. ACM Press.

[93] A. S. Szalay, P. Z. Kunszt, A. Thakar, J. Gray, D. Slutz, and R. J. Brunner. Designing and mining multi-terabyte astronomy archives: The Sloan digital sky survey. *SIGMOD Rec.*, 29(2):451–462, 2000.

[94] Y. Theodoridis. The R-Tree-portal, 2003.

[95] L. A. Treinish. Scientific data models for large-scale applications, 1995.

[96] J. S. Vitter. External memory algorithms and data structures: Dealing with massive data. *ACM Computing Surveys*, 33(2):209–271, 2001.

[97] R. Weber, H.-J. Schek, and S. Blott. A quantitative analysis and performance study for similarity-search methods in high-dimensional spaces. In *VLDB'98*, pp. 194–205. Morgan Kaufmann, 1998.

[98] A. Whitney. Abridged kDB+ database manual. http://kx.com/q/d/a/kdb+.htm, 2007.

[99] I. H. Witten, A. Moffat, and T. C. Bell. *Managing gigabytes: compressing, and indexing documents and images*. Van Nostrand Reinhold, International Thomson Publ. Co., New York, 1994.

[100] H. K. T. Wong, H.-F. Liu, F. Olken, D. Rotem, and L. Wong. Bit transposed files. In *Proceedings of VLDB 85, Stockholm*, pp. 448–457, 1985.

[101] K. Wu, J. Gu, J. Lauret, A. M. Poskanzer, A. Shoshani, A. Sim, and W.-M. Zhang. Grid collector: Facilitating efficient selective access from

datagrids. In *International Supercomputer Conference, Heidelberg, Germany, June 21–24, 2005*, May 2005.

[102] K. Wu, E. J. Otoo, and A. Shoshani. On the performance of bitmap indices for high cardinality attributes. In *VLDB'2004*, 2004.

[103] K. Wu, E. J. Otoo, and A. Shoshani. Optimizing bitmap indices with efficient compression. *ACM Trans. Database Syst.*, 31(1):1–38, 2006.

[104] K. Wu, W.-M. Zhang, V. Perevoztchikov, J. Lauret, and A. Shoshani. The grid collector: Using an event catalog to speed-up user analysis in distributed environment. In *Computing in High Energy and Nuclear Physics (CHEP) 2004*, Interlaken, Switzerland, September 2004.

[105] K.-L. Wu and P.S. Yu. Range-based bitmap indexing for high-cardinality attributes with skew. Technical report, IBM Watson Research Center, May 1996.

[106] K. Wu. FastBit reference guide. Technical Report LBNL/PUB-3192, Lawrence Berkeley National Laboratory, Berkeley, CA, 2007.

[107] K. Wu, E. Otoo, and A. Shoshani. A performance comparison of bitmap indices. In *CIKM*. ACM Press, November 2001.

[108] K. Wu, E. J. Otoo, and A. Shoshani. Compressed bitmap indices for efficient query processing. Technical Report LBNL-47807, LBL, Berkeley, CA, 2001.

[109] K. Wu, E. J. Otoo, and A. Shoshani. Compressing bitmap indexes for faster search operations. In *SSDBM*, pp. 99–108, 2002.

[110] K. Wu, K. Stockinger, and A. Shoshani. Performance of multi-level and multi-component compressed bitmap indexes. Technical Report LBNL-60891, Lawrence Berkeley National Laboratory, Berkeley, CA, 2007.

[111] K. Wu, K. Stockinger, and A. Shoshani. Breaking curse of cardinality on bitmap indexes. In *SSDBM'08*, pp. 348–365, 2008.

[112] M.-C. Wu and A. P. Buchmann. Encoded bitmap indexing for data warehouses. In *International Conference on Data Engineering*, Orlando, FL, February 1998. IEEE Computer Society Press.

Chapter 7

Emerging Database Systems in Support of Scientific Data

Per Svensson,[1] Peter Boncz,[2] Milena Ivanova,[2] Martin Kersten,[2] Niels Nes,[2] and Doron Rotem[3]

[1] Swedish Defence Research Agency, Stockholm, Sweden
[2] Centrum Wiskunde & Informatica (CWI), The National Research Institute for Mathematics and Computer Science, The Netherlands
[3] Lawrence Berkeley National Laboratory, Berkeley, California

Contents

- 7.1 Introduction to Vertical Databases 236
  - 7.1.1 Basic Concepts 236
  - 7.1.2 Design Rules and User Needs 237
  - 7.1.3 Architectural Opportunities 240
- 7.2 Architectural Principles of Vertical Databases 241
  - 7.2.1 Transposed Files and the Decomposed Storage Model 241
  - 7.2.2 The Impact of Modern Processor Architectures 243
  - 7.2.3 Vectorization and the Data-Flow Execution Model 245
  - 7.2.4 Data Compression 247
  - 7.2.5 Buffering Techniques for Accessing Metadata 250
  - 7.2.6 Querying Compressed, Fully Transposed Files 251
  - 7.2.7 Compression-Aware Optimizations for the Equi-Join Operator 252
  - 7.2.8 Two Recent Benchmark Studies 253
  - 7.2.9 Scalability 253
- 7.3 Two Contemporary Vertical Database Systems: MonetDB and C-Store 254
  - 7.3.1 MonetDB 254
  - 7.3.2 C-Store 255
- 7.4 The Architecture and Evolution of MonetDB 256
  - 7.4.1 Design Principles 256
  - 7.4.2 The Binary Association Table Algebra 257

        7.4.3   Efficiency Advantages of Using the BAT Algebra 259
        7.4.4   Further Improvements 261
        7.4.5   Assessment of the Benefits of Vertical Organization 261
7.5   Experience with SkyServer Data Warehouse Using MonetDB 263
        7.5.1   Application Description and Planned Experiments 263
        7.5.2   Efficient Vertical Data Access for Disk-Bound Queries 264
        7.5.3   Improved Performance 265
        7.5.4   Reduced Redundancy and Storage Needs 266
        7.5.5   Flexibility 267
        7.5.6   Use Cases Where Vertical Databases
                May Not Be Appropriate 267
        7.5.7   Conclusions and Future Work 268
7.6   Extremely Large Databases and SciDB 268
        7.6.1   Differences between the Requirements of Scientific and
                Commercial Databases 268
        7.6.2   The Array Data Model in SciDB 269
                7.6.2.1   Definition and Creation 269
                7.6.2.2   Operators 270
        7.6.3   Data Overwrite and Provenance 271
        7.6.4   Uncertainty 271
        7.6.5   Storage Layout 272
Acknowledgments 272
References 272

7.1 Introduction to Vertical Databases

7.1.1 Basic Concepts

Consider a high-energy physics experiment, where elementary particles are accelerated to nearly the speed of light and made to collide. These collisions generate a large number of additional particles. For each collision, called an *event*, about 1–10 MB of raw data are collected. The rate of these collisions is about 10 per second, corresponding with hundreds of millions or a few billion events per year. Such events are also generated by large-scale simulations. After the raw data are collected they undergo a *reconstruction* phase, where each event is analyzed to determine the particles it produced and to extract hundreds of summary properties (such as the total energy of the event, momentum, and number of particles of each type).

To illustrate the concept of vertical versus horizontal organization of data, consider a dataset of a billion events, each having 200 properties, with values labeled $V_{0,1}$, $V_{0,2}$, and so on. Conceptually, the entire collection of summary data can be represented as a table with a billion rows and 200 columns as

Emerging Database Systems in Support of Scientific Data

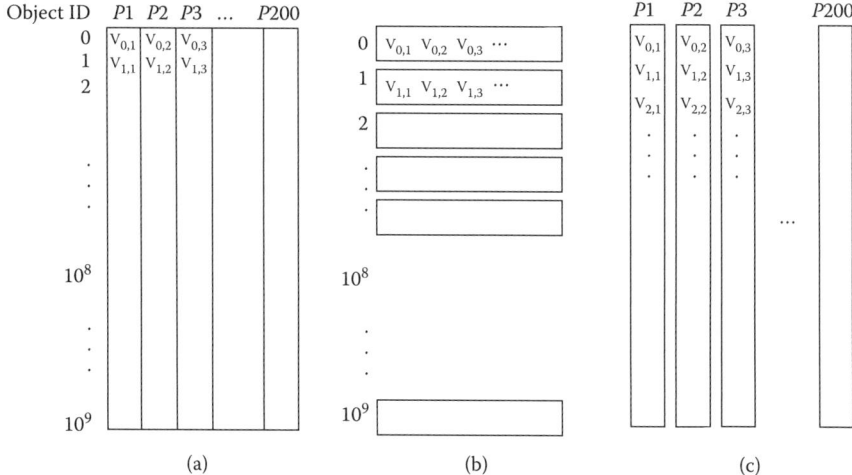

Figure 7.1 Horizontal vs. vertical organization of tabular (relational) data.

shown in Figure 7.1(a). A horizontal organization of the table simply means that the physical layout of the data is row-wise, one row following its predecessor, as shown in Figure 7.1(b). Usually the entire table is stored into disk pages or files, each containing multiple rows. A vertical organization means that the layout of the data is column-wise as shown in Figure 7.1(c). Note that the entire column containing a billion values is usually stored in multiple disk pages or multiple files.

Suppose that a user wishes to get the event IDs that have energy, E, greater than 10 MeV (million electron-volts) and that have number of pions, N_p, between 100 and 200, where *pion* is a specific type of particle. This predicate can be written as: $((E > 10) \setminus (100 < Np < 200))$. It is obvious that in this case searching over the vertically organized data is likely to be faster, since only the data in the two columns for E and N_p have to be brought into memory and searched. In contrast, the horizontal organization will require reading the entire table. Given this simple observation, why were relational database systems typically built with a horizontal organization? As will be discussed next, the majority of database systems were designed for transaction processing, where frequent updates of randomly requested rows were expected, which is the reason for choosing the horizontal organization. In this chapter we discuss the class of applications that benefit greatly from a vertical organization, which includes most scientific data applications.

7.1.2 Design Rules and User Needs

A recent contribution to the literature on database design is a description of the design rationale for Sybase IQ Multiplex, what its authors call *complex analytics*[1]: a parallel, multi-node, shared-storage vertical database system[2] whose major design goal is to efficiently manage large-scale data warehousing workloads. It is argued in that paper that the consequences of adopting the primary design criterion for transactional, write-oriented databases "minimize the portion of stored data that must be locked for exclusive access and the length of time that locks are held." Thus, according to Reference 1, these consequences have generally led to the following set of design rules for transactional, write-oriented databases:

- Since data are usually accessed and modified one record at a time, data should be stored row-wise to allow each record to be updated by a single write operation. Also, data should be stored in small disk pages to minimize the amount of data transferred between memory and disk and to minimize the part of the disk file that needs to be locked during a transaction.
- Indexes should be restricted to a few attributes to avoid locking the entire index tree structures on disk and thereby denying access to whole sets of rows, which might otherwise become necessary when indexes are updated.
- Compression of data is usually not profitable because there is often a mix of different data types and unrelated data values in each row. The CPU time required for compression and decompression will therefore not be recovered by reduced data transfer volume.
- Adding or deleting attributes and indexes is likely to be expensive since all or a large part of the data pages used by the parent table may be affected.
- Finally, updates of an attribute according to even a simple predicate are likely to be costly because the entire row must be read and written when a single attribute is to be updated.

Once the primary criterion for the internal design of a database system becomes the achievement of high performance of complex analytics tasks, this set of rules should be changed as follows:

- Since by storing data column-wise instead of row-wise it is possible to avoid touching those disk pages of a table that are not at all affected by a query, considerable performance improvements may be achieved. Cache efficiency will be enhanced because commonly accessed columns will tend to stay in the cache.
- Data are likely to be read many more times than they are written or updated, making CPU time "investment" in the creation of efficient storage structures more likely to be profitable. Also, data should be stored in large pages so that a large number of relevant data items can be retrieved in a single read operation, resulting in a high overall "hit ratio." Row-wise storage, on the other hand, tends to disfavor large page sizes since each read operation also drags into memory attributes that are not relevant to the query in question, resulting in a low overall hit ratio.

By using version management instead of record-wise locking techniques (which becomes possible largely because of the typically much smaller number of concurrent updates), each query would see a consistent database state in which no locks, or very few locks, ever occur. Also, version management is what complex analytics users need to be able to keep track of their many different analysis paths across the database.

It is possible (although probably often not necessary) to index every attribute since searches greatly dominate over updates and because adding an index to an attribute requires only that attribute to be read, not the entire row.

Data compression is likely to be profitable because the data belonging to one attribute are highly likely to be homogeneous and even auto-correlated.

Adding or deleting attributes of a table is likely to be cheap since only relevant data would be accessed. Updates of an attribute are likely to be relatively cheap because no irrelevant attribute values need to be read or written.

Similar observations were made much earlier,[3] but were for a long time considered irrelevant in mainstream database research. In fact, these alternative design principles have only comparatively recently received serious attention. This renewed interest is at least partly due to the fact that today, many very large databases are actually used for data warehousing, decision support, or business or security intelligence applications, areas where similar characteristics apply as those claimed above. In Svensson,[3] the following additional observations were made:

In scientific data analysis, the number of simultaneous users is typically much smaller than in large-scale commercial applications; but on the other hand the users tend to put more complex, usually unanticipated, queries to the system.

A more automatic response to complex user requests is required from the system components, since in scientific applications no systems or database specialists are usually available.

A system should be transportable to many computer models, including medium-sized computers, because in scientific data analysis applications many users prefer to work with an in-house computer dedicated to the acquisition and analysis of data.

One of the first systems to be developed based on the above principles was the system *Cantor*.[4–7] The Cantor project pioneered the analysis and coordinated application of many of the above-mentioned techniques and concepts in relational systems.

7.1.3 Architectural Opportunities

Today, there is a growing interest in what has been called *read-optimized* database systems,[8,9] that is, systems that are oriented toward *ad hoc* querying of large amounts of data that require little or no updating. Data warehouses represent one class of read-optimized systems, in which bulk loads of new data are periodically carried out, followed by a relatively long period of read-only querying. Early interest in this class of systems came from various statistical and scientific applications, such as epidemiological, pharmacological, and other data analytical studies in medicine,[10] as well as intelligence analysis applications.[11] *Transposed files*, as vertical storage schemes were usually called at the time, were used in a number of early nonrelational read-optimized database systems. A fairly comprehensive list of such systems was given in Copeland and Khoshafian,[12] which asserted that the standard tabular scheme for storage of relations is not necessarily the best, and that transposed files can offer many advantages.

In the field of database technology during the 1970s and early 80s, there was little consensus on how to perform experiments, or even on what to measure while performing them. Today's experiments and analyses are usually far better planned and executed, and the accumulated scientific knowledge in database technology is vastly greater. There is now very good evidence that vertical database systems can offer substantial performance advantages, in particular when used in those statistical and analytical kinds of applications for which the concept was originally developed, (cf. Section 7.5).

An important conceptual step associated with the use of transposed files is that a whole range of new architectural opportunities opens. Below is a partial list of such architectural opportunities. We note, however, that only a subset of these techniques has been widely used in systems currently available:

- *column-wise storage* of data, in place of the conventional row-wise data storage layout used in most relational database management systems (RDBMS), can eliminate unnecessary data access if only a subset of the columns is involved in the query
- *clustering*, in particular *sort ordering*, of attribute values, can speed up searches over column data
- various kinds of lightweight data compression: *minimum byte size, dictionary encoding, differencing* of attribute value sequences, can reduce the amount of data accessed from disk to memory
- *run-length encoding (RLE)* data compression for columns that are ordered can reduce the amount of data fetched from disk into main memory
- *dynamically optimized sequential combinations* of different compression techniques can reduce processing time

B-tree variants or other indexing techniques *designed to efficiently store and retrieve variable-length data* in columns, a requirement for profitable exploitation of many data compression techniques

conjunctive search, join, and set algebra *algorithms exploiting the column-wise storage structure and working directly on compressed data*

lazy decompression of data, that is, data are decompressed only as needed instead of as soon as having been brought into main memory, is required if such algorithms are to be used

compressed lists of tuple *ID*s to represent intermediate and final results in such algorithms

vectorized operations on data streams and the *vectorized dataflow network architecture* paradigm, to reduce call overhead costs and allow efficient query evaluation by interpretation of algebraic expressions rather than by compilation to low-level code

specially designed buffering techniques for storing and accessing metadata and results of simple-transaction-type queries, which are in general not well suited to column storage schemes

Next, we will discuss several of these approaches, with an emphasis on those techniques that have been claimed in the literature to be of particular importance in high-performance systems.

7.2 Architectural Principles of Vertical Databases

The architectural principles discussed in this section were proposed by several groups who have designed different vertical databases over the years. Bringing them together in this way does not mean that these principles can be arbitrarily combined with each other. However, they form a collection of ideas one should probably be aware of when designing or acquiring such systems.

The literature review presented next shows that most of the advantages of vertical storage in databases for analytical purposes have been known and exploited since the early 80s at least, but recently there is renewed, widespread market and research interest in the matter. The lack of interest in the past now seems to reverse into what might be construed as a canonical vertical storage architecture, replacing the previous consensus that the "flat file with indexes" approach is always preferable.

7.2.1 Transposed Files and the Decomposed Storage Model

A number of early papers deal with issues related to how to *group*, *cluster*, or *partition* the attributes of a database table. For example, Navathe et al.[13]

state: "Partitioning in database design is the process of assigning a logical object (relation) from the logical schema of the database to several physical objects (files) in a stored database. *Vertical partitioning* subdivides attributes into groups and assigns each group to a physical object." That paper, however, was not concerned with such analytical applications in which *ad hoc* queries are dominant. Instead, it discusses how tables may be partitioned in order to exploit known correlations between attribute hit rates, to obtain better average query performance. This is worthwhile mainly in routinely repeating processes where access patterns that display such correlations dominate and change slowly.

The term *[fully] transposed file* was used in early papers,[3,14–16] to denote what is today called "vertically fragmented" or "vertically decomposed" data structures,[17] "vertical partitioning,"[9] "column-oriented" databases,[18] or "column store" databases.[8]

Copeland and Khoshafian[12] is the first published paper on transposed files and related structures that is widely referenced in recent database literature. While the authors note that some early database systems used a fully transposed storage model, for example, RM,[19] TOD,[10] RAPID,[14] ALDS,[20] Delta[21] and Tanaka,[22] they described in that paper the advantages of a *fully decomposed storage model* (DSM). A DSM is a "[fully] transposed storage model with surrogates* included." In a DSM each column of a relational table is stored in a separate *binary association table* (BAT), as an array of fixed-size two-field records (TID, value), where TID refers to Tuple ID. According to Khoshafian et al.,[23] the DSM further assumes that two copies of each binary relation are stored, one clustered (i.e., sorted or hashed with an index) on each of the two attributes (TID, value). Copeland and Khoshafian[12] conclude that there seems to be a general consensus among the database community that the conventional N-ary Storage Model (NSM) is better. They suggest that the consensus opinion is not well founded and that neither is clearly better until a closer analysis is made.

In Khoshafian et al.,[23] a parallel query-processing strategy for the DSM is presented, called the *pivot algorithm*. The algorithm makes use of the *join index* concept.[24] An informal description of a generic select-join-project query is given in the paper, where all selects are assumed to be range restriction operations and all joins are equi-joins. The initial *select phase* executes a select operation for every predicate binding in the query, using the appropriate value-clustered BAT as index. The output of this phase is a collection of temporary index lists, each containing the TIDs of selected tuples from conceptual relation tables. All these operations are done in parallel. During the *pivot phase* of the algorithm, the main m-way join operation of the query is executed. A "pivot" TID column is chosen from those attributes that

---

*The term *surrogate* is used by some researchers to denote object identifiers, or as here in relational databases, *tuple identifiers* or *TIDs*. We will use the latter term unless we make a direct quotation.

appear in the join expression. The result of this phase is another collection of temporary index lists indicating which tuples in each conceptual relation satisfy the query. Since a join index clustered on the desired TID exists for all entity-based equi-joins, a full scan can always be avoided. During the *value materialization phase* several independent joins are evaluated, preferably in parallel. The join operands are small binary relations containing only TIDs. The final *composition phase* executes an *m*-way merge join, which permits a large degree of parallelism. Its operands are all small binary relations containing only TID lists whose cardinality has been maximally reduced due to the select operations.

The practical conclusions from this work, reported in Valduriez et al.[25] and cited in Khoshafian et al.,[23] are (1) that DSM with join indexes provides better retrieval performance than NSM when the number of retrieved attributes is low or the number of retrieved records is medium to high, but NSM provides better retrieval performance when the number of retrieved attributes is high and the number of retrieved records is low; and (2) that the performance of single attribute modification is the same for both DSM and NSM, but NSM provides better record insert/delete performance.

This approach is similar to those used in MonetDB[17,26] and in Cantor,[27] with the following main differences: (1) DSM provides two predefined join indexes for each attribute, one clustered on each of the two attributes (attribute value, TID), while Cantor and MonetDB both use indexes that are created as needed during query evaluation; (2) Cantor stores these indices using RLE compression; MonetDB introduces a novel radix cluster algorithm for hash join; (3) although potentially important, parallelism has not been presented as a key design issue for MonetDB, nor was it one for Cantor; (4) the algorithms used in MonetDB and Cantor were both presented as simple two-way joins, corresponding mainly to the composition phase in the DSM algorithm, which is presented as an m-way join.

7.2.2 The Impact of Modern Processor Architectures

Research has shown that DBMS performance may be strongly affected by "cache misses"[28] and can be much improved by use of *cache-conscious* data structures, including column-wise storage layouts such as DSM and within-page vertical partitioning techniques.[29] In Ailamaki et al.[28] (p. 266) this observation is summarized as follows: "Due to the sophisticated techniques used for hiding I/O latency and the complexity of modern database applications, DBMSs are becoming compute and memory bound." In Boncz et al.,[30] it is noted that past research on main-memory databases has shown that main-memory execution needs different optimization criteria than those used in I/O-dominated systems.

On the other hand, it was a common goal early on for scientific database management systems (SDBMS) development projects to exploit the superior CPU power of computers used for scientific applications, which in the early

70s could be orders of magnitude higher than those of processors designed for commercial workloads. The main purpose was to make query evaluation compute and memory bound rather than I/O bound whenever possible.

The MonetDB developers[17,26,31] have conducted thorough analyses of the effect of modern computer hardware architectures on database performance. As advances in CPU speed far outpace advances in dynamic random access memory (DRAM) latency, the effect of optimal use of the memory caches is becoming ever more important. In Manegold et al.[17] a detailed discussion is presented of the impact of modern computer architectures, in particular with respect to their use of multilevel cache memories to alleviate the continually widening gap between DRAM and CPU speeds that has been a characteristic for computer hardware evolution since the late 70s. Memory access speed has stayed almost constant (within a factor of 2), while CPU speed has increased by almost a factor of 1,000 from 1979 to 1999. Cache memories, which have been introduced on several levels to reduce memory latency, can do so effectively only when the requested data are found in the cache.

Manegold et al.[17] claim that it is no longer appropriate to think of the main memory of a computer system as "random access" memory, and show that accessing data sequentially also in main memory may provide significant performance advantages. They furthermore show that, unless special care is taken, a database server running even a simple sequential scan on a table may spend 95% of its cycles waiting for memory to be accessed. This memory-access bottleneck is even more difficult to avoid in more complex database operations such as sorting, aggregation, and join, which exhibit a random access pattern. The performance advantages of exploiting sequential data access patterns during query processing have thus become progressively more significant as faster processor hardware has become available.

Based on results from a detailed analytical cost model, Manegold et al.[17] discuss the consequences of this bottleneck for data structures and algorithms to be used in database systems and identify vertical fragmentation as the storage layout that leads to optimal memory cache usage.

A key tool whose utilization the MonetDB developers pioneered in database performance research is the use of detailed access cost models based on input from hardware event counters that are available in modern CPUs. Use of such models has enabled them, among other things, to identify a significant bottleneck in the implementation of the partitioned hash-join and hence to improve it using *perfect hashing*. Another contribution is their creation of a *calibration tool*, which allows relevant performance characteristics (cache sizes, cache line sizes, cache miss latencies) of the cache memory system to be extracted from the operating system for use in cost models, in order to predict the performance of, and to automatically tune, memory-conscious query processing algorithms on any standard processor.

It is the experience of the MonetDB developers that virtual-memory advice on modern operating systems can be effectively utilized in a way that makes a *single-level storage software architecture* approach feasible. Thus, the

MonetDB database software architecture does not feature secondary storage structures or explicit I/O operations, whereas the underlying "physical" storage architecture is multilevel hierarchical, consisting of CPU registers on the lowest level, two levels of hardware cache memory (L1 and L2), main memory, and virtual swap memory on disk.

A conclusion of these studies is that database algorithms and data structures should be designed and optimized for efficient multilevel memory access from the outset. Careless implementation of the key algorithms can lead to a performance disaster that even faster CPUs will not be able to rescue, whereas careful design can lead to an order of magnitude performance improvement. The authors claim, on very good analytical and experimental grounds, that the vertical decomposition storage feature is in fact the basis of achieving such high performance.

One final achievement of these design studies and subsequent implementation improvements is a data mining benchmark result, which is two orders of magnitude better than that of some commercial database products.

7.2.3 Vectorization and the Data-Flow Execution Model

The use of vector operators (vectorization) in interpretive query evaluation aims at distributing the (usually heavy) interpretation overhead over many elementary CPU operations. It is the same basic idea that is exploited in a vectorized CPU, although used in this context primarily to improve the efficiency of a software process. It turns out, however, that it is not straightforward to devise an efficient vectorized query evaluation process. This is a topic discussed at some length by Boncz et al.[31] It would seem that the MonetDB developers are more or less alone in taking advantage of this approach today.

In Karasalo and Svensson,[5] a vectorized interpretation technique for query evaluation was presented, claiming to be analogous to the operation of a vectorized dataflow computer.[32] This architectural approach was chosen to make sense of the transposed file architecture when extended to support general relational queries from the basic access patterns and conjunctive queries previously studied[3,16] and briefly discussed in Section 7.2.6. It has in fact several key features in common with those of MonetDB/X100 described in Section 7.4.4. Karasalo and Svensson[5] survey the methods that were developed for translating the parsed and syntactically optimized expression of a relational query into an execution plan in the form of one or more hierarchies of static dataflow networks. Each network in the execution plan is a *bipartite graph*, that is, if in such a network two nodes are connected by an arc, then one is a data buffer node and the other an operator node.

Network generation is followed by an execution phase, which proceeds in two stages: (1) when initializing a network hierarchy for evaluation, space is assigned to its buffers from a buffer pool common to all networks; and (2) when evaluating a network, initially all but the upstream boundary buffer nodes are empty. Evaluation proceeds by executing operator nodes in some order until all downstream boundary buffer nodes contain a value (usually a vector value). An operator node may execute whenever none of its inbuffer nodes are empty, and none of its outbuffer nodes are full. In the last system version of Cantor (1991), 31 different vectorized *stream operators* were available to the dataflow network generator. They are software analogues of the machine instructions of a vectorized dataflow computer. On modern "multicore" computers as well as on shared-nothing multinode computer systems, Cantor's vectorized dataflow query evaluation process could quite easily be parallelized.

Boncz et al.[31] argue that database systems usually execute less than one instruction per cycle (IPC), while in scientific computation, such as matrix multiplication, or in multimedia processing, IPCs of two or more are not uncommon on modern CPUs. The authors claim that database systems do not need to perform so badly relative to scientific computing workloads. Based on experimental results they conclude that there are interpretation techniques that, if exploited, would allow DBMS compute performance to approach that of scientific computing workloads. A key technique by which this may be achieved is *loop pipelining*, whereby interpretation overhead is distributed over many elementary operations. This technique is central to the vectorized prototype query processor X100, recently designed and evaluated by the MonetDB developers. According to Boncz et al.,[31] its goal is to:

1. execute high-volume queries at high CPU efficiency,
2. be extensible to other application domains like data mining and multimedia retrieval, and
3. scale with the size of the lowest storage hierarchy (disk).

To achieve these goals, X100 must manage bottlenecks throughout the computer architecture:

Disk. The columnBM I/O subsystem of X100 is geared toward efficient sequential data access. To reduce bandwidth requirements, it uses a vertical storage layout that in some cases is enhanced with lightweight data compression.

RAM. Like I/O, RAM access is carried out through explicit memory-to-cache routines, which contain platform-specific optimizations. The same vertically partitioned and even compressed disk data layout is used in RAM to save space and bandwidth.

Cache. A Volcano-like[33] execution pipeline with a vectorized processing model is used. Small vertical chunks (e.g., 1,000 values) of cache-resident data items, called *vectors*, are the unit of operation for X100 execution primitives. The CPU cache is the only place where bandwidth does not matter, and therefore (de)compression happens on the boundary between RAM and cache.

CPU. Vectorized primitives expose to the compiler that processing a tuple is independent of the previous and next tuples. Vectorized primitives for projections (expression calculation) do this easily, but Boncz et al.[31] try to achieve the same for other query processing operators as well (e.g., aggregation). This allows compilers to produce efficient loop-pipelined code.

7.2.4 Data Compression

There are two obvious ways a DBMS can trade CPU cycles to save disk space and thereby I/O bandwidth, a precious resource.[17] In this context, we are concerned with I/O bandwidth only, since disk space itself has recently become so cheap that its cost rarely matters at all except perhaps in extreme applications, such as large-scale Web search.[34] First, data may be coded into a more compact form.[8] For example, if one is storing an attribute that is a U.S. customer's state of residence, the state can be coded directly into 6 bits, whereas the standard two-character abbreviation requires 16 bits and a variable-length character string for the full name of the state requires many more. Second, data values may be packed compactly by storing N values in K*N bits, where K is the smallest byte size that can hold any value in the column (*bit packing*). Of course, more sophisticated schemes may be used, which can save even more space while allowing for flexible updates with data items which require more than K bits. It is also possible to use additional techniques, in particular sort order, to save additional I/O bandwidth. Note that these simple data compression techniques are equally applicable to row-wise as to column-wise storage schemes. Therefore, when a query is processed using a column-wise storage scheme, what makes the basic difference with respect to I/O bandwidth is the fact that there is no need to transfer data from irrelevant columns into main memory. As we will see, however, additional, more sophisticated compression techniques may also be exploited for column-wise storage.

Although Abadi et al.[18] state that "it was not until the 90s when researchers began to concentrate on how compression affects database performance," the two early papers[16,35] both emphasize in different ways that fast data access may be achieved through judicious use of data compression in fully transposed (*a.k.a.* vertically partitioned) files. Early tutorials on the subject include Severance,[36] Bassiouni,[37] and Roth and Van Horn;[38] however, none of them specifically discusses how to achieve query evaluation-speed improvements. Recent papers on the use of data compression in relational databases are Westmann et al.,[39] Abadi et al.,[18] and Raman and Swart.[40] The most significant advantages are typically obtained when combining data compression with fully transposed file or DSM storage, but there are also proposals to use compression in row-oriented storage schemes.

Abadi et al.[18] discuss an extension of the column-store storage and access subsystem of the C-Store system,[8] while also addressing the issue of querying compressed data. The extended column storage exploits the fact that "sorted data is usually quite compressible"[18] (p. 671), and suggests storing columns in multiple sort orders to maximize query performance, rather than to minimize storage space. The authors propose an architecture that allows for direct operation on compressed data while minimizing the complexity of adding new compression algorithms. They state: "Compression in traditional databases is known to improve performance significantly. It reduces the size of the data and improves I/O performance by reducing seek times (the data are stored nearer to each other), reducing transfer times (there is less data to transfer), and increasing buffer hit rate (a larger fraction of the DBMS fits in the buffer pool). For queries that are I/O limited, the CPU overhead of decompression is often compensated for by the I/O improvements"[18] (p. 671).

Compression techniques for row-wise stored data often employ *dictionary schemes* to code attribute values in fewer bits. Sometimes Huffman encoding is used, whereby varying symbol frequencies may be exploited to gain a more compact total encoding, at the price of having to use varying length codes. Also, the idea of *frame of reference* encoding (*FOR*) is considered, where values are expressed as small differences from some frame-of-reference value, such as the minimum, in a block of data.[18] RLE, where repeats of the same element are expressed as (value, run-length) pairs, is presented as an attractive approach for compressing sorted data in a column-wise store.

An important point made is that if one wants to exploit different compression techniques depending on local properties of data, it is important to find ways to avoid an associated increase in code complexity, where each combination of compression types used to represent the arguments of a join operation would otherwise require its own piece of code.[18] The authors give several examples showing how, by using compressed blocks as an intermediate representation of data, operators can operate directly on compressed data whenever possible, degenerating to a lazy decompression scheme when not possible. Also, by abstracting general properties of compression techniques and letting operators check these properties, operator code may be shielded from having to know details of the way data are encoded.

Harizopoulos et al.[9] discuss techniques for performance improvements in read-optimized databases. The authors report how they have studied performance effects of compressing data using three commonly used lightweight compression techniques: dictionary, bit packing, and *FOR-delta*, the latter a variation of FOR where the value stored is the difference of a value from the previous one, instead of from the same base value.

Database compression techniques are discussed from a more general perspective, pointing out that certain statistical properties of data in a DBMS, in particular skew, correlation, and lack of tuple order, may be used to achieve additional compression.[40] They present a new compression method based on a mix of column and tuple coding, while employing Huffman coding, lexicographical sorting, and delta coding. The paper provides a deeper performance analysis of its methods than has been customary, making it an important contribution also in this context, although it does not specifically focus on column-wise storage. Finally, it briefly discusses how to perform certain operations, specifically *index scan, hash join, group by with aggregation,* and *sort-merge join,* directly on compressed data.

In the MonetDB system,[17] two space optimizations have been applied that reduce the per-tuple memory in BATs (binary association tables, see Section 7.2.1):

Virtual TIDs. Generally, when decomposing a relational table, MonetDB avoids allocating the 4-byte field for the TID since it can be inferred from the data sequence itself. MonetDB's (and DSM's) approach leaves the possibility open for nonvirtual TIDs as well, a feature that may be useful, for example, when performing a hash lookup.

Byte encodings. Database columns often have low domain cardinality. For such columns, MonetDB uses fixed-size encodings in 1- or 2-byte integer values.

In Karasalo and Svensson,[5] the approach to data compression used in Cantor and its testbed is described. It presupposes the existence of an efficient way to organize attribute subfiles containing varying length data. The so-called *b-list* structure developed for this purpose is a B-tree variant suitable for storing linear lists that exploit an observation made in Knuth[41] and also shows similarity to a structure discussed in Maruyama.[42]

When data are read from or written into a b-list node, data values are accessed one block at a time. When a block is written, a compression algorithm is first applied to its data, working as follows: First, from all values in a given sequence, its minimum value is subtracted and stored in a sequence header (cf. the FOR technique discussed above). Then, to store an arbitrary-length subsequence of integers compactly, four alternatives are considered:

1. Use the smallest possible common byte length for the subsequence and store the byte length in the header (bit packing)
2. Use the smallest possible common byte length for the difference subsequence and store the first element of the original subsequence as well as the byte length in the header (FOR-delta)
3. If the sequence is a *run* of equal values, store the value and length of the subsequence in the subsequence header (RLE)
4. If the subsequence is a run of equal differences, store the first element of the subsequence, the common difference, and the subsequence length in the header (delta-RLE)

To combine these alternatives optimally to store a given data sequence (of length n) as compactly as possible, a dynamic programming, branch-and-bound algorithm was developed. This algorithm subdivides the sequence into subsequences, each characterized by *storage alternative, byte size, cardinality,* and *size of header,* so as to represent the entire sequence using as few bits as possible, given the above-mentioned constraints. The algorithm solved this problem in time $O(n)$.

For reading and writing data in b-lists, six access procedures are available, allowing sequential as well as direct-addressed read and write access of data segments. One of the sequential read procedures does not unpack run-compressed data sequences, enabling fast access to such data. This facility is used by the conjunctive query search algorithm as well as by the merge-join algorithm.

7.2.5 Buffering Techniques for Accessing Metadata

In vertical databases, metadata usually play an important role, not only to support users with information about database contents, but also internally to support parameter and algorithm selection during runtime. In order to exploit the advantages of vertical segmentation to achieve the best possible runtime performance, algorithms for searching and accessing data need to frequently consult the metadatabase (MDB) for information about the current status of database contents. For example, to find the fastest way to search a run-length compressed, fully transposed, ordered file (CFTOF search, see Section 7.2.6), it is important to be able to quickly access at runtime certain properties of the attributes involved in the query, in particular their sort key position, as well as their cardinality and value range. The latter information is needed to obtain good estimates of the selectivity of search clauses, to be used when determining which attribute access sequence the query should choose.

However, the vertical storage structure is not well suited for the management of metadata, since the access pattern of system-generated, runtime metadata lookup queries and updates is much more "horizontally" clustered than typical user queries. In fact, the access pattern of such queries is quite similar to that of a random sequence of simple-transaction queries being issued during a short time interval. To a first approximation the individual items in such a sequence can be assumed to be uncorrelated, that is, when the system brings into memory a vertically structured MDB buffer load to access one data item in an MDB attribute, the likelihood that it will access another data item from the same buffer load in the near future is no greater than that of a random hit into the attribute. Therefore, the next time a data item is needed from the same MDB attribute, it is quite likely that a new buffer load will have to be fetched from disk. On the other hand, the likelihood is quite high that the next few MDB accesses will involve other properties associated with the same MDB relation, which should favor horizontal clustering. In the first case, the hit rate for the buffered data will be low and the performance of the system will suffer (unless the entire column fits into the buffer and can therefore be assumed to stay in memory permanently).

A simple solution to this "impedance mismatch" problem could be to store the metadata as a collection of hash tables or linked lists; but if one wants the system to be able to answer general queries that involve metadata relation tables, and perhaps other relation tables as well, a more elaborate solution is needed, possibly one similar to that used in C-Store to manage simple-transaction-type queries (see Section 7.3.2). In Cantor, this issue was handled by designing a separate cache memory for MDB relation tables with a least recently used (LRU) replacement regime, structured as a collection of linked lists of metadata records, one for each relation, attribute, and b-list storage structure. This amounts to adopting an NSM architecture for the cache, to be used for internal queries only. When a user query involves an MDB relation, the entire MDB is first "unified," that is, all updates made to the cache since the previous unification are written out to the transposed files used to permanently store the columns of MDB tables.

7.2.6 Querying Compressed, Fully Transposed Files

Stonebraker et al.,[8] while describing the architectural properties of the C-Store system, acknowledge previous work on using compressed data in databases, stating (p. 564) that "Roth and Van Horn[38] provide an excellent summary of many of the techniques that have been developed. Our coding schemes are similar to some of these techniques, all of which are derived from a long history of work on the topic in the broader field of computer science ... Our observation that it is possible to operate directly on compressed data has been made before."[39] Indeed, the capability to represent multiple values in a single field to simultaneously apply an operation on all the values at once was exploited earlier in the algorithm for CFTOF search described in Svensson.[16]

Batory[15] showed that search algorithms designed for use with transposed files could outperform commonly used techniques such as the use of inverted files (indexes) in a large proportion of cases. In Svensson,[16] theoretical and empirical results were presented, showing that conjunctive queries may be evaluated even more efficiently if the transposed file structure is combined with sorting with respect to the primary key followed by column-wise RLE data compression, forming a CFTOF organization. The performance of a testbed system was measured and compared with a commercially available database system and with results from an analytical performance model. The results showed that order-of-magnitude performance gains could indeed be achieved by combining transposed file storage and data compression techniques.

In Anderson and Svensson[7] the authors later refined these results by combining interpolation search, sequential search, and binary search into a polyalgorithm which dynamically selects the appropriate method, given known metadata. It is shown how this "modified CFTOF interpolation search" significantly improves search performance over sequential CFTOF search in critical cases. The only situation where inverted file range search retains a clear advantage in a CFTOF-structured database is in highly selective queries over nonkey attributes. To add the complex and costly machinery of updatable inverted files to handle that (in a read-optimized database) fairly uncommon special case seems unwarranted in most analytic DBMS applications.

One feature of RLE-compressed vertical data storage architectures that allows for direct operation on compressed data is that not only can they be exploited in conjunctive query searches, but they can be used profitably also in join and set operations on relations, as well as in duplicate removal operations. An important design feature of Cantor's search and sort subsystem,[5] which contains algorithms for internal and external sorting, duplicate tuple detection, conjunctive query search, key lookup, merge-join, set union, set difference, and set intersection, is that all these algorithms are designed to work one (or a small, fixed number of) attribute(s) at a time, to match the transposed file principle. In most of these algorithms, scanning a compressed transposed file is done by using the sequential read interface which retains run-compressed data. Internal sorting is carried out using a modified Quicksort algorithm, which for each (group of) attribute(s) produces a stream of TIDs as output, according to which subsequent attributes are to be permuted.

7.2.7 Compression-Aware Optimizations for the Equi-Join Operator

In Abadi et al.,[18] a discussion of a nested loop–join algorithm capable of operating directly on compressed data is presented, and the paper also contains pseudo-code showing how the join operator may take into account the compression state of the input columns. Combinations of the three cases *uncompressed*, *RLE*, and *bit-vector encoded* data are considered. For example, if one of the input columns is bit-vector encoded and the other is uncompressed, then the resulting column of positions for the uncompressed column can be represented using RLE coding, and the resulting column of positions for the bit-vector column can be copied from the appropriate bit-vector for the value that matched the predicate. The authors also present results from several benchmark tests, showing clear, often order-of-magnitude, performance improvements from judicious use of data compression, in particular the use of RLE on ordered data.

In Svensson,[27] the main equi-join algorithm used in Cantor, based on merging CFTOF-represented relations, was presented by way of a simple example. The paper states that "the search [scanning] phase of the equi-join operation ... is so similar to conjunctive query search that it is to be expected that analogous results hold," but no performance measurement data are given. The end result of the scanning phase is a compressed, ordered list of qualifying TID pairs, that is, a compressed join index. When a join search is to be done on nonkey attributes, one or both factors have to be re-sorted first. In such cases, sorting will dominate in the equi-join process, unless the resulting Cartesian product has much greater cardinality than its factors. However, in situations where no re-sorting of operands is necessary, this algorithm provides a fast way of allowing the equi-join operator to work directly on run-length compressed data.

7.2.8 Two Recent Benchmark Studies

In Stonebraker et al.,[43] results from a benchmarking study are presented, and performance comparisons are made between commercial implementations based on what these authors call "specialized architectures" and conventional relational databases. The tests involve a range of DBMS applications, including both a standard data warehouse benchmark (TPC-H) and several unconventional ones, namely a text database application, message stream processing, and some computational scientific applications. The "specialized architecture" system used in the data warehouse benchmarks was Vertica, a recently released parallel multinode, shared-nothing, vertical database product[44] designed along the lines of C-Store. It utilizes a DSM data model, data compression, and sorting/indexing. On these examples, Vertica spent between one and two orders of magnitude less time than the comparison system, running in a big and expensive RDBMS installation.

Another database design and benchmarking study using semantic Web text data was reported in Abadi et al.[45] The vertical database used in this study was an extension of C-Store capable of dealing with Semantic Web applications, while the row-store system used for comparison was the open source RDBMS PostgreSQL,[67] which has been found more efficient when dealing with sparse data than typical commercial database products (in this application, NULL data values are abundant). The authors showed that storing and processing Semantic Web data in resource description framework (RDF) format efficiently in a conventional RDBMS requires creative representation of the data in relations. But, more importantly, they showed that RDF data may be most successfully realized by vertically partitioning the data that obey logically a fully DSM. The authors demonstrated an average performance advantage for C-Store of at least an order of magnitude over PostgreSQL, even when data are structured optimally for the latter system.

7.2.9 Scalability

Over the last decade, the largest data warehouses have increased from 5 to 100 terabytes, and by 2010, most of today's data warehouses may be 10 times larger than today. Since there are limits to the performance of any individual processor or disk, all high-performance computers include multiple processors and disks. Accordingly, a high-performance DBMS must take advantage of multiple disks and multiple processors. In Dewitt et al.,[46] three approaches to achieving the required scalability are briefly discussed.

In a *shared-memory* computer system, all processors share a single memory and a single set of disks. Distributed locking and commit protocols are not needed, since the lock manager and buffer pool are both stored in the memory system where they can be accessed by all processors. However, since all I/O and memory requests have to be transferred over the same bus that all processors share, the bandwidth of this bus rapidly becomes a bottleneck so there is very limited capacity for a shared-memory system to scale.

In a *shared-disk* architecture, there are a number of independent processor nodes, each with its own memory. Such architectures also have a number of drawbacks that limit scalability. The interconnection network that connects each processor to the shared-disk subsystem can become a bottleneck. Since there is no pool of memory that is shared by the processors, there is no obvious place for the lock table or buffer pool to reside. To set locks, one must either centralize the lock manager on one processor or introduce a distributed locking protocol. Both are likely to become bottlenecks as the system is scaled up.

In a *shared-nothing* approach, each processor has its own set of disks. Every node maintains its own lock table and buffer pool, eliminating the need for complicated locking and consistency mechanisms. Data are "horizontally partitioned" across nodes, such that each node has a subset of the rows (and in vertical databases, maybe also a subset of the columns) from each big table in the database. According to these authors, shared-nothing is generally regarded as the best-scaling architecture (see also Dewitt and Gray[47]).

7.3 Two Contemporary Vertical Database Systems: MonetDB and C-Store

We give next a brief overview of two recently developed vertical database systems to contrast their styles.

7.3.1 MonetDB

MonetDB[48,49] uses the DSM storage model. A commonly perceived drawback of the DSM is that queries must spend "tremendous additional time" doing extra joins to recombine fragmented data. This was, for example, explicitly claimed Ailamaki et al.[29] in p. 169. According to Boncz and Kersten,[30] for this reason the DSM was for a long time not taken seriously by the database research community. However, as these authors observe (and as was known and exploited long ago[16,27]) vertical fragments of the same table contain different attribute values from identical tuple sequences; and if the join operator is aware of this, it does not need to spend significant effort on finding matching tuples. MonetDB maintains fragmentation information as *properties* (metadata) on each binary association table and propagates these across operations. The choice of algorithms is typically deferred until runtime and is done on the basis of such properties.

With respect to query-processing algorithms, the MonetDB developers have shown that a novel *radix-cluster algorithm* for hash-join is better than standard bucket-chained alternatives. In a radix-cluster algorithm, both relations are first partitioned on hash number into a number of separate clusters which each fit the memory cache, before appropriately selected pairs of clusters are hash-joined together (see Manegold et al.[17,26] for details). The result of a hash-join is a binary association table that contains the (TID1, TID2) combinations of matching tuples, that is, a join index. As indicated above, subsequent tuple reconstruction is a cheap operation that does not need to be included in the analysis.

The architectural design and key features of the MonetDB system are presented in Section 7.4, and experience from using it in a large-scale scientific database application is presented in Section 7.5.

7.3.2 C-Store

C-Store[8,50] features a two-level store with one writable part and one, typically much larger, read-only part. Both storage levels are column oriented. The use of this principle, called a *differential file*, in data management was studied in Severance and Lohman,[51] although its full realization in a relational database may perhaps have been first achieved in C-Store much later. This way, C-Store attempts to resolve the conflicting requirements for a fast and safe parallel-writable store (WS) and a powerful read-only query processor and storage system (RS). Tuples are periodically moved from WS to RS by a batch update process. Although the storage model of C-Store is more complex than the DSM in order to also manage queries of simple-transaction type efficiently, most of its basic design principles are similar to those of DSM and hence best suited for read-optimized databases.

To store data, C-Store implements *projections*. A C-Store projection is *anchored* on a given relation table, and contains one or more attributes from this table, retaining any duplicate rows. In addition, a projection can contain any number of attributes from other tables as long as there is a complete sequence of foreign key relationships from the anchor table to the table from which an attribute is obtained. Hence, a projection has the same number of rows as its anchor table. The attributes in a projection are stored column-wise, using one separate storage structure per attribute. Tuples in a projection are sorted with respect to the same *sort key*, that is, any attribute or sequence of distinct attributes of the projection. Finally, every projection is horizontally partitioned into one or more *segments*, each of which is given a segment identifier value *sid*, based on the sort key of the projection. Hence, each segment of a given projection is associated with a *key range* of the sort key for the projection.

To answer SQL queries in C-Store, there has to exist at least one covering set of projections for every table in the logical schema. In addition, it must be possible to reconstruct complete rows of all tables from the collection of stored segments. To do this, C-Store has to join segments from different projections, which is accomplished using *storage keys* and join indexes. Each segment associates every data value of every column with a storage key SK. Values from different columns in the same segment with matching storage keys belong to the same row. Storage keys in RS are implicit, while storage keys in WS are explicitly represented as integers, larger than the largest storage key in RS. Assuming that T1 and T2 are projections that together *cover* the attributes of a table T, an entry in the join index for a given tuple in a segment of T1 contains the segment ID and storage key of the joining tuple in T2. Since all join indexes are between projections anchored at the same table, this is always a one-to-one mapping. This strategy contributes to efficient join operations in C-Store.

7.4 The Architecture and Evolution of MonetDB

7.4.1 Design Principles

MonetDB has been introduced in the previous section as a database system that uses vertical decomposition in order to reduce disk I/O. However, the principal motivation to use this storage model was not so much to reduce disk I/O in scientific or business intelligence query loads—though that is certainly one of its effects. Rather, the MonetDB architecture was based on other considerations given in the original Decomposition Storage Model (DSM)[12] paper, namely it focused on data storage layout *and* query algebra, with the purpose of achieving higher CPU efficiency.

When the original relational databases appeared, CPU hardware followed an in-order, single-pipeline, one-at-a-time design, using a low clock frequency, such that RAM latency took just a few cycles, and disk I/O was the strongest performance factor in database performance. In modern multi-GHz computers, however, the cycle cost of a CPU instruction is highly variable and in pipelined CPU designs with multiple instructions per clock, it depends on CPU cache hit ratio, on branch misprediction ratio (see Ross[52]), and on dependencies on other instructions (less dependencies leading to faster execution). In other words, the difference in throughput between "good" and "bad" program code has been increasing significantly with newer CPU designs, and it had become clear that traditional database code turned out to fit mostly in the "bad" basket.[28] Therefore, MonetDB attempted to follow a query execution strategy radically different from the prevalent tuple-at-a-time pull-based iterator approach (where each operator gets its input by calling the operators of its children in the operator tree), as that can be linked to the "bad" performance characteristics of database code. MonetDB aimed at mimicking the success of scientific computation programs in extracting efficiency from modern CPUs by expressing its calculations typically in tight loops over arrays, which are well supported by compiler technology to extract maximum performance from CPUs through techniques such as strength reduction (replacing an operation with an equivalent, less costly operation), array blocking (grouping subsets of an array to increase cache locality), and loop pipelining (mapping loops into optimized pipeline executions).

7.4.2 The Binary Association Table Algebra

The distinctive feature of MonetDB thus is the so-called Binary Association Table (BAT) Algebra, which offers operations that work only on a handful of BATs. The term *binary association table* refers to a two-column <surrogate, value> table as proposed in DSM. The left column (often the surrogate of the record identity) is called the *head* column, and the right column, the *tail*. The BAT Algebra is closed on BATs, that is, its operators get BATs (or constants) as parameters, and produce a BAT (or constant) result. Data in execution is always stored in (intermediate) BATs, and even the result of a query is a collection of BATs. Some database systems using vertical fragmentation typically use a relational algebra that adopts the table data model, that is, horizontal records inside the execution engine. In the implementation, this leads to query processing strategies where relational tuples (i.e., horizontal structures) are reconstructed early in the plan, typically in the Scan operator. This is not the case in MonetDB; data always remain vertically fragmented.

BAT storage takes the form of two simple memory arrays, one for the head column and one for the tail column (variable-width types are split into two arrays: one with offsets, and the other with all concatenated data). MonetDB allows direct access to these entire arrays by the BAT Algebra operators. In case the relations are large, it uses memory-mapped files to store these arrays. This was in line with the philosophy of exploiting hardware features as much as possible. In this case, allowing array lookup as a way to locate tuples in an entire table in effect means that MonetDB exploits the MMU (memory management unit) hardware in a CPU to offer a very-fast $O(1)$ lookup mechanism by position—where the common case is that the DSM surrogate columns correspond to positions.

As shown in Figure 7.2, MonetDB follows a front-end/back-end architecture, where the front end is responsible for maintaining the illusion of data stored in some end-user format (i.e., relational tables or objects, or XML trees or RDF graphs in SQL, ODMG, XQuery, and SPARQL front ends, respectively). In the MonetDB back end, there is no concept of relational tables (nor concepts of objects); there are only BATs. The front ends translate end-user queries in SQL, OQL, XQuery, and SPARQL into BAT Algebra, execute the plan, and use the resulting BATs to present results. A core fragment of the language is presented below:

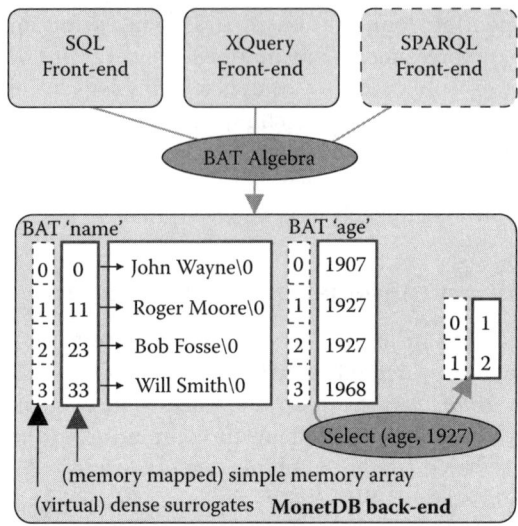

Figure 7.2 MonetDB architecture: The front end translates queries into BAT Algebra expressions; the back end executes the BAT plan.

reverse(bat[t1,t2] B) : bat[t2,t1]   = [<B[i]. tail ,B[i]. head>|i<|B|] "swap columns"
mirror(bat[t1,t2] B) : bat[t1,t1]   = [<B[i].head,B[i]. head>|i<|B|] "make tail equal to head"
mark(bat[t1,t2] B) : bat[t1,TID]   = [<B[i].head,i>|i<|B|] "number tail"

join(bat[t1,t2] L, bat[t2,t3] R) : bat[t1,t3]   = [<L[i].head,R[j]. tail > |i<|L|,j<|R}, L[i]. tail =R[j].head]
"inner join"
uselect(bat[t1,t2] B, t2 v) : bat[t1,void]   = [<B[i].head,nil>|i<|B|, B[I]. tail =v] " selection on tail "
[+](bat[t1,t2] L, bat[t1,t2] R) : bat[t1,t2]   = [<L[i].head,L[i]. tail +R[i]. tail >|i<|L|, j<|R|,
L[i]. head=R[i].head]
" map [op]"
group(bat[t1,t2] L, bat[t1,t2] R) : bat[t1,t2]   = [<L[i].head,unique(L[i]. tail ,R[j]. tail)>|i<|L|,j<|R|,
L[i]. head=R[i].head] "groupby"
unique(bat[t1,t2] B) : bat[t1,t2]   = {<B[i].head,B[i]. tail >|i<|B|} "duplicate elimination "
{sum}(bat[t1,t2] B) : bat[t1,t2]   = [<U[i].head,sum(select(reverse(L)),h.head)>|U= unique(mirror(B))] "aggr{op}"

The BAT-algebra notation used above consists of an operation applied to an operand on the left of the ":", and the result in the right. For example, the "join" operator is applied to left (L) and right (R) BATs, and for entries where tail of L is equal to head of R, it generates a result with columns corresponding to head of L and tail of R.

The reverse(), mirror(), and mark() operators all produce a result in which at least one of the input columns appears unchanged. In the MonetDB implementation, these operations have a constant-cost implementation that just manipulates some information in the column descriptor, since the result BAT shares the (large) array data-structures holding the column data with its input BAT. The [op]() and {op}() are second-order operators that take an operator name *op* and construct for it a map operator (that works on the natural join of all input BATs on head column) and a grouped aggregate function, respectively.

7.4.3 Efficiency Advantages of Using the BAT Algebra

The main advantage of the BAT Algebra is its hard-coded semantics, causing all operators to be predicateless. For comparison, in relational algebra, the join and select operators take an arbitrary Boolean column expression that determines which tuples must be joined and selected. The fact that this Boolean expression is arbitrary, and specified at query time only, means that the RDBMS must include some expression interpreter in the critical runtime code-path of the join and select operators. Such predicates do not occur in BAT Algebra; therefore, we also say it has a "zero degree of freedom." For instance, the hard-coded semantics of join (L,R) is that the predicate is a simple equality between the inner columns of the left BAT, L, and right BAT, R; and its output is the outer columns for the matching tuples. In case of select, the predicate is equality on the tail column. This absence of freedom allows the implementation of the query algebra to forsake an expression interpreting engine; rather, all BAT algebra operations in the implementation map onto array operations. For instance, the expression "select(bat[TID,int] B, int V): bat[TID,TID] R" in BAT Algebra can be represented at the C-level code as something like:

```
for(i=j=0; i<n; i++)
   if (B.tail[i] == V) R.tail[i] = j++;
```

Note the following in the "select" BAT statement above:

The select operator has two parameters B and V and one result R.

B is a BAT with a head-column of type TID, and a tail-column of type int.

V is a constant (a single) value of type int.

R is a BAT with both head and tail column of type TID, where the head represents a surrogate sequence (=0,1,2, ...) and the tail contains the qualifying row-IDs of the rows that matched the int=V condition (in the example shown in Figure 7.2, these are rows 1,2).

Such simple loops are amenable to compiler optimization and CPU out-of-order speculation, which lead to high performance. The philosophy behind BAT Algebra can be paraphrased as "the RISC approach to database query languages": By making the algebra simple, the opportunities are created for implementations that execute the common case very fast.

Note that the above code is only correct if the head column of a BAT, B, contains a densely ascending TID (tuple identifier, that is, surrogate) sequence starting with 0 (that is, B.head = 0,1,2, ...). This happens to be a common case, and MonetDB recognizes this as the *dense* property. Dense TID columns are in fact not stored at all in the implementation, as they are the same as the array index in the column. Many head columns are dense, so MonetDB BAT processing often equates to simple array processing. In addition to denseness, MonetDB keeps a series of other runtime properties on columns (uniqueness, sortedness, min/max) that are exploited at runtime under various circumstances. We show below an example of translating an SQL query into expressions in the BAT Algebra to illustrate the advantages of executing the BAT Algebra expressions. The details of the algebra are not important, but rather this example is intended to illustrate the operators used that execute this query. Note that the algebraic expressions represent an execution plan.

Example: The SQL Query

SELECT DISTINCT P.firstname,P.lastname, SUM(I.price)
FROM Person P, Item I
WHERE P.id = I.buyer and I.year = 2007
GROUP BY P.Firstname,P.lastname

translates into BAT algebra:

s := reverse(mark(uselect(Item_year, 2007)))
b := join(s, Item_buyer)
p := join(b, reverse(Person_id))
r := reverse(mark(reverse(p)))
g := group(join(r, Person_firstname), join(r, Person_lastname))
a := {sum}(join(join(reverse(g), r), Item_price)
[print](join(g, Person_firstname), join(g, Person_lastname), a)

A potential disadvantage of the DSM model is the large number of joins needed to relate columns, also visible in the above plan, which has eight join operators. However, note that only a single join operation is a real value-based join (shown in boldface); all other joins are cases where a TID tail-column from a known min/max range is joined into a dense head-column that spans that range. The property detection in MonetDB successfully derives all such joins into a fetch-join algorithm which for each left input fetches a single tail result from the right input using a positional array lookup. Note that fetch-join is a linear operation at very low CPU cost (a single load instruction). Therefore, MonetDB turns out to perform no expensive additional joins relative to N-ary storage model (NSM) execution engines that store tuple records contiguously in disk pages.

The BAT Algebra core process arrays directly and thus forgoes locking and other transaction processing operations. Rather, a separate module with explicit locking primitives and WAL (write ahead log) functionality is offered. Thus, it is up to the various front ends to ensure that queries do not conflict (referred to as ACID properties), if needed. Note that some front ends do not perform online updates (which is typical in scientific applications and data-mining tools) and therefore do not need to use any transaction management. The advantage in MonetDB is that such applications do not suffer any overhead from transaction facilities that they do not use. The SQL and XQuery front ends both offer full ACID properties, showing that a separation of execution and transaction processing enforcement can indeed be achieved in database architecture.

7.4.4 Further Improvements

The original design of MonetDB had two main weaknesses. First, the reliance on virtual memory for disk storage means that the buffer manager is removed from the system architecture. While removing this layer makes it easier to write efficient data processing algorithms, it means that MonetDB relies on virtual memory advice calls to perform buffering policies. The downside is mainly practical, that is, the implementation of such virtual memory advice can often be incomplete or ineffective, depending on the OS (version). Furthermore, virtual memory prefetching is configured at the OS kernel level and tuned for different access patterns than those that MonetDB targets. This often leads to I/O prefetch sizes that are too small (and thus, lower bandwidth is achieved). The second main problem in the design is that the BAT Algebra implementation follows a design of full materialization. An algebra operator fully consumes its input BATs, producing a full-result BAT. Again, while such loop code is simple and efficient, problems may occur if the result arrays are large. If these are huge, which is often the case with queries on scientific data, output flows via virtual memory to disk, and swapping may happen, deteriorating performance. Both of these problems have been fixed in the subsequent MonetDB/X100 system, which introduces a pipelined model operating on small BAT pieces (vectors) and a buffer manager that can perform efficient asynchronous I/O. The use of a buffer manager also means that compression techniques, which work well with vertical storage, can be exploited. Furthermore, vertically oriented compressed indexes, such as FastBit (described in Chapter 6) can be exploited as well.

7.4.5 Assessment of the Benefits of Vertical Organization

The vertical organization of storage in MonetDB led to the achievement of the original goal of high-performance and CPU efficiency and was shown to outpace relational competitors on many query-intensive workloads, especially when data fits into RAM (see case study in the next subsection). Because of the vertical data layout, it was possible to develop a series of architecture-conscious query processing algorithms, such as for instance radix-partitioned hash-joins and radix cluster/decluster (cache-efficient permutation). Also, pioneering work in architecture-conscious cost modeling and automatic cost calibration were done in this context.

The approach taken by MonetDB of using a front-end/back-end architecture provides practical advantages as well. It is relatively easy to extend with new modules that introduce new BAT Algebra operators. This ease can be attributed to the direct array interface to data in MonetDB, which basically implies that no API is needed to access data (therefore, database extenders do not have to familiarize themselves with a complex API).

The use of a vertical data layout not only on disk but also throughout query processing turned out to be beneficial, especially when operators access data sequentially. Random data access, even if data fits into RAM, is difficult to make efficient, especially if the accessed region does not fit into the CPU cache. In fact, random access does not exploit all the RAM bandwidth optimally; this is typically only achieved if the CPU detects a sequential pattern and the hardware prefetcher is activated. Therefore main-memory algorithms that have predominantly sequential access tend to outpace random-access algorithms, even if they do more CPU work. Sequential algorithms, in turn, strongly favor vertical storage, as memory accesses are dense regardless of whether a query touches all table columns. Also, sequentially processing densely packed data allows compilers to generate single instruction, multiple data (SIMD) code, which further accelerates processing on modern machines.

Finally, the idea articulated in the DSM paper,[12] that DSM could be the physical data model building block that can power many more complex user-level data models, was validated in the case of MonetDB, where a number of diverse front ends were built. We describe briefly below the way BATs were used for processing of different front-end data models and their query languages.

SQL. The relational front-end decomposes tables by column, in BATs with a dense (nonstored) TID head, and a tail column with values. For each table, a BAT with deleted positions is kept. For each column, an additional BAT with insert value is kept. These delta BATs are designed to delay updates to the main columns and allow a relatively cheap snapshot isolation mechanism (only the delta BATs are copied). MonetDB/SQL also keeps additional BATs for join indexes, and value indexes are created on-the-fly.

XQuery. The work in the Pathfinder project[53] makes it possible to store XML tree structures in relational tables as <pre,post> coordinates, represented in MonetDB as a collection of BATs. In fact, the pre-numbers are densely ascending, hence can be represented as a (nonstored) dense TID column, saving storage space and allowing fast O(1) lookups. Only slight extensions to the BAT Algebra were needed, in particular a series of region-joins called "staircase joins" was added to the system for the purpose of accelerating XPath predicates. MonetDB/XQuery provides comprehensive support for the XQuery language, the XQuery Update facility, and a host of specific extensions.

Arrays. The Sparse Relational Array Mapping (SRAM) project maps large (scientific) array-based datasets into MonetDB BATs, and offers a high-level, comprehension-based query language.[54] This language is subsequently optimized on various levels before being translated into BAT Algebra. Array front ends are particularly useful in scientific applications.

SPARQL. The MonetDB team started work in 2008 to offer scalable RDF storage and support for the W3C query language SPARQL to the system.

7.5 Experience with SkyServer Data Warehouse Using MonetDB

7.5.1 Application Description and Planned Experiments

To illustrate the advantages of vertical databases for scientific data management we summarize the experiences from porting the SkyServer application[55] onto MonetDB. The SkyServer application is a good example of a read-optimized database system with long periods of *ad hoc* querying of large data volumes, and periodic bulk-loading of new data. In these settings a column-store architecture offers more efficient data access patterns for disk-bound queries, flexibility in the presence of changing workloads, and reduced storage needs. The MonetDB/SkyServer project[56] started with the purpose of providing an experimentation platform to develop new techniques addressing the challenges posed by scientific data management. Our intent was to examine and demonstrate the maturity of column-store technology by providing the functionality required by this real-world astronomy application. The project shows the advantages of vertical storage architectures for scientific applications in a broader perspective. It goes way beyond micro benchmarks and simulations typically used to examine individual algorithms and techniques. MonetDB/SkyServer allows testing the performance of the entire software stack.

The SkyServer application gives public access to data from the Sloan Digital Sky Survey[57], an astronomy survey with the ambition to map one-quarter of the entire sky in detail. The survey has already collected several terabytes of data. The sky object catalog stored in a relational database reached the volume of 4 TB for data release 6 in 2007. The database schema is organized in several sections among which *Photo* and *Spectro* contain the most important photometric and spectroscopic factual data from the survey. The *Photo* section has a structure centered in the *PhotoObjAll* table. The table contains more than 440 columns and more than 270 million rows, which already stresses the capabilities of most DBMSs. A single record in a row-store representation occupies almost 2 KB, and the majority of the fields are real numbers representing CCD measurements.

Porting of the SkyServer application to MonetDB was organized in three phases. The goal of the first phase was to develop and enhance MonetDB's features to handle the functionality requirements of the SkyServer application. The target dataset during this phase was the so-called Personal SkyServer, a 1% subset of the archive with a size of approximately 1.5 GB. Since this dataset fits entirely in memory, there were no scalability issues with the main-memory orientation of MonetDB. The large vendor-specific schema (consisting of 91 tables, 51 views, and 203 functions, of which 42 are table-valued) and its extensive use of the SQL persistent storage module functionality required an engineering effort. We had to cast vendor-specific syntax (such as identifiers "datetime" vs. "timestamp") in the schema definition into the SQL:2003 standard supported by MonetDB/SQL. We also adapted the application to the column-store architecture and slightly modified the schema, reducing data redundancy.

The challenge addressed in the second phase was to scale the application to sizes beyond the main memory limit. The target dataset was a 10% subset of approximately 150 GB. At the time of writing, the project is in its third phase aiming to support the full-size 4 TB database. Some interesting techniques yet to be investigated that may increase system efficiency are exploring parallel load, interleaving of column I/O with query processing, self-organizing indexing schemes, and exploitation of commonalities in query batches.

7.5.2 Efficient Vertical Data Access for Disk-Bound Queries

As explained in the introductory section, the major advantage of column-wise storage comes from minimizing the data flow from disk through memory into the CPU caches. Many scientific analytical applications involve examination of an entire table, or a big portion of it, while at the same time spanning just a few attributes at a time. The immediate benefit the column-wise storage brings is that only data columns relevant for processing are fetched from disk.

In contrast, the access pattern in a row-wise storage of wide tables, such as the *PhotoObjAll* table, might require hundreds of columns to be transferred from disk, where many of the columns are irrelevant to the query. This becomes the major performance bottleneck for analytical queries. To illustrate the problem, consider the following SQL query searching for moving asteroids (Q15 in Gray et al.[58]).

SELECT objID, sqrt(power(rowv,2) + power(colv,2)) as
    velocity
FROM PhotoObj
WHERE (power(rowv,2) + power(colv,2)) > 50
    and rowv >= 0 and colv >= 0;

The execution plan for a row-wise storage organization involves a full table scan, which leads to transferring entire records of 440+ columns in order to process the four columns referred to in the query. For the 150 GB dataset, the volume transferred is almost 50 GB. The execution plan in MonetDB involves scans strictly limited to the columns directly referenced in the query, which amounts to 370 MB for the example query above.

The access pattern problem in row-wise storage systems has already been addressed by a variety of techniques, such as indexes, materialized views, and replicated tables. For example, if all the columns in a query are indexed, the query can be substantially sped up by scanning the shorter index records instead of touching the wide records of the main table. We illustrate this with the next query example. It extracts celestial objects that are low-z quasar candidates, a property specified through correlations between the objects' magnitudes in different color bands (query SX11 in Gray et al.[58]).

SELECT g, run, rerun, camcol, field, objID}
FROM Galaxy
WHERE ((g <= 22)
  and (u − g >= −0.27) and (u − g < 0.71)
  and (g − r >= −0.24) and (g − r < 0.35)
  and (r − i >= −0.27) and (r − i < 0.57)
  and (i − z >= −0.35) and (i − z < 0.70))

The query predicates do not allow efficient index search of the qualifying rows; instead scanning of all the rows is needed. However, a full table scan can be avoided using available indexes that contain all the necessary columns. The data volume transferred for the 150 GB dataset is 1.8 GB, a substantial reduction with respect to the full table scan, but still twice as large as the 850 MB transferred in MonetDB for the same query. The reason is that the indexes chosen for the query execution contain several additional columns irrelevant for this query.

7.5.3 Improved Performance

In addition to the efficient vertical access pattern, MonetDB employs a number of techniques to provide high performance for analytical applications. Among these are runtime optimization, such as choosing the best algorithm fitting the argument properties, and efficient cache-conscious algorithms exploiting modern computer architecture. To demonstrate the net effect of these techniques on the performance experienced by the end user, we performed a few experiments with the above table- and index-scan queries against both the 1.5 GB and 150 GB datasets. The elapsed times in seconds are shown in Table 7.1. The performance of the vertical database for index-supported queries is comparable for the small dataset, and 30% better for the large dataset. Queries involving full table scans are sped up by a factor of 5 for the large dataset.

TABLE 7.1 Elapsed times in seconds for two types of queries against a "small" (1.5 GB) and a "large" (150 GB) dataset

	Table Scan 1.5 GB	Index Scan 1.5 GB	Table Scan 150 GB	Index Scan 150 GB
Row-store	6.6	0.4	245	24
Column-store	0.4	0.47	53	16

7.5.4 Reduced Redundancy and Storage Needs

The original SkyServer system utilizes indexes and replication to speed up important disk-bound queries. All tables have primary and foreign key constraints supported by B-tree indexes, and many tables have covering indexes created after careful workload analysis. Replicated tables are also used to speed up some frequent classes of queries. For instance, the *PhotoTag* table is a vertical partition of the *PhotoObjAll* table that stores redundantly its most popular 100+ columns. The *SpecPhotoAll* table stores the most popular of the columns from the precomputed join of photo and spectrum tables.

In order to support the original queries we replaced the *PhotoTag* and *SpecPhotoAll* tables with views exploiting the advantages of the column-wise storage of MonetDB. This replacement had little impact on the performance of queries that involve those tables because of the column-wise storage organization. However, generating the views was still worthwhile, saving approximately 10% of the storage needs.

The index support in MonetDB is limited to primary and foreign keys. The system generates indexes on-the-fly when columns are touched for the first time. The net effect of reducing data volume is that the storage needs of MonetDB database image decreased by approximately 30%.

7.5.5 Flexibility

Although secondary access structures in row-wise storage systems improve performance substantially in comparison to full table scans, they exhibit relatively static behaviors with respect to changing workloads. Modern DBMSs come with advanced database design-tuning wizards, which derive design recommendations using representative workloads. Due to its complexity, the workload analysis is mostly performed offline and requires database administrator (DBA) competence to decide on the final database design. When the workload changes, it is probable that the new, unanticipated queries are not supported (or partially supported) by the existing indexes, which leads to suboptimal system performance. The typical solution is that the DBA monitors the system functionality and periodically reruns the workload analysis and modifies the supporting secondary structures.

Recently, online tuning tools have been proposed[59] that take the burden from the DBA but still incur overhead for monitoring and creation of secondary structures. Dealing with this issue is completely avoided in MonetDB. When the query load changes to incorporate new attributes, the execution plans simply transfer to memory only the new columns of interest. This is achieved without any storage, creation, or monitoring overhead for secondary structures, but simply based on the architectural principles of the column-wise storage systems.

7.5.6 Use Cases Where Vertical Databases May Not Be Appropriate

There are a number of situations where column-wise storage is comparable to or slower than row-wise systems. The category of point-and-range queries is usually efficiently supported in the row-store databases since the available indexes enable quick retrieval of qualifying rows. For a small number of qualifying rows, the data transfer is sufficiently efficient and is not perceivable by the end user. For the same query category MonetDB often uses a sequential scan, which might be slower than searching with a B-tree index. However, for append-only data, which is the case for scientific data, new types of compressed bitmap indexes (described in Chapter 6) require a relatively small space overhead of only 30% of the original data. If this overhead is not prohibitive, then all columns (or columns searched often) can be indexed to provide efficient point-and-range queries in vertical databases.

Another source of performance overhead in vertical databases is tuple reconstruction joins. Despite their efficient implementation, they may still contribute a substantial cost for queries that request all attributes (referred to as "SELECT *" queries), or queries with a large number of attributes. Here again, using compressed bitmap indexes can mitigate this overhead, since joining the results of qualifying tuples from each column can be done by logical operations (AND, OR, NOT) over multiple bitmaps, where each bitmap represents the result of searching the index of each column.

There are some uncommon applications where all (or most) columns are needed in every query. In such cases there is no value to using column-wise organization, and row-wise organization with appropriate indexing (for selecting the desired tuples given predicate conditions) may prove more efficient. Also, row-wise organization may be more appropriate in applications where very few rows are selected, and several columns are involved. An extensive analysis of which organization is best was conducted in.[60] Given a characterization of the query patterns, a formula was developed in order to determine which organization is better. By and large, for applications where a large number of rows is selected, and only a subset of the columns is involved in the query, column-wise organization is superior. Furthermore, in practical experiments described in O'Neil et al.,[60] it was shown that when sequential reads (which are much faster than random read operations) are considered as a possible strategy, column-wise organization is even more favorable because it is much easier to utilize sequential read operations with the vertical data organization.

Although we prefer to reduce data redundancy, in some cases it may prove useful to store derived data when generated, for instance, by expensive computations. For example, the *Neighbors* table groups together pairs of SDSS objects within an a-priori distance bound of 0.5 arc-minutes. Our attempt to replace this table with a view computing the distances was shown to be less efficient than accessing the precomputed table.

7.5.7 Conclusions and Future Work

Our experiences with MonetDB/SkyServer application confirm the advantages of column-wise storage systems for scientific applications with analytical disk-bound processing. To improve the performance for point-and-range queries several techniques for workload-driven self-organization of columns have been developed in MonetDB, such as cracking (continuous physical organization based on access patterns),[61] and adaptive segmentation and replication (splitting columns into segments or replicating segments).[62] We intend to integrate those techniques in support of the SkyServer application. Since compression has shown to be particularly efficient in combination with column-wise storage,[18] we also intend to investigate and utilize appropriate compression schemes for the SkyServer application.

The MonetDB execution engine differs in a fundamental way from state-of-the-art commercial systems. The execution paradigm is based on full materialization of all intermediate results in a query plan. This opens another direction of research exploiting commonalities in query batches by carefully preserving and reusing common intermediate results.

7.6 Extremely Large Databases and SciDB

In this section we describe very recent developments in the area of scientific databases that may lead to yet another type of database architecture that is neither horizontal nor vertical but rather based on array structures. These developments were initiated in two successive workshops called Extremely Large Databases (XLDB)[63,64] that were organized in order to address the challenge of designing databases that can support the complexity and scale involved in scientific applications. An important outcome of these workshops was the foundation of an organization consisting of researchers and implementers from a variety of disciplines dedicated to the design and implementation of a new open source science database called SciDB.

7.6.1 Differences between the Requirements of Scientific and Commercial Databases

Major differences between the requirements of large scientific databases and current commercial DBMS offerings were noted.[65] This led to the conclusion that the new database system should not just consist of incremental improvements to existing commercial DBMSs but requires a complete new design from the ground up. The most important differences between these two types of databases are summarized in Table 7.2.

TABLE 7.2 Main differences between requirements for commercial DBMS and SciDB

Feature	Commercial DBMS	Scientific Databases
Transaction Processing	Commonly used, requires optimization	Less important, no optimization needed
Insertion of New Data	One row at a time	Bulk loading
Main Usage	Multiple usage modes	Large-scale scientific analysis
Data Model	Relational tables	Arrays, vectors, graphs, sequences, meshes
Language Binding	SQL	MATLAB, R package, Python, C++, IDL
Scalability	Usually single machine	Must run on incrementally scalable clusters or clouds of industry-standard hardware
Provenance	Seldom required	Required, data derivation process must be reproducible
Uncertainty	No support	Scientific data is inherently uncertain, support required (error bars)
Data Overwrite	Old values are replaced with new values	Old values cannot be discarded; queries on historical data must be supported
Administrative Resources	Commonly one or more DBAs may be employed	Minimal, preferably none

In the following sections we provide more details about the requirements from SciDB and the planned implementations of some of the required features.

7.6.2 The Array Data Model in SciDB

7.6.2.1 Definition and Creation

One of the first tasks taken by members of the SciDB organization was the incorporation of an array data model. In this section, more details are provided about this model as presented in a recent document.[66] The plan is to support multidimensional arrays that can have any number of dimensions. Arrays can be nested, that is, array cells may contain records (i.e., tuples), which in turn can contain components that are by themselves multidimensional arrays. An array structure must be defined first, and after that multiple instances of it can be created by supplying high watermarks (dimension boundaries) for each instance. The basic syntax for defining an array (based on Reference 66) is as follows:

define ArrayName ({VaribleName=VariableType}) ({DimensionName})
where "{}" means one or more instances

The definition consists of two pairs of brackets. The first pair of brackets describes the array cell contents. Each element has a name and a data type, which can be either an array or a scalar. The second pair of brackets describes the dimensions of the array; these must be integer-valued. For example, given below is a definition of a three-dimensional array (based on Reference 66) called climate_array, where each cell consists of two floating numbers.

define climate_array (temperature = float,
  pressure = float) (I, J, K)

A physical array can be created from this definition by specifying sizes of each dimension. For example, the statement

create climate_region **as** climate_array [1024,1024,1024]

will create a new instance of climate_array, where each of the respective three dimensions has a size of 1024. In addition, arrays can be enhanced with user-defined functions (UDFs) in order to perform transposition, scaling, translation, and other coordinate transformations.

7.6.2.2 Operators

Several operators on arrays are defined. They fall into two main categories: structural operators and content-dependent operators. Structural operators can operate on the structure of the array independent of the data, and content-dependent operators perform some content-based operation, such as aggregation, based on data values stored in the array.

Examples of structural operators are **Subsample** and **Reshape**. The **Subsample** operator takes as its input an array A and a predicate specified on the dimensions of A. It then generates a new array with the same number of dimensions where the dimension values must satisfy the predicate (e.g., every 10th value of the dimension). **Reshape** is a more advanced structural operator that can convert an array to a new one with more or fewer dimensions possibly with new dimension names, but the same number of cells. Other structural operators include **Structural-Join (SJoin), Add Dimension, Remove Dimension, Concatenate, and Cross Product.**

Content-dependent operators are those whose result depends on the data that is stored in the input array. A simple example of this kind of operator is

Filter, which takes as input an array A and a predicate P over the data values that are stored in the cells of A and returns an array with the same dimensions as A. The predicate P is applied to each cell of the array, and if the predicate is satisfied, the cell value is unchanged; otherwise it is set to NULL. Other operators are **Aggregate, CJoin** (content-based join), **Apply**, and **Project**.

7.6.3 Data Overwrite and Provenance

In commercial database systems old values are overwritten with new values once data is updated. In contrast, scientific work often requires that no data be discarded, as the old values are needed for provenance (lineage) purposes. This requires that a history dimension be added to every updatable array. The history dimension may use integers to denote the sequence of updates, or it can be enhanced with a mapping between these integers and a wall clock allowing the array to be addressed using conventional time.

A related requirement from scientific users is the ability to repeat data derivations. In case the derivation was done inside SciDB, a simple log can do the job. However, in many cases arrays will be loaded from external sources. In these cases a metadata repository will be needed where scientists will enter information about the programs that were used to derive the data alongside with their run-time parameters, so that a record of provenance is available. A provenance query language is also planned such that queries about derivation steps for a given data element or finding other data elements whose values are impacted by a given data element can be easily found. The management of provenance for scientific data is of major importance, and is covered in detail in Chapter 12.

7.6.4 Uncertainty

As scientific data is inherently imprecise, an important requirement from SciDB is the ability to support uncertain data. The current preferred model of uncertainty used by many scientists involves the representation of data elements by appropriate normal distributions. SciDB will support "uncertain x" for any data type x using two values (mean and standard deviation). This potentially may lead to increase in the size of the database, as two values are stored per each element rather than one. However, in many practical cases some compression is possible for arrays that have the same error bounds for all values. Appropriate operators on uncertain data that compute the aggregated error will also be supported.

Another source of errors and uncertainty in scientific databases is a result of location calculation for observed objects commonly used in astronomy and GIS databases. These calculations may contain some approximation errors (whose bounds are known) due to hardware calibration. Federating or combining astronomical datasets may sometimes require spatial join operators that match objects between several datasets based on their location calculation.

This may require specialized algorithms and data structures to ensure that objects are correctly matched, taking into account that their location information may contain some bounded error.

7.6.5 Storage Layout

As mentioned before, it is anticipated that most data will be bulk loaded into SciDB from input sources such as scientific measurement devices or sensors. The data will be linearized based on some ordering of the dimensions (typically, time will be the most dominant dimension) and will be initially stored in the compute node's memory. Due to the large volume of data, it will need to be written into disk buckets that will contain rectangular chunks of the array. Keeping track of the location on disk and contents of these buckets will require a data structure such as R-tree. Algorithms for determining optimal shapes of buckets (size of stride in each dimension), bucket compression, as well as deciding when to merge several buckets into a larger one, are still open research issues.

Acknowledgments

The work on the MonetDB/SkyServer project was supported by the Bsik-Bricks and MultimediaN programs. We would like to thank all members of the CWI database team in the past decade for their joint efforts that made MonetDB into a successful open-source database product. The work on the section describing XLDB/SciDB was supported by the Director, Office of Science, Office of Advanced Scientific Computing Research, of the U.S. Department of Energy, under Contract No. DE-AC02-05CH11231.

References

[1] MacNicol, R., French, B.: Sybase IQ multiplex—designed for analytics. In *Proc. of the 30th Int. Conf. on Very Large Databases*, Toronto, Canada (2004).

[2] http://www.sybase.com/products/datawarehousing/sybaseiq. Accessed May 22, 2008.

[3] Svensson, P.: *Contributions to the design of efficient relational data base systems. Summary of the author's doctoral thesis.* Report TRITA-NA-7909, Royal Institute of Technology, Stockholm (1979).

[4] Karasalo, I., Svensson, P.: An overview of Cantor—a new system for data analysis. In *Proc. 2nd Int. Workshop on Statistical Database Management (SSDBM)* (1983).

[5] Karasalo, I., Svensson, P.: The design of Cantor—a new system for data analysis. In *Proc. 3rd International Workshop on Statistical and Scientific Database Management (SSDBM)* (1986).

[6] Svensson, P.: Database management systems for statistical and scientific applications: Are commercially available DBMS good enough? In *Proc. 4th Int. Working Conf. on Statistical and Scientific Database Management (SSDBM)*, Rome, Italy, June 21–23. LNCS 339, Springer Verlag (1988).

[7] Andersson, M., Svensson, P.: A study of modified interpolation search in compressed, fully transposed, ordered files. In *Proc. 4th Int. Working Conf. on Statistical and Scientific Database Management (SSDBM)*, Rome, Italy, June 21–23. LNCS 339, Springer-Verlag (1988).

[8] Stonebraker, M., Abadi, D.J., Batkin, A., et al.: C-Store: A column-oriented DBMS. In *Proc. 31st Int. Conf. on Very Large Databases*, Trondheim, Norway, August 30–September 2, 553–564 (2005).

[9] Harizopoulos, S., Liang, V., Abadi, D.J., Madden, S.: Performance tradeoffs in read-optimized databases. In *Proc. 32nd Int. Conf. on Very Large Databases*, September 12–15, Seoul, South Korea (2006).

[10] Wiederhold, G., Fries, J.F., Weyl, S.: Structured organization of clinical data bases, In *Proc. of the National Computer Conference*, AFIPS Press (1975).

[11] Bergsten, U., Schubert, J., Svensson, P.: Applying data mining and machine learning techniques to submarine intelligence analysis. In *Proc. 3rd Int. Conf. on Knowledge Discovery and Data Mining*. AAAI Press, Menlo Park, CA, USA (1997).

[12] Copeland, G.P., Khoshafian, S.N.: A decomposition storage model. In *Proc. 1985 SIGMOD Conf.*, ACM, New York (1985).

[13] Navathe, S., Ceri, S., Wiederhold, G., Dou, J.: Vertical partitioning algorithms for database design. *ACM Trans. on Database Systems*, 9(4): 680–710 (1984).

[14] Turner, M.J., Hammond, R., Cotton, P.: A DBMS for large statistical databases. In *Proc. 5th Int. Conf. on Very Large Databases*, 319–327 (1979).

[15] Batory, D.S.: On searching transposed files. *ACM Trans. on Database Systems* (TODS), 4(4): 531–544 (1979).

[16] Svensson, P.: On search performance for conjunctive queries in compressed, fully transposed ordered files. In *Proc. 5th Int. Conf. on Very Large Databases*, 155–163 (1979).

[17] Manegold, S., Boncz, P.A., Kersten, M.L.: Optimizing database architecture for the new bottleneck: Memory access. *The VLDB Journal* 9(3): 231–246 (2000).

[18] Abadi, D.J., Madden, S., Ferreira, M.C.: Integrating compression and execution in column-oriented database systems. In *Proc. 2006 SIGMOD Conf.*, June 27–29, Chicago, IL, USA. ACM, New York (2006).

[19] Lorie, R.A., Symonds, A.J: *A Relational Access Method for Interactive Applications. Data Base Systems*, Courant Computer Science Symposia, vol. 6. Prentice-Hall (1971).

[20] Burnett, R., Thomas, J.: Data management support for statistical data editing. In *Proc. 1st Lawrence Berkeley Laboratory Workshop on Statistical Database Management* (1981).

[21] Shibayama, S., Kakuta, T., Miyazaki, N., Yokota, H., Murakami, K.: A relational database machine with large semiconductor disk and hardware relational algebra processor. *New Generation Computing* vol. 2 (1984).

[22] Tanaka, Y.: A data-stream database machine with large capacity. In *Advanced Database Machine Architectures*, Hsiao, D.K. (ed.), Prentice-Hall (1983).

[23] Khoshafian, S.N., Copeland, G.P., Jagodis, T., Boral, H., Valduriez, P.: A query processing strategy for the decomposed storage model. In *Proc. ICDE*, 636–643, IEEE Computer Society (1987).

[24] Valduriez, P.: Join indices. *ACM Trans. on Database Systems* (TODS), 12(2): 218–246 (1987).

[25] Valduriez, P., Khoshafian, S., Copeland, G.: Implementation techniques of complex objects, *Proc. 12th Int. Conf. on Very Large Databases*, Kyoto, Japan (1986).

[26] Manegold, S., Boncz, P.A., Kersten, M.L.: What happens during a Join? Dissecting CPU and memory optimization effects. In *Proc. of the 26th Int. Conf. on Very Large Databases*, Cairo, Egypt (2000).

[27] Svensson, P.: Highlights of a new system for data analysis. In *Proc. CERN Workshop on Software in High Energy Physics* (invited paper), 4–6 Oct., 119–146 (1982).

[28] Ailamaki, A., DeWitt, D., Hill, M., Wood, D.A.: DBMSs on a modern processor: Where does time go? In *Proc. 25th Int. Conf. on Very Large Databases*, Edinburgh, Scotland (1999).

[29] Ailamaki, A., DeWitt, D., Hill, M., Skounakis, M.: Weaving relations for high performance. In *Proc. of the 27th Int. Conf. on Very Large Databases*, Rome, Italy (2001).

[30] Boncz, P.-A., Kersten, M.L.: MIL primitives for querying a fragmented world. *The VLDB Journal* 8(2):101–119 (1999).

[31] Boncz, P., Zukowski, M., Nes, N.: MonetDB/X100: Hyper-pipelining query execution. In *Proc. 2nd Biennial Conference on Innovative Data Systems Research (CIDR)*, VLDB Endowment (2005).

[32] Giloi, W.K.: Towards a taxonomy of computer architecture based on the machine data type view. In *Proc. 10th Ann. Symp. on Computer Architecture*, Stockholm. IEEE Inc., New York (1983).

[33] Graefe, G.: Volcano—an extensible and parallel query evaluation system. *IEEE Trans. Knowl. Data Eng.* 6(1):120–135 (1994).

[34] Chang, F., Dean, J., Ghemawat, S., Hsieh, W.C., Wallach, D.A., Burrows, M., Chandra, T., Fikes, A., Gruber, R.E.: Bigtable: A distributed storage system for structured data. In *Proc. 7th Symposium on Operating Systems Design and Implementation* (OSDI '06), November 6–8, Seattle, USA (2006).

[35] Eggers, S.J., Olken, F., Shoshani, A.: A compression technique for large statistical databases. In *Proc. 7th Int. Conf. on Very Large Databases* (1981).

[36] Severance, D.G.: A practitioner's guide to database compression—A tutorial. *Inf. Syst.* 8(1):51–62 (1983).

[37] Bassiouni, M.A.: Data compression in scientific and statistical databases. *IEEE Trans. on Software Eng.*, SE-11(10), Oct., 1047–1058 (1985).

[38] Roth, M.A., Van Horn, S.J.: Database compression. *SIGMOD Record* 22(3) (1993).

[39] Westmann, T., Kossmann, D., Helmer, S., Moerkotte, G.: The implementation and performance of compressed databases. *SIGMOD Record* 29(3): 55–67 (2000).

[40] Raman, V., Swart, G.: How to wring a table dry: Entropy compression of relations and querying of compressed relations. In *Proc. 32nd Int. Conf. on Very Large Databases*, September 12–15, Seoul, South Korea (2006).

[41] Knuth, D.E.: *The Art of Computer Programming. Vol 3: Sorting and Searching.* Addison-Wesley (1973).

[42] Maruyama, K., Smith, S.E: Analysis of design alternatives for virtual memory indexes. *Comm. of the ACM* 20(4) (1977).

[43] Stonebraker, M., Bear, C., Cetintemel, U., Cherniack, M., Ge, T., Hachem, N., Harizopoulos, S., Lifter, J., Rogers, J., Zdonik, S.B.: One size fits all?—Part 2: Benchmarking results. In *Proc. 3rd Biennial Conf. on Innovative Data Systems Research (CIDR)*, VLDB Endowment (2007).

[44] http://www.vertica.com/product/relational-database-management-system-overview. Accessed May 22, 2008.

[45] Abadi, D.J., Marcus, A., Madden, S., Hollenbach, K.: Scalable Semantic Web data management using vertical partitioning. In *Proc. 33rd Int. Conf. on Very Large Databases*, September 23–28, Vienna, Austria (2007).

[46] DeWitt, D., Madden, S., Stonebraker, M.: How to build a high-performance data warehouse. http://db.csail.mit.edu/madden/high_perf.pdf. Accessed May 22, 2008.

[47] DeWitt, D., Gray, J.: Parallel database systems: The future of high performance database processing. *Comm. ACM*, 35(6): 85–98 (1992).

[48] http://MonetDB.cwi.nl/. Accessed July 20, 2009.

[49] *Monet: A next-generation DBMS kernel for query-intensive applications.* PhD Thesis, Univ. of Amsterdam, The Netherlands, May 2002.

[50] http://db.csail.mit.edu/projects/cstore/ Accessed May 22, 2008.

[51] Severance, D.G., Lohman, G.M.: Differential files: Their application to the maintenance of large databases. *ACM Trans. Database Syst.* 1(3): 256–267, Sept. 1976.

[52] Ross, K.A.: Conjunctive selection conditions in main memory. ACM SIGMOD 2002.

[53] Boncz, P.A., Grust, T., Van Keulen, M., Manegold, S., Rittinger, J., Teubner, J.: MonetDB/XQuery: A fast XQuery processor powered by a relational engine. SIGMOD Conference 2006: 479–490.

[54] Cornacchia, R., Héman, S., Zukowski, M., de Vries, A.P., Boncz, P.A.: Flexible and efficient IR using array databases. *VLDB Journal*, 17(1): 151–168 (2008).

[55] Szalay, A.S., Gray, J., Thakar, A.R., et al.: The SDSS SkyServer: Public access to the Sloan digital sky server data. In *Proc. 2002 SIGMOD Conf.*, 570–581 (2002).

[56] M. Ivanova, N. Nes, R. Goncalves, M.L. Kersten: MonetDB/SQL meets SkyServer: The challenges of a scientific database. In *Proc. 19th Int. Conf. on Statistical and Scientific Database Management (SSDBM)* (2007).

[57] Sloan Digital Sky Survey / SkyServer: http://cas.sdss.org/. Accessed July 20, 2009.

[58] Gray, J., Szalay, A.S., Thakar, A.R., et al.: *Data mining the SDSS SkyServer database*. Microsoft publication MSR-TR-2002-01, January 2002.

[59] Bruno, N., Chaudhuri, S.: An online approach to physical design tuning. In *Proc. ICDE*, 826–835, IEEE Computer Society (2007).

[60] O'Neil, E., O'Neil, P.E., Wu, K.: Bitmap index design choices and their performance implications. *Proc. of IDEAS* 2007,72–84.

[61] Idreos, S., Kersten, M.L., Manegold, S.: Database cracking. In *Proc. 3rd Biennial Conference on Innovative Data Systems Research (CIDR)* 68–78, VLDB Endowment (2007).

[62] Ivanova, M., Kersten, M.L., Nes, N. Self-organizing strategies for a column-store database. *Proc. 11th International Conference on Extending Database Technology*, March 25–30, Nantes, France (2008).

[63] http://www-conf.slac.stanford.edu/xldb07/. Accessed on July 20, 2009.

[64] http://www-conf.slac.stanford.edu/xldb08/. Accessed on July 20, 2009.

[65] http://scidb.org/. Accessed July 20, 2009.

[66] Stonebraker, M., Becla, J., Dewitt, D., Lim, K.-T., Maier, D., Ratzesberger, O., Zdonik, S.: Requirements for science data bases and SciDB, in CIDR Persepectives 2009.

[67] http://www.postgresql.org/. Accessed May 22, 2008.

Part IV

Data Analysis, Integration, and Visualization Methods

Chapter 8

*Scientific Data Analysis**

Chandrika Kamath,[1] Nikil Wale,[2] George Karypis,[2] Gaurav Pandey,[2] Vipin Kumar,[2] Krishna Rajan,[3] Nagiza F. Samatova,[4] Paul Breimyer,[4] Guruprasad Kora,[4] Chongle Pan,[4] and Srikanth Yoginath[4]

[1] *Lawrence Livermore National Laboratory*
[2] *University of Minnesota*
[3] *Iowa State University*
[4] *Oak Ridge National Laboratory and North Carolina State University*

Contents

8.1 Introduction 282
8.2 The Process of Scientific Data Analysis 283
8.3 Analysis of Cheminformatics Data 285
    8.3.1 Trends in Cheminformatics Data Mining and Modeling 286
    8.3.2 Structural Descriptors for Chemical Compounds 287
        8.3.2.1 Descriptors Based on Bounded-Size Subgraphs 287
    8.3.3 Indirect Similarity Measures for Similarity Searching
        and Scaffold Hopping 288
    8.3.4 Classification Algorithms for Chemical Compounds 290
    8.3.5 Future Direction of Cheminformatics Data Analysis 291
8.4 Computational Prediction of Protein Function:
Survey and Enhancements 291
    8.4.1 Preprocessing of Biological Datasets to Enhance
        Function Prediction 295
    8.4.2 Future Directions 297
8.5 Materials Informatics 299
    8.5.1 Data-Dimensionality Reduction: Classification
        and Clustering Applications 300
    8.5.2 Prediction via Data Mining 303
    8.5.3 Data Mining for Descriptor Development:
        Enhancing Databases for Predictions 304
    8.5.4 Future Challenges 305

---

*Sections 8.1 and 8.2 were authored by Chandrika Kamath, 8.3 by Nikil Wale and George Karypis, 8.4 by Gaurav Pandey and Vipin Kumar, 8.5 by Krishna Rajan, and 8.6 by Nagiza F. Samatova, Paul Breimyer, Guruprasad Kora, Chongle Pan, and Srikanth Yoginath.

8.6 Parallel R for High-Performance Analytics: Applications
     to Biology 306
     8.6.1 Advanced Analytics for High-Throughput
            Quantitative Proteomics 308
            8.6.1.1 Parallel Processing of Core Analysis Steps
                     in ProRata 310
            8.6.1.2 Estimation of Peptide Abundance Ratios
                     and Scoring of their Variability and Bias 312
            8.6.1.3 Protein Abundance Ratio Estimation with
                     Confidence Interval Evaluation 313
8.7 Summary 315
Acknowledgment 316
References 316

8.1 Introduction

The analysis of data is a key part of any scientific endeavor, as it leads to a better understanding of the world around us. With scientific data now being measured in terabytes and petabytes, this analysis is becoming quite challenging. In addition, the complexity of the data is increasing as well due to several factors such as improved sensor technologies and increased computing power. This complexity can take various forms such as multisensor, multispectral, multiresolution data, spatio-temporal data, high-dimensional data, structured and unstructured mesh data from simulations, data contaminated with different types of noise, three-dimensional data, and so on.

Over the last decade, techniques from machine learning, image and signal processing, and high-performance computing have gained acceptance as viable approaches for finding useful information in science data. These techniques complement the more established approaches from statistics and pattern recognition to provide solutions to a diverse set of problems in a variety of application domains.

This chapter is organized as follows: first, we describe a typical data flow diagram used in scientific data analysis. It shows how one might start with various forms of scientific data and process them iteratively to extract useful information. This is followed by several specific examples of how this general process is applied to problems in domains ranging from materials science to biology and cheminformatics. Finally, we conclude with a brief summary of some challenges in the analysis of scientific datasets.

8.2 The Process of Scientific Data Analysis

The process of scientific data analysis is usually a multistep process, with the details of each step being driven by the type of data and the type of analysis being performed. As scientific data is often in the form of meshes, images, or other sensor outputs, the data has to be first preprocessed to identify the objects of interest and extract characteristics for these objects. One approach, which is a simplified version of the approach used in the Sapphire project (http://computation.llnl.gov/casc/sapphire), is to split the process into five steps, as shown in Figure 8.1.

The first step is to identify and extract the objects or items of interest in the data. This can be relatively easy in some cases, for example, when the objects are chemical compounds or proteins and we are provided data on each of the compounds or proteins. In other cases, it can be very complicated, for example, when the data is in the form of images and we need to identify the objects (say, galaxies in astronomical images or various organs in medical images) and extract them from the background. Once the objects have been identified, we need to extract features or descriptors that describe them. These reflect the analysis task at hand. For example, if the task focuses on the structure or the shape of the objects, the descriptors must reflect the structure or the shape, respectively. In many cases, one may extract far more features than is necessary, requiring a reduction in the number of features or the dimension of the problem. These key features are then used in the pattern recognition step, and the patterns extracted are visualized for validation by the domain scientists.

The data analysis process is iterative and interactive; any step may lead to a refinement of one or more of the previous steps, and not all steps may be used during the analysis of a particular dataset. The domain scientist is actively involved in all steps, starting from the initial description of the data and the problem, the extraction of potentially relevant features, the identification of the training set (where necessary), and the validation of the results.

The rest of this chapter describes analysis approaches to different problems in various scientific domains. They illustrate the diversity of domains and problems, as well as the similarities in the solution approaches. But first, a few words of caution. Data analysis, especially in the context of scientific data, is a very broad and multidisciplinary subject. There are many different solution

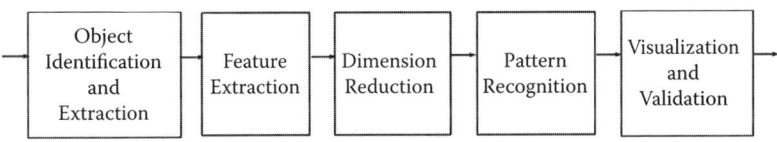

Figure 8.1 Scientific data analysis: an iterative and interactive process.

techniques in each of these disciplines, and one can solve a problem in many different ways. The terminology used in the different analysis disciplines, as well as the different application domains, is not always the same. Further, the process of data analysis is not a blind application of techniques to the data, but a careful and considered process, where any interpretation of the knowledge in the data should take into account the subtleties of the data, the algorithms, and the domain. These issues make it difficult to capture all aspects of scientific data analysis in a single chapter, especially one in a book whose focus is on data management and whose readers are unlikely to be familiar with data analysis. As a result, this chapter should be considered as providing the briefest of glimpses into the intricacies of scientific data analysis; the interested reader is referred to,[1] and the references therein, for further details.

The next four sections describe different facets of scientific data analysis. First, Section 8.3 describes the use of analysis techniques in cheminformatics. The authors provide a brief overview of the field, followed by a summary of some of their work in graph-based descriptors to represent chemical compounds. These descriptors are used in similarity searches as well as classification to detect the patterns among the compounds. In this example, the focus of the work is on the pattern recognition step from Figure 8.1, with the added emphasis that the way the object is described, in this case as graphs, is an important aspect of the analysis. Section 8.4 focuses on the biology domain, specifically, the prediction of protein function. Following a brief introduction to the field, the authors describe how their work in pre-processing the data to reduce noise is substantially enhancing the performance of standard function prediction algorithms. The focus is again on the pattern recognition step, with an emphasis on the role noise in the data can play in our abilities to make accurate inferences. The presence of noise in the data is common in many scientific domains and can adversely affect the results of the analysis.

A rather different aspect of scientific data analysis is highlighted in Section 8.5. The author is interested in combining information from two different datasets in materials science, one with crystallographic information on various materials, and the other a compilation of fundamental thermochemical information. The section describes how techniques such as principal component analysis for dimension reduction and partial least squares for regression can be used to uncover relationships between the structure and the properties of materials. The focus in this section is on the dimension reduction and pattern recognition steps from Figure 8.1. Interestingly, as the author points out, sometimes in materials science, a few careful measurements may be of great value. This is in contrast to the commonly held belief that massive-sized datasets are critical to any informatics endeavor. While large amounts of data might not be necessary, they are becoming more common. Section 8.6 considers the situation when domain scientists, who are accustomed to using a tool on small-to moderate-sized data, find themselves in a situation where the size of the data is such that the tool cannot handle it or the response from the tool is very slow. This section describes an approach to parallelizing a well-known statistical software package **R** and discusses its performance in the context of a biology problem.

8.3 Analysis of Cheminformatics Data

In this first application, we use cheminformatics as a problem domain to illustrate the importance of descriptors that are used to describe the objects being considered in the pattern recognition step of the analysis. *Cheminformatics* is a term used for the broad field of organizing, storing, retrieving, mining, and analyzing vast amounts of chemical information. It has extensive applications in the field of drug discovery and development.[2] The problems in cheminformatics entail the development of well-organized chemical databases to store chemical information like chemical structures and properties, the development of algorithms that can operate effectively and efficiently on these databases to extract relevant information for a given problem, and the development of effective ways of visualizing the information to make informed decisions during the drug discovery process.

A large number of public and proprietary databases such as PubChem, ChemBank, DrugBank, ChemDB, and MDDR have been developed that organize basic chemical information such as the 2D/3D structure(s) of chemical entities or compounds, as well as their physical properties such as molecular weight, polarity, water solubility, and lipophilicity.[3] Many of these databases contain advanced information such as structural descriptors, key functional groups (hydrogen bond donors/acceptors), references to known biological targets, and references to relevant literature.[3]

Once all the pertinent information is organized in a suitable database format, the next step is to develop algorithms to generate, retrieve, mine, and analyze this information in the context of a particular problem in drug discovery. The typical algorithmic tasks performed on chemical structure(s) include calculation of physical and biological characteristics of chemical compounds from first principles, searching, clustering, classification/regression, and docking.[2,4]

Most of these algorithms operate on the assumption that the properties and biological activity of a chemical compound are related to its structure.[2,4] Hansch et al.[5] demonstrated that the biological activity of a chemical compound can be mathematically expressed as a function of its physiochemical properties, which lead to the development of quantitative methods for modeling structure-activity relationships (QSAR). Since that work, many different approaches have been developed for building such structure-activity-relationship (SAR) models. These models have become an essential tool for predicting biological activity from the structural properties of a molecule.

The extensive use of cheminformatics methods has led to the development of a number of large commercial software packages such as Daylight's Toolkit,[6]

SciTegic's Pipeline Pilot,[7] Tripos' SYBYL,[8] and Chemaxon's SCREEN[9] that provide a wide range of capabilities including database management and searching, compound filtering, physical–chemical property calculations, SAR modeling, and visualization. Visualization techniques are covered in detail in Chapter 9.

In the rest of this section we first review some of the current trends in cheminformatics and then highlight some of the techniques that we developed for representing chemical compounds, determining their similarity, and building classification models. Finally, we outline some of the future research directions in cheminformatics.

8.3.1 Trends in Cheminformatics Data Mining and Modeling

Calculation of similarity between chemical compounds is a fundamental task in order to analyze cheminformatics data. The analysis includes, but is not limited to, retrieving, mining, and building SAR models on the data. To perform this analysis effectively and efficiently, many algorithms first convert the 2D/3D structure into *descriptor space* or *descriptor representation*[2,4] and then apply various information retrieval, data-mining, statistical, and machine-learning approaches on the transformed data. The descriptors employed range from physiochemical property descriptors,[2,4] to topological descriptors derived from the compound's molecular graph,[2,6,10,11] to 2D and 3D pharmacophore descriptors that capture interactions important to protein-ligand binding.[2,4] Among them, hashed 2D descriptors corresponding to subgraphs of various sizes and types (e.g., paths, trees, rings) are the most common and include the extensively used Daylight fingerprints,[6] Chemaxon fingerprints[9] and the extended connectivity fingerprints,[2,7] that have been recently implemented in Scitegic's Pipeline Pilot.[7]

Over the years, the approaches that have been employed to learn SAR models have evolved from the initial regression-based techniques used by Hansch et al., to approaches that utilize more complex statistical model-estimation procedures. These procedures include partial least squares (PLS), linear discriminant analysis, Bayesian models, and approaches that employ various machine-learning/pattern recognition methods such as recursive partitioning, neural networks, and support vector machines.[2,4] Another class of methods for building SAR models operate directly on the structure of the chemical compound. These methods employ inductive logic programming (ILP) (such as the WARMR system[12]) or heuristics (such as the MultiCASE system[2]) to automatically identify a small number of chemical substructures that relate to their biological activity. These substructures are used as descriptors. Finally, in recent years, a new class of machine-learning techniques has been developed that builds SAR models that measure the similarity between two compounds by operating directly on their molecular graphs.[13,14] These techniques measure the similarity by using powers of adjacency matrices,[13] calculating Markov random walks on the underlying graphs,[13] finding the maximum common subgraph,[15] or using weighted substructure matching between two graphs.[14] These similarity measures are then used as kernels, and the SAR models are built using kernel-based learning algorithms that include kernel PCA,[16] kernel Fisher discriminant analysis,[16] and support vector machines.[2] The advantage of these techniques is that they directly compute similarity values for all pairs of compounds and eliminate the step of generating descriptors for chemical compounds. However, such methods are inherently less descriptive as it is not possible to analyze the different features of the molecular graph that might be important for activity.

8.3.2 Structural Descriptors for Chemical Compounds

Descriptor-based representations of chemical compounds are used extensively in cheminformatics, as they represent a convenient and computationally efficient way to capture key characteristics of the structure of the compounds. Such representations have extensive applications to similarity search and various structure-driven prediction problems (e.g., activity, toxicity, absorption, distribution, metabolism, and excretion). As part of our research, we have developed novel structural descriptors corresponding to the set of frequent topological subgraphs[17] and all bounded-size subgraphs[11] that are present in a compound library. The key advantage of these descriptors is that unlike structural descriptors that use path-based (e.g., Daylight[6]) or extended connectivity fingerprints,[2] they impose no limit on the complexity of the descriptor's structure.[11,18]

8.3.2.1 Descriptors Based on Bounded-Size Subgraphs

We have developed a new descriptor space that consists of all connected subgraphs up to a given length l that exist in a compound library.[11] We refer to this descriptor space as *graph fragments* (GF). This descriptor space is determined dynamically from the library (i.e., a database of chemical compounds) and consists of features that have complex topologies. We have also developed an algorithm to generate the GF descriptors of a library by efficiently enumerating all bounded-size subgraphs using a recursive technique initially developed for generating all the spanning trees in a graph.[11] Moreover, the algorithm uses the efficient canonical labeling algorithm that we developed[17] to identify isomorphic fragments in the same graph. In addition, if desired, the fragment's canonical labeling is also used to count the number of occurrences (embeddings) of a fragment in a molecular graph.

We have also identified the key characteristics that make certain two-dimensional descriptors better than others. In order to establish these characteristics, we compared the GF descriptors and three of its subsets (AF, TF, and PF)[11] to five previously developed descriptors (Chemaxon's fingerprints (fp),[9] extended connectivity fingerprints (ECFP),[2] Maccs keys (MK),[10] Cycles and Trees (CT),[19] and frequent subgraph-based descriptors (FS).[17,18]

TABLE 8.1 SAR performance of different descriptors

Datasets	GF	ECFP	fp	MK	FS
NCI1	**0.33**	0.32	0.30	0.29	0.27
NCI109	**0.32**	**0.32**	0.27	0.24	0.26
NCI123	**0.27**	**0.27**	0.25	0.24	0.23
NCI145	**0.37**	0.35	0.30	0.28	0.30
NCI167	**0.07**	0.06	0.06	0.04	0.06
NCI220	0.29	0.28	**0.33**	0.26	0.21
NCI33	**0.33**	0.31	0.26	0.26	0.25
NCI330	**0.36**	**0.36**	0.34	0.31	0.24
NCI41	**0.36**	**0.36**	0.25	0.28	0.30
NCI47	**0.31**	**0.31**	0.26	0.26	0.24
NCI81	**0.28**	**0.28**	0.27	0.25	0.24
NCI83	**0.31**	**0.31**	0.26	0.26	0.25

The numbers correspond to the ROC_{50} values of SVM-based SAR models for twelve screening assays obtained from the National Cancer Institute (NCI). The ROC_{50} value is the area under the receiver-operating characteristic curve (ROC) up to the first 50 false positives. These values were computed using a five-fold cross-validation approach. The descriptors being evaluated are graph fragments (GF),[11] extended connectivity fingerprints (ECFP),[2] Chemaxon's fingerprints (fp) (Chemaxon Inc.),[9] Maccs keys (MK) (MDL Information Systems Inc.),[10] and frequent subgraphs (FS).[18]

We observed that descriptors that are determined dynamically from the dataset and use fragments with simple and complex topologies lead to precise representations. In addition, they have a high degree of coverage and may be expected to perform better in the context of chemical compound classification and retrieval as they allow for a better representation of the underlying compounds.[11] The descriptor space that satisfies all the desirable characteristics is GF.[11] ECFP virtually satisfies all of the characteristics except precise representation since there is the possibility of collisions, although in practice it is quite low. The quantitative and statistical results on the performance of each of these descriptors on 28 datasets were found to be consistent with our qualitative analysis.[11] Table 8.1 shows a subset of our results for the NCI datasets obtained from the PubChem Project.[20] These results show that the GF descriptor space achieves a performance that is either better or comparable to that achieved by currently used descriptors, indicating that the GF descriptors can effectively capture the structural characteristics of the compounds.

8.3.3 Indirect Similarity Measures for Similarity Searching and Scaffold Hopping

The task of searching a library to find compounds that are similar to a query is extensively performed in cheminformatics. These compounds are termed as *hit compounds* or *hits*. In order to identify these hits, the methods employed typically utilize similarity values between a pair of compounds. This similarity value is usually computed over a suitable descriptor-space representation[4] of chemical compounds, which is typically derived from the two-dimensional topological molecular graph of the chemical compounds. It has been shown that when this similarity is high, these two-dimensional descriptor-based methods are very effective in finding compounds that share similar activity against a biomolecular target.[2] However, the task of identifying hit compounds is complicated by the fact that the query might have undesirable properties such as toxicity, bad ADME (absorption, distribution, metabolism, and excretion) properties, or may be promiscuous.[2] These properties will also be shared by most of the compounds similar to the query, as they will correspond to very similar structures. In order to overcome this problem, it is important to identify (i.e., rank high) as many chemical compounds as possible that not only show the desired activity for the biomolecular target but also have different structures (come from diverse chemical classes or chemotypes). Finding novel chemotypes using the information of already known bioactive small molecules is termed *scaffold hopping*.[2]

We developed techniques,[21] inspired by research in social network analysis, that measure the similarity between the query and a compound by taking into account additional information beyond their *direct* descriptor-space-based representation. These techniques derive indirect similarities by analyzing the network connecting the query and the library compounds. This network is determined using an undirected k-*nearest-neighbor* graph (NG) and an undirected k-*mutual-nearest-neighbor* graph (MG). Both of these graphs contain a node for each of the compounds as well as a node for the query. However, they differ on the set of edges that they contain. In the k-nearest-neighbor graph there is an edge between a pair of nodes corresponding to compounds c_i and c_j, if c_i is in the k-nearest-neighbor list of c_j or vice-versa. In the k-mutual-nearest-neighbor graph, an edge exists only when c_i is in the k-nearest-neighbor list of c_j and c_j is in the k-nearest-neighbor list of c_i. The indirect similarity between a pair of nodes is computed as the Tanimoto coefficient of their adjacency lists, which assigns a high similarity value to a pair of compounds if they have a large number of common similar compounds. Thus, the indirect similarity between a pair of compounds will be high if there are a large number of size-two paths connecting them in the network.

The performance of indirect similarity-based retrieval strategies based on the NG as well as MG graph was compared with direct similarity based on the Tanimoto coefficient.[21] The compounds were represented using different descriptor-spaces (GF, ECFP, etc.). The quantitative results showed that indirect similarity is consistently, and in many cases substantially, better than direct similarity. Figure 8.2 shows a part of our results in which we compare MG-based indirect similarity to direct Tanimoto-coefficient similarity searching using ECFP descriptors. It can be observed from the figure that indirect similarity outperforms direct similarity for scaffold-hopping active retrieval in five out of six datasets (COX2, A1A, CDK2, FXa, MAO, and PDE5) on

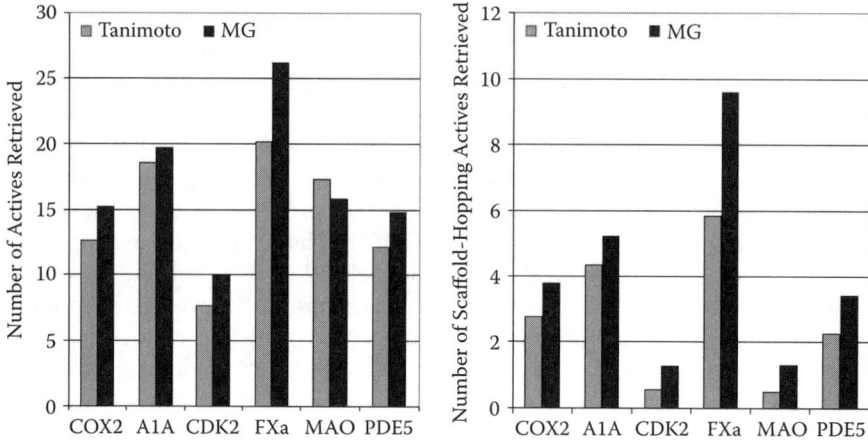

Figure 8.2 Performance of indirect similarity measures (MG) as compared to similarity searching using the Tanimoto coefficient.

which we tested our method. It can also be observed that indirect similarity outperforms direct similarity for active compound retrieval in all datasets except MAO. Moreover, the relative gains achieved by indirect similarity for the task of identifying active compounds with different scaffolds is much higher, indicating that it performs well in identifying compounds that have similar biomolecule activity even when their direct similarity is low.

8.3.4 Classification Algorithms for Chemical Compounds

We have developed structure-based prediction models for classifying compounds into various classes of interest (e.g., active/inactive, toxic/non-toxic). These models were based on support vector machines (SVMs) and utilized novel kernel functions to determine the similarity between a pair of compounds.[11,18] These kernel functions were developed by representing the structure of the compound as a vector in a high-dimensional descriptor space whose dimensions corresponded to two- or three-dimensional structures present in the compounds. The descriptor spaces that we developed and studied include the FS[18] and GF descriptors[11] and a descriptor based on frequently occurring geometric subgraphs that were discovered automatically from the predicted three-dimensional structure of the compounds.[22] We studied the performance of different kernel functions including linear, radial basis functions, and Tanimoto.[11,18] We also developed novel extensions to existing kernel functions for the GF descriptors that take into account the size of the different descriptors. This function calculates a different similarity value for fragments belonging to each of the different sizes and then combines them to yield a single similarity value. This approach leads to better results for GF descriptors in the context of SVM.[11] Although SVM is quite effective for high-dimensional datasets, the selection of the most relevant features has been shown to yield superior results in practice. We have developed feature selection techniques for the problem of chemical compound classification in the context of FS descriptors that take into account the class distribution of a feature (belonging to a predominantly active or inactive set of compounds) and their prevalence (support) in the dataset. Our results showed that using selected features derived from this technique leads to better models for SVM-based classification.[18]

Overall, the outcome of these studies has been that support vector machines are a powerful and flexible methodology for building predictive models and lead to models that often outperform other supervised learning approaches.

8.3.5 Future Direction of Cheminformatics Data Analysis

Mining and retrieving data for a single biomolecular target, and building SAR models on it, has been traditionally used to analyze the structure–activity relationships, which play a key role in drug discovery. However, in recent years the widespread use of high-throughput screening (HTS) technologies by the pharmaceutical industry has generated a wealth of protein-ligand activity data for large compound libraries. These data have been systematically collected and stored in centralized databases.[23] At the same time, the completion of the human genome sequencing project has provided a large number of "druggable" protein targets[24] that can be used for therapeutic purposes. Additionally, a large fraction of the protein targets that have been or are currently being investigated for therapeutic purposes belong to a small number of gene families.[25] The combination of these three factors has led to the development of methods that utilize information that goes beyond the traditional single biomolecular target data analysis. Recently, the trend has been to integrate cheminformatics data with protein and genetic data (bioinformatics data) and analyze the problem over multiple proteins or different protein families. Various approaches such as the identification of active compounds for a given new target within the same family (chemogenomics),[26] discovering new biology (chemical genetics),[27] discovering new targets for well-characterized chemical compound(s) (target fishing),[28] and establishing promiscuity or selectivity of chemical compound(s) (poly-pharmacology)[23] are being developed to aid the drug discovery process.

8.4 Computational Prediction of Protein Function: Survey and Enhancements

In our second example, we consider a different application domain, namely, bioinformatics, where we focus on the important problem of predicting the function of proteins. We also discuss the role that noise can play in our ability to identify the patterns in the data.

The macromolecules known as proteins are responsible for some of the most important functions in an organism, such as constitution of the organs (structural proteins), the catalysis of biochemical reactions necessary for metabolism (enzymes), and the maintenance of the cellular environment (transmembrane proteins). Thus, proteins are the most essential and versatile macromolecules of life, and the knowledge of their functions is a crucial link in the development of new drugs, better crops, and even the development of synthetic biochemicals such as biofuels.

Traditional approaches for predicting protein function are experimental and usually focus on a specific target gene or protein, or a small set of proteins forming natural groups such as protein complexes. These approaches are low-throughput because of the huge experimental and human effort required in analyzing a single gene or protein. As a result, even large-scale experimental annotation initiatives, such as the EUROFAN project,[29,30] are able to annotate only a small fraction of the proteins that are becoming available due to rapid advances in genome sequencing technology. This has resulted in a continually expanding sequence–function gap for discovered proteins.[31] Figure 8.3 shows that the gap between the number of available gene sequences in GenBank and that of manually annotated protein sequences in SwissProt is continuously increasing and currently stands at about three orders of magnitude.[32]

In an attempt to close this gap, numerous high-throughput experimental procedures have been invented to investigate the mechanisms leading to the

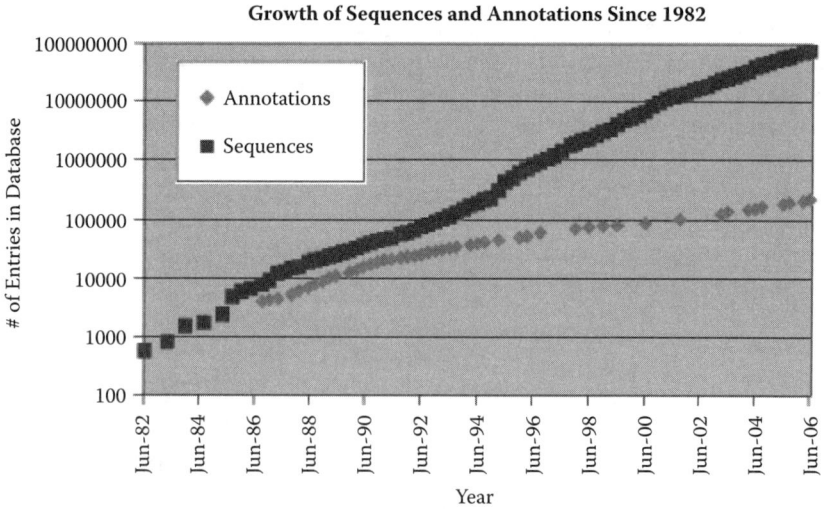

Figure 8.3 Growth of sequences and annotations on a log scale using numbers from GenBank and SwissProt (Figure adapted from Mayer, F. 2006. Genome sequencing vs. Moore, Law: Cyber challenges for the next decade. *CTWatch Quarterly*. Used with permission).

accomplishment of a protein's function. These procedures have generated a wide variety of useful data that range from simple protein sequences to complex high-throughput data, such as gene expression datasets and protein interaction networks. These data offer different types of insights into a protein's function and related concepts. For instance, protein interaction data shows which proteins come together to perform a particular function, while the three-dimensional structure of a protein determines the precise sites to which the interacting protein binds itself. Due to its utility, recent years have seen the recording of this data in very standardized and professionally maintained databases such as SWISS-PROT,[33] MIPS,[34] DIP[35] and PDB.[36]

However, the huge amount of data that has accumulated through these experiments over the years has made biological discovery via manual analysis tedious and cumbersome, and has led to the emergence of the field of bioinformatics. Indeed, an increasingly accepted path for biological research is the creation of hypotheses by generating results from an appropriate bioinformatics algorithm in order to narrow the search space and the subsequent experimental validation of these hypotheses to reach the final conclusion.[31,37] Owing to this change in ideology and the importance of understanding protein function, numerous researchers have applied computational and mathematical techniques to predict protein function and attempt to close the sequence–function gap. Early approaches used sequence similarity tools such as BLAST[38,39] to transfer functional annotation from the most similar proteins. Subsequently, several other approaches have been proposed that utilize other types of biological data for computational protein function prediction, such as gene expression data, protein interaction networks, and phylogenetic profiles. These techniques enable the prediction of the functions of those proteins that cannot be reliably annotated using sequence similarity-based techniques.[40,41] Table 8.2 summarizes the general ideas used by several of these approaches, categorized by the type of biological data they utilize. Additional details on several hundred such techniques that have been published in this area in the last few years are available in Pandy et al.[42] and other reviews on this topic.[43–46]

As can be seen, many of the techniques listed in Table 8.2, such as classification, clustering, and association analysis,[47] are drawn from the fields of data mining and machine learning. Indeed, these techniques have produced among the best results for this problem, some of which have even been experimentally verified.[48,49]

However, despite this progress, the gap between the number of known proteins and those that have been functionally annotated is astounding, as illustrated by Figure 8.3. This gap is possibly due to several outstanding issues that need to be convincingly addressed. Some of the general issues that have to be addressed are the possibility of a protein's performing multiple functions, and thus having multiple functional labels; the widely varying sizes of functional classes with most classes being very small; hierarchical arrangement of functional labels, such as in gene ontology;[50] and incompleteness and various types and extents of noise in biological data.

TABLE 8.2 Types of biological data and analysis techniques used for predicting protein function

Data Type	Analysis Techniques
Amino acid sequences	Flexible sequence similarity measures and clustering Classification based on subsequences such as motifs and domains Classification based on biological features derived from sequences
Protein structure	Structural similarity-based inference Inference using 3D structural motifs Inference using features of the 3D surface of a protein Classification using structure kernels and frequent substructures
Genome sequences	Proximity of two genes and their orthologs in multiple genomes Fusion of two proteins into a single protein in other genomes
Evolutionary data	Co-occurrence of two genes in multiple genomes Functional inference based on duplication and speciation events
Microarray data	Clustering for finding functional groups Classification to infer functions of individual proteins Classification of temporal microarray data
Protein interaction networks	Annotation transfer from neighboring proteins in network Global annotation transfer from the whole network Clustering to find densely connected regions Association analysis to find frequently occurring subnetworks
Biomedical literature	Information retrieval using word frequency-based statistics Text mining using classification and clustering of documents Natural language processing-based approaches
Multiple data types	Combination of multiple data types into a single graph Combination of predictions from different data types Intelligent fusion of multiple datasets of different types

In this section, we discuss some approaches that have been proposed for handling one of the above issues, namely preprocessing of biological datasets to reduce their incompleteness and the noise contained in them. We also discuss some further research issues that should be addressed in this direction.

8.4.1 Preprocessing of Biological Datasets to Enhance Function Prediction

Biological datasets usually contain significant amounts of noise,[51] which may hamper their use for making accurate inferences about biological processes and protein function. This noise may arise from inaccuracies in the experimental methods used to generate the data, or in the subsequent data analysis methods to process the data generated into a more usable form. In particular, the problem of noise is further pronounced for data generated by high-throughput experimental methods, such as protein interaction and microarray data. For instance, it has been reported in a recent survey of the yeast and human interactomes that most of the available protein interaction datasets have an exceedingly high fraction of false positive interactions—up to almost 80% for some datasets.[52] Similarly, a slightly less acknowledged but equally important problem with biological data is that of incompleteness. Even for the most well studied organisms, experimental data for several biological processes is generally unavailable. For instance, Hart et al. have estimated that for the commonly used model organism *S. cerevisiae*, only 50% of all viable protein–protein interactions are known, while for the human genome, this number is as small as 11%.[52] This incompleteness may delay the discovery of proteins involved in the processes represented by the missing data. This illustrates that the problem of noise and incompleteness in biological datasets needs to be adequately addressed in order to ensure that accurate inferences about protein function are drawn from them.

We illustrate the efforts in preprocessing of biological data through the work done on noise quantification and elimination in protein interaction datasets. Several methods have been proposed for estimating the quality of a dataset consisting of direct protein–protein interactions, such as the EPR (expression profile reliability) index.[53]* This method estimates the reliability of the input interaction dataset by comparing the distribution of the correlations between the expression values of the constituent proteins with those of the proteins constituting the DIPCore dataset,[53] which is a set of about 5,000 highly reliable interactions in the DIP database[35] and is treated here as the reference dataset. Although this estimation of the reliability of an entire dataset is useful, it is often the case that some interactions in the dataset are more reliable than others. Hence, it is very useful to estimate the reliabilities of individual interactions. A popular tool, known as PVM (paralogous verification

---

*Available online at http://dip.doe-mbi.ucla.edu/dip/Services.cgi?SM=1. Accessed July 12, 2008.

method),[53][†] computes this reliability by estimating the likelihood that the two proteins also have paralogs that are known to interact.

In many cases, the structure of the protein–protein interaction network is itself able to provide useful information about the likelihood of an interaction between two proteins. Such information may be in the form of the interconnection patterns of these proteins with other proteins in the network. An important structural feature of networks that provides a more robust estimate of the reliability is the concept of *common neighborhood*, that is, the set of proteins that are interaction partners of both of the proteins whose interaction is being evaluated. Intuitively, the larger the number of common neighbors two proteins have, the more likely they are to have a direct interaction between them. Some approaches have used measures for evaluating the reliability of a given interaction in terms of this common neighborhood-based similarity measure, such as Jaccard similarity or a variation thereof.[54–56] A more systematic approach has been proposed recently,[57] where the *h-confidence* measure[58] from the field of association analysis[47] in data mining was used to quantify this likelihood. For interaction networks, the *h-confidence* measure may be defined as shown in Equation (8.1). Here, N_{P_1} and N_{P_2} denote the sets of neighbors of P_1 and P_2 respectively.

$$h-confidence(P_1, P_2) = \min\left(\frac{|N_{P_1} \cap N_{P_2}|}{|N_{P_1}|}, \frac{|N_{P_1} \cap N_{P_2}|}{|N_{P_2}|}\right) \quad (8.1)$$

As defined above, *h-confidence* is only applicable to binary data or, in the context of protein interaction graphs, to unweighted graphs. However, the notion of *h-confidence* can be readily generalized to networks where the edges carry real-valued weights indicating their reliability. In this case, Equation (8.1) can be conveniently modified to calculate $h\text{-}confidence(P_1, P_2)$ by making the following substitutions: (1) $|N_{P_1}| \rightarrow$ sum of weights of edges incident on P_1 (similarly for P_2) and (2) $|N_{P_1} \cap N_{P_2}| \rightarrow$ sum of minimum of weights of each pair of edges that are incident on a protein P from both P_1 and P_2. In both of these cases, the *h-confidence* measure is guaranteed to fall in the range [0, 1].

Now, with the hypothesis that a function of the number of common neighbors indicates the reliability of an interaction between two proteins, two types of interactions, namely *spurious* and *potential* interactions, can be characterized in an interaction network as follows:

- Interactions in the network that have a low *h-confidence* score are likely to be *spurious*.
- A pair of proteins that have a high pairwise *h-confidence* score are likely to interact, even if the network currently does not contain an interaction between them, and these are termed as *potential* interactions.

[†]Available online at http://dip.doe-mbi.ucla.edu/dip/Services.cgi?SM=2. Accessed July 12, 2008.

Using these definitions, Pandey et al.[57] proposed the following graph transformation approach for purifying available interaction datasets. Consider the input interaction network $G = (V, E)$, where V is the set of nodes representing the proteins in the network, and E is the set of edges representing the protein–protein interactions constituting the network. First, the *h-confidence* measure is computed between each pair of constituent proteins, whether connected or unconnected by an edge in the input network. Next, a threshold is applied to drop the protein pairs with a low *h-confidence* to remove spurious interactions and control the density of the network. The resultant graph $G' = (V, E')$ is hypothesized to be the less noisy and more complete version of G, since it is expected to contain fewer noisy edges, some biologically viable edges that were not present in the original graph, and more accurate weights on the remaining edges.

In order to evaluate the efficacy of the resultant networks for protein function prediction, we provided the original and the transformed graphs as input to the *FunctionalFlow* algorithm.[59] *FunctionalFlow* is a graph-theory-based algorithm that enables insufficiently connected proteins to obtain functional annotations from distant proteins in the network and has produced much better results than several other function prediction algorithms operating on protein interaction networks. We also tested several transformed versions of the input network generated using our graph transformation approach in conjunction with some other common neighbor-based similarity measures, such as the number of common neighbors, and Samanta et al.'s p-value measure.[55] Figure 8.4 shows the performance of the *FunctionalFlow* algorithm on these transformed versions of two standard interaction networks, measured in terms of the accuracy of the top scoring 1,000 predictions of the functions of the constituent proteins.

The significant improvement in the accuracy of the predictions derived from the *h-confidence*-based transformations of standard interaction networks, one of which is constructed by combining several popular yeast interaction datasets (combined) and weighted using the EPR index tool, and the other being a confident subset of the DIP database[35] (DIPCore), shows that this association analysis-based graph transformation approach is indeed able to reduce noise, enhance completeness, and assign more reliable weights to the constituent edges. The other similarity measures were also substantially outperformed by *h-confidence*. This result is in coherence with those of an earlier study, where *h-confidence* and *hypercliques* were used to eliminate noisy objects from datasets.[60]

8.4.2 Future Directions

The above discussion shows that the preprocessing of biological data can enhance the performance of standard function prediction algorithms substantially, and thus should be considered as an integral step of the process of

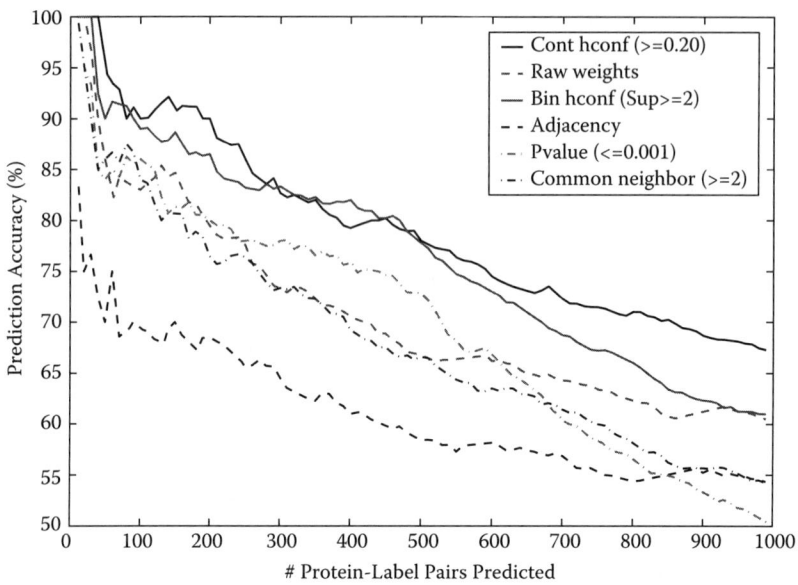

(a) Accuracy of top 1,000 predictions on the combined network

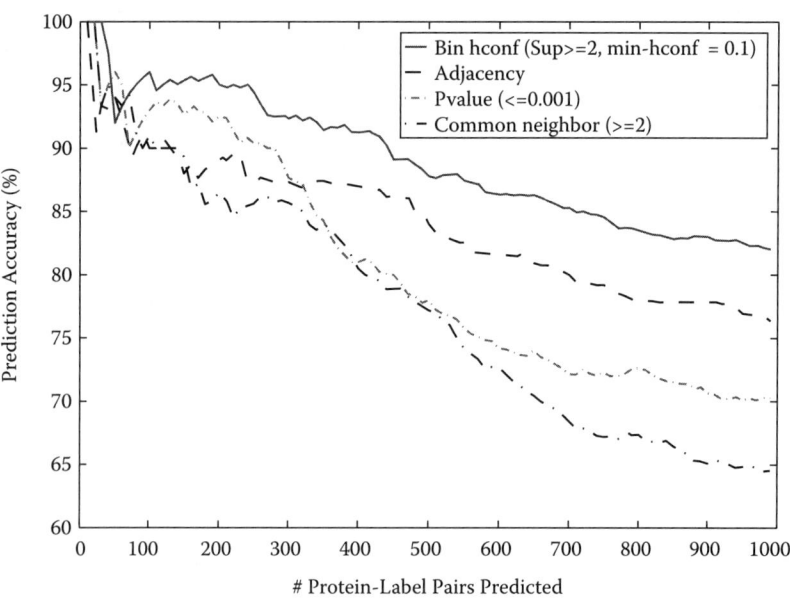

(b) Accuracy of top 1,000 predictions on the DIPCore network

Figure 8.4 (See color insert following page 224.) Comparison of performance of various transformed networks and the input networks.

functional discovery from different datasets. However, it should be noted that the techniques used for this task have to be aware of the nature and distribution of the dataset being processed. For instance, while techniques for microarray data preprocessing should take into account the *log*-transformed nature of the data from each experiment, techniques for interaction data should not ignore the network structure of the complete dataset. Another complexity that may have to be handled occurs when a dataset poses several preprocessing issues simultaneously. For example, microarray data produced through experiments often suffers from the problems of missing values and a difference in the scales of values produced by the constituent experiments. Often, the effect of one preprocessing operation may have implications for the subsequent operations. Thus, a systematic preprocessing pipeline needs to be developed for each data type, as has been done for microarray data.[61] Finally, an effort should be made toward finding biological justifications for the changes made to the original dataset during the preprocessing step(s).

8.5 Materials Informatics

Our third example is from the domain of materials science and focuses on a different step in the data analysis process, namely, dimension reduction. It illustrates how inferences from even small datasets can be made with a careful application of the analysis techniques, coupled with domain expertise.

One may naturally assume that having large amounts of data is critical for any serious informatics studies. However, what constitutes "enough" data in materials science applications can vary significantly. In studying structural ceramics, for instance, fracture toughness measurements are difficult to make, and in some of the more complex materials, just a few careful measurements can be of great value. Similarly, having reliable measurements on fundamental constants or properties for a given material involves very detailed measurement and/or computational techniques. In essence, datasets in materials science fall into two broad categories. The first is datasets on the behavior of a given material as related to mechanical or physical properties. The other is datasets related to intrinsic information based on the chemical characteristic of the material such as thermodynamic datasets.

Historically, in the materials science community, crystallographic and thermochemical databases have been two of the most well established datasets. The former serves as the foundation for interpreting crystal structure data of metals, alloys, and inorganic materials. The latter involves the compilation of fundamental thermochemical information in terms of heat capacity and calorimetric data. While crystallographic databases are primarily used as a reference source, thermodynamic databases were actually one of the first early examples of informatics as these databases were integrated into thermochemical computations to map phase stability in binary and ternary alloys.[62–72] This led the development of computationally derived phase diagrams, which is a classic example of integrating information in databases to data models. The evolution of both databases has occurred independently, although in terms of their scientific value, they are extraordinarily intertwined. Phase diagrams map out regimes of crystal structure in temperatures-composition space or temperature-pressure space. Yet, crystal structure databases have been developed totally independently. At present, the community has to work with each database separately, making information searches cumbersome and the interpretation of data analysis involving both databases very difficult. Researchers only integrate such information on their own for a very specific system at a time, based on their individual interests. Hence, there is at present no unified way to explore patterns of behavior across both databases, which are so scientifically related.

One of the more systematic efforts to address this challenge has been that of Ashby[73–77] who showed how, by merging phenomenological relationships in materials properties with discrete data on specific materials characteristics, one can begin to develop patterns of classification of materials behavior. The visualization of multivariate data was managed by using normalization schemes that permitted the development of "maps" that provided a means of capturing the clustering of materials properties. It also provided a methodology to establish common structure–property relationships across seemingly different classes of materials. This approach, while very valuable, is limited in its predictive value and is ultimately based on utilizing prior models to build and seek relationships.

In the *informatics* strategy of studying materials behavior, we are approaching the problem from a broader perspective. By exploring all types of data that may have varying degrees of influence on a given property (or properties) with no prior assumptions, one utilizes data-mining techniques to establish both classification and predictive assessments in materials behavior. This is not done, however, from a purely statistical perspective. Instead, we carefully integrate a physics-driven approach to data collection with data mining, and then validate or analyze with theory-based computation and/or experiments. The origins of the data can be either from experiment or computation; the former, when organized in terms of combinatorial experiments, can provide an opportunity to screen large amounts of data in a high throughput fashion.[78–82] In the following discussion, we provide an example of both the classification and predictive uses of data mining in materials science.

8.5.1 Data-Dimensionality Reduction: Classification and Clustering Applications

Figure 8.5 shows a three-dimensional principal component analysis (PCA) plot of a multivariate database of high-temperature (Tc) superconductors. The initial dataset consisted of 600 compounds (that is, the rows of our database)

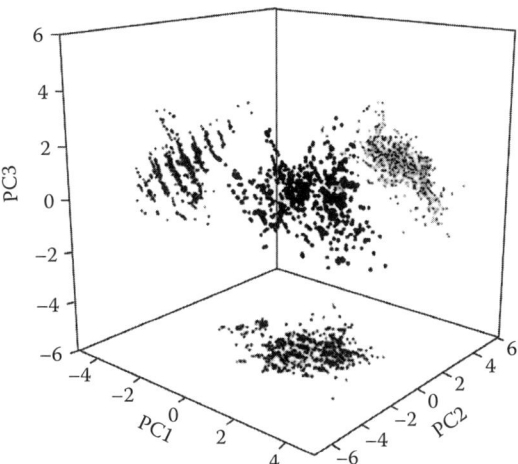

Figure 8.5 (See color insert following page 224.) A three-dimensional PCA plot of a multivariate database of high-temperature superconductors. The initial dataset has 600 compounds with 8 attributes each. The projections in the PC1, PC2, and PC3 space are shown.

with data on 8 different attributes or variables associated with each compound (i.e., the columns). These attributes include cohesive energy, pseudo-potential radii, electronegativity, ionization energy, and valency. The PCA analysis indicates that no clear pattern emerges (from visual inspection at least) unless one looks at the PC3-PC2 projection, where a clear clustering pattern is seen. By inspection with the original dataset, the linear clusters were found to be associated with systematic valency changes among the compounds studied. It should be noted that this process of inspection and data association has to be a deliberate and careful process of understanding what data was input and how to infer interpretations from these patterns. Automatic methods of "unsupervised learning" and data interpretation are possible, and that is where data-mining methods become very valuable.[83]

It was a systematic association of trends in transition temperature with the original datasets that led to the discovery that each linear cluster was associated with a given "average valency," a term proposed by Villars and Phillips[84] nearly two decades ago. In other words, all the compounds in this score plot cluster according to this hypothesis. Given that the data used by Villars and Phillips just precedes the discovery of ceramic superconductors, our work, which includes data since then, demonstrates the broader impact of the valency-clustering criterion in superconductors. As shown in Figure 8.6, that also helps in providing a physical interpretation of trajectories in PCA space. For instance PC2 and PC3 represent orthogonal projections of the "valency vector" effect on high Tc superconductors as it is approximately at 45 degrees to the principal component axes. While the PC axes themselves in

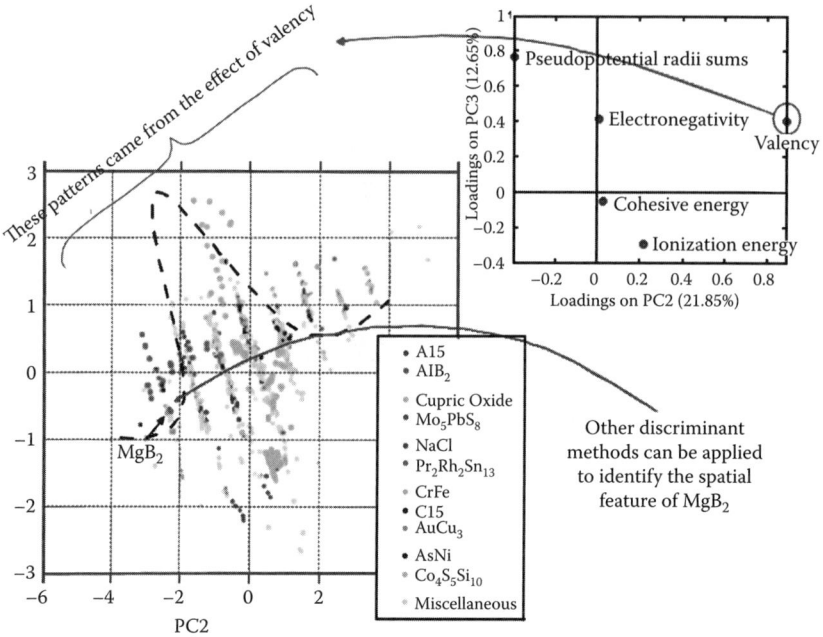

Figure 8.6 (See color insert following page 224.) Interpreting results by comparing the scoring plot on the left and the loading plot on the right. The dashed curve shows the trend in transition temperature across the linear clusters. The peak corresponds to those with the highest recorded temperatures. By comparison to the earlier scoring plot, and using a different visualization scheme, we can now capture a more complete perspective of trends in this multivariate dataset.

this case do not have a discrete physical meaning, other directions in the PC projections can. We will later show examples of how the PC axes can correlate directly to trends in a physical parameter. It is also interesting to note that the more recent discovery of MgB2 as a high-temperature superconductor actually shows up in this plot, indicating how this multidimensional analysis appears to capture the critical physics governing high-Tc materials, where Tc is the transition temperature of the superconductor.

Up to this point, we have been examining the scoring plot looking at how the compound chemistries correlate to each other. Let us now examine the loading plot of the variables used in the calculation to examine how they correlate with each other in terms of their influence on superconducting transition temperature.

The loading plot in the top right corner of Figure 8.6 suggests that the attribute *cohesive energy* does not play a major role in discriminating among compounds in terms of the linear clustering projection, as it is near the origin

(0, 0) position of the loading plot. However, attributes such as *pseudo-potential radii* are a very dominant factor, while *electronegativity* and *ionization energy* play an important but lesser role. The loading plot also indicates the noticeable negative correlation between ionization energy and pseudo-potential radii in terms of their influence on average valency clustering of the compounds as they reside in opposite quadrants. We can further gather more information by juxtaposition of information from both scoring and loading plots and by different visualization schemes.

The strong effect of valency on the linear pattern of clustering in the scoring plot is consistent with its large distance from the origin on the loading plot. In Figure 8.6, we have presented the same scoring plot as before, except now each compound is labeled according to structure type rather than transition temperature. The highest transition compounds are the cupric oxides (marked in light green).

To summarize, when we start with a multivariate data matrix, PCA analysis permits us to reduce the dimensionality of that dataset. This reduction in dimensionality now offers us better opportunities to

- identify the strongest patterns in the data,
- capture most of the variability of the data by a small fraction of the total set of dimensions, and
- eliminate much of the noise in the data, making it beneficial for both data mining and other data analysis algorithms.

8.5.2 Prediction via Data Mining

While a fundamental tenet in materials science is to establish structure–property relationships, it is the life sciences and organic chemistry communities that have formally introduced the concept of *quantitative structure–activity* (or also termed *property*) *relationships* (QSAR or QSPR), as discussed in Section 8.3. Unlike classical materials science approaches, which is relating structure and function through physically based models, QSARs are derived from a model-independent approach, sometimes referred to as *soft modeling*. These data-driven dimensionality reduction techniques help to guide links between structure and properties. The partial least squares (PLS) technique expresses a dependent variable (target property) in terms of linear combinations of the principal components. The PLS method can be applied to rationalize the materials attributes relevant to materials function or property; this permits one to use PLS methods to develop explicit quantitative relationships that identify the relative contributions of different data descriptors, and the resulting relationship between all these descriptors as a linear combination, to the final property.

For instance, Suh and Rajan[85] explored the attributes used in electronic structure calculations and their influence on predicting bulk modulus. Using PLS, a QSAR was developed relating bulk modulus with a variety of electronic structure parameters:

$$\text{Bulk modulus} = -1.00096EN - 0.35682x - 0.77228BL_{A-N} - 0.83367BL_{B-N}$$
$$+ 0.03296Q^*_{tet} + 0.18484Q^*_{oct} - 0.13503Q^*_N$$

where EN is the weighted electronegativity difference, x is the internal anion parameter, BL_{A-N} is the A–N bond length, BL_{B-N} is the B–N bond length, Q^*_{tet} is the Mulliken effective charge for tetrahedral site ion, Q^*_{oct} is the Mulliken effective charge for octahedral site ion, and Q^*_N is the Mulliken effective charge for N ion.

By systematically exploring the number and type of variables needed, Suh and Rajan found very strong agreement in being to able to predict properties consistent with ab-initio calculations based strictly on a data-driven analysis. Based on our QSAR formulation, the role of the effective charge (Q^*) in enhancing modulus is particularly notable. This is consistent with theoretical studies, which show that it is the effective charge parameter that helps to define the degree of charge transfer and the level of covalency associated with the specific site occupancy of a given species. Ab-initio calculations of this effective charge can then be used as a major screening parameter in identifying promising crystal chemistries for promoting the modulus. Hence, using PLS to develop a QSAR formulation, combined with an interpretation of the physics governing these materials, can indeed be valuable. Our predictions fit well with systems of similar electronic structure and allow us to clearly identify outliers based on these quantum mechanical calculations. Based on these predictions, we can now seriously and effectively accelerate materials design by focusing on promising candidate chemistries. Those selected can then be subjected to further analysis via experimentation and computational methods to validate crystal-structure-level properties. The data generated by these selective experiments and computations also serve to refine the next generation of "training" data for another iterative round of data mining, which permits a further refinement of high-throughput predictions.

8.5.3 Data Mining for Descriptor Development: Enhancing Databases for Predictions

Standard materials databases containing experimental and theoretical data about properties and structures can be enhanced by the addition of chemistry-structure-property descriptors for specific applications such as materials design. While these descriptors are usually statistically derived, they must incorporate the physics of the problem. In this case study, the starting database is the zeolite framework database of the Structure Commission of the International Zeolite Association (http://www.iza-sc.ethz.ch/IZA-SC/). However this database is not useful to predict the mesopore sizes of the framework, suggesting the need for additional descriptors or types of data. Using principal component analysis and PLS, we have developed secondary descriptors

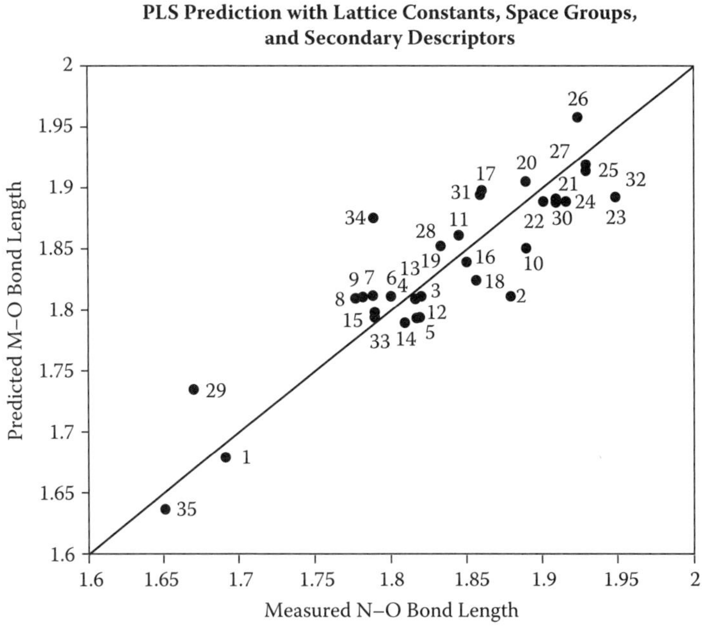

Figure 8.7 The PLS prediction with 12 space group descriptors, 4 lattice constants, and 4 secondary descriptors. M–O bond lengths are given in Angstroms.

that describe the local topology of the frameworks and have used these statistically derived variables to enhance the prediction of structural properties. In our study of zeolites, we first use PLS as a predictive tool based on only the primary descriptors from the database.[86,87] We then added secondary topological descriptors such as c/a ratio (chosen using principal components) and compared the results (Figure 8.7). This illustrates how we can enhance a materials database by adding statistical secondary descriptors tailored for specific applications.

8.5.4 Future Challenges

With crystallographic and thermodynamic databases as an exception, in materials science, the building of databases is still largely an ad hoc process. What data is collected, compiled, and managed is primarily defined by the community that may use that data. As such databases in materials science are not set up to easily integrate, correlate, and process the diversity in data that needs to be considered for solving many complex problems. Crystallographic and thermodynamic databases have a well-established formalism and scientific rationale for organization of data that lend themselves well to be used and applied to experiments and models. For instance, group theory provides the mathematical foundation for establishing a hierarchical database for crystallography. While this is not possible for all types of data bases in materials science, data mining and informatics provide a framework for identifying, searching, and organizing descriptors that form the foundation of databases; and hence provide the key for data analysis of databases in materials science.

8.6 Parallel R for High-Performance Analytics: Applications to Biology

This last section focuses on a different aspect of data analysis, namely, the role of parallel processing in the analysis of massive amounts of data. We describe how a commonly used statistical analysis package can be enhanced to apply it to large-data problems in biology.

R[88] is an open-source software platform for statistical computing and graphics. It is broadly used by the statistics, bioinformatics, engineering, and other communities. R supports diverse statistical analysis tasks such as linear regression, classic statistical tests, time-series analysis, and clustering. It also provides a variety of graphical functions such as histograms, pie charts, and 3D surface plots. More importantly, R provides easy-to-use hooks for adding extension packages by external developers. The major drawback of R is the lack of scalability to massive datasets, which are quite common in scientific domains. For instance, a typical output from mass spectrometry proteomics measurements for a single bacterial genome easily reaches gigabytes, while the output from a climate simulation reaches terabytes. The most straightforward approach to address this challenge is to equip R with high-performance, scalable, parallel processing capabilities.

There are a number of requirements that should be met for any solution to the parallelization of R to be practical and to be easily adapted by a broad community of users. First, it would be ideal to attain the performance comparable with the performance of parallel solutions for compiled languages, like C or Fortran. While R is a scripting language, it is written on top of C and provides mechanisms for calling functions written in such languages. Second, the interface to calling parallel analysis routines should mimic the original R interface, and ideally, should not require users to change their R code to run in a parallel mode. Third, it should carry enough intelligence to detect on the end-user's behalf which parts of the code are parallelizable and which parts are not. Finally, it should require no knowledge or very minimal knowledge of parallel computing from the end user.

Two major approaches to enable parallel processing with R have been introduced (Figure 8.8). The first approach offers message-passing capabilities

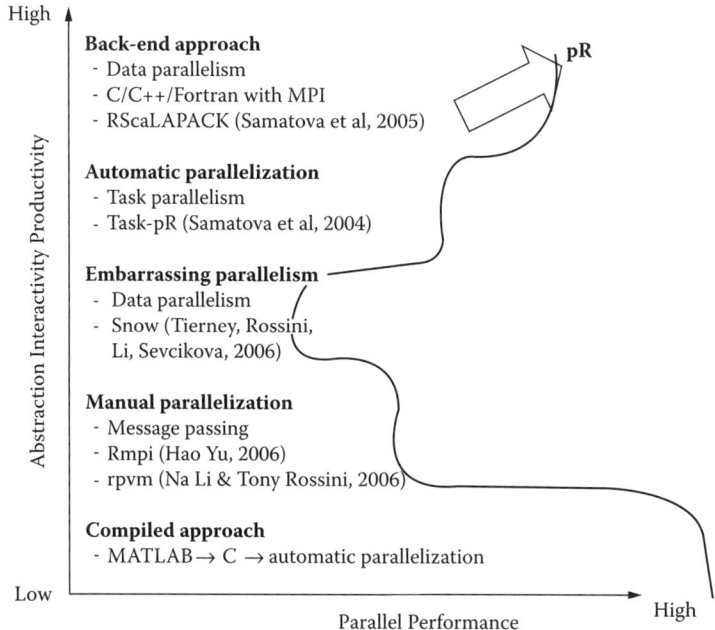

Figure 8.8 Evolution of parallel computing with R.

between **R** processes. It essentially provides wrappers for message passing interface (MPI) and parallel virtual machine (PVM) routines realized through R add-on libraries like $Rmpi$[89] and $rpvm$.[90] While flexible for writing almost any parallel program in **R**, these approaches lack efficiency due to the interpreted nature of the underlying programming language. These libraries are not transparent to the users with limited knowledge of parallel computing. The second approach, such as the *snow* library,[91] offers a capability to address embarrassingly parallel statistical computations. While simple to use, it is limited to computations that require no coordination between R processes: All processes work on their local data, perform exactly the same function, and return the result to the parent process.

In contrast, our approach with ***pR*** addresses the above issues by providing a framework that aims to automatically and efficiently exploit both *data parallelism* (e.g., cluster analysis, principal component analysis, correlation) and *task parallelism* (e.g., likelihood maximization, bootstrap sampling, Markov Chain Monte Carlo, animations) (Figure 8.9). With a data parallel approach, the given data is divided among various processes that perform more or less the same task on their data chunks to obtain the desired result. Demonstration of this capability is realized through an *RScaLAPACK* library,[92,93] which provides hooks to the ScaLAPACK[94] routines for parallel, optimized, and portable linear algebra solvers. *RScaLAPACK* manages

308   *Scientific Data Management*

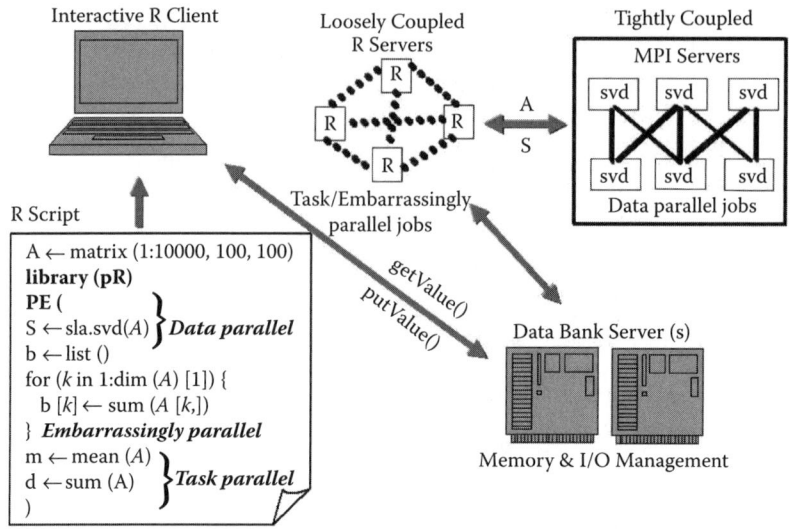

Figure 8.9 The architecture of pR in use.

parallel computation through a single-function call from the R environment. Through *RScaLAPACK*, the user can configure the parallel environment, distribute data and carry out the required parallel computation. While the interface maintains the look and feel of the R system, *RScaLAPACK* allows users to execute analyses that scale in both problem size and the number of processors. *RScaLAPACK* is developed using the C and FORTRAN languages and is distributed as an add-on library to the R statistical package at http://cran.r-project.org/src/contrib/Descriptions/RScaLAPACK.html. With task parallelism, *pR* divides a given job into various tasks that are executed in parallel to obtain the expected result. It provides a way to detect the out-of-order independent R tasks and delegates each task to remote worker processes to execute them concurrently. Again, the changes introduced to the interface are minimal to provide simple user-level control over task parallel versus serial execution (Figure 8.9, Algorithm 8.1). In the next section we describe an application from the biology domain that deals with data analytics challenges and takes advantages of *pR*.

8.6.1 Advanced Analytics for High-Throughput Quantitative Proteomics

Organisms often respond to environmental or physiological stimuli by adjusting the type and abundance of proteins in their cells. Measurement of the *relative* abundances of proteins in treatment cells subjected to stimuli, compared with that in reference cells, provides valuable insights about protein function and regulation. Quantitative "shotgun" proteomics has recently emerged as a high-throughput technique for measuring the relative abundances of thousands of proteins between two cellular conditions. In quantitative shotgun proteomics, proteolysis-derived peptides are measured with liquid chromatography-tandem mass spectrometry (LC-MS/MS) and used as surrogates of their parent proteins for relative quantification. By employing stable isotope labeling, the proteomes under comparison are mixed together and analyzed in one LC-MS/MS run. Each peptide in the mixture of two isotopically labeled proteomes has two mass-different isotopic variants, the light isotopologue (e.g., ^{14}N) from one proteome and the heavy isotopologue (e.g., ^{15}N) from the other. Because sample handling in proteome measurements is highly complex, proteome quantification requires rigorous statistical approaches. Moreover, the size of the experimental data easily reaches several gigabytes, thus challenging data analysis algorithms even further. Below, we describe some of these challenges and approaches to address them for metabolically labeled proteome quantification.

Figure 8.10 depicts the overall pipeline for proteome quantification from data generation through data processing and analysis to visualization. The resulting software system is called ProRata,[95,96] which is available as open source from http://www.MSProRata.org. The analytical engine of ProRata is available both as a serial C++ code and as a parallel *pR* code, and the graphical user interface allows manual data interrogation for visualization and

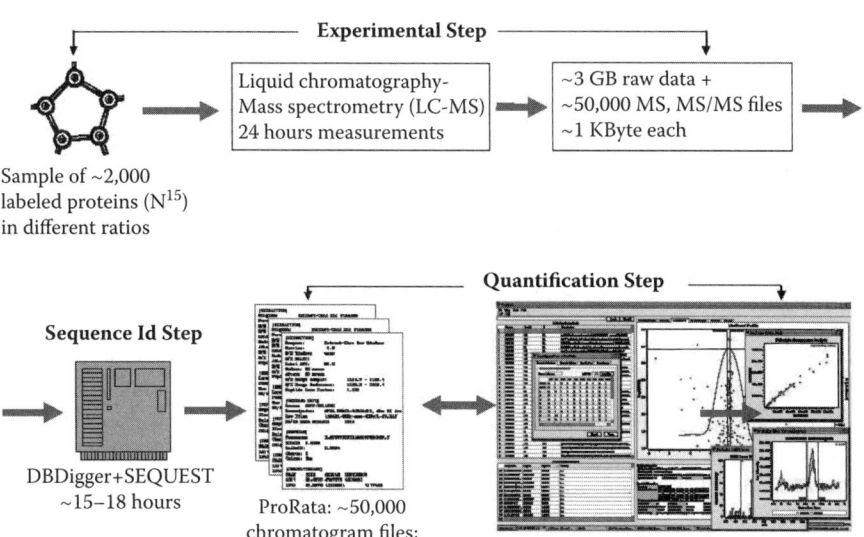

Figure 8.10 Data processing pipeline in mass spectrometry proteomics for proteome quantification.

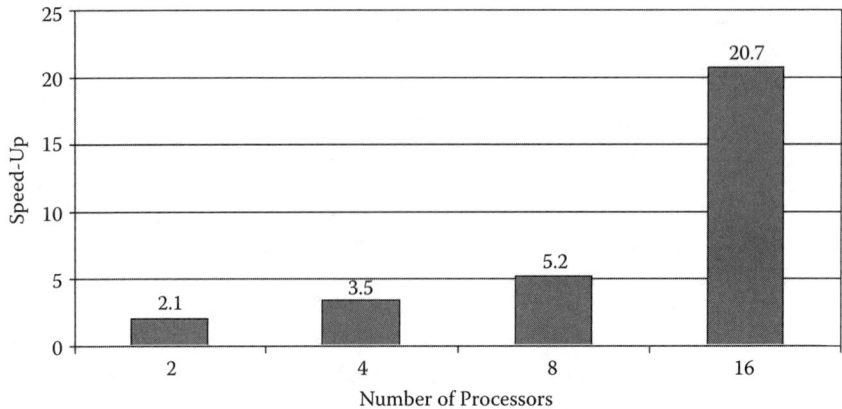

Figure 8.11 Scalability of processing task-parallel jobs in ProRata.

validation of results. Next we describe how ProRata addresses the problem of efficiency in data processing and the problem of data noise.

8.6.1.1 Parallel Processing of Core Analysis Steps in ProRata

The number of files that are typically generated by mass spectrometry devices for a whole proteome experiment easily reaches several thousands. Although each individual file is relatively small in size, processing them all collectively is time-consuming. Since initial processing of individual files does not depend on the other files, the ***R*** version of ProRata allows one to employ the task-parallelism feature of ***pR*** for concurrent processing of all these files on multiple processors. Algorithm 8.1 depicts the fragment of the ProRata code with the slight changes that were introduced to the serial ***R*** code to enable such a task-parallel processing. The modifications are via **PE()** highlighted in boldface. As a result, the linear speed-up has been gained, as shown in Figure 8.11. Note that a superlinear speed-up has been observed for 16 processors, partly due to the fact that each processor had its own copy of the file stored on a local disk (see Algorithm 8.1).

Likewise, one of the key steps in ProRata is the use of principal component analysis (PCA) (see details in Section 8.6.1.2). ***R*** supports this kind of analysis through its *prcomp()* function, which underneath calls a serial singular value decomposition function, called *svd()*. Since *svd()* calculation is a matrix calculation, for large matrices, this calculation is computationally demanding. The parallel and optimized *svd()* calculation is, however, available through the ScaLAPACK parallel linear algebra package. To invoke a parallel version of the *prcomp()* library, one needs to load the ***RScaLAPACK*** library and replace the call to *prcomp()* function with the call to *sla.prcomp()* function. These slight modifications for getting access to the data-parallelism feature of

```
 1:  chroList ← list.files(pattern="*.chro");
 2:  cat ("Chro", "samSN", "refSN", "PPCSN", "HR", "PCA", "PCASN",
        file="Pratio-Peptide.txt");
 3:  PE (for (i in 1:length(chroList))
 4:  {
 5:     currResult [i] = Pratio(filename=chroList[i]);
 6:  } )
 7:  for (i in 1:length(chroList))
 8:  {
 9:     cat (chroList[i], currResult$samSN, file="Pratio-Peptide.txt");
10:  }
```

Algorithm 8.1: Code fragment from ProRata with enabled task-parallelism feature.

pR are not mandatory and are provided for better user-driven control in the choice of parallel and serial routines. Since the overhead introduced by pR for these kinds of function calls is less than 10% and reduces with the size of the matrix, the scalability in terms of the number of processors is largely determined by the scalability of the underlying ScaLAPACK's routines (scalability benchmarks are not shown here, but described elsewhere).[92,93]

Dealing with the second challenge—the noise in the data—to improve both quantification accuracy and quantification confidence requires optimization of the core analysis steps described below: chromatographic peaks detection, peptide relative abundance estimation, and protein relative abundance estimation (Figure 8.12).

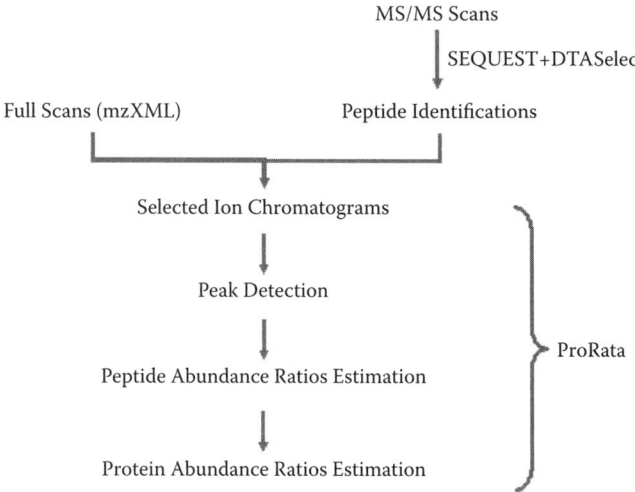

Figure 8.12 Key data analysis steps in ProRata.

8.6.1.2 Estimation of Peptide Abundance Ratios and Scoring of Their Variability and Bias

The abundance ratio between the light and heavy isotopologues of an isotopically labeled peptide can be estimated from their selected ion chromatograms. However, quantitative shotgun proteomics measurements yield selected ion chromatograms at highly variable signal-to-noise ratios (S/Ns) for tens of thousands of peptides. This challenge calls for algorithms that not only robustly estimate the abundance ratios of different peptides but also rigorously score each abundance ratio for the expected estimation bias and variability. Scoring of the abundance ratios enables filtering of unreliable peptide quantification and use of formal statistical inference in the subsequent protein abundance ratio estimation.

The general computational procedure for estimating the abundance ratio between the light isotopologue and the heavy isotopologue of a peptide includes the following steps (Figure 8.12). First, *peak detection* is performed in the selected ion chromatograms for the two isotopologues. Since a large fraction of chromatographic peaks have a very low chromatographic S/N, incorrect assignments of their peak boundaries are often observed. We have improved the robustness of peak detection by employing a parallel paired covariance, which is the product of the background subtracted ion intensities at that full scan in the two selected ion chromatograms. The parallel paired covariance algorithm has largely enhanced the S/N of the two isotopologues' chromatograms and, as a result, has enabled much more accurate peak detection.

Next, the peptide abundance ratios from the detected chromatographic peaks are estimated. Existing algorithms that evaluate peptide abundance ratios do not formally score the abundance ratio estimates for their expected bias and variability. In quantitative shotgun proteomics, the abundance ratios for tens of thousands of identified peptides can be estimated, but with dramatically varying error. We proposed to employ principal component analysis (PCA) of the peak profile to estimate the peptide abundance ratio and to score the estimation with the signal-to-noise ratio of the peak profile (profile-S/N) (Figure 8.13). Specifically, we observed that the peptide abundance ratio correlates with the slope of the first eigenvector, and the profile-S/N correlates with the the square root of the ratio between the first and the second eigenvalues. We demonstrate that the profile-SN is inversely correlated with the variability and bias of peptide abundance ratio estimation. Thus, the profile-S/N allows stratification of the peptide abundance ratios into those with greater or lesser estimation accuracy and precision. As a result, we observed superior estimation accuracy in peptide abundance ratios compared to the traditional methods based on peak height and peak area. In addition, the efficiency of data processing has been achieved due to ProRata's support of parallel PCA processing through the RScaLA-PACK's sla.prcomp() function, which scaled almost linearly with the number

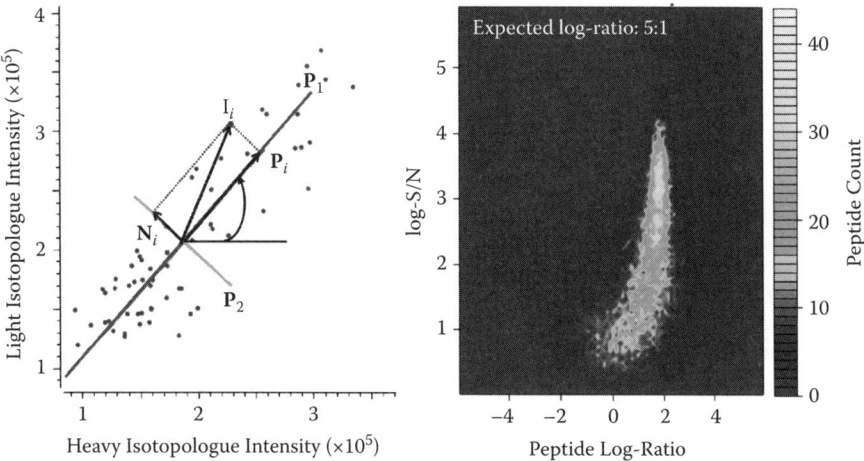

Figure 8.13 (See color insert following page 224.) Peptide abundance ratio and signal-to-noise ratio estimation via PCA (left). Signal-to-noise ratio is inversely correlated with the variability and bias of peptide abundance ratio (right).

of processors used (up to 32) after less than 10% overhead as described in Section 8.6.1.1.

8.6.1.3 Protein Abundance Ratio Estimation with Confidence Interval Evaluation

To evaluate protein abundance ratios from quantitative proteomics measurements, two types of statistical estimation should be employed: *point estimation* and *interval estimation*. The point estimation gives an abundance ratio for every quantified protein, which best approximates the true abundance ratio. Unfortunately, the point estimation provides no information about protein quantification precision, which can significantly vary across different proteins. Generally, a protein should have better quantification precision if it has more peptides quantified from mass spectral data of higher signal-to-noise ratio. It is misleading in quantitative proteomics to treat all proteins' abundance ratios identically, regardless of their estimation precision.

The interval estimation complements the point estimation by providing confidence intervals for protein abundance ratios. If 90% of quantified proteins have confidence intervals that contain their true abundance ratios, then confidence intervals are estimated at a 90% confidence level. The confidence level for the interval estimation in quantitative proteomics is analogous to the true positive rate for protein identification in qualitative proteomics. More importantly, at a given confidence level, the confidence interval intuitively reflects the quantification precision for each protein as an error bar of the abundance ratio estimate.

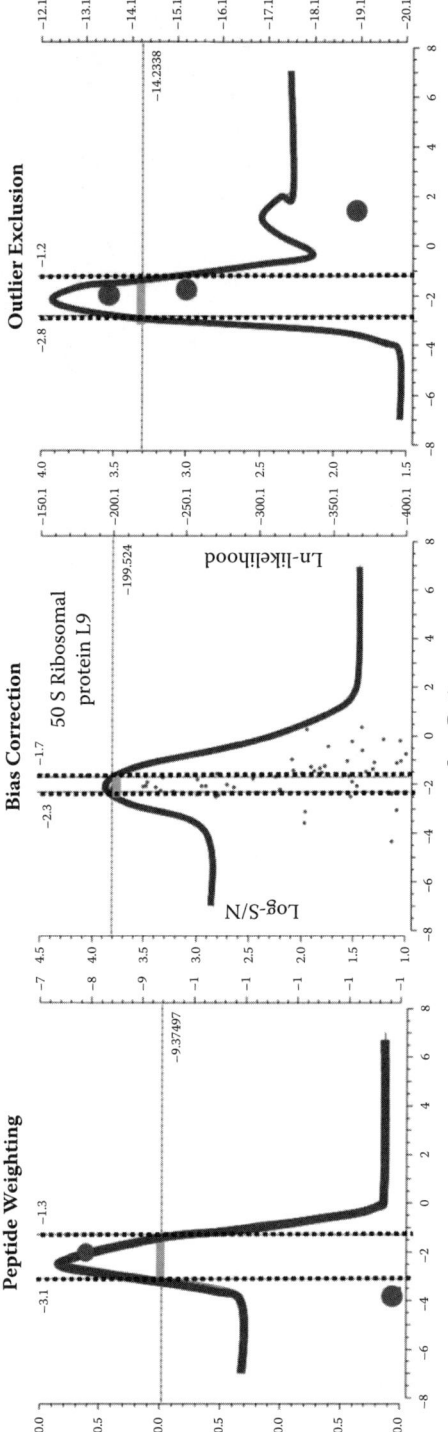

Figure 8.14 (See color insert following page 224.) Protein abundance ratio and confidence interval estimation via profile likelihood algorithm.

In quantitative shotgun proteomics, each protein abundance ratio is estimated by combining multiple peptide abundance ratios and often using the average value. The standard deviations of peptide abundance ratios are used to measure the variation of protein abundance ratio estimates. However, without assuming the normality of the peptide abundance ratio distribution, the standard deviation is not directly related to the confidence interval of the protein abundance ratio.

We devised a profile likelihood algorithm to infer the abundance ratios of proteins from the abundance ratios of isotopically labeled peptides. Given multiple quantified peptides for a protein, the profile likelihood algorithm probabilistically weighs the peptide abundance ratios by their inferred estimation variability, accounts for their expected estimation bias, and suppresses contribution from outliers (Figure 8.14). This algorithm yields maximum likelihood point estimation and profile likelihood confidence interval estimation of protein abundance ratios. This point estimator is more accurate than an estimator based on the average of peptide abundance ratios. The confidence interval estimation provides an error bar for each protein abundance ratio that reflects its estimation precision and statistical uncertainty. The profile likelihood algorithm not only showed more accurate protein quantification and better coverage than the widely used programs (e.g., RelEx) but also a more robust estimate of a confidence interval for each differential protein expression ratio.

8.7 Summary

In this chapter, we described the application of various data analysis algorithms to find useful information in datasets from several different scientific domains, ranging from biology to materials science and cheminformatics. Though these domains, and the problems being solved, are very different, we observe several similarities. For example, the presence of noise in the data is frequently an issue as it can affect the analysis done using the data. The representation of the objects in the data is also important—if the representation captures the key features that are critical to the analysis problem at hand, the results of the analysis can be improved. We also saw that some techniques, such as principal component analysis, are used in several different domains, including biology and materials science. And finally, it was observed that analysis in scientific domains is not just the application of statistical or data mining algorithms, but a careful integration of such techniques with domain expertise, along with a careful and deliberate process of understanding the data input to the algorithms and interpreting the patterns found in the data.

Acknowledgment

N. Wale and G. Karypis were supported by NSF IIS-0431135 and NIH RLM008713A and by the Digital Technology Center at the University of Minnesota.

The research of G. Pandey and V. Kumar was partially supported by NSF grants CNS-0551551 and IIS-0713227 the University of Minnesota.

KR acknowledges support from the National Science Foundation International Materials Institute program for the Combinatorial Sciences and Materials Informatics Collaboratory (CoSMIC-IMI) grant no. DMR-0603644; National Science Foundation Materials Digital Library (MatDL) Pathway Project grant no. DUE 0532831; DARPA Center for Interfacial Engineering for MEMS (CIEMS) grant no. 1891874036790B; the Air Force Office of Scientific Research grant no. FA95500610501; Office of Naval Research (ONR) and DARPA as part of the Dynamic 3-D Digital Structures (Contracts N00014-07-WX-2-0381 and N00014-07-WX-2-0382); and the Office of Naval Research MURI program for Novel Vaccines: Targeting and Exploiting the Bacterial Quorum Sensing Pathway, award no. N00014-06-1-1176.

The work of NFS, PB, GK, CP, and SY is performed as part of the Scientific Data Management Center (http://sdmcenter.lbl.gov) under the Department of Energy's Scientific Discovery through Advanced Computing program (http://www.scidac.org) and conducted at Oak Ridge National Laboratory, which is managed by UTBattelle for the LLC U.S. D.O.E. under contract no. DEAC05-00OR22725.

LLNL-MI-402814: The work of Chandrika Kamath is performed under the auspices of the U.S. Department of Energy by Lawrence Livermore National Laboratory under Contract DE-AC52-07NA27344.

References

[1] Kamath, C. *Scientific Data Mining: A Practical Perspective.* SIAM, Philadelphia, PA, 2009.

[2] Leach, A. R., and Gillet, V. J. *An Introduction to Chemoinformatics.* Dordorecht, The Netherlands: Springer, 2003.

[3] Southan, C., Varkonyi, P., and Muresan, S. Complementarity between public and commercial databases: new opportunities in medicinal chemistry informatics. *Current Topics in Medicinal Chemistry* 7, 15 (2007), 1502–1508.

[4] Bohm, H., and Schneider, G. *Virtual Screening for Bioactive Molecules.* Weinheim, Germany Wiley-VCH, 2000.

[5] Hansch, C., Maolney, P. P., Fujita, T., and Muir, R. M. Correlation of biological activity of phenoxyacetic acids with hammett substituent constants and partition coefficients. *Nature 194* (1962), 178–180.

[6] Daylight Inc, Mission Viejo, CA, USA. http://www.daylight.com. Accessed July 12, 2009.

[7] Pipeline project, Scitegic inc. http://www.scitegic.com. Accessed July 12, 2009.

[8] Sybyl, Tripos inc. http://www.tripos.com/. Accessed July 12, 2009.

[9] Screen, Chemaxon inc. http://www.chemaxon.com. Accessed July 12, 2009.

[10] MDL Information Systems Inc., San Leandro, CA, USA. http://www.mdl.com. Accessed July 12, 2009.

[11] Wale, N., Watson, I. A., and Karypis, G. Comparison of descriptor spaces for chemical compound retrieval and classification. *Knowledge and Information Systems (in press)* (2007).

[12] King, R. D., Srinivasan, A., and Dehaspe, L. Warmr: A data mining tool for chemical data. *Journal of Computer Aided Molecular Design 15* (2001), 173–181.

[13] Ralaivola, L., Swamidassa, S. J., Saigo, H., and Baldi., P. Graph kernels for chemical informatics. *Neural Networks 18*, 8 (2005), 1093–1110.

[14] Menchetti, S., Costa, F., and Frasconi, P. Weighted decomposition kernels. *Proceedings of the 22nd International Conference in Machine Learning. 119* (2005), 585–592.

[15] Raymond, J. W., and Willett, P. Maximum common subgraph isomorphism algorithms for the matching of chemical structures. *J. Comp. Aided Mol. Des. 16*, 7 (2002), 521–533.

[16] Muller, K. R., Mika, S., Ratsch, G., Tsuda, K., and Scholkopf., B. An introduction to kernel-based learning algorithms. *IEEE Trans. Neural. Net. 12*, 2 (2001), 181–201.

[17] Kuramochi, M., and Karypis, G. An efficient algorithm for discovering frequent subgraphs. *IEEE TKDE. 16*, 9 (2004.), 1038–1051.

[18] Deshpande, M., Kuramochi, M., Wale, N., and Karypis., G. Frequent substructure-based approaches for classifying chemical compounds. *IEEE TKDE. 17*, 8 (2005), 1036–1050.

[19] Horvath, T., Gartner, T., and Wrobel, S. Cyclic pattern kernels for predictive graph mining. *Proceedings of the SIGKDD* (2004), 158–167.

[20] The Pubchem project. http://pubchem.ncbi.nlm.nih.gov. Accessed July 12, 2009.

[21] Wale, N., Karypis, G., and Watson, I. A. Method for effective virtual screening and scaffold-hopping in chemical compounds. *Comput Syst Bioinformatics Conf 6* (2007), 403–414.

[22] Kuramochi, M., and Karypis, G. Discovering frequent geometric subgraphs. *Information Systems 32*, 8 (2007), 1101–1120.

[23] Paolini, G. V., Shapland, R. H., Hoorn, W. P. V., Mason, J. S., and Hopkins., A. Global mapping of pharmacological space. *Nature Biotechnology 24* (2006), 805–815.

[24] Russ, A. P., and Lampel, S. The druggable genome: an update. *Drug Discov Today 10*, 23–24 (2005), 1607–1610.

[25] Yildirim, M. A., Goh, K.-I., Cusick, M. E., Barabási, A.-L., and Vidal, M. Drug-target network. *Nat Biotechnol 25*, 10 (Oct 2007), 1119–1126.

[26] Rognan, D. Chemogenomic approaches to rational drug design. *Br J Pharmacol 152*, 1 (Sep 2007), 38–52.

[27] Stockwell, B. R. Exploring biology with small organic molecules. *Nature 432*, 7019 (Dec 2004), 846–854.

[28] Jenkins, J. L., Bender, A., and Davies, J. W. In silico target fishing: predicting biological targets from chemical structure. *Drug Discovery Today 3*, 4 (2006), 413–421.

[29] Oliver, S. A network approach to the systematic analysis of yeast gene function. *Trends in Genetics 12*, 7 (1996), 241–242.

[30] *EUROFAN 2000: The Final Meeting* (2000).

[31] Roberts, R. J. Identifying protein function, a call for community action. *PLoS Biology 2*, 3 (2004), 293–294.

[32] Meyer, F. Genome sequencing vs. Moore's law: Cyber challenges for the next decade. *CTWatch Quarterly*, August 2006.

[33] Boeckmann, B., Bairoch, A., Apweiler, R., Blatter, M.-C., Estreicher, A., et al. The SWISS-PROT protein knowledgebase and its supplement TrEMBL in 2003. *Nucleic Acids Research 31*, 1 (2003), 365–370.

[34] Mewes, H. W., Frishman, D., Guldener, U., Mannhaupt, G., Mayer, K., et al. MIPS: a database for genomes and protein sequences. *Nucleic Acids Research 30*, 1 (2002), 31–34.

[35] Xenarios, I., Salwinski, L., Duan, X. J., Higney, P., Kim, S.-M., and Eisenberg, D. DIP and the Database of Interacting Proteins: a research tool for studying cellular networks of protein interactions. *Nucleic Acids Research 30*, 1 (2002), 303–305.

[36] Berman, H. M., Westbrook, J., Feng, Z., Gilliland, G., Bhat, T. N., Weissig, H., Shindyalov, I. N., and Bourne, P. E. The protein data bank. *Nucleic Acids Research 28*, 1 (2000), 235–242.

[37] Rastan, S., and Beeley, L. J. Functional genomics: going forwards from the databases. *Curr Opin Genet Dev.* 7, 6 (1997), 777–783.

[38] Altschul, S. F., Gish, W., Miller, W., Meyers, E. W., and Lipman, D. J. Basic local alignment search tool. *J Mol Biol.* 215, 3 (1990), 403–410.

[39] Altschul, S. F., Madden, T. L., Schiffer, A. A., Zhang, J., Zhang, Z., Miller, W., and Lipman, D. J. Gapped BLAST and PSI-BLAST: a new generation of protein database search programs. *Nucleic Acids Research* 25, 17 (1997), 3389–3402.

[40] Devos, D., and Valencia, A. Practical limits of function prediction. *Proteins* 41, 1 (2000), 98–107.

[41] Whisstock, J. C., and Lesk, A. M. Prediction of protein function from protein sequence and structure. *Q Rev Biophys.* 36, 3 (2003), 307–340.

[42] Pandey, G., Kumar, V., and Steinbach, M. Computational approaches for protein function prediction: A survey. Tech. Rep. 06-028, Department of Computer Science and Engineering, University of Minnesota, Twin Cities, 2006.

[43] Bork, P., Dandekar, T., Diaz-Lazcoz, Y., Eisenhaber, F., Huynen, M., and Yuan, Y. Predicting function: from genes to genomes and back. *J Mol Biol.* 283, 4 (1998), 707–725.

[44] Rost, B., Liu, J., Nair, R., Wrzeszczynski, K. O., and Ofran, Y. Automatic prediction of protein function. *Cell Mol Life Sci.* 60, 12 (2003), 2637–2650.

[45] Gabaldon, T., and Huynen, M. A. Prediction of protein function and pathways in the genome era. *Cell Mol Life Sci.* 61, 7–8 (2004), 930–944.

[46] Seshasayee, A. S. N., and Babu, M. M. Contextual inference of protein function. In *Encyclopaedia of Genetics and Genomics and Proteomics and Bioinformatics*, M. J. Dunn, L. B. Jorde, P. F. R. Little, and S. Subramaniam (Eds.), Hoboken, NJ: John Wiley and Sons, 2005.

[47] Tan, P.-N., Steinbach, M., and Kumar, V. *Introduction to Data Mining.* Pearson Addison-Wesley, 2006.

[48] Huynen, M. A., Snel, B., Bork, P., and Gibson, T. J. The phylogenetic distribution of frataxin indicates a role in iron-sulfur cluster protein assembly. *Hum Mol Genet.* 10, 21 (2001), 2463–2468.

[49] Myers, C. L., Robson, D., Wible, A., Hibbs, M. A., Chiriac, C., et al. Discovery of biological networks from diverse functional genomic data. *Genome Biology* 6 (2005), R114.

[50] Ashburner, M., Ball, C. A., Blake, J. A., Botstein, D., Butler, H., et al. Gene ontology: tool for the unification of biology. *Nature Genetics* 25, 1 (2000), 25–29.

[51] Brusic, V., and Zeleznikow, J. Knowledge discovery and data mining in biological databases. *Knowl. Eng. Rev. 14*, 3 (1999), 257–277.

[52] Hart, G. T., Ramani, A. K., and Marcotte, E. M. How complete are current yeast and human protein-interaction networks? *Genome Biology 7*, 11 (2006), 120.

[53] Deane, C. M., Salwinski, L., Xenarios, I., and Eisenberg, D. Protein interactions: two methods for assessment of the reliability of high throughput observations. *Mol Cell Proteomics 1*, 5 (2002), 349–356.

[54] Brun, C., Chevenet, F., Martin, D., Wojcik, J., Guenoche, A., and Jacq, B. Functional classification of proteins for the prediction of cellular function from a protein-protein interaction network. *Genome Biology 5*, 1 (2003), R6.

[55] Samanta, M. P., and Liang, S. Predicting protein functions from redundancies in large-scale protein interaction networks. *Proc Natl Acad Sci U.S.A. 100*, 22 (2003), 12579–12583.

[56] Chen, J., Chua, H. N., Hsu, W., Lee, M.-L., Ng, S.-K., Saito, R., Sung, W.-K., and Wong, L. Increasing confidence of protein-protein inteactomes. In *Proceedings of 17th International Conference on Genome Informatics (GIW)* (2006), pp. 284–297.

[57] Pandey, G., Steinbach, M., Gupta, R., Garg, T., and Kumar, V. Association analysis-based transformations for protein interaction networks: a function prediction case study. In *KDD '07: Proceedings of the 13th ACM SIGKDD International Conference on Knowledge Discovery and Data Mining* (2007), pp. 540–549.

[58] Xiong, H., Tan, P.-N., and Kumar, V. Hyperclique pattern discovery. *Data Min. Knowl. Discov. 13*, 2 (2006), 219–242.

[59] Nabieva, E., Jim, K., Agarwal, A., Chazelle, B., and Singh, M. Whole-proteome prediction of protein function via graph-theoretic analysis of interaction maps. *Bioinformatics 21*, Suppl. 1 (2005), i1–i9.

[60] Xiong, H., Pandey, G., Steinbach, M., and Kumar, V. Enhancing data analysis with noise removal. *IEEE Transactions on Knowledge and Data Engineering 18*, 3 (2006), 304–319.

[61] Smyth, G. K., and Speed, T. Normalization of cDNA microarray data. *Methods 31*, 4 (2003), 265–273.

[62] Grimvall, G. *Thermophysical Properties of Materials*. Elsevier, Amsterdam, 1999.

[63] Ledbetter, H., and Kim, S. Bulk moduli systematics in oxides, including superconductors. In *Handbook of Elastic Properties of Solids, Liquids and Gases*, H. B. M. Levy and R. R. Stern, Eds. San Diego, CA: Academic Press, 2000.

[64] Iwata, S., et al. Materials database for materials design. *Journal of Nuclear Materials 179–181* (1992), 1135.

[65] Davis, J. W. Development of an international fusion materials database. *Journal of Nuclear Materials 179–181* (1992), 1139.

[66] Saxena, S., et al. *Thermodynamic Data on Oxides and Silicates*. Springer, New York, 1993, page 428.

[67] Fabrichnaya, O. B., and Sundman, B. The assessment of thermodynamic parameters in the Fe-O and Fe-Si-O systems. *Geochimica et Cosmochimica ACTA 61*, 21 (1997), 4539–4555.

[68] Fabrichnaya, O. B., et al. *Thermodynamic Data, Models, and Phase Diagrams in Multicomponent Oxide Systems*. Springer, New York, 2003.

[69] Bale, C. W., et al. Factsage thermochemical software and databases. *Calphad 26*, 2 (2002), 189–228.

[70] Cox, J., et al., Eds. *CODATA Key Values for Thermodynamics*. Hemisphere Publishing Corp., New York, 1989.

[71] Soligo, D., et al. Non-generic concentrations for shape-memory alloys; the case of CuZnAl. *Acta Materialia 47* (1999), 2741.

[72] Villars, P., et al. Binary, ternary and quaternary compound former/non-former prediction via mendeleev number. *Journal of Alloys and Compounds 317–318* (2001), 26.

[73] Ashby, M. F. *Materials Selection in Mechanical Design*. Butterworth-Heinemann, Oxford, 1999.

[74] Shercliff, H. R., and Lsvatt, A. M. Selection of manufacturing processes in design and the role of process modeling. *Progress in Materials Science 46* (2001), 429–459.

[75] Lovatt, A. M., and Shercliff, H. R. Manufacturing process selection in engineering design. *Materials and Design 19* (1998), 205–230.

[76] Ashby, M. F. Checks and estimates for materials properties : I. ranges and simple correlations. *Proc Royal Soc. London A 454*, 1873 (1998), 1301–1321.

[77] Bassetti, D., et al. Checks and estimates for materials properties : I. the method of multiple correlations. *Proc Royal Soc. London A 454*, 1873 (1998), 1323–1336.

[78] Zaki, M., and Rajan, K. Data mining: a tool for materials discovery. In *Proceedings of 17th CODATA meeting* (2002). http://www.cs.rpi.edu/zaki/ps/codata.00.ps.gz. Accessed on July 12, 2009.

[79] Bajorath, J. Integration of virtual and high-throughput screening. *Nature Reviews Drug Discovery 1* (2002), 882–894.

[80] Quakenbush, J. Compuational analysis of microarray data. *Nature Reviews Genetics 2* (2001), 418–427.

[81] Li, G., et al. High dimensional model representations. *J. Physical Chemistry 104* (2001), 7765.

[82] Rajan, K., Suh, C., and Narasimhan, B. Informatics methods for combinatorial materials science. In *Combinatorial Materials Science*, S. Mallapragada, B. Narasimhan, and M. Porter (Eds.), John Wiley & Sons, New York, 2007.

[83] Suh, C., and Rajan, K. Invited review: Data mining and information for crystal chemistry: Establishing measurement techniques for mapping structure-property relationships. *J. Materials Science and Technology* Volume 25(4) April 2009 pp. 466–471.

[84] Willars, P., and Phillips, J. C. Quantum structural diagrams and high Tc superconductivity. *Phys. Rev Letters 37* (1988), 2345–2348.

[85] Suh, C., and Rajan, K. Virtual screening and QSAR formulations for crystal chemistry. *QSAR & Combinatorial Science Journal 24* (2005), 114.

[86] Rajagopalan, A., and Rajan, K. Informatics based optimization of crystallographic descriptors for framework structures. In *Combinatorial and High Throughput Discovery and Optimization of Catalysts and Materials*, R. A. Potyrailo and W. F. Maier (Eds.), CRC Press, Boca Raton, FL, 2006.

[87] Rajagopalan, A., et al. Secondary descriptor development for zeolite framework design: an informatics approach. *Applied Catalysis A 254* (2003), 147–160.

[88] R Development Core Team. *R: A Language and Environment for Statistical Computing*. R Foundation for Statistical Computing, Vienna, Austria, 2005. ISBN 3-900051-07-0.

[89] Yu, H. *The Rmpi Package*. R Foundation for Statistical Computing, Vienna, Austria, 2006.

[90] Li N., and Rossini, A. J. *The rpvm Package*. R Foundation for Statistical Computing, Vienna, Austria, 2005.

[91] Tierney, L., Rossini, A. J., Li, N., and Sevcikova, H. *The Snow Package*. R Foundation for Statistical Computing, Vienna, Austria, 2006.

[92] Yoginath, S., Samatova, N., Bauer, D., Kora, G., Fann, G., and Geist, A. RScaLAPACK: High performance parallel statistical computing with R and ScaLAPACK. In *Proceedings of the 18th International Conference on Parallel and Distributed Computing Systems* (2005), pp. 61–67.

[93] Samatova, N., et al. High performance statistical computing with parallel R: applications to biology and climate modeling. *Journal of Physics: Conference Series*, 46 (2006).

[94] L. S. Blackford, J. Choi, A. Cleary, E. D'Azevedo, J. Demmel, I. Dhillon, J. Dongarra, S. Hammarling, G. Henry, A. Petitet, K. Stanley, D. Walker, and R. C. Whaley. *ScaLAPACK User's Guide*, 1997.

[95] Pan, C., Kora, G., Tabb, D. L., Pelletier, D. A., McDonald, W. H., Hurst, G. B., Hettich, R. L. and Samatova, N. F. Robust estimation of peptide abundance ratios and rigorous scoring of their variability and bias in quantitative shotgun proteomics. *Analytical Chemistry 78(20)* (2006), 7110–7120.

[96] Pan, C., Kora, G., McDonald, W. H., Tabb, D. L., VerBerkmoes, N. C., Hurst, G. B., Pelletier, D. A., Samatova, N. F., and Hettich, R. L. Prorata: a quantitative proteomics program for accurate protein abundance ratio estimation with confidence interval evaluation. *Analytical Chemistry 78(20)* (2006), 7121–7131.

Chapter 9

Scientific Data Management Challenges in High-Performance Visual Data Analysis

E. Wes Bethel,[1] Prabhat,[1] Hank Childs,[2] Ajith Mascarenhas,[3] and Valerio Pascucci[4]

[1] High Performance Computing Research Department, Lawrence Berkeley National Laboratory, Berkeley, California
[2] Computing Applications and Research, Lawrence Livermore National Laboratory, Livermore, California
[3] Center for Applied Scientific Computing, Lawrence Livermore National Laboratory, Livermore, California
[4] Scientific Computing and Imaging (SCI) Institute, School of Computing, University of Utah, Salt Lake City, Utah

Contents

9.1 Introduction 326
9.2 Production-Level, Parallel Visualization Tool Perspective on SDM 327
    9.2.1 How Data Is Processed 328
        9.2.1.1 I/O 328
        9.2.1.2 Processing 329
        9.2.1.3 Rendering and Remote/Distributed Visualization .. 330
    9.2.2 How Metadata Can Enable Optimizations 332
    9.2.3 Data Models and Semantics 334
    9.2.4 A Real-World Production Parallel Visualization Tool 335
9.3 Multiresolution Data Layout for Large-Scale Data Analysis 336
    9.3.1 Background 336
    9.3.2 Hierarchical Indexing for Out-of-Core Access to Multiresolution Data 337
        9.3.2.1 Hierarchical Subsampling Framework 338
        9.3.2.2 Binary Trees and the Lebesgue Space-Filling Curve 340
        9.3.2.3 Performance 344

9.4 File Formats for High-Performance Visualization
    and Analytics 345
    9.4.1 H5Part 347
          9.4.1.1 Motivation 347
          9.4.1.2 File Organization and API 349
          9.4.1.3 Parallel I/O 350
    9.4.2 HDF5_ FastQuery 352
          9.4.2.1 Functionality 352
          9.4.2.2 Architectural Layout 353
    9.4.3 Status and Sample Applications 354
9.5 Query-Driven Visualization 358
    9.5.1 Implementing Query-Driven Visualization 359
    9.5.2 Case Study—Network Traffic Analysis 360
9.6 Summary and Conclusion 363
Acknowledgments 364
References 365

9.1 Introduction

Scientific visualization, which is the transformation of abstract data into readily comprehensible images, and visual data analysis/analytics, which combines visualization with analysis, play a central role in the modern scientific process. We use the umbrella term *visualization* to refer to this broad set of investigatory techniques aimed at enabling knowledge discovery—gaining insight—from large, complex collections of scientific data.

The term *scientific visualization* was coined in 1987 in a landmark report,[1] which said:

"Visualization is a method of computing. It transforms the symbolic into the geometric, enabling researchers to observe their simulations and computations. Visualization offers a method for seeing the unseen. It enriches the process of scientific discovery and fosters profound and unexpected insights. In many fields, it is already revolutionizing the way scientists do science... The goal of visualization is to leverage existing scientific methods by providing new scientific insight through visual methods."

Various data analysis methods covered in Chapter 8 can also benefit from techniques covered in this chapter.

Although the term was coined in 1987, the art and science of visualization dates back hundreds of years to DaVinci's illustrations, or even earlier, to Cicero's written account of an early orrery constructed by the Greek philosopher Posidonious to exhibit the diurnal motions of the sun, moon, and five known planets.* A relatively recent overview in Adelmann et al. provides a good survey of the field's breadth and depth: algorithms for visualizing scalar, vector, and tensor fields, geometric modeling, virtual environments for visualization, large data visualization, perceptual and cognitive issues in visualization, visualization software and frameworks, software architecture, and so forth.

Visualization is a very data-intensive science: visualization algorithms take as input vast amounts of data produced by simulation or experiment, and then transform that data into imagery. It turns out, as we shall explore in this chapter, that visualization reveals a somewhat different view of scientific data management challenges than are examined elsewhere in this book. For example, a data ordering and storage layout that works well for saving data from memory to disk may not be the best thing for subsequent visual data analysis algorithms.

This chapter will present four broad topic areas under this general rubric: (1) a view of SDM-related issues from the perspective of implementing a production-quality, parallel capable visual data analysis infrastructure; (2) novel data storage formats for multiresolution, streaming data movement, access and use by postprocessing tools; (3) data models, formats and APIs for performing efficient I/O for both simulations and postprocessing tools, discussion of issues, and previous work in this space; (4) how combining state-of-the-art techniques from scientific data management and visualization enables visual data analysis of truly massive datasets.

9.2 Production-Level, Parallel Visualization Tool Perspective on SDM

A production-level, parallel visualization tool is a robust program that is used by a potentially large population of users to perform diverse visualizations and analyses, normally on data from many different types of file formats and with varying types of data models. As such, these tools are somewhat different from scientific simulation codes in that

- They "unify" many different data models. For example, they support many mesh types, field types, and various centerings for those fields (point centered, cell centered, etc).

- They are not the originators of the semantics placed on the data. Therefore, the meaning of each of the arrays of data must somehow be provided to the visualization tool.

---

*See http://en.wikipedia.org/wiki/Orrery

- When run in a parallel environment, visualization tools are expected to adapt to available resources (e.g., number of processors) and partition the data for processing in a way that achieves good load balance.

In the following subsections, we describe how these visualization tools use the data from scientific simulations. In particular, we will discuss

- How a production-level, parallel visualization tool loads data, processes it, and produces results
- How a production-level, parallel visualization tool can optimize its data management and processing with the presence of metadata
- The importance of data semantics from the perspective of a production-level, parallel visualization tool

9.2.1 How Data Is Processed

The three major parallelized, production-level visualization tools—EnSight,[3] VisIt,[4] and ParaView[5]—all employ similar strategies. They use a client-server design, where the client provides a user interface on the user's desktop, and the server runs where the data is located, which is assumed to have resources for parallel processing. The general data management strategy for the parallel server can essentially be described as a *scatter–gather* algorithm. The process can be characterized in three steps:

1. I/O (scatter): load data (in parallel) onto the server
2. Processing: employ visualization and analysis algorithms; transform the data to geometry
3. Rendering (gather): transform the geometry into images

9.2.1.1 I/O

Since visualization is a data-intensive endeavor, I/O is frequently the slowest and most expensive part of the entire visualization pipeline. As such, it is advantageous to parallelize the data loading. A typical design pattern is for each processor of the parallel server to read a portion of the input dataset, which is the mechanism that "scatters" the dataset across each of the processors. The key question during the I/O phase is how to assign portions of the input dataset to the processors of the server. We simplify the discussion below, by assuming that the data is being read from disk, that is, *not* being processed in situ as part of a single program with the simulation code.

When the visualization server processes a portion of the dataset, the input dataset must be partitioned and distributed across the server's processors. When the simulation outputs data, it may impose restrictions on data partitioning. There are two typical scenarios:

1. The underlying I/O infrastructure only supports a partitioning scheme fixed by the simulation when the file(s) were created. Most of the time, this scenario corresponds to having one atomic chunk of data for each processor.* Examples of I/O libraries of this type are Silo[6] and Exodus.[7] Other examples include file-per-processor output, which may be a good way to achieve I/O performance for the simulation's data-write phase, but has undesirable consequences for processing tools that read their data. Those processing tools are then forced to reconcile between the simulation's degree of parallelism and their own. For example, the simulation may decompose a three-dimensional space into 1,000 pieces, but the processing tool may be running with only five processors. In this case, the processing tool must find a way to partition the pieces across its processors, either by combining all of the pieces on a given processor into one large piece or by respecting piece layout and supporting multiple pieces per processor.

2. The underlying I/O infrastructure supports re-partitioning during read. Most of the time, this scenario corresponds to having all of the data in one large file, with the I/O infrastructure supporting operations like hyperslab reads, collective I/O, and so forth. Examples of formats that can repartition data in this manner are ViSUS,[8] SAF,[9] and HDF5.[10]

These two scenarios are well supported by the major, parallel, production visualization tools, although the scenarios are supported differently in terms of how the tools do parallel partitioning. For the first case (imposed partitioning), each subset of the partition normally consists of *domains*, where each domain consists of the portion operated on by a single processor. In this case, the visualization tool distributes the domains across its processors. For the second case (adaptive partitioning), the visualization tool forms its own partition of the dataset by having each processor read in a unique piece. In both cases, it is important that each processor has an approximately equal amount of data to read, which correlates strongly with work to be performed in subsequent stages. From a scientific data management (SDM) perspective, the summary is that both ways of writing data are acceptable.

9.2.1.2 Processing

The modern parallel visualization tools all use a data flow network processing design.[11–13] Data flow networks have base types of *data objects* and *components* (sometimes called process objects). The components can be *filters*, *sources*, or *sinks*. Filters have an input and an output, both of which are data objects. Sources have only data object outputs, while sinks have only data object inputs. A *pipeline* is an ordered collection of components. Each pipeline has a source (typically a file reader) followed by one or more filters (for example

---

*Atomic in the sense that partial reads of the chunk of data are not possible.

slicing or contouring algorithms) followed by a sink (typically a rendering algorithm). When a pipeline is executed, data comes from the source and flows from filter to filter until it reaches the sink. There are many variations on this general design that include caching, how the execution takes place (push versus pull), multiplicity in terms of sources and sinks, feedback loops in the filters, reusing arrays from data object to data object to reduce memory footprint, and optimizations like parallel-pipelined operation so that different stages of the pipeline may operate concurrently.

When the client asks the server to perform some operations on a data object, each processor of the server sets up an identical data flow network. They only differ in the portion of the dataset that they process. The majority of visualization operations are "embarrassingly parallel"—the processing can occur in parallel with no communication between the parallel processes. For these operations, the only concern is artifacts that can occur along the boundaries of a chunk. For example, a stencil-based algorithm that is run in parallel may require data from adjacent grid points that are owned by another processor. The typical way to resolve this problem is using redundant data located at the boundary, which is often referred to as *ghost data*.

9.2.1.3 Rendering and Remote/Distributed Visualization

Within the context of visualization software architectures, the majority of SDM-related concerns reside in I/O and processing stages. The later stage of visualization—rendering, where visualization results (geometry, 3D volumes, etc.) are transformed into images—has its own unique set of visualization-centric SDM-related issues.

As context, remote and distributed visualization applications can use one of three general types of architectures as shown in Figure 9.1. A discussion of the relative performance and usability merits of these different configurations is presented in Reference.[14] The important point here, within the context of SDM-related issues, is that moving data across machine boundaries can be a nontrivial task. The data might be raw data, as in the *desktop-only* configuration; it might be geometric output produced by visualization tools, as in the *cluster isosurface* configuration; or it might be raw image pixels, as in the *cluster render* configuration. Unlike "traditional" data movement applications (e.g., ftp and its variants), the visualization use model often dictates which pipeline partitioning will work best given a particular problem size and set of machines/networks. For instance, if maximizing rendering interactivity of static data is the desired target, then one of the configurations that uses desktop graphics hardware for rendering is the best choice, assuming the problem will fit onto the desktop machine. If maximizing throughput is the objective, for example, cycling through large, time-varying data, then the configuration where data I/O and processing is performed on a parallel machine and image sent to the remote viewer is the best choice. The trend we see in

Scientific Data Management Challenges

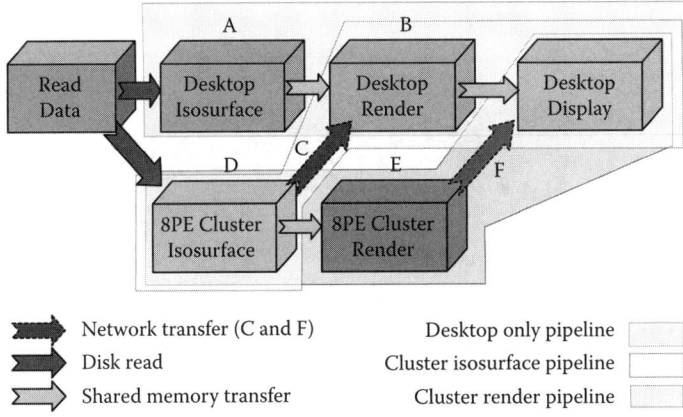

Figure 9.1 (See color insert following page 224.) This image shows three different partitionings of a simple visualization pipeline consisting of four stages: data I/O, computing an isosurface, rendering isosurface triangles, and image display. In one partitioning, all operations happen on a single desktop machine (A-B, blue background). In another, data is loaded onto an eight-processor cluster for isosurface processing, and the resulting isosurface triangles are sent to the desktop for rendering (D-C-B, yellow background). In the third, data is loaded onto the cluster for isosurface processing and rendering, and the resulting image is sent to the desktop for display (D-E-F, magenta background).

high-performance visualization for many problem domains is more toward this latter configuration, which exhibits favorable scaling characteristics as data sizes grow larger.

In the case when all rendering occurs on the desktop, all data representation and data transfer issues are encapsulated inside the graphics library (though there may be substantial SDM issues to consider in the visualization pipeline prior to the rendering stage). In the case where rendering occurs on one machine and image pixels are transmitted to one or more remote machines for viewing, several interrelated issues appear: security (authorization and authentication), compression (lossless vs. lossy), efficient data movement (lossless vs. lossy, multistreamed, multicast), data formats, and models for the pixel data.

For this latter issue, a widely adopted approach is the remote framebuffer protocol (RFB), which is part of the popular virtual network computing (VNC) client/server application for remote desktop access.[15] Here, the remote client connects to a central server where the data-intensive application is run; the resulting imagery is "harvested" by the VNC server, encoded into RFB format, and transmitted to the VNC client for display via a standard transmission control protocol (TCP) connection. In its native form, VNC is not capable of capturing image pixels created by graphics hardware. Other

Figure 9.2 (See color insert following page 224.) Shown is a 36-domain dataset. The domains have thick black lines and are colored red or green. Mesh lines for the elements are also shown. To create the dataset sliced by the transparent gray plane, only the red domains need to be processed. The green domains can be eliminated before ever being read in.

recent work rectifies this shortcoming, as well as provides a solution layered atop the RFB protocol to capture and deliver image pixels produced by hardware-accelerated, distributed memory-rendering infrastructure.[16] In both cases, the rendering infrastructure, particularly the image capture and remote delivery, is transparent to the visualization application.

9.2.2 How Metadata Can Enable Optimizations

In this section, we give an example that motivates how the presence of metadata* can lead to extensive optimizations for the visualization tool. I/O is the most expensive portion of a pipeline execution for almost every operation a visualization tool performs. We can reduce I/O and processing load by reading only the domains that are relevant to any given pipeline operation. This performance gain propagates through the pipeline, since the domains not read in do not have to be processed downstream.

Consider the example of slicing a three-dimensional dataset by a plane (Figure 9.2). In this case, most of the domains will not intersect the plane— loading and processing those domains that do not intersect the slice plane is

---

*We define *metadata* as data about the total dataset to be processed, whose size is small relative to the total dataset itself.

wasted effort. By using metadata, we can dramatically reduce both I/O and processing load: We can limit I/O and processing only to those domains needed to complete the task at hand. For example, if the slice filter had access to the spatial extents for each domain, it could calculate the list of domains whose bounding boxes intersect the slice and only process that list (note that false positives can potentially be generated by considering only the bounding box).

The performance gains one can realize from metadata optimizations can be extensive. From a theoretical perspective, if D is the total number of domains, then the number of domains intersected by the slice is typically $O(D^{2/3})$. Using this fact, we observe that we might expect an order-of-magnitude improvement in performance by using metadata to optimize visualization processing. From a practical perspective, we ran some performance experiments to show exactly how much performance gain can result from metadata optimizations.

Table 9.1 presents the results of the study where we run a pair of visualization algorithms—slicing and isocontouring—and measure the I/O and processing costs with and without metadata configurations. In the slicing case, we use spatial metadata to limit the subsets of data that are loaded to only those that intersect the slice plane. In the isocontouring case, we use metadata describing the data range for each block to limit I/O and processing only to those blocks containing data ranges of interest—those that intersect the isosurface.

The data for this study was produced by a Rayleigh-Taylor Instability simulation, which models fluid instability between heavy fluid and light fluid. The simulation was performed on a 1152 × 1152 × 1152 rectilinear grid, for a total of more than 1.5 billion elements. The data was decomposed into 729 domains, with each domain containing more than 2 million elements. All timings were taken on a cluster of 1.4 GHz Intel Itanium2 processors, each with access to 2 gigabytes of memory.

The processing time includes the time to read in a dataset from disk, perform operations to it, and prepare it for rendering. Rendering was not included

TABLE 9.1 Performance results of visualization processing—slicing and isocontouring—with and without metadata optimization. For slicing and early-time isocontouring, we see an order-of-magnitude performance gain resulting from the metadata optimization. For the late-time isocontouring, the performance gain is still substantial, but not as profound due to the fact that late-stage isocontour is more complex and spans more blocks of data

		Processing Time (sec)		Data Processed (MB)	
Algorithm	Processors	Without Metadata	With Metadata	Without Metadata	With Metadata
Slicing	32	25.3	3.2	6,375.6	708.4
Contouring (early time)	32	41.1	5.8	6,375.6	708.4
Contouring (late time)	32	185.0	97.2	6,375.6	3,948.0

because it can be highly dependent on screen size. We note that using spatial metadata typically yields a consistent performance improvement, whereas performance gains resulting from metadata about fields defined on the mesh (e.g., pressure, density, etc.) can be highly problem specific. To illustrate this effect, we show results from running the contouring algorithm on simulation data from both early and late timesteps. In earlier timesteps, the fluids have not mixed much, so the shape of the contour approximates a planar slice and does not intersect many domains. In the later timestep, the fluid has undergone substantial mixing, and the contour has much greater surface area due to folding, so it intersects many more data domains.

In summary, the presence of metadata can improve performance to the point of being interactive for certain algorithms. And interactivity is widely regarded to be a key component for scientific discovery.

9.2.3 Data Models and Semantics

Visualization tools devote a substantial amount of code to representing data (i.e., data structures), importing data, and translating it into those data structures. Anecdotal evidence suggests that as much as 80% of any given visualization application is dedicated to these very activities. The large amount of SDM-related code typically comes from the fact that various simulation tools have many different ways to represent their data, and the visualization often must support them all. By way of example, the VisIt visualization tool devotes approximately 40,000 lines of code to various data structures. This does not include the portion of the data model that is incorporated from a third-party library (the Visualization ToolKit), which in fact forms the core of the data model (the portion for mesh and field representations). In addition, VisIt has over 80 separate file format readers, each of which ranges from 800 lines of code to 12,000 lines of code. Approximately 150,000 lines of code in VisIt is devoted to file format readers.

In addition to a basic data model for representing standard mesh types (e.g., rectilinear, curvilinear, unstructured, adaptive mesh refinement (AMR), and point meshes), and fields (e.g., scalars, vectors, and tensors), production visualization tools must understand many types of metadata about the dataset to perform certain operations. We list a subset of this metadata to give a feel for how deep the visualization tool must go:

- For each array in the file that corresponds to data that should be visualized, the tool must understand what this array is and how it should be interpreted. For example, the visualization tool must understand that an array in a file labeled "den" is in fact a scalar field defined on a mesh. It is often not necessary to know that "den" is actually the density field, but it may be necessary to know if the field values are explicit (i.e., density of some material per unit of volume) or implicit (i.e., density is directly related to the volume of the cell).

- The tool must understand which cells in the mesh, if any, are ghost cells.
- If the mesh is hierarchical, as is the case with AMR meshes, then the visualization tool must understand metadata describing how the patches of the mesh nest from coarse to fine resolution.
- If the mesh is from a multidomain dataset, the metadata defines how the domains abut. This information is required for many operations, such as computation of ghost data.
- The tool must understand metadata for optimizations, such as the per-domain bounding boxes discussed in the previous section.
- The tool must understand metadata about temporal characteristics, such as the simulation time and cycle identifier.
- The tool must understand information such as volume fractions for Eulerian calculations, including maintaining sparse matrix structures for efficient representation of this information.

9.2.4 A Real-World Production Parallel Visualization Tool

For concreteness, we describe the architecture and operation of a specific, production-quality, parallel visualization, VisIt, which implements many of the concepts described in the previous sections. In terms of I/O, it supports both imposed and adaptive partitioning (see Section 9.2.1.1). It also caches all I/O, which is frequently the dominant portion of execution time. In terms of processing, it uses a pipeline design, although its user interface does not expose pipeline constructs to users. VisIt's pipeline design makes full use of *contracts*, which enable the components of a data flow network to specify and communicate optimizations to achieve higher levels of performance efficiency (see Section 9.2.2). Many filters utilize metadata to limit the amount of data being processed, and many file format readers produce metadata, such as per-domain bounding boxes. VisIt uses two approaches for rendering. First, surfaces with a relatively small number of geometric primitives (e.g., triangles, line segments, etc.) are sent to the client and rendered locally using the local desktop's graphics hardware. Surfaces with a relatively large amount of geometric primitives remain on the server where VisIt renders them in parallel, and then the VisIt server sends the resulting imagery to the remote client. VisIt automatically decides which rendering approach to use based on the number of geometric primitives, but this decision can be overridden by users. In terms of VisIt's data model, VisIt processes all of the mesh types and field types described in Section 9.2.3. It also pays special attention to preserving information about data layout and ordering, so that users can ask debugging-type questions such as, "What is the value of the 110th element in the 41st domain?" This design represents a large amount of effort. VisIt has over 200 filters, 20 different ways to render data, and 90 different file readers, adding up to over 1.5 million lines of C++ code.

Summarizing the entire section, data representation issues and designs of file formats are a critical issue for visualization tools. First, the visualization tool needs to be aware of most of the data that the simulation code itself is aware of, simply because much of that information is directly visualized or needed for proper visualization. Second, additional metadata can enable optimizations and greatly improve the performance of a visualization tool. Third, data layout issues, such as the way data can be partitioned for parallelization, are very important and can have a profound impact on end-to-end performance and usability.

9.3 Multiresolution Data Layout for Large-Scale Data Analysis

In recent years, computational scientists with access to powerful supercomputers have successfully simulated fundamental physical processes with the goal of shedding new light on our understanding of nature. Such simulations often produce massive amounts of data: grids of size 1024^3 to 4096^3 at multiple timesteps and dozens of variables per grid point are not uncommon. This data must be visualized and analyzed to verify and validate the underlying model, to understand the phenomenon in detail, and to develop new insights into fundamental physics. Both data visualization and data analysis are vibrant research areas, and much effort is being spent on developing advanced, new techniques to process the massive amounts of data produced by scientists. In this section, we describe a multiresolution data layout, which provides the ability for quick access to data at varying levels of resolution, from coarse to fine.

9.3.1 Background

To provide context, we highlight these two components in a typical visualization and analysis pipeline shown in Figure 9.3. We assume that raw data from simulations is available as real-valued, regular samples of space-time. Due to the large size of datasets, we emphasize that all data samples cannot all be loaded into main memory at once; it is not feasible to use standard implementations of visualization and analysis algorithms on these large datasets.

Reordering this raw data into a suitable multiresolution data layout can improve the efficiency of both visualization and analysis. Multiresolution layouts enable interactive visualization by allowing the user to first load the data at a coarse level, then progressively refine by adding more samples to obtain a more detailed view. Classical schemes, for example, those based on bricking or chunking, do not readily support the type of data access required for progressive or multiresolution techniques. In the following, we describe our hierarchical Z-order data layout scheme. It builds on the coherent layout

Scientific Data Management Challenges

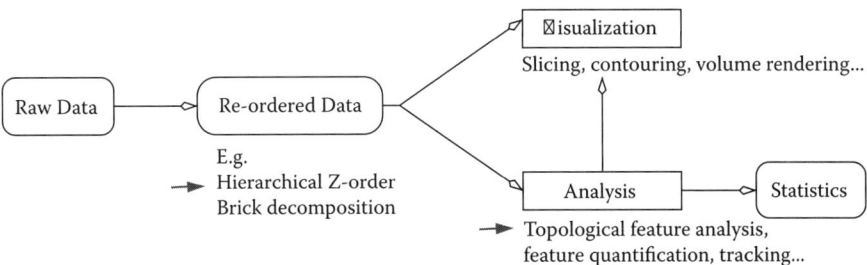

Figure 9.3 Hierarchical Z-ordered data layout and topological analysis components highlighted in the context of a typical visualization and analysis pipeline.

provided by the Z-order space filling curve by incorporating a coarse-to-fine hierarchy on the ordering. Our system is very simple to implement and has been used as a core technology and applied to a variety of visualization algorithms, such as slicing, isosurfacing, and volume rendering on massive amounts of scientific simulation data.

A multiresolution data layout is a key technology that plays a central role in advanced analysis algorithms that go beyond simple images or movies. Topological analysis is one such technique that is useful for providing a deeper understanding of scientific phenomena. In this type of application, the analysis takes the form of defining, detecting, and quantifying features in data. Features can and do exist at multiple scales in data. For this reason, an efficient, multiresolution data layout and model is an integral part of high-performance implementations of such algorithms. For more information on state-of the art topological analysis, see,[17–22] and for application of such techniques to the analysis of simulation data, including hydrodynamic instability, and comparative analysis.[23–25]

9.3.2 Hierarchical Indexing for Out-of-Core Access to Multiresolution Data

Out-of-core computing[26] specifically addresses the issues of algorithm redesign and data layout restructuring that are necessary to enable data access patterns having minimal out-of-core processing performance degradation. Research in this area is also valuable in parallel and distributed computing, where one has to deal with the similar issue of balancing processing time with the time required for data access and movement among elements of a distributed or parallel application.

The solution to the out-of-core processing problem is typically divided into two parts: (1) algorithm analysis, to understand data access patterns and, when possible, redesign to maximize data locality; (2) storage of data in secondary memory using a layout consistent with the access patterns of the algorithm, amortizing the cost of individual I/O operations over several memory access operations.

In the case of hierarchical visualization algorithms for volumetric data, the 3D input hierarchy is traversed from a coarse grid to the fine-grid levels to build derived geometric models having adaptive levels of detail. The shapes of the output models are then modified dynamically with incremental updates of their level of detail. The parameters that govern this continuous modification of the output geometry are dependent on runtime user interaction, making it impossible to determine, *a priori*, what levels of detail will be constructed. For example, parameters can be external, such as the viewpoint of the current display window, or internal, such as the isovalue of a contour or the position of a slice plane. The general structure of the access pattern can be summarized into two main points: (1) the input hierarchy is traversed from coarse to fine and level by level so that data in the same level of resolution is accessed at the same time, and (2) within each level of resolution, the regions that are in close geometric proximity are stored as much as possible in close memory locations and also traversed at the same time.

In this section, we describe a static indexing scheme that induces a data layout satisfying both requirements (1) and (2) for the hierarchical traversal of n-dimensional regular grids. The scheme has three key features that make it particularly attractive. First, the order of the data is independent of the out-of-core block structure, so that its use in different settings (e.g., local disk access or transmission over a network) does not require any large data reorganization. Second, conversion from the Z-order indexing[27] used in classical database approaches to the new indexing scheme can be implemented with a simple sequence of bit-string manipulations, making it appealing for a possible hardware implementation. Third, since there is no data replication, we avoid the performance penalties associated with dynamic updates as well as increased storage requirements typically associated with most hierarchical and out-of-core schemes.

Beyond the theoretical interest in developing hierarchical indexing schemes for n-dimensional space-filling curves, our approach targets practical applications in out-of-core visualization algorithms. For details on related work, algorithmic analysis, and experimental results see Pascucci and Frank.[28]

9.3.2.1 Hierarchical Subsampling Framework

This section discusses the general framework for an efficient definition of a hierarchy over the samples of a dataset.

Consider a set S of n elements decomposed into a hierarchy $\mathcal{H}$ of k levels of resolution $\mathcal{H} = \{S_0, S_1, \ldots, S_{k-1}\}$ such that:

$$S_0 \subset S_1 \subset \cdots \subset S_{k-1} = S$$

where S_i is said to be coarser than S_j if $i < j$. The order of the elements in S is defined by a cardinality function $I : S \to \{0 \ldots n-1\}$. This means that the following identity always holds:

$$S[I(s)] \equiv s$$

where square brackets are used to index an element in a set.

One can define a derived sequence $\mathcal{H}'$ of sets S'_i as follows:

$$S'_i = S_i \setminus S_{i-1} \qquad i = 0, \ldots, k-1$$

where formally $S_{-1} = \emptyset$. The sequence $\mathcal{H}' = \{S'_0, S'_1, \ldots, S'_{k-1}\}$ is a partitioning of S. A derived cardinality function $I' : S \to \{0 \ldots n-1\}$ can be defined on the basis of the following two properties:

- $\forall s, t \in S'_i : I'(s) < I'(t) \Leftrightarrow I(s) < I(t)$
- $\forall s \in S'_i, \forall t \in S'_j : i < j \Rightarrow I'(s) < I'(t)$

If the original function I has strong locality properties when restricted to any level of resolution S_i, then the cardinality function I' generates the desired global index for hierarchical and out-of-core traversal. The scheme has strong locality if elements with close indexes are also close in geometric position. These locality properties are well studied in Moon et al.[28]

The construction of function I' can be achieved as follows: (1) determine the number of elements in each derived set S'_i and (2) determine a cardinality function $I''_i = I'|_{S'_i}$ restriction of I' to each set S'_i. In particular, if c_i is the number of elements of S'_i, one can predetermine the starting index of the elements in a given level of resolution by building the sequence of constants $C_0, \ldots, C_{k-1}$ with

$$C_i = \sum_{j=0}^{i-1} c_j. \qquad (9.1)$$

Next, one must determine a set of local cardinality functions $I''_i : S'_i \to \{0 \ldots c_i - 1\}$ so that

$$\forall s \in S'_i : I'(s) = C_i + I''_i(s). \qquad (9.2)$$

The computation of the constants C_i can be performed in a preprocessing stage so that the computation of I' is reduced to the following two steps:

- given s determine its level of resolution i (that is the i such that $s \in S'_i$)
- compute $I''_i(s)$ and add it to C_i

These two steps must be performed very efficiently as they will be executed repeatedly at runtime. The following section reports a practical realization of this scheme for rectilinear cube grids in any dimension.

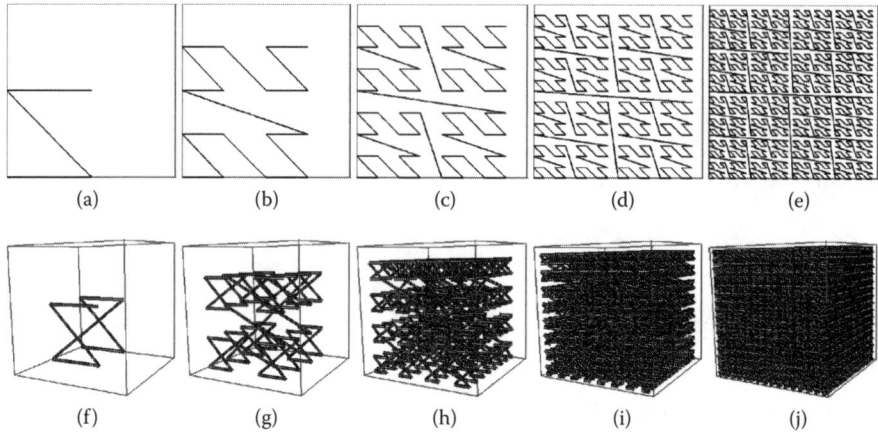

Figure 9.4 (See color insert following page 224.) (a–e) The first five levels of resolution of the 2D Lebesgue's space-filling curve. (f–j) The first five levels of resolution of the 3D Lebesgue's space filling curve.

9.3.2.2 Binary Trees and the Lebesgue Space-Filling Curve

This section reports the details on how to derive from the Z-order space-filling curve the local cardinality functions l_i'' for a binary tree hierarchy in any dimension and its remapping to the new index I'.

Indexing the Lebesgue Space-Filling Curve. The Lebesgue space-filling curve, also called Z-order space-filling curve for its shape in the 2D case, is depicted in Figure 9.4(a–e). The Z-order space-filling curve can be defined inductively by a base Z shape of size 1 (Figure 9.4(a)), that is by the vertices of a square of side 1 that are connected along a Z pattern. Such vertices can then be replaced each by a Z shape of size $\frac{1}{2}$ as in Figure 9.4(b). The vertices obtained in this way are then replaced by Z shapes of size $\frac{1}{4}$ as in Figure 9.4(c), and so on. In general, the ith level of resolution is defined as the curve obtained by replacing the vertices of the $(i-1)$th level of resolution with Z shapes of size $\frac{1}{2^i}$. The 3D version of this space-filling curve has the same hierarchical structure with the only difference being that the basic Z shape is replaced by a connected pair of Z shapes lying on the opposite faces of a cube as shown in Figure 9.4(f). Figure 9.4(f–j) shows five successive refinements of the 3D Lebesgue space-filling curve. The d-dimensional version of the space-filling curve also has the same hierarchical structure, where the basic shape (the Z of the 2D case) is defined as a connected pair of $(d-1)$-dimensional basic shapes lying on the opposite faces of a d-dimensional cube.

The property that makes the Lebesgue's space-filling curve particularly attractive is the easy conversion from the d indexes of a d-dimensional matrix to the 1D index along the curve. If one element e has d-dimensional reference $(i_1, \ldots, i_d)$, its 1D reference is built by interleaving the bits of the binary representations of the indexes $i_1, \ldots, i_d$. In particular if i_j is represented by the

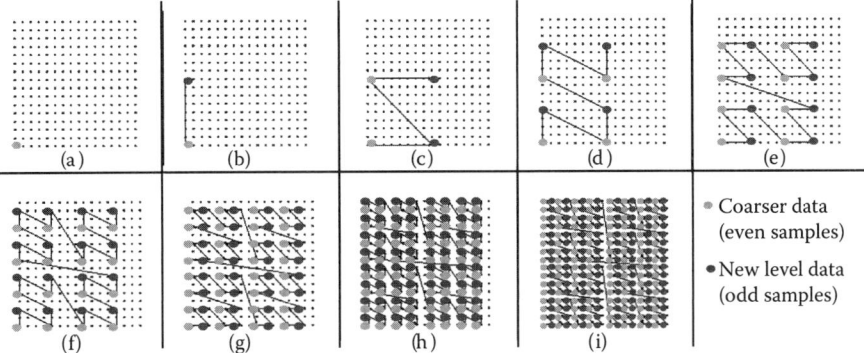

Figure 9.5 (See color insert following page 224.) The nine levels of resolution of the binary tree hierarchy defined by the 2D space-filling curve applied on 16×16 rectilinear grid. The coarsest level of resolution (a) is a single point. The number of points that belong to the curve at any level of resolution (b) to (i) is double the number of points of the previous level.

string of h bits "$b_j^1 b_j^2 \cdots b_j^h$" (with $j = 1, \ldots, d$) then the 1D reference I of e is represented by the string of hd bits $I = \text{"}b_1^1 b_2^1 \cdots b_d^1 b_1^2 b_2^2 \cdots b_d^2 \cdots b_1^h b_2^h \cdots b_d^h\text{"}$.

The 1D order can be structured in a binary tree by considering elements of level i, those that have the last i bits all equal to 0. This yields a hierarchy where each level of resolution has twice as many points as the previous level. From a geometric point of view this means that the density of the points in the d-dimensional grid is doubled alternating along each coordinate axis. Figure 9.5 shows the binary hierarchy in the 2D case where the resolution of the space-filling curve is doubled alternately along the x and y axes. The coarsest level (a) is a single point, the second level (b) has two points, the third level (c) has four points (forming the Z shape), and so on.

Index Remapping. The cardinality function discussed in Section 9.3.2.1 for the binary tree case has the structure shown in Table 9.2. For example, the element of index 0 is always at the top of the tree (level 0) for any granularity. If the index has 5 granularity levels (the 4th row in the table), node 8 is at the second level of the tree; the nodes 4 and 12 are at the next level of the

TABLE 9.2 Structure of the hierarchical indexing scheme for binary tree combined with the order defined by the Lebesgue space-filling curve

Level of Tree	0	1	2	3				4								
Z-order index (2 levels)	0	1														
Z-order index (3 levels)	0	2	1	3												
Z-order index (4 levels)	0	4	2	6	1	3	5	7								
Z-order index (5 levels)	0	8	4	12	2	6	10	14	1	3	5	7	9	11	13	15
hierarchical index	0	1	2	3	4	5	6	7	8	9	10	11	12	13	14	15

tree; the node 4 has nodes 2 and 6 below it at the next level of the tree, while the node 12 has 10 and 14 below it at the next level of the tree. The last level has nodes 1 and 3 below 2, 5 and 7 below 6, and so on. Note that this is a general structure suitable for out-of-core storage of static binary trees. It is independent of the dimension d of the grid of points or of the Z-order space-filling curve.

The structure of the binary tree defined on the Z-order space-filling curve allows one to easily determine the three elements necessary for the computation of the cardinality. They are (1) the level i of an element, (2) the constants C_i of Equation (9.1), and (3) the local indexes I_i''.

i — if the binary tree hierarchy has k levels then the element of Z-order index j in the Z-order belongs to the level $k - h$, where h is the number of trailing zeros in the binary representation of j.

C_i — the total number of elements in the levels coarser than i, with $i > 0$, is $C_i = 2^{i-1}$ with $C_0 = 0$.

I_i'' — if an element has index j and belongs to the set S_i', then $\frac{j}{2^{k-i}}$ must be an odd number, by definition of i. Its local index is then:

$$I_i''(j) = \left\lfloor \frac{j}{2^{k-i+1}} \right\rfloor.$$

The computation of the local index I_i'' can be explained easily by looking at the bottom right part of Table 9.2 where the sequence of indexes (1, 3, 5, 7, 9, 11, 13, 15) needs to be remapped to the local index (0, 1, 2, 3, 4, 5, 6, 7). The original sequence is made of a consecutive series of odd numbers. A right shift of one bit (or rounded division by two) turns them into the desired index.

These three elements can be put together to build an efficient algorithm that computes the hierarchical index $I'(s) = C_i + I_i''(s)$ in the two steps shown in Figure 9.6(a):

1. Set the bit in position $k + 1$ to 1.
2. Shift to the right until a 1 comes out of the bit string.

This algorithm could have a very simple and efficient hardware implementation. The software C++ version can be implemented as follows:

```
inline adhocindex remap(register adhocindex i){
    i |= last_bit_mask;   // set leftmost one
    i /= i&-i;            // remove trailing zeros
    return (i>>1);        // remove rightmost one
}
```

This code would work only on machines with two's complement representation of numbers. In a more portable version, one needs to replace `i /= i&-i` with `i /= i&((~i)+1)`.

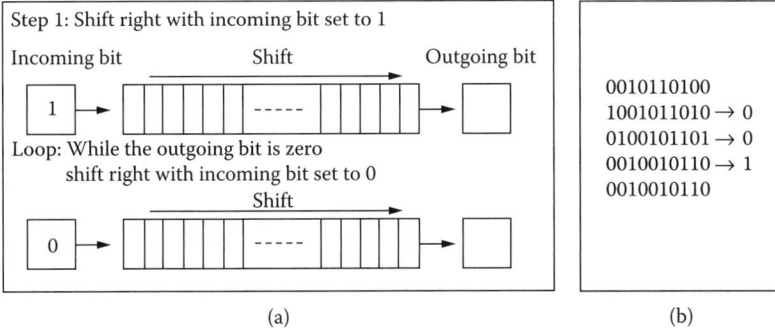

(a) (b)

Figure 9.6 (a) Diagram of the algorithm for index remapping from Z-order to the hierarchical out-of-core binary tree order. (b) Example of the sequence of shift operations necessary to remap an index. The top element is the original index and the bottom is the remapped, output index.

Figure 9.7 shows the data layout obtained for a 2D matrix when its elements are reordered following the index I'. The data is stored in this order and divided into blocks of constant size. The 2D image of such decomposition has the first block corresponding to the coarsest level of resolution of the data. The subsequent blocks correspond to finer and finer resolution data, which is distributed more and more locally.

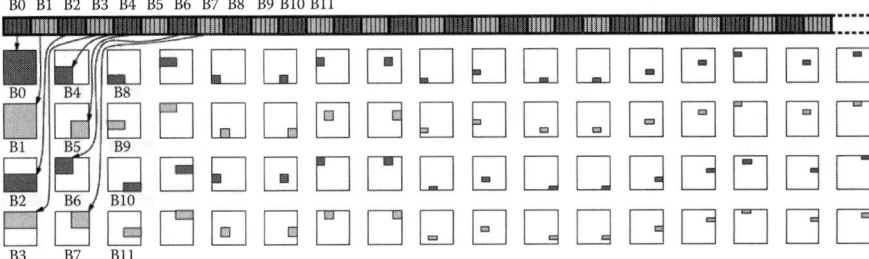

Figure 9.7 Data layout obtained for a 2D matrix reorganized using the index I' (1D array at the top). The 2D image of each block in the decomposition of the 1D array is shown. Each gray region (odd blocks dark gray, even blocks light gray) shows where the block of data is distributed in the 2D array. In particular the first block is the set of coarsest levels of the data distributed uniformly on the 2D array. The next block is the next level of resolution still covering the entire matrix. The next two levels are finer data covering each half of the array. The subsequent blocks represent finer resolution data distributed with increasing locality in the 2D array.

9.3.2.3 Performance

In this section, we describe experimental results for a simple, fundamental visualization technique: orthogonal slicing of a 3D rectilinear grid. Slices can be at different resolutions to allow interactivity: As the user manipulates the slice parameters, we compute and display a coarse resolution slice, then refine it progressively. We compare our layout with two common array layouts: row major, and $h \times h \times h$ brick decomposition.

Data I/O Requirements. As we shall see, the amount of data required to be read from disk varies substantially from one array layout to another. By way of example, consider the case of an 8 GB dataset (a 2048^3 mesh of `unsigned char` data values). An orthogonal slice of this mesh consists of 2048×2048, or 4,194,304 points/bytes. In this example, disk pages are 32 KB in size (see Figure 9.8(a)). For the brick decomposition case, one would use $32 \times 32 \times 32$ blocks of 32 KB for the entire dataset. The data loaded from disk for a slice is 32 times larger than the output, or 128 MB. As the subsampling increases up to a value of 32 (one sample out of 32), the amount of data loaded does not decrease because each $32 \times 32 \times 32$ brick needs to be completely loaded. At lower subsampling rates, the data overhead remains the same: The data loaded is 32,768 times larger than the data needed. In the binary tree with Z-order remapping, the data layout is equivalent to a KD-tree, constructing the same subdivision as an octree. For a 2D slice, the KD-tree mapping is equivalent to a quadtree layout. The data loaded is grouped into blocks along the hierarchy that gives an overhead factor in number of blocks of $1 + \frac{1}{2} + \frac{1}{4} + \frac{1}{16} + \cdots < 2$ (as for one added to a geometric series), while each block is 32 KB.

Tests with Memory-Mapped Files. A series of basic tests was performed to verify the performance of the approach using a general purpose paging system. The out-of-core component of the scheme was implemented simply by mapping a 1D array of data to a file on disk using the `mmap` function. In this way, the I/O layer is implemented by the operating system virtual memory subsystem, paging in and out a portion of the data array as needed. No multithreaded component is used to avoid blocking the application while retrieving the data. The blocks of data defined by the system are typically 4 KB. Figure 9.8(b) shows performance tests executed on a Pentium III laptop. The proposed scheme shows the best scalability in performance. The brick decomposition scheme with 16^3 chunks of regular grids shows the next best performance. The (i, j, k) row-major storage scheme has the worst performance because of its dependency on the slicing direction: best for (j, k) plane slices and worst for (j, i) plane slices. Figure 9.8(c) shows the performance results for a test on a larger, 8 GB dataset, run on an SGI Octane. The results are similar.

The hierarchical Z-order is our layout of choice for efficient data management for visualizing and analyzing large-scale scientific simulation data. In the next section, we describe some of the fundamental mathematical techniques that we use for analyzing such datasets.

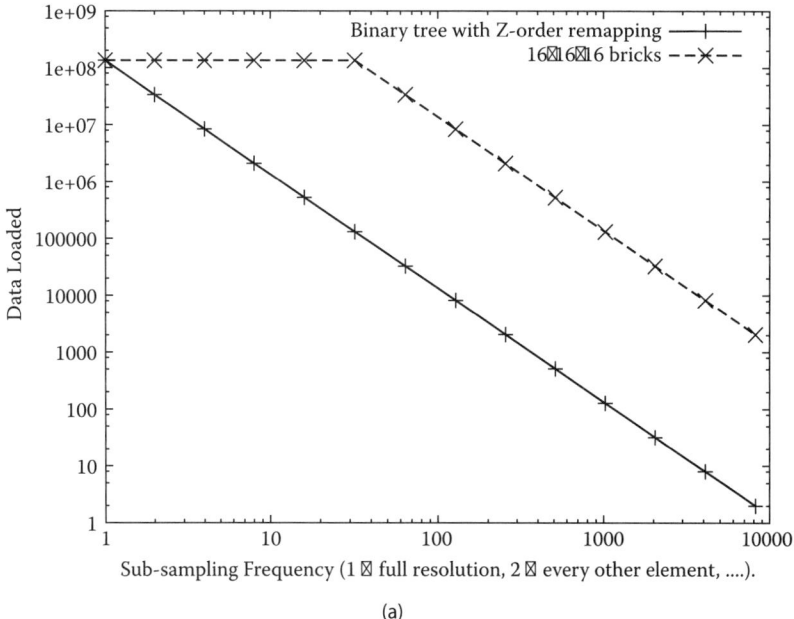

(a)

Figure 9.8 Comparison of static analysis and real performance in different conditions. For an 8 GB dataset, (a) compares the amount of data loaded from disk (vertical axis) per slice while varying the level of subsampling and using two different access patterns/storage layouts: Z-order remapping and brick decomposition. The values on the vertical axis are reported using a logarithmic scale to highlight the performance difference—orders of magnitude—at any level of resolution. (b–c) Two comparisons of slice computations times (log scale) of four different data layout schemes with slices parallel to the (j, k) plane (orthogonal to the x axis). The horizontal axis is the level of subsampling of the slicing scheme, where values on the left are finer resolution. Note how the practical performance of the Z-order versus brick layout is predicted very well by the static analysis. The only layout that can compete with the hierarchical Z-order is the row-major array layout optimized for (j, k) slices (orthogonal to the x axis). Of course, the row-major layout performs poorly for (j, i) slices (orthogonal to the z axis), while the hierarchical Z-order layout would maintain the same performance for access at any orientation.

9.4 File Formats for High-Performance Visualization and Analytics

The notion of data models and formats is a central focal point in the nexus between producers and consumers of data. As indicated earlier in Section 9.2, production visualization applications are expected to be able to read and

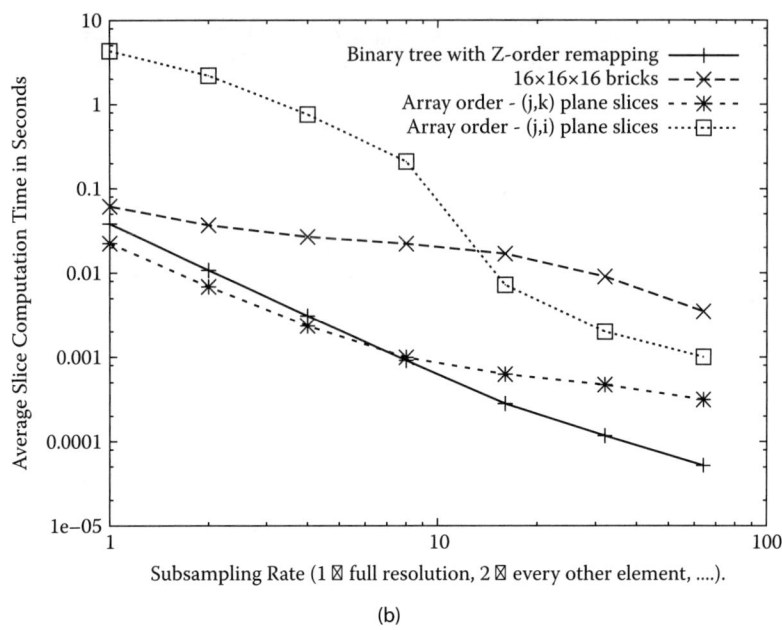

(b)

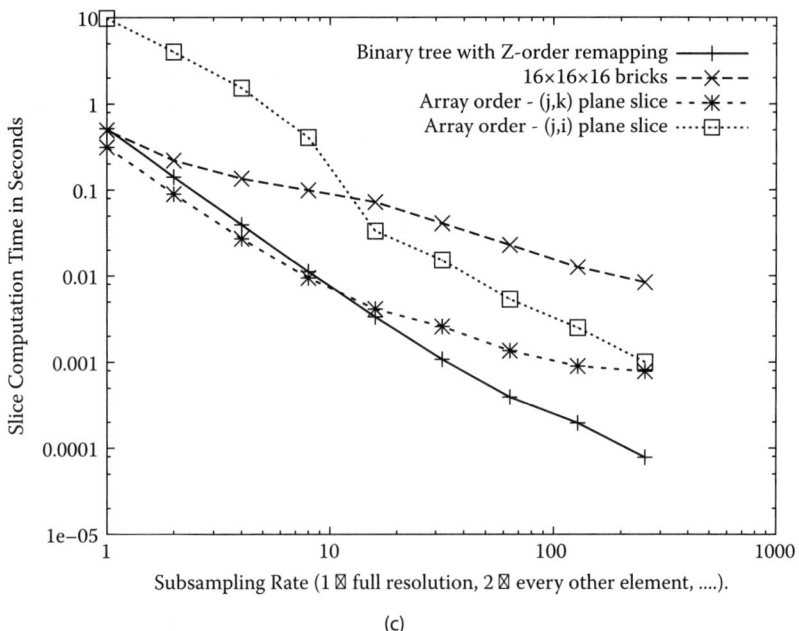

(c)

Figure 9.8 (*Continued*)

process a large number of different formats. Many of these formats are built atop underlying I/O libraries, like HDF5 and netCDF. Even with well-established I/O libraries, which implement an on-disk data format for storing and retrieving arrays of data, there exist several issues that perennially affect producers and consumers of data. One is the fact that the APIs for these I/O libraries can be complex: They provide a great deal of functionality, which is exposed through an API. Another is that it is possible to "misuse" the I/O library in a way that will result in less-than-optimal I/O performance. Yet another is that semantic conventions are not consistent across and within disciplines. The existence of well-established I/O libraries is of huge benefit to developers, for these technologies accelerate development by not reinventing the wheel. However, they can be complex to use, and they don't solve the semantic gap problem that exists between producers and consumers of data.

This section addresses these topics from a visualization-centric perspective. In our work, we often are the ones who end up having to reconcile semantic as well as format discrepancies between simulation, experiment, and visualization technologies. As it does not seem practical for a panacea solution that will solve these problems for all disciplines, we have adopted a bottom-up approach that focuses on addressing these problems for communities and for specific new capabilities at the crossover point between the fields of visualization and scientific data management.

The first subsection below presents H5Part, which is a high-level API that provides a solution to the semantic gap for use in computational accelerator modeling. Concurrently, H5Part (which stands for HDF5 format for particle data) is engineered to provide good I/O performance in a way that is reasonably "immune to misuse." It encapsulates the complexity of an underlying I/O library, thereby providing advanced data I/O capabilities to both simulation and visualization developers. The relative simplicity of its API lowers the developer cost of taking advantage of such technology. Next, we present another high-level API aimed at encapsulating both data I/O and index/query capabilities. Again, our motivation is to encapsulate the complexity of both of these technologies. Finally, we present some examples of how these technologies are used in practice.

9.4.1 H5Part

9.4.1.1 Motivation

Modern science—both computational and experimental—is typically performed by a number of researchers from different organizations who collectively collaborate on a challenging research problem. In the case of particle accelerator modeling, different groups collaborate on different aspects of modeling the entire beamline. One group works on modeling injection of particles into the beam, another works on modeling magnetic confinement and beam focusing along the beamline, while yet another works on modeling the impact of the energized particles with the target. Each of these different models is studied with one or more different simulation codes. One grand challenge in accelerator modeling is to simulate the entire beamline, from injection to impact. This ambitious project is being approached by the development and use of a number of different codes, each of which models a different portion of the beamline.* Accurate end-to-end modeling of accelerators is crucial to create an optimized design prior to building the multi-billion-dollar instrument.

Unfortunately, there has historically been little coordination between these code teams: The individual codes often output particle and field data in different, incompatible formats. Worse yet, some of these codes output multiple terabytes of simulation data in ASCII format because it is easy to do so. Many of these codes use serial I/O; for example, ASCII, HDF4. As a result, there is data format incompatibility between different modeling stages, I/O is slow when performed in serial, and it is wasteful of space in the case of ASCII formats. Of more concern is the impediment to scientific progress that results from data format incompatibility. It becomes extremely difficult to compare results from different simulations that model the same part of the beamline due to different data formats (e.g., different units of measure, different coordinate systems, different data layouts, and so forth).

H5Part† was motivated by the desire of the accelerator modeling community to address these problems. The vision is to have a community-centric data format and API that would enable codes from all processing stages to: (1) read/write particle and field data in a single format; (2) allow legacy codes—simulations, visualization and analysis tools—to quickly take advantage of modern parallel I/O capabilities on HPC platforms; (3) accelerate software development and engineering in the area of I/O for accelerator modeling codes; (4) facilitate code-to-code and code-to-experiment data comparison and analysis; (5) enable the accelerator modeling community to more easily share data, simulation, and analysis code; and (6) facilitate migrating toward community standards for data formats and analysis/visualization. The intended result is to accelerate scientific discovery by reducing software development time, by fostering best practices in data management within the particle accelerator modeling community, and to improve the I/O efficiency of simulation and analysis tools.

H5Part is a very simple data storage schema and provides an API that simplifies the reading/writing of the data to the HDF5 file format.[10] An important foundation for a stable visualization and data analysis environment is a stable and portable file storage format and its associated APIs. The presence of a common file storage format, including associated APIs, fosters a fundamental level of interoperability across the project's software infrastructure. It

---

*See http://compass.fnal.gov for an example of a large, community-based accelerator modeling project with exactly these objectives.
†We coined the term H5Part for the API to reflect use of HDF5 for particle-based datasets common in high-energy physics.

also ensures that key data analysis capabilities are present during the earliest phases of the software development effort. The H5Part file format and APIs enable disparate research groups with different simulation implementations to transparently share datasets and data analysis tools. For instance, the common file format enables groups that depend on completely different simulation implementations to share data analysis tools.

H5Part is built on top of HDF5 (hierarchical data format). HDF5 offers a number of advantages: It is a self-describing, machine-independent binary file format that supports scalable parallel I/O performance for MPI codes on a variety of supercomputing systems and works equally well on laptop computers. HDF5 is available for C, C++, and Fortran codes. The primary disadvantage of HDF5 is in the complexity of the API. Because of the rich set of functionality in HDF5, it can be challenging for domain scientists to write out a simple 1D array of data using raw HDF5 calls. Worse, doing parallel I/O is further complicated by the variety of data layout and I/O tuning options available in HDF5.

By restricting the usage scenario to particle accelerator data, H5Part encapsulates much of the complexity of HDF5 to present a simple interface for data I/O to accelerator scientists that is much easier to use than the HDF5 API. Compared with code that calls HDF5 directly, code that calls H5Part is more terse, less complex, and easier to maintain. For example, code that uses H5Part needs to make any HDF5 calls that set up organization for data groups inside the HDF5 file since it encapsulates such functionality. The internal layout of data groups inside the HDF5/H5Part file has proven to be effective for both efficiently writing data from simulation code as well as for efficiently reading data, either serially or in parallel, into visual data analysis tools.

9.4.1.2 File Organization and API

The H5Part file storage format uses HDF5 for the low-level file storage and a simple API to provide a high-level interface to that file format. A programmer can use the H5Part API to access the data files or to write directly to the file format using some simple conventions for organizing and naming the objects stored in the file.

In order to store particle data in the HDF5 file format, we have formalized the hierarchical arrangement of the datasets and naming conventions for the groups and associated datasets. The H5Part API formally encodes these conventions in order to provide a simple and uniform way to access these files from C, C++, and Fortran codes. The API makes it easy to write very portable data adaptors for visualization tools in order to expand the number of tools available to access the data. Users may write their own HDF5-based interface for reading and writing the file format, or may use the *h5ls* and *h5dump* command-line utilities, which are included with the HDF5 distribution, to display the organization and contents of H5Part files. The standards

TABLE 9.3 Sample H5Part code to write multiple fields from a time-varying simulation to a single file

---
```
if(serial)
   handle=H5PartOpenFile(filename, mode);
else
   handle=H5PartOpenFileParallel(filename, mode, mpi_comm);
H5PartSetNumParticles(handle, num_particles);
loop(step=1,2)
   // compute data
   H5PartSetStep(handle, step);
   H5PartWriteDataFloat64(handle,"px",data_px);
   H5PartWriteDataFloat64(handle,"py",data_py);
   H5PartWriteDataFloat64(handle,"pz",data_pz);
   H5PartWriteDataFloat64(handle, "x",data_x);
   H5PartWriteDataFloat64(handle, "y",data_y);
   H5PartWriteDataFloat64(handle, "z",data_z);
H5PartCloseFile(handle);
```
---

offered by the sample API are completely independent of the standard for organizing data within the file. The file format supports the storage of multiple timesteps of datasets that contain multiple fields.

The data model for particle data allows storing multiple timesteps where each timestep can contain several datasets of the same length. Typical particle data consists of the three-dimensional Cartesian positions of particles (x, y, z) as well as the corresponding three-dimensional momenta (px, py, pz). These six variables are stored as six HDF5 datasets. The type of the dataset can be either integer or real. H5Part also allows storing attribute information for the file and timesteps.

Table 9.3 presents sample H5Part code for storing particle data with two timesteps. The resulting HDF5 file with two timesteps is shown in Table 9.4. These examples show the simplicity of an application that uses the H5Part API to write or read H5Part files. One point is that there is basically a one-line difference between serial and parallel code. Another is that the H5Part application is much simpler than an HDF5-only counterpart: This example code need not worry about setting up data groups inside HDF5; that task is performed inside the H5Part library.

9.4.1.3 Parallel I/O

A naïve approach to writing data from a parallel program is to write one file per processor. Although this approach is simple to implement and very efficient on most cluster file systems, it leads to file management difficulties when the data needs to be analyzed. One must either recombine these separate files into a single file or create unwieldy user interfaces that allow a

TABLE 9.4 Contents of the H5Part file generated in Table 9.3

```
GROUP "/" {
  GROUP "Step#0" {
    DATASET "px" {
      DATATYPE H5T_IEEE_F64LE
      DATASPACE SIMPLE { ( 1000 ) / ( 1000 ) }
    }
    DATASET "py" {
      DATATYPE H5T_IEEE_F64LE
      DATASPACE SIMPLE { ( 1000 ) / ( 1000 ) }
    }
    DATASET "pz" {
      DATATYPE H5T_IEEE_F64LE
      DATASPACE SIMPLE { ( 1000 ) / ( 1000 ) }
    }
    DATASET "x" {
      DATATYPE H5T_IEEE_F64LE
      DATASPACE SIMPLE { ( 1000 ) / ( 1000 ) }
    }
    DATASET "y" {
      DATATYPE H5T_IEEE_F64LE
      DATASPACE SIMPLE { ( 1000 ) / ( 1000 ) }
    }
    DATASET "z" {
      DATATYPE H5T_IEEE_F64LE
      DATASPACE SIMPLE { ( 1000 ) / ( 1000 ) }
    }
  }
  GROUP "Step#1" {
    ...information for 6 datasets...
  }
}
```

data analysis application to read from a directory full of files instead of just one file. An arguably better approach is to provide the means for a parallel application to write data into a single file from all processing elements (PEs), which is known as *collective I/O*. Collective I/O performance is typically (but not always) lower than that of writing one file per processor, but it makes data management much simpler after the program has finished. No additional recombinination steps are required to make the file accessible by visualization tools or for restarting a simulation using a different number of processors.

Parallel HDF5 uses MPI-IO for its low-level implementation. The mechanics of using MPI-IO are hidden from the user by the H5Part API (the code looks nearly identical to reading/writing the data from a serial program). While the performance is not always as good as writing one file per processor, we have shown that writing files with Parallel HDF5 is consistently faster than writing the data in raw/native binary using the MPI-IO library.[29] This efficiency is made possible through sophisticated HDF5 tuning directives, which are transparent to the parallel application, that control data alignment and caching within the HDF5 layer. Therefore, we argue that it would be difficult to match HDF5 performance even using a home-grown binary file format.

9.4.2 HDF5_FastQuery

Large-scale scientific data is often stored in scientific data formats like FITS, netCDF, and HDF. These storage formats are of particular interest to the scientific user community since they provide multidimensional storage and retrieval capabilities. However, one of the drawbacks of these storage formats is that they do not support the ability to extract subsets of data that meet multidimensional, compound range conditions. Such multidimensional range conditions are often the basis for defining "features of interest," which are the focus of scientific inquiry and study.

HDF5_FastQuery[30] is a high-level API that provides the ability to perform multidimensional indexing and searching on large HDF5 files. It leverages an efficient bitmap indexing technology called FastBit[31-33] (described in Chapter 6) that has been widely used in the database community. Bitmap indexes are especially well suited for interactive exploration of large-scale read-only data. Storing the bitmap indexes into the HDF5 file has the following advantages: (1) significant performance speed-up of accessing subsets of multidimensional data and (2) portability of the indexes across multiple computer platforms. The HDF5_FastQuery API simplifies the execution of queries on HDF5 files for general scientific applications and data analysis. The design is flexible enough to accommodate the use of arbitrary indexing technology for semantic range queries.

HDF5_FastQuery provides an interface to support semantic indexing for HDF5 via a query API. HDF5_FastQuery allows users to efficiently generate complex selections on HDF5 datasets using compound range queries like ($energy > 10^5$) AND ($70 < pressure < 90$) and retrieve only the subset of data elements that meet the query conditions. The FastBit technology generates the compressed bitmap indexes that accelerate searches on HDF5 datasets, as well as the raw indexes (the compressed bitmap indexes), which are stored together with the datasets in an HDF5 file. Compared with other indexing schemes, compressed bitmap indices are compact and very well suited for searching over multidimensional data even for arbitrarily complex combinations of range conditions.

9.4.2.1 Functionality

HDF5 supports slab and hyperslab selections of n-dimensional datasets. HDF5_FastQuery extends the HDF5 selection mechanism to allow subset selection based upon arbitrary range conditions on the data values contained in the datasets using the bitmap indexes. As a result, HDF5_FastQuery supports fast execution of searches based upon compound queries that span multiple datasets. The API also allows us to seamlessly integrate the FastBit query mechanism for data selection with HDF5's standard hyperslab selection mechanism. Using the HDF5_FastQuery API, one can quickly select subsets of data from an HDF5 file using text-string queries.

The bitmap indexes are created and stored through a single call to the HDF5_FastQuery API. The storage of these indexes uses separate arrays in the same file as the datasets they refer to and are opaque to the general HDF5 functions. It is important to note that all such indexes must be built before any queries are posed to the API. Once the bitmap indexes have been built and stored in the data file, queries are posed to the API as a text string such as ($temperature > 1000$) AND ($70 < pressure < 90$), where the names specified in the range query correspond to the names of the datasets in the HDF5 file. The HDF5_FastQuery interface uses the stored bitmap indexes that correspond to the specified dataset to accelerate the selection of elements in the datasets that meet the search criteria. An accelerated query on the contents of a dataset requires only small portions of the compressed bitmap indexes to be read into memory, so extremely large datasets can be searched with little memory overhead. The query engine then generates an HDF5 selection that is used to read only the elements from the dataset that are specified by the query string.

9.4.2.2 Architectural Layout

In this section, we present a high-level view of the HDF5_FastQuery architectural layout. We begin by defining relevant terms used throughout the architectural layout as well as the HDF5_FastQuery API.

Groups are the logical way to organize data in an HDF5 file. We use the term *group* or *grouping* to refer to this logical structuring. These groups act as containers of various types of metadata, which in our approach are specific to a given dataset. Note that these groups may be assigned type information (float, int, string, etc.) to uniquely describe these datasets.

Variables vs. Attributes. The properties assigned to a specific group (i.e., group metadata) are called attributes or group attributes. For all datasets, the specific physical properties that the dataset quantizes (density, pressure, helicity, etc.) will be referred to as dataset variables. To organize a given multivariate dataset consisting of a discrete range of time steps, a division is made between the raw data and the attributes that describe the data. This division is represented in the architectural layout by the separation and formation of two classes of groups: the *TimeStep* groups for the raw data, and the *VariableDescriptor* groups for the metadata used to describe the dataset variables.

For the dataset variables, one *VariableDescriptor* group is created for each variable (pressure, velocity, etc.). The metadata saved under these groups usually includes the size of the dataset; name of the dataset variable; coordinate system used in the dataset (spherical, Cartesian, etc.); the schema (structured, unstructured, etc.); centering (cell centered, vertex centered, edge centered, etc.); and number of coordinates that must exist per centering element (each vertex, each face, etc.).

The various *VariableDescriptor* groups are then organized under one TOC (table of contents) group that retains common global information about the file's variables (the names of all variables, bitmap indexes, metadata information). For the raw datasets, a unique *TimeStep* group is created for each timestep in the discrete time range. Under each *TimeStep* group exists one HDF5 dataset that contains the raw data for a given variable at that timestep. The bitmap dataset corresponding to the variable is also stored under the same *TimeStep* group.

This division between data and metadata is essential for the primary reason that variable metadata for a given dataset is relevant and accurate across all timesteps for that dataset variable (there is no need to store redundant metadata). Figure 9.9 illustrates the HDF5_FastQuery architectural layout.

9.4.3 Status and Sample Applications

H5Part research and development is an active, international collaborative effort involving researchers from high-performance computing and accelerator modeling. It has evolved from focusing on I/O for particle-based datasets

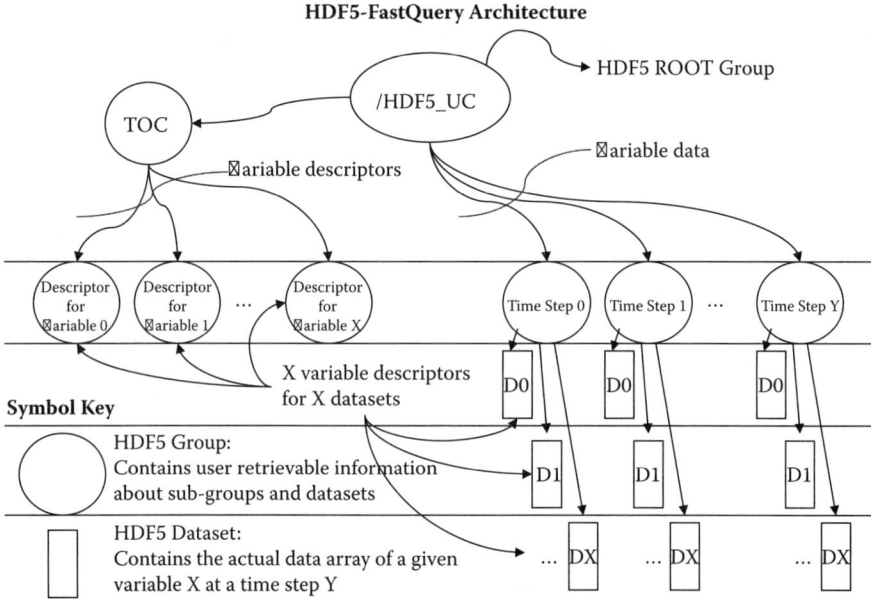

Figure 9.9 Architectural layout of HDF5_FastQuery.

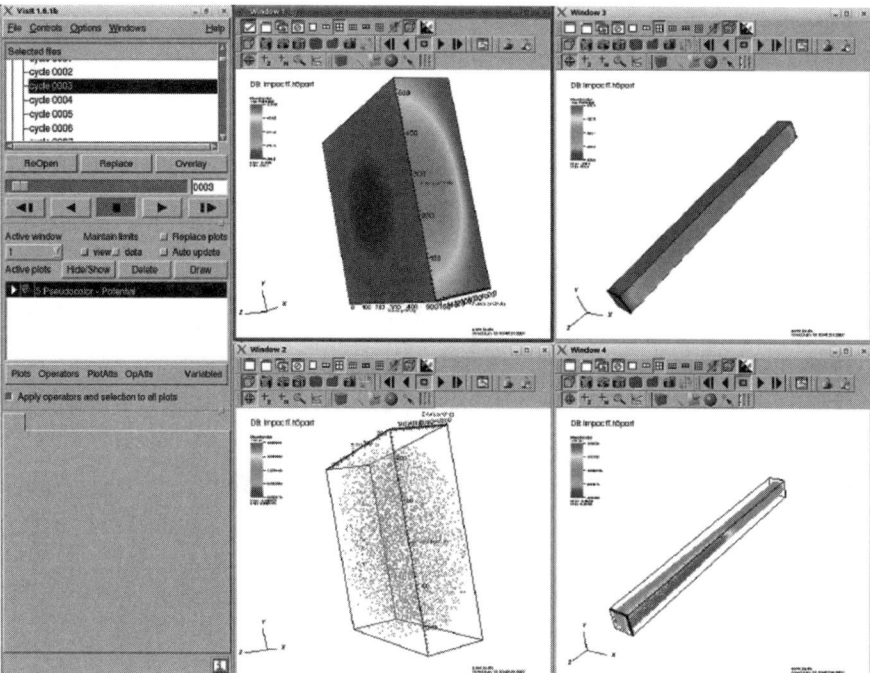

Figure 9.10 (See color insert following page 224.) H5Part readers are included with visualization and data analysis tools in use by the accelerator modeling community. This image shows H5Part data loaded into VisIt for comparative analysis of multiple variables at different simulation timesteps.

to include support for block-structured fields and unstructured meshes. It is an open source project, and source code may be downloaded from https://codeforge.lbl.gov/projects/h5part. Several accelerator modeling codes use H5Part for writing data in parallel, and H5Part readers have been created and integrated into visual data analysis applications in widespread use by the accelerator modeling community (see Figures 9.10 and 9.11).

Recent efforts have focused on merging the capabilities of HDF5_FastQuery, which provides a high-level API interface for advanced index/query capabilities, into H5Part. The objective here is to encapsulate the complexity of index/query APIs and to integrate that capability into what appears to the application as an I/O layer. This work has proven very useful in enabling rapid visual data exploration in several projects, including advanced accelerator design using laser wakefield acceleration (see Figure 9.12). Here, the tightly integrated index/query and I/O provides the fundamental scientific data management infrastructure needed to implement query-driven visualization: A visual interface enables rapid exploration of high-level data characteristics, and enables visual specification of "interesting" data subsets, which are then quickly extracted from a very large dataset and used by downstream visual

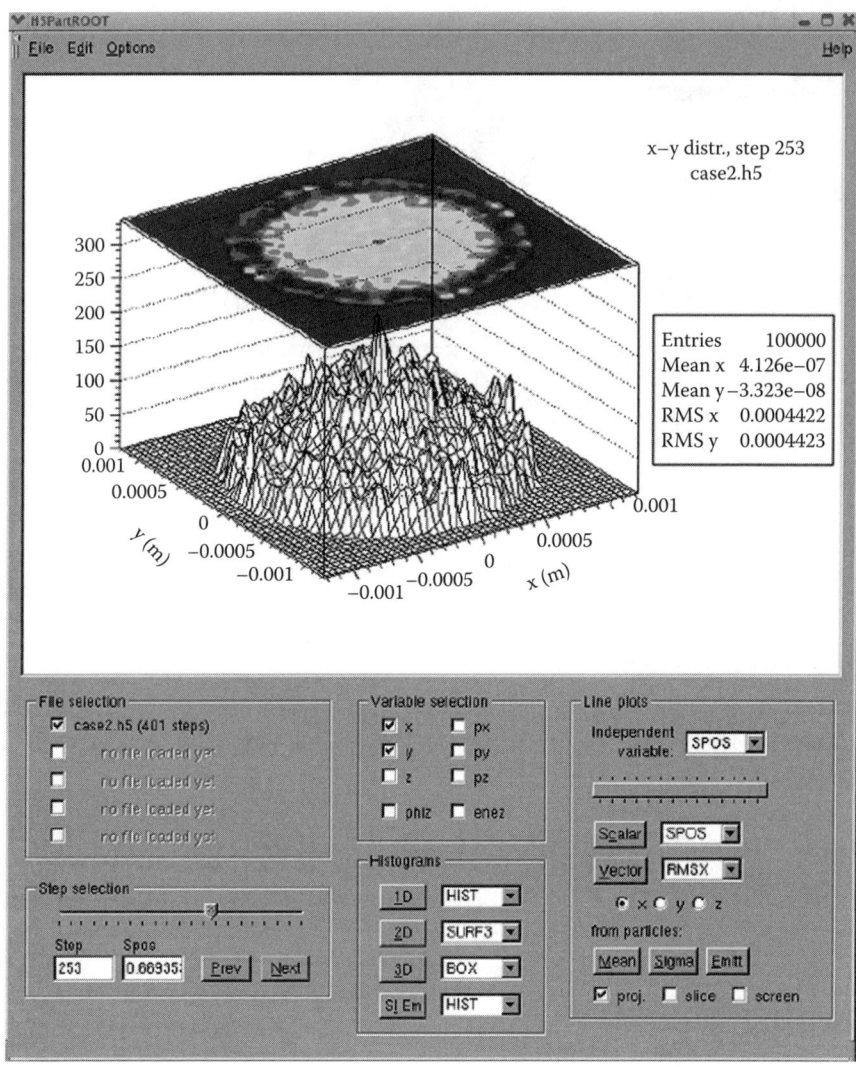

Figure 9.11 ROOT, a freely available, open source system for data management and analysis, is in widespread use by the high-energy physics community. An H5Part reader is included with ROOT. This image shows an example of visual data analysis of an H5Part dataset produced by a particle accelerator modeling code.

data analysis tools. The index/query capability has been implemented inside the H5Part layer to run in parallel to take advantage of high-performance computing platforms to accelerate I/O and range-based subsetting to support interactive exploration.

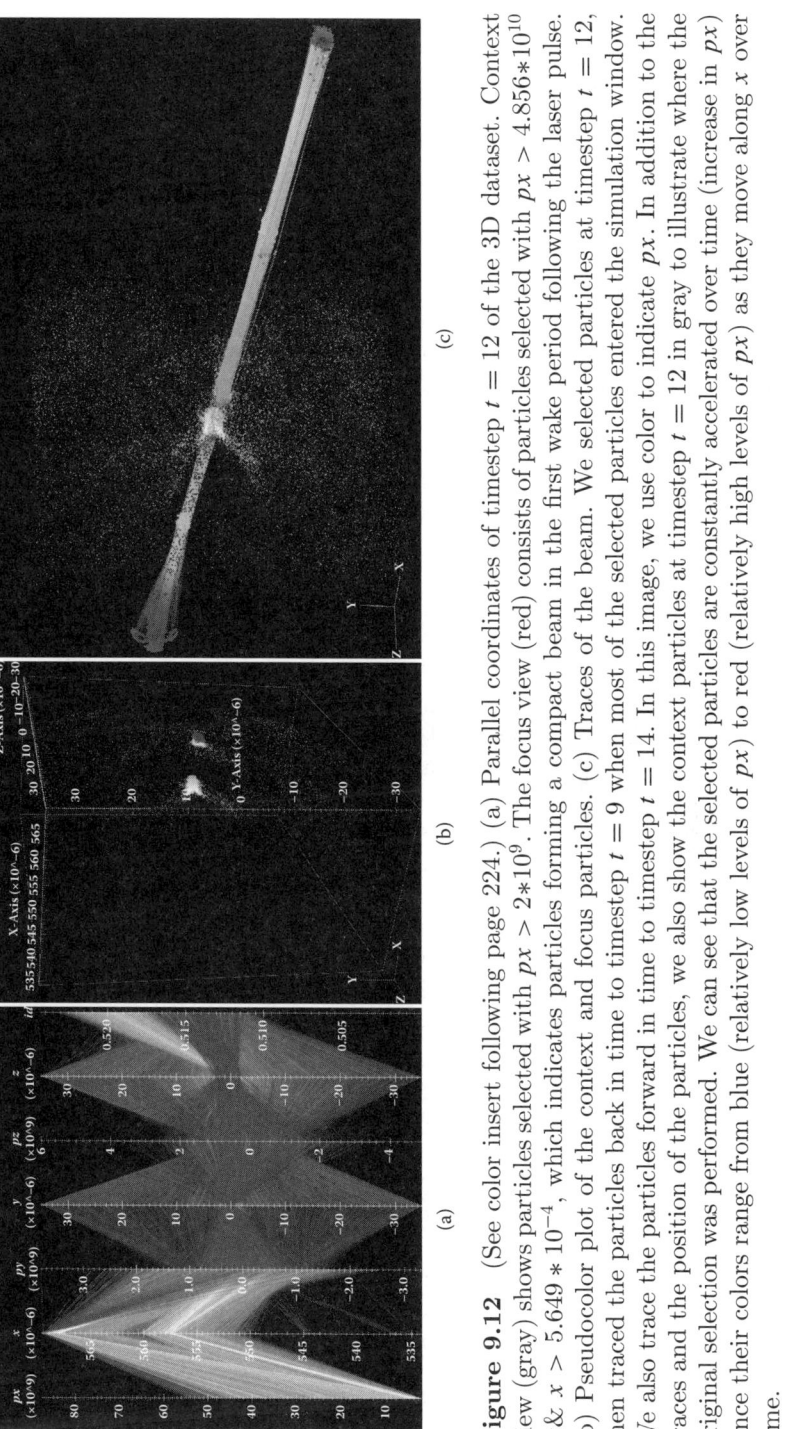

Figure 9.12 (See color insert following page 224.) (a) Parallel coordinates of timestep $t = 12$ of the 3D dataset. Context view (gray) shows particles selected with $px > 2*10^9$. The focus view (red) consists of particles selected with $px > 4.856*10^{10}$ && $x > 5.649 * 10^{-4}$, which indicates particles forming a compact beam in the first wake period following the laser pulse. (b) Pseudocolor plot of the context and focus particles. (c) Traces of the beam. We selected particles at timestep $t = 12$, then traced the particles back in time to timestep $t = 9$ when most of the selected particles entered the simulation window. We also trace the particles forward in time to timestep $t = 14$. In this image, we use color to indicate px. In addition to the traces and the position of the particles, we also show the context particles at timestep $t = 12$ in gray to illustrate where the original selection was performed. We can see that the selected particles are constantly accelerated over time (increase in px) since their colors range from blue (relatively low levels of px) to red (relatively high levels of px) as they move along x over time.

9.5 Query-Driven Visualization

The term *query-driven visualization* (QDV) refers to the process of limiting visual data analysis processing only to "data of interest."[34] In brief, QDV is about using software machinery combined with flexible and highly useful interfaces to help reduce the amount of information that needs to be analyzed. The basis for the reduction varies from domain to domain but boils down to what subset of the large dataset is really of interest for the problem being studied. This notion is closely related to that of feature detection and analysis, where *features* can be thought of as subsets of a larger population that exhibits some characteristics that are either intrinsic to individuals within the population (e.g., data points where there is high pressure and high velocity) or that are defined as relations between individuals within the population (e.g., the temperature gradient changes sign at a given data point).

QDV is one approach to visual data analysis of problems that are of massive scale in size. Other common approaches focus on increasing capacity of the visualization processing pipeline through increasing levels of parallelism to scale up existing techniques to accommodate larger data sizes. While effective in the primary objective—increase capacity to accommodate larger problem sizes—there is a fundamental problem with these approaches: They do not necessarily increase the likelihood of scientific insight. By processing more data and creating an image that is more complex, such an approach can actually impede scientific understanding.

Let's examine the first question a bit more closely. First, let's assume that we're operating on a gigabyte-sized dataset (10^9 data points), and we're displaying the results on a monitor that has, say, 2 million pixels ($2*10^6$ pixels). For the sake of discussion, let's assume we're going to create and display an isosurface of this dataset. Studies have shown that on the order of about $N^{2/3}$ grid cells in a dataset of size N^3 will contain any given isosurface.[35] In our own work, we have found this estimate to be somewhat low—our results have shown the number to be closer to $N^{0.8}$ for N^3 data. Also, we have found an average of about 2.4 triangles per grid cell will result from the isocontouring algorithm.[36] If we use these two figures as lower and upper bounds, then for our gigabyte-sized dataset, we can reasonably expect on the order of between about 2.1 and 40 million triangles for many isocontouring levels. At a display resolution of about 2 million pixels, the result is a depth complexity—the number of objects at each pixel along all depths—of between 1 and 20.

With increasing depth complexity come at least two types of problems. First, more information is "hidden from view." In other words, the nearest object at each pixel hides all the other objects that are further away. Second, if we do use a form of visualization and rendering that supports transparency—so that we can, in principle, see all the objects along all depths at each pixel—we are assuming that a human observer will be capable of distinguishing among the objects in depth. At best, this latter assumption does not always hold true, and at worst, we are virtually guaranteed the viewer will not be able to gain any meaningful information from the visual information overload.

If we scale up our dataset from gigabyte (10^9) to terabyte (10^{12}), we then can expect on the order of between 199 million and 9.5 billion triangles representing a depth complexity ranging between about 80 and 4,700, respectively. Regardless of which estimate of the number of triangles we use, we end up drawing the same conclusion: depth complexity, and correspondingly scene complexity and human cognitive workload, all grow at a rate that is a function of the size of the source data. Even if we are able to somehow display all those triangles, we would be placing an incredibly difficult burden on the user. They will be facing the impossible task of trying to visually locate smaller needles in a larger haystack.

The multifaceted approach we are adopting takes square aim at the fundamental objective: help scientific researchers more quickly and efficiently do science. In one view, one primary tactical approach that seems promising is to help focus user attention on easily consumable images from the large data collection. We do not have enough space in this chapter to cover all aspects in this regard. Instead, we provide a few details about a couple of especially interesting challenge areas.

9.5.1 Implementing Query-Driven Visualization

In principle, the QDV idea is conceptually quite simple: restrict visualization and analysis processing only to data of interest. In practice, implementing this capability can be quite a challenge. Our approach is to take advantage of the state-of-the-art index/query technology mentioned earlier in this chapter, FastBit, and use it as the basis for data subsetting in query-driven visualization applications. In principle, any type of data-subsetting technology can be used to provide the same functionality. In practice, FastBit has proven to be very efficient and effective at problems of scale.

Some of our early work in this space focused on comparing the performance of FastBit as the basis for index/query in QDV applications with some of the industry standard algorithms for isosurface computation.[34] In computing isosurfaces, one must first find all the grid cells that contain the surface of interest, then for each such cell, compute the isosurface that intersects the cell. Our approach, which leverages FastBit, shows a performance gain of between 25% to 300% over the best isocontouring algorithms created by the visualization community. Of greater significance is the fact that our approach extends to *n*-dimensional queries (i.e., queries of *n* different variables), whereas the indexing structures created for use in isocontouring are applicable only to univariate queries. This approach was demonstrated on datasets created by a combustion simulation (see Figure 9.13, which shows three different query conditions performed on a single dataset; see Stockinger et al.[34] for more details).

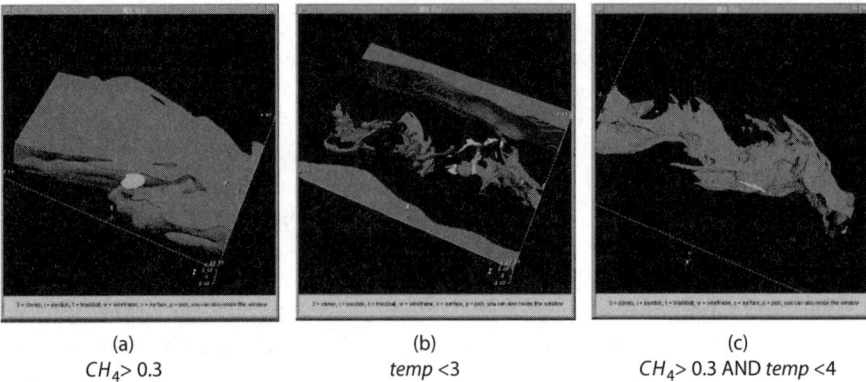

(a)	(b)	(c)
$CH_4 > 0.3$	$temp < 3$	$CH_4 > 0.3$ AND $temp < 4$

Figure 9.13 (See color insert following page 224.) A visualization of flames in a high-fidelity simulation of methane-air jet. The images show the cells in a 3D block-structured dataset that were returned by three different queries.

9.5.2 Case Study—Network Traffic Analysis

While our earlier work established the viability of the approach, particularly when compared with the best search algorithms from the visualization community, more recent work extends and applies these techniques to a "hero-sized" problem. In this application, our objective is to perform interactive visual data analysis of one year's worth of network connection data. The case study in this work focuses on rapid drill-down using multiresolution histograms computed by FastBit for the purposes of identifying the existence of a distributed network scan attack, then for identifying the set of hosts participating in the attack. The results of that study[37,38] show that this approach performs up to four orders of magnitude faster than conventional techniques commonly used in the field of network traffic analysis.

The basic use model for this application, which is shown pictorially in Figure 9.14, is as follows: first, compute and display a histogram of traffic levels at a coarse granularity (each day over a 365-day period). Histogram display is augmented with statistical analysis to help highlight anomalous behavior. Next, through a visual user interface, "drill into" the data by allowing the user to specify a temporal window of finer resolution. FastBit computes a new histogram over the specified temporal window and at finer resolution, allowing more details of the data to emerge. This process repeats until coherent temporal patterns of the attack begin to emerge. Once the attack signature is identified and confirmed to be a network scan (see Figure 9.15), FastBit can quickly locate and return the network traffic records that contain information about hosts participating in the attack (see Stockinger et al.[37] for more details).

In this work, FastBit was extended to support the rapid creation of multidimensional, conditional histograms. The implementation and performance study was conducted on a parallel, shared-memory platform. The results of

Scientific Data Management Challenges

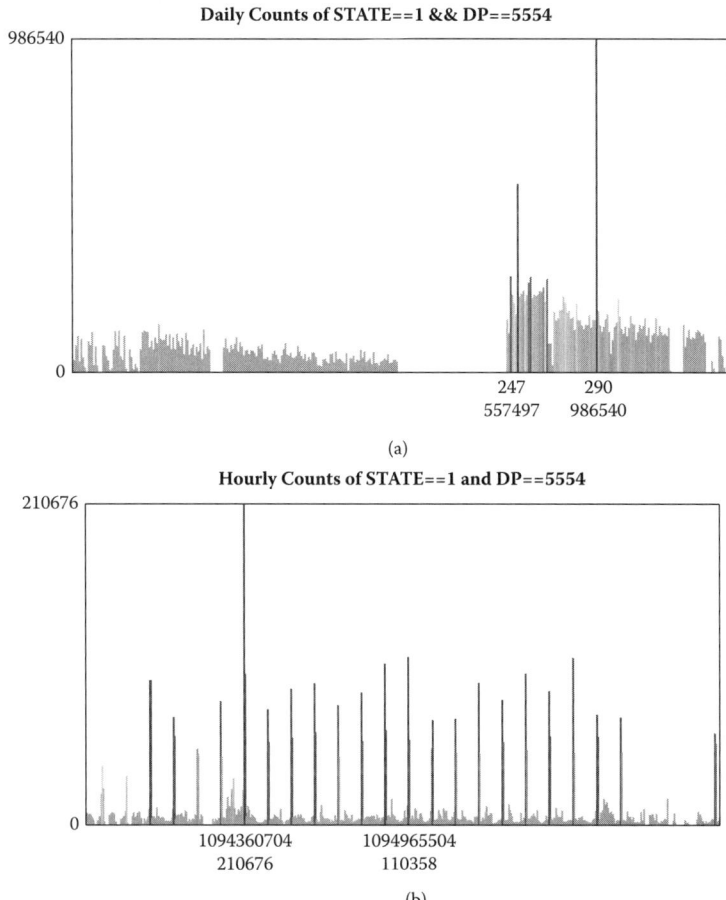

Figure 9.14 (See color insert following page 224.) (a) Histogram of suspicious activity levels over a one-year period at one-day temporal resolution. (b) Suspicious activity levels over a four week period at one-hour temporal resolution. (c) Suspicious activity levels over a one-day period at one-hour temporal resolution. Forensic network traffic analysis is conducted by examining histograms of suspicious traffic activity at varying temporal resolution. These examples go from coarse, per-day resolution over a one-year time window down to per-minute resolution over a five-day window and show a regular pattern of systematic network attacks that occur with temporal regularity.

the study show that forensic investigation of such massive datasets can be conducted in an interactive fashion. Previous approaches would require hours or days to conduct a similar investigation. The significance of this work is that it shows the potential of coupling state-of-the-art scientific data management

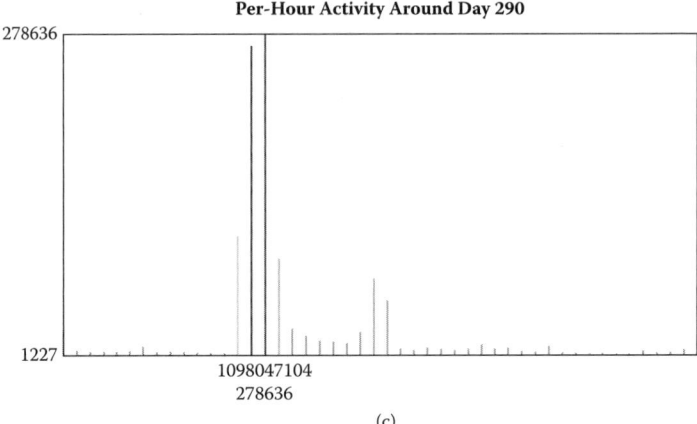

Figure 9.14 (*Continued*)

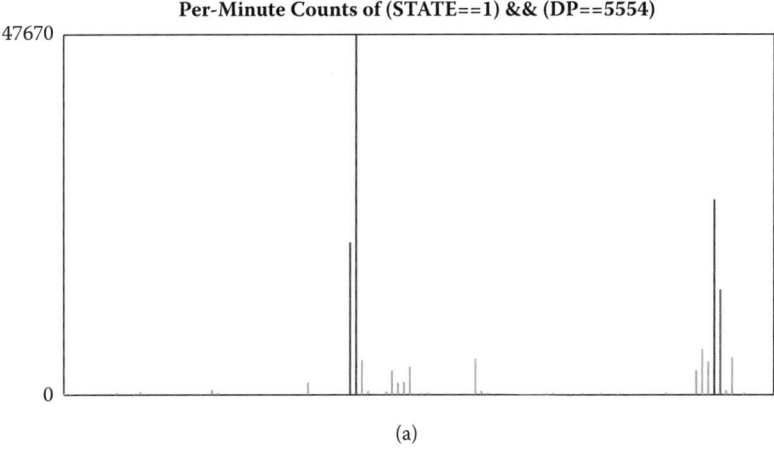

Figure 9.15 Two- and three-dimensional histograms are the building blocks for visual data exploration of network traffic analysis. Here we see evidence of an organized scan: One or more remote hosts are probing sequential IP addresses within a block of addresses, hoping to find a vulnerability. (a) This histogram shows suspicious activity over a two-hour period at one-minute temporal resolution. The spikes in this histogram correspond to the "sheets" in the adjacent image. (b) A 3D histogram; the vertical axis is time, the other two axes are the C and D octets of the destination host address. The "sheets" indicate that the remote host(s) are performing a scan of all IP addresses within a given IP address space, indicating the attack is a scan.

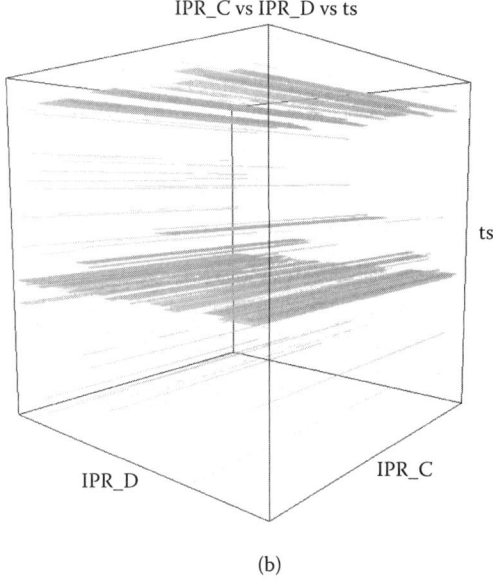

Figure 9.15 (*Continued*)

technology with visual data analysis technology to tackle problems of scale in a manner that virtually guarantees scientific insight and knowledge discovery.

9.6 Summary and Conclusion

Visualization, which is the transformation of abstract data into images, plays a central role in virtually all fields of scientific endeavor. It is an indispensable part of hypothesis testing and knowledge discovery. Like most other fields discussed in this book, visualization faces substantial scientific data management challenges that are the result of growth in size and complexity of data being produced by simulations and collected from experiments. Our objective in this chapter has been to reveal some of the scientific data management issues, challenges, and solutions that are somewhat unique to the field of visualization.

Production visualization applications, namely those that run on large-scale parallel machines, can derive great benefit from close attention to scientific data management issues. Staple operations, like slicing and isosurfacing, can be vastly accelerated by taking advantage of metadata. In some cases, the simulation or experiment produces such metadata as part of the data production process. In other cases, we must generate that data ourselves. A significant fraction of the code in these production visualization applications is dedicated to scientific data management: they must support a plethora of input data formats, and therefore, they contain a number of data loader modules; they must create an internal data structure that is suitable for use by a potentially large collection of visualization, analysis, and rendering modules, all of which may potentially run in parallel on shared or distributed memory machines. An open problem is one of data models and semantics, where meaning is assigned to arrays of data stored in data files.

With data of massive scale, it is often useful to perform a multiresolution analysis, working first with a smaller, coarser version of data, then progressively refine the analysis as interesting features are revealed. We saw that a space-filling curve model has proven to be highly efficient for interactive analysis of massive data. However, such a data model and layout is unlikely to be output directly from a simulation. This is a good example of how a data model and layout that works very well for multiresolution analysis is unlikely to be used by simulations for output. Multiresolution, quantitative feature detection and analysis methods were demonstrated on two different datasets from the field of turbulent mixing. This quantitative analysis approach is very useful for enabling scientific knowledge discovery by focusing on features rather than on the machinery for creating potentially incomprehensible images of large-scale scientific data.

A significant barrier faced by many computational and experimental science projects is the complexity of using state-of-the art technology from scientific data management. We discussed an approach for encapsulating complexity in the form of a high-level API for data storage and retrieval that lowers the entry point for using such technology. This concept was also applied to index/query technology. We have applied these concepts to multiple application areas to produce results showing that visual data analysis, as a field, can benefit from a close collaboration with the field of scientific data management. The benefit is improved performance for many data intensive operations, like data I/O and data subsetting, as well as the potential to conceive and create completely new, paradigm-changing approaches to solve the problem of scientific knowledge discovery for massive, complex datasets.

Acknowledgments

This work was supported by the Director, Office of Advanced Scientific Computing Research, Office of Science, of the U.S. Department of Energy, under Contract No. DE-AC02-05CH11231 through the Scientific Discovery through Advanced Computing (SciDAC) program's Visualization and Analytics Center for Enabling Technologies (VACET). This research used resources of the

National Energy Research Scientific Computing Center, which is supported by the Office of Science of the U.S. Department of Energy under Contract No. DE-AC02-05CH11231.

References

[1] B. McCormick, T. Defanti, and M. Brown (eds.). Visualization in scientific computing. *Computer Graphics*, 21(6), 1987.

[2] C. Johnson and C. Hansen (eds.). *Visualization Handbook*. Academic Press, Inc., Orlando, FL, USA, 2004.

[3] Computational Engineering International, Inc. *Ensight Visualization Software*, 2008. http://www.ensight.com/. Accessed July 10, 2009.

[4] Lawrence Livermore National Laboratory. *VisIt Visualization Software*, 2008. http://www.llnl.gov/visit/. Accessed July 10, 2009.

[5] Kitware, Inc., Los Alamos National Laboratory, Sandia National Laboratory, and CSimSoft. *ParaView Visualization Software*, 2008. http://www.paraview.org/. Accessed July 10, 2009.

[6] L. Roberts. *Silo User's Guide*. Lawrence Livermore National Laboratory, 2000. UCRL-MA-118751-REV-1.

[7] L. Schoof and V. Yarberry. *EXODUS II: A Finite Element Data Model*. Sandia National Laboratory, 1994. Tech Rep. SAND92-2137.

[8] V. Pascucci and R. J. Frank. Global static indexing for real-time exploration of very large regular grids. In *Supercomputing '01: Proceedings of the 2001 ACM/IEEE conference on Supercomputing (CDROM)*, pages 2–2, New York, NY, USA, 2001. ACM Press.

[9] M. C. Miller, J. F. Reus, R. P. Matzke, W. J. Arrighi, L. A. Schoof, R. T. Hitt, and P. K. Espen. Enabling interoperation of high performance, scientific computing applications: Modeling scientific data with the sets & fields (saf) modeling system. In *ICCS '01: Proceedings of the International Conference on Computational Science—Part II*, pages 158–170, London, UK, 2001. Springer-Verlag.

[10] HDF Group. Hierarchical Data Format—HDF5, 2008. http://www.hdfgroup.com. Accessed July 10, 2009.

[11] G. Abram and L. A. Treinish. An extended data-flow architecture for data analysis and visualization. Research report RC 20001 (88338), IBM T. J. Watson Research Center, Yorktown Heights, NY, USA, February 1995.

[12] W. J. Schroeder, K. M. Martin, and W. E. Lorensen. The design and implementation of an object-oriented toolkit for 3D graphics and visualization. In *VIS '96: Proceedings of the Conference on Visualization '96*, pages 93–100. IEEE Computer Society Press, 1996.

[13] C. Upson, T. Faulhaber Jr., D. Kamins, D. H. Laidlow, D. Schlegel, J. Vroom, R. Gurwitz, and A. van Dam. The application visualization system: A computational environment for scientific visualization. *Computer Graphics and Applications*, 9(4):30–42, July 1989.

[14] J. Shalf and E. W. Bethel. How the Grid Will Affect the Architecture of Future Visualization Systems. *IEEE Computer Graphics and Applications*, 23(2):6–9, May/June 2003.

[15] T. Richardson, Q. Stafford-Fraser, K. R. Wood, and A. Hopper. Virtual Network Computing. *IEEE Internet Computing*, 2(1):33–38, 1998.

[16] B. Paul, S. Ahern, E. W. Bethel, E. Brugger, R. Cook, J. Daniel, K. Lewis, J. Owen, and D. Southard. Chromium Renderserver: Scalable and Open Remote Rendering Infrastructure. *IEEE Transactions on Visualization and Computer Graphics*, 14(3), May/June 2008. LBNL-63693.

[17] K. Cole-McLaughlin, H. Edelsbrunner, J. Harer, V. Natarajan, and V. Pascucci. Loops in reeb graphs of 2-manifolds. In *Proceedings of the 19th Annual Symposium on Computational Geometry*, pages 344–350. ACM Press, 2003.

[18] H. Edelsbrunner, J. Harer, and A. Zomorodian. Hierarchical Morse-Smale complexes for piecewise linear 2-manifolds. *Discrete Comput. Geom.*, 30:87–107, 2003.

[19] H. Edelsbrunner, D. Letscher, and A. Zomorodian. Topological persistence and simplification. In *FOCS '00: Proceedings of the 41st Annual Symposium on Foundations of Computer Science*, page 454, Washington, D.C., USA, 2000. IEEE Computer Society.

[20] P.-T. Bremer, H. Edelsbrunner, B. Hamann, and V. Pascucci. A multiresolution data structure for two-dimensional Morse-Smale functions. In G. Turk, J. J. van Wijk, and R. Moorhead, editors, *Proc. IEEE Visualization '03*, pages 139–146, Los Alamitos, CA, 2003. IEEE, IEEE Computer Society Press.

[21] P.-T. Bremer, H. Edelsbrunner, B. Hamann, and V. Pascucci. A topological hierarchy for functions on triangulated surfaces. *IEEE Trans. on Visualization and Computer Graphics*, 10(4):385–396, 2004.

[22] V. Natarajan, Y. Wang, P.-T. Bremer, V. Pascucci, and B. Hamann. Segmenting molecular surfaces. *Comput. Aided Geom. Des.*, 23(6):495–509, 2006.

[23] D. Laney, P. T. Bremer, A. Mascarenhas, P. Miller, and V. Pascucci. Understanding the structure of the turbulent mixing layer in hydrodynamic instabilities. *IEEE Transactions on Visualization and Computer Graphics*, 12(5):1053–1060, 2006.

[24] A. Gyulassy, M. Duchaineau, V. Natarajan, V. Pascucci, E. Bringa, A. Higginbotham, and B. Hamann. Topologically Clean Distance fields. *IEEE Transactions on Visualization and Computer Graphics*, 13(6):1432–1439, November/December 2007.

[25] P. Miller, P.-T. Bremer, W. Cabot, A. Cook, D. Laney, A. Mascarenhas, and V. Pascucci. Application of Morse theory to analysis of Rayleigh-Taylor topology. In *Proceedings of the 10th International Workshop on the Physics of Compressible Turbulent Mixing*, 2006.

[26] J. S. Vitter. External memory algorithms and data structures: Dealing with massive data. *ACM Computing Surveys*, March 2000.

[27] J. K. Lawder and P. J. H. King. Using space-filling curves for multidimensional indexing. In *Lecture Notes in Computer Science*, pages 20–35, 2000.

[28] B. Moon, H. Jagadish, C. Faloutsos, and J. Saltz. Analysis of the clustering properties of Hilbert spacefilling curve. *IEEE Transactions on knowledge and data engeneering*, 13(1):124–141, 2001.

[29] A. Adelmann, A. Gsell, B. Oswald, T. Schietinger, E. W. Bethel, J. Shalf, C. Siegerist, and K. Stockinger. Progress on H5Part: A Portable High Performance Parallel Data Interface for Electromagnetic Simulations. In *Particle Accelerator Conference PAC07* June 25–29, 2007, Albuquerque, NM, 2007. http://vis.lbl.gov/Publications/2007/LBNL-63042.pdf. Accessed July 10, 2009.

[30] L. Gosink, J. Shalf, K. Stockinger, K. Wu, and E. W. Bethel. HDF5_FastQuery: Accelerating Complex Queries on HDF Datasets Using Fast Bitmap Indices. In *Proceedings of the 18th International Conference on Scientific and Statistical Database Management*. IEEE Computer Society Press, July 2006. LBNL-59602.

[31] K. Wu, E. Otoo, and A. Shoshani. Optimizing bitmap indices with efficient compression. *ACM Transactions on Database Systems*, 31:1–38, 2006.

[32] LBNL Scientific Data Management Research Group. FastBit: An Efficient Compressed Bitmap Index Technology, 2008. http://sdm.lbl.gov/fastbit/. Accessed July 10, 2009.

[33] K. Wu, E. Otoo, and A. Shoshani. On the performance of bitmap indices for high cardinality attributes. In *International Conference on Very Large Databases*, pages 24–35, 2004.

[34] K. Stockinger, J. Shalf, K. Wu, and E. W. Bethel. Query-Driven Visualization of Large datasets. In *Proceedings of IEEE Visualization 2005*, pages 167–174. IEEE Computer Society Press, October 2005. LBNL-57511.

[35] C. L. Bajaj, V. Pascucci, and D. R. Schikore. Fast isocontouring for improved interactivity. In *VVS '96: Proceedings of the 1996 symposium on volume visualization*, pages 39–46, Piscataway, NJ, USA, 1996. IEEE Press.

[36] I. Bowman, J. Shalf, K.-L. Ma, and E. W. Bethel. Performance Modeling for 3D Visualization in a Heterogenous Computing Environment. Technical Report LBNL-56977, Lawrence Berkeley National Laboratory, 2004.

[37] K. Stockinger, E. W. Bethel, S. Campbell, E. Dart, and K. Wu. Detecting Distributed Scans Using High-Performance Query-Driven Visualization. In *SC '06: Proceedings of the 2006 ACM/IEEE Conference on High Performance Computing, Networking, Storage and Analysis*. IEEE Computer Society Press, October 2006, pages 82–94.

[38] E. W. Bethel, S. Campbell, E. Dart, K. Stockinger, and K. Wu. Accelerating Network Traffic Analysis Using Query-Driven Visualization. In *Proceedings of 2006 IEEE Symposium on Visual Analytics Science and Technology*, pages 115–122. IEEE Computer Society Press, October 2006. LBNL-59891.

Chapter 10

Interoperability and Data Integration in the Geosciences

Michael Gertz,[1] Carlos Rueda,[2] and Jianting Zhang[3]

[1] Institute of Computer Science, University of Heidelberg, Heidelberg, Germany
[2] Monterey Bay Aquarium Research Institute, Moss Landing, California
[3] Department of Computer Science, The City College of New York, New York, New York

Contents

10.1 Introduction 370
10.2 Geospatial Data Management and Integration 372
    10.2.1 Geospatial Data Models and Representations 373
    10.2.2 Geospatial Data Management Systems and Formats 374
    10.2.3 Schemas and Metadata 375
        10.2.3.1 GML Application Schemas 375
        10.2.3.2 Metadata Standards 376
    10.2.4 Approaches to Integrating Geospatial Data 377
10.3 Service-Based Data and Application Integration 379
    10.3.1 Approaching Integration through Interoperability 379
    10.3.2 Service-Oriented Architectures 380
    10.3.3 Registry and Catalog Services 380
    10.3.4 Geospatial Web-Services and Standards 381
    10.3.5 Sensor Web Enablement 383
    10.3.6 OPeNDAP 386
10.4 An Example of an Integration and Interoperability Scenario 386
    10.4.1 Environmental Modeling Task—Evapotranspiration 386
    10.4.2 Integration Platform 387
    10.4.3 Overall Integration and Results 389
10.5 Conclusions 391
Acknowledgements 392
References 392

10.1 Introduction

The past decade has witnessed a dramatic increase in scientific data being generated in the physical, earth, and life sciences. This development is primarily a result of major advancements in sensor technology, surveying techniques, computer-based simulations, and instrumentation of experiments. As stated by Szalay and Gray,[1] it is estimated that the amount of scientific data generated in these disciplines is now doubling every year. Organizations in government, industry, as well as academic and private sectors, have made significant investments in infrastructures to collect and maintain scientific data and make them accessible to the public. Good examples of such efforts are the Sloan Digital Sky Survey in astronomy,[2] the GDB Human Genome Database and Entrez Genome Database in genomics,[3,4] and the Global Biodiversity Information Facility in ecology,[5] to name only a few.

More and more such domain-specific data management infrastructures are built to allow users easy access to scientific data, often in a Web-based fashion through comprehensive Web portals. However, a key challenge is to provide users with effective means to *integrate* data from diverse sources to facilitate data exploration and analysis tasks. Data integration is one of the more traditional yet still very active fields in the area of databases and data management. It is concerned with models, techniques, and architectures that provide users with a uniform logical view of and transparent access to physically distributed and often heterogeneous data sources.[6–9] Data integration is a key theme in many e-commerce and e-business IT infrastructures, often called enterprise information integration.[10] In these application domains, the objective is to integrate business and consumer data from different transactional databases in order to obtain new information that drives business activities and decisions. Nowadays, several commercial and open-source data integration platforms exist that help businesses to integrate (typically relational) data from transactional databases, leading to data warehouse and federated database architectures.

It seems natural to apply similar techniques realized in those business-oriented data integration platforms to scientific data collections as well. However, because of the complexity, unprecedented quantities, and diversity of scientific data, traditional schema-based approaches to data integration are in general not applicable. In many scientific application domains, there often is no single conceptual schema that can be developed from the data and schemas associated with the individual data sources to be integrated. Furthermore, scientific data integration often occurs in an ad hoc fashion. For example, data relevant to evaluate a scientific hypothesis needs to be discovered and dynamically integrated into often complex data analysis and exploration tasks without requiring to persistently store the data used in these tasks. The problem many scientists are facing nowadays is how to easily make use of the ever-increasing number of data repositories in an effective way.

A prominent domain where these problems become more and more apparent and pressing is in the geosciences. Geospatial data, that is, data that is spatially referenced to Earth, have become ubiquitous. This is primarily due to major advancements in remote-sensing technology, surveying techniques, and computer-based simulations. As an example, the satellites operated by NASA and NOAA generate dozens of terabytes of imagery and derived data products per day, leading to one of the fastest growing repositories with petabytes of science data. In the year 2003, NOAA already maintained about 1,300 databases containing more than 2,500 environmental variables.[11] The diverse types of geospatial data collected by federal and local governments as well as organizations in industry and academia play a significant role in developing mission-critical spatial data infrastructures.[12,13]

The use of geospatial data obtained through observations and simulations and their management in spatial data infrastructures have become essential in many application domains. These include environmental monitoring, climate research, disaster prevention, natural resource management, transportation, and decision support at various levels of local and state governments. The types of geospatial data considered in these domains come in a variety of types. Common types include maps and imagery from air and space-borne instruments, vector data describing geographic objects and features, outputs from simulations, and numerous types of real-time sensor data. In particular, the latter are an emerging data source, driven by large-scale environmental observation networks such as those envisioned by NEON.[14]

With such a proliferation of a wide range of geospatial data repositories, many of which are readily accessible through the Web, it is imperative to achieve a high degree of interoperability among these systems as a prerequisite to facilitating data-integration tasks. By realizing this objective, geospatial data that is managed in specialized repositories in support of specific domains and tasks can serve whole communities and scientists in different disciplines.

In this chapter, we present the current trends and technologies in support of developing interoperable geospatial data sources and management architectures that *enable the efficient sharing, use, and integration of physically distributed and heterogeneous geospatial data collections*. Our primary focus is on emerging technologies that facilitate true interoperability among geospatial data repositories, such as the development and implementation of standards for geospatial content and services promoted by the Open Geospatial Consortium (OGC).[15] A key concept underlying this approach is (geospatial) Web services, which realize a standard way to interoperate with diverse geospatial data management infrastructures and to access heterogeneous forms of geospatial data in a uniform and transparent fashion.

Such type of interoperability, of course, is only one ingredient in effective data-integration approaches. Compared with data-integration techniques for traditional relational databases, there are several special properties pertinent to geospatial data. For example, a complicating factor in integrating geospatial data is the variety of formats in which the data is managed, ranging from flat files to specialized geographic information systems (GIS). As we will illustrate in the following sections, geospatial Web services provide an effective means to request geospatial data from heterogeneous repositories in a format suitable for data integration tasks. Such services help greatly in dealing with data heterogeneity and conflict resolution aspects in data integration. Further data integration challenges, such as heterogeneity of the data in terms of structure and semantics are often dealt with by employing standard representation formats and taxonomies, respectively. In particular, for these two aspects, we show that there have been significant achievements in various subdisciplines of the geosciences, especially in the development of schema frameworks for describing geospatial data and metadata/taxonomy frameworks that focus on the semantics of geospatial data components.

A novel aspect we focus on in this chapter is the integration of streaming real-time data, which is becoming a predominant source of geospatial data, for example, in remote-sensing and sensor observation networks. We describe recent technologies that have been developed for the service-based management and consumption of streaming geospatial data and show how computing infrastructures can be built that effectively consume diverse static as well as dynamic (streaming) geospatial data from heterogeneous and distributed data sources. General techniques for managing streaming data are discussed in Chapter 11.

The remainder of this chapter is organized as follows. In Section 10.2, we review basic geospatial data management and integration concepts, including data formats, metadata standards, and existing approaches and techniques to geospatial data integration. In Section 10.3, we discuss in detail the technologies surrounding geospatial Web services. Furthermore, we outline emerging technologies in the context of sensor Web enablement architectures. In Section 10.4, we use a relevant practical scenario from the environmental sciences to demonstrate how the different techniques presented in this chapter can effectively be deployed to perform geospatial data integration tasks, including the integration of real-time sensor data. We conclude the chapter in Section 10.5 with a summary.

10.2 Geospatial Data Management and Integration

In this section, we review some fundamental concepts and techniques for the management of geospatial data. In Section 10.2.1, we give a brief overview of geospatial data models and representation formats. Section 10.2.2 outlines some standard approaches to managing geospatial data. Our particular focus in these two sections is on issues relevant to interoperability and integration aspects. After a discussion of schema and metadata concepts in Section 10.2.3, we discuss in Section 10.2.4 some existing approaches for integrating geospatial data.

10.2.1 Geospatial Data Models and Representations

Depending on the application domain and collected geospatial information, geospatial data can be modeled and represented in different ways. The two most common approaches to model geographic information are using either an *object-based model* or a *field-based model*.[16,17] In an object-based model, geographic objects correspond to real-world entities (also called *features*) about which information needs to be managed. A feature typically has two parts: (a) a spatial component (or *spatial extent*), which specifies the shape and location of the object in the embedding space; and (b) a descriptive component that describes the *nonspatial properties* of the feature in the form of attributes. The spatial extent of an object is typically modeled as a point, polyline, or polygon, depending on the required spatial granularity and scale of the data to be managed. For the representation of a collection of features, different approaches exist, such as the network model, spaghetti model, or topological model.[16] The left part of Figure 10.1 shows an example of an object-based presentation of geographic information (a road network).

In field-based approaches, the space to be modeled is partitioned (tessellated) into two- or multidimensional cells, a cell having a spatial extent. With each cell one or more attribute values are associated, each attribute describing a continuous function in space. A typical example of field-based data are multispectral or hyperspectral raster imagery obtained from remote-sensing instruments. Field-based data are also common as outputs of simulations where with each point in space a set of attribute values (measurements) is associated. Note that in a field-based model, there is no notion of objects but observations of phenomena in space, which are described by attribute values (measurements) that vary with the location in space. The right part of Figure 10.1 shows an example of a field-based representation (estimated temperature over an area).

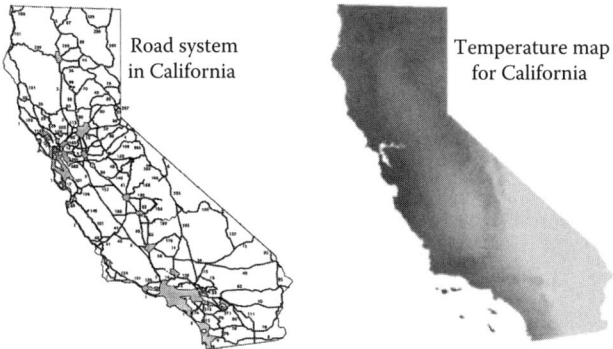

Figure 10.1 (See color insert following page 224.) Examples of object-based (left) and field-based (right) geospatial data representation.

In order to precisely describe the spatial extent of geographic objects or cells in a raster image, it is important to have a *spatial reference system (SRS)* (or *coordinate reference system (CRS)*) underlying the space in which features and phenomena are modeled. A reference system is a particular *map projection* that represents the two-dimensional curved surface of the Earth. There are numerous such map projections used in practice, ranging from global projections such as latitude/longitude or Universal Transverse Mercator (UTM) to parameterized local ones tailored to specific regions on the Earth's surface, such as the State Plane Coordinate System [18] used in the United States. From a spatial data management point of view and in particular for the integration of diverse datasets, an important aspect is to be able to re-project geospatial data from one reference system to another one.[19,20]

10.2.2 Geospatial Data Management Systems and Formats

Compared with data management systems for relational data, which are all based on the same model (the relational model) and make use of the same language (SQL), there is a plethora of commercial and open-source systems for managing geospatial data. In the following, we give a brief overview of the different types of systems and focus on aspects that are relevant to data integration approaches.

GISs are the predominant type of systems to manage, store, analyze, and display geographic data and associated attributes that are spatially referenced to the Earth.[21] A widely used type of GIS is ESRI's (Environmental Systems Research Institute's) ArcGIS products, such as ArcView to view spatial data, create maps, and perform basic spatial analysis operations. ArcInfo is an advanced version of the ArcGIS product line that also includes functions for manipulating, editing, and analyzing geospatial data[22,23] and services for geoprocessing and geocoding.[24] ArcGIS also provides different types of Web services to access geospatial data.

There are also traditional relational database management system vendors that offer spatial extensions to their relational engines. For example, Oracle Spatial provides several functions for storing, querying, and indexing spatial data, including raster and gridded data.[25] The spatial extension models a majority of the spatial types and operations described in the SQL/MM spatial standard.[26] IBM's DB2 product line also offers spatial extensions to their relational DB2 core system such as the DB2 Spatial Extender and the DB2 Geodetic Extender.[27] Also here, the spatial extensions implement types and functions specified in the SQL/MM standard.

Prominent open-source GIS type systems are PostGIS,[28] the spatial extension of the object-relational database management system PostgreSQL, and the Geographic Resources Analysis Support System (GRASS).[29,30] Like the spatial extensions for Oracle and DB2, PostGIS follows the *Simple Features for SQL specification* developed as an implementation specification by the OGC.[31] This standard specifies the storage of different types of geographic objects (points, lines, polygons, etc.) and includes specifications for various spatial operators to derive new objects from existing ones. PostGIS makes use of the proj.4 library[32] for converting geographic data between different map projections, an important functionality to integrate geospatial objects that are based on different reference systems.

GRASS provides a variety of functions to manage raster data and topological vector data. It natively uses and supports a number of vector and raster formats, which are expanded with several other formats using the Geospatial Data Abstraction Library (GDAL).[33] GRASS offers the option to manage nonspatial attributes associated with geographic objects and raster images in either files or an SQL-based database management system.

Besides the above GIS type of data management infrastructures, geospatial data are also often managed just at the file level. That is, applications generate geospatial data and simply record them in standard file formats for consumption by and exchange with other programs. One can basically distinguish between file formats for vector data (object-based data) and file formats for raster or gridded data (field-based data). One of the most common formats for vector data are *shapefiles*, which have been developed by ESRI and are used to exchange data among ESRI products and other software.[34] Another important, although less widely used, format for vector data is the Topologically Integrated Geographic Encoding and Referencing (TIGER) format used by the U.S. Census Bureau. It is employed for modeling geographic information such as roads, rivers, lakes and census tracts.[35]

For raster and gridded data, widely used file formats are the Network Common Data Form (NetCDF),[36] the Hierarchical Data Format (HDF5),[37] and GeoTIFF.[38] These file formats only represent a small but important portion of a large collection of scientific data formats (many of which also come in an XML framework) that have been developed over the past decades in different disciplines.

The above discussions about the variety of commercial and open-source geospatial data management software as well as file formats for the exchange of complex (geo)spatial data clearly illustrate that achieving interoperability among heterogeneous geospatial data sources is a great challenge.

10.2.3 Schemas and Metadata

An essential ingredient to any data integration approach is to have information about the schemas as well as metadata for schema components and the data managed in heterogeneous scientific data repositories. In the following, we first discuss an emerging standard for geospatial data to represent both schema information and data and then detail some prominent metadata frameworks used in the context of geospatial data.

10.2.3.1 GML Application Schemas

The Geography Markup Language (GML) is an XML-based specification developed by the OGC for representing geographic features.[39,40] GML serves as an open interchange format for geospatial data as well as a modeling language for geographic information. In GML, real-world objects are called *features* and have a spatial component (geometry) and nonspatial properties. The most recent GML version, 3.1, is being standardized as ISO 19136. While earlier GML versions used Document Type Definitions (DTDs), the later versions are based on XML-Schema. GML version 3.x also includes support for two-dimensional complex geometries and topology, three-dimensional geometries, spatial and temporal reference systems, and visualization.

Because GML is based on XML-Schema, it allows users to create their own application schemas by making use of GML (core) schema components such as geometry, topology, and time, and follow the simple, structured rules of the GML encoding specification. GML application schemas are very flexible in that they allow users to tailor and extend predefined GML data types (mostly geometrical and topological) to specific needs in an application domain. GML also serves as data exchange format for geospatial data, an aspect that is particularly important to achieve a high degree of interoperability among geospatial data repositories through geospatial Web services.

10.2.3.2 Metadata Standards

There are many metadata frameworks for spatial data and applications that have geospatial components. Most of these frameworks and initiatives are driven by individual science communities. Metadata frameworks can be found at all data management levels, ranging from metadata associated with traditional database and GIS schemas to approaches where metadata is simply encoded as part of a file format containing the (geo)spatial data.

The most widely used geospatial metadata standard in GIS products is the standard developed and maintained by the Federal Geographic Data Committee (FGDC).[41] The FGDC developed the Content Standard for Digital Geospatial Metadata (CSDGM) in 1994, which is often simply referred to as the FGDC metadata standard.[42] This standard has components to describe the availability, fitness for use, and access and transfer information of geospatial datasets. According to the CSDGM version 2 published in 1998, Section 1 has entries to describe the geographical area a geospatial dataset covers; Section 3 describes the spatial data model that is used to encode the spatial data (vector/raster) or other possible methods for indirect georeferencing; and Section 4 describes the information about the spatial reference system.

In addition to the CSDGM, several other metadata standards have been developed over the past few years for different application domains in the geosciences and environmental sciences. For example, the Ecological Metadata Language (EML) developed by the National Center for Ecological Analysis and Synthesis (NCEAS) has been widely adopted in the ecological data management community.[43–45] EML has been designed as a collection of modules and has an extensible architecture. For the data module, EML has detailed structures to describe tabular, raster, and vector data. In EML, the metadata is much more tightly coupled with data, compared with that of the FGDC metadata standard. Such a coupling is an important aspect in metadata-driven data integration.[45] Another extensive data description framework for Earth science initiatives is the Semantic Web for Earth and Environmental Terminology (SWEET) developed by NASA.[46] SWEET is a standard vocabulary rather than a full-fledged metadata framework, and it includes a variety of data description components in the context of physical phenomena; processes; and properties, sensors, space, time, and units.

10.2.4 Approaches to Integrating Geospatial Data

Traditional data integration basically follows a schema-matching approach in which related schema components (relations and attributes) from the different sources are identified, homogenized, and suitably integrated to provide the user with a single conceptual view over the data managed at the sources.[6–9] Using such a view, the distributed data then can either be physically integrated at a single site or queried in a uniform and transparent fashion. The former approach then leads to some kind of data warehouse that physically stores the integrated data, but now in a homogeneous representation and format, leading to the *physical integration* of data. The latter approach, on the other hand, results in a federated or multidatabase system, realizing a so-called *logical integration*. Key to the integration is resolving the various types of structural and semantic heterogeneities that occur due to differences in data representation and meaning.[47]

For integrating geospatial data sources, the approach can be significantly more complicated, especially because there is a wider variety of geospatial data types (compared to just relational data), including various vector data representations and formats for field-based data, as discussed above. But, what is actually meant by *integrating geospatial data*? In practice, the most common view of this is to have a GIS that allows users to overlay different *themes* (or *layers*). That is, for a given geographic area, there are several georeferenced themes that represent different characteristics of that area. Figure 10.2 illustrates an example in which several themes based on vector and field-based data

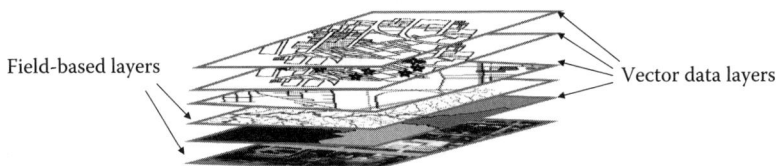

Figure 10.2 (See color insert following page 224.) Illustration of an overlay of themes in a GIS. Geo-referenced and aligned layers include both vector data and field-based data.

are overlayed. Theme overlays allow users to view and explore geographic data in different contexts. Being able to visualize data in context is an important functionality in integrating diverse types of geospatial data.

A theme can be represented by either vector data or field-based data. A road network with roads being individual features, for example, would be represented as vector data, whereas a vegetation index would be represented as field-based data (more specifically a raster image). If the data for the layers come from different sources, two problems can occur. First, the system used to integrate the data has to be *interoperable* with the other systems the geospatial data are retrieved from. Here, *interoperability* means that systems can exchange information and data using standard protocols and formats. As we will discuss below, a high degree of interoperability can be achieved when distributed and heterogeneous geospatial data sources can be uniformly accessed using geospatial Web services.

Second, the geospatial data may come in different formats with conflicting structures and semantics. For example, if two sources provide vector data for the same theme and region, the data might conflict in terms of their spatial components as well as their descriptive components (see Section 10.2.1). Such a situation can even occur if both datasets are based on the same projection (spatial reference system), have been georeferenced/aligned, and have the same scale (spatial resolution). Re-projection, georeferencing, and scaling are tasks that are frequently used in the context of remotely sensed imagery and are typically performed on the datasets prior to their overlay or integration. Another typical example often occurring in practice is when some raster imagery is overlayed with vector data. Phenomena in the image might not align or match up with the features modeled by the vector data. Approaches to resolving these types of conflicts are known as *conflation*, meaning to "replace two or more versions of the same information with a single version that reflects the pooling, weighted averaging, of the sources."[21]

The key in dealing with conflicting spatial components of two or more datasets to be integrated is to make use of the location information associated with geospatial objects (and cells/pixels in a raster image), something unique to spatial datasets. Several approaches have been proposed that deal with the integration of vector data and road maps in particular and the combination of imagery with vector data.[48–50] More fine-grained approaches have been developed for finding corresponding objects in datasets to be integrated. Corresponding objects (features) represent the same real-world entity but are possibly misaligned across different data sources. Approaches for point-based data have been presented by Beeri et al.[51,52] referred to as *location-based join*. Related approaches are so-called entity resolution techniques, which try to determine the true location of a real-world entity in case geospatial data about the entity comes from a collection of data sources.[53]

For resolving spatial conflicts, that is, if the same real-world entity has conflicting feature location information in the different sources, nonspatial attributes associated with the features can help in resolving such conflicts. For example, if it is known that a feature has been updated recently at a source, the feature at this source might be more likely to represent the correct location information about the real-world entity. As with any other data integration approach, the quality of the data plays a crucial role in resolving individual spatial and nonspatial data conflicts.[54]

Once different features have been matched to the same real-world entity, the next step is to resolve conflicts that might exist among the descriptive attributes. As these are ordinary attributes such as in relational databases, respective approaches can be used. In the context of geospatial data such attributes are typically based on metadata standards and application schemas described in Section 10.2.3, which are likely to produce a more coherent data description in terms of semantics.

10.3 Service-Based Data and Application Integration

In the following, we present emerging standards, techniques, and architectures that enable interoperability among distributed and heterogeneous geospatial data sources. In Section 10.3.1, we outline the relationships between interoperability and data integration aspects. An overall framework for data integration employed by the techniques presented in this chapter is the service-oriented architecture, which is described in Section 10.3.2. In Sections 10.3.3 and 10.3.4, we give an overview of service registry and geospatial Web services, respectively. We place a particular focus on services that deal with real-time sensor data, described in Section 10.3.5. We conclude the section with a brief overview of a practically relevant alternative to geospatial Web services.

10.3.1 Approaching Integration through Interoperability

Interoperability among heterogeneous and distributed data sources is a fundamental requirement not only in the context of scientific data management, but in any type of distributed computing infrastructure. *Interoperability* is generally defined as "the ability of two or more systems or components to exchange information and to use the information that has been exchanged."[55] Interoperability can be achieved at different levels of network protocols and data exchange formats. In several scientific application domains, interoperability among data repositories and applications has become a main driver to facilitate scientific data management and exploration on a large scale. Grid computing infrastructures have significantly contributed to this development[56] and are widely employed in science domains, such as in Earth observation,[57] climate modeling,[58] and physics,[59] to name only a few. A more recent trend in these science initiatives is to increase interoperability aspects through *service-oriented science*.[60] A well-known early example that realizes such an approach is the WorldWide Telescope.[61]

One major driver in the area of geoprocessing and geospatial data management technologies is the OGC Interoperability Institute,[62] where the OGC is also developing and promoting diverse types of geospatial Web services. Such type of Web services play an increasing role in geospatial data integration frameworks. Services do not just provide easy access to diverse types of geospatial data using standard protocols and interfaces, but they also often offer functionality that helps in resolving data conflicts. For example, requesting data in a particular projection or at a particular scale are important data preprocessing steps that already can be accomplished by services rather than at the data integration site. In this sense, such services provide some application functionality too. There are a few geospatial Web services approaches that address both interoperability and integration aspects, for example, based on mediation,[63] services,[64] or a combination of service and mediation-based techniques.[65–67] In the following, we describe how integration infrastructures can be built based on such services and architectures.

10.3.2 Service-Oriented Architectures

In the past few years, there have been significant developments in terms of architectures and standards that help developers build Web-based services that allow for a uniform and transparent access to data managed at different sources. One such development is the service-oriented architecture (SOA),[68] which allows an effective cooperation among data sources and data processing components hosted at different organizational units. In particular, SOA supports reusability and interoperability of software and service components on the Web, thus increasing the efficiency of developing and composing new services. In an SOA-based system, all data and process components are modeled as Web services.

10.3.3 Registry and Catalog Services

Of particular interest in this chapter are catalog and registry services. As scientific data are accumulating at an ever-increasing speed, it is very difficult if not impossible for users to know exactly the details of all the data that might be relevant to their project. As such, repositories that provide catalog services and allow users to interactively or programmatically search and retrieve metadata that are related to the use of the datasets are playing an inreasingly important role in scientific data management. In the context of geospatial applications, OGC's Catalog Service for the Web (CSW) Implementation Standard[69] provides this functionality in the form of several operations: the mandatory GetCapabilities operation returns metadata about the specific repository server (ServiceIdentification), the operations supported by the service including the URL(s) for operation requests (OperationMetadata), the type of resource cataloged by the repository server (Content), and the query language and its functionality supported by the repository. The GetRecords operation allows users to specify query constraints and metadata to be retrieved and returns the number of items in the result set and/or selected metadata for the result set. The DescribeRecord operation allows a client to discover elements of the information model supported by the target catalog service. The optional GetDomain operation is used to obtain runtime information about the range of values of a metadata record element. Finally, the mandatory GetRecordByID request retrieves the default representation of catalog records using their identifier. Through the GetCapabilities → GetRecords → DescribeRecord → GetDomain → GetRecordByID sequence, users are able to probe the repository server's capabilities, search the repository, negotiate the format of the metadata and finally retrieve the metadata of the dataset(s) of interest.

10.3.4 Geospatial Web-Services and Standards

The OGC was founded with the mission of advancing the "development of international standards for geospatial interoperability."[15] The OGC currently comprises, at time of writing, over 350 companies, universities, and government agencies from around the world. In the Earth sciences in particular, the role of standard data and interface protocols is crucial in the context of climate monitoring and forecasting. The National Weather Service,[70] for example, has recently started to make forecast data available to users using Web Feature Service (WFS) and Geography Markup Language (GML), two of the open standards developed by OGC.

In this Web service framework, the concepts of coverage, feature, and layer play a key role in publishing and accessing diverse types of geospatial datasets through OGC Web services. Both coverage and feature provide associations between observed or measured values with a geographical domain, such as a particular region or spatial extent (see also Section 10.2.4). A coverage can be thought of as a measurement that varies over space, while a feature is a spatial object that has associated measurements. In the context of remotely-sensed data, for example, a satellite image covering an area can be represented as a coverage. On the other hand, the observation values of a (point-based) weather station can be represented as a feature. A layer, which basically corresponds to the concept of a theme, can be either a gridded coverage or a collection of similar features.

The ability to map heterogeneous forms of geospatial datasets to a few simple types (such as features, coverages, and layers) greatly reduces the complexity of diverse data types in application and data integration scenarios in particular, and it makes it possible to standardize publishing datasets using Web services. Although it is beyond the scope of this chapter to give a detailed technical description of OGC Web standards, Figure 10.3 shows the

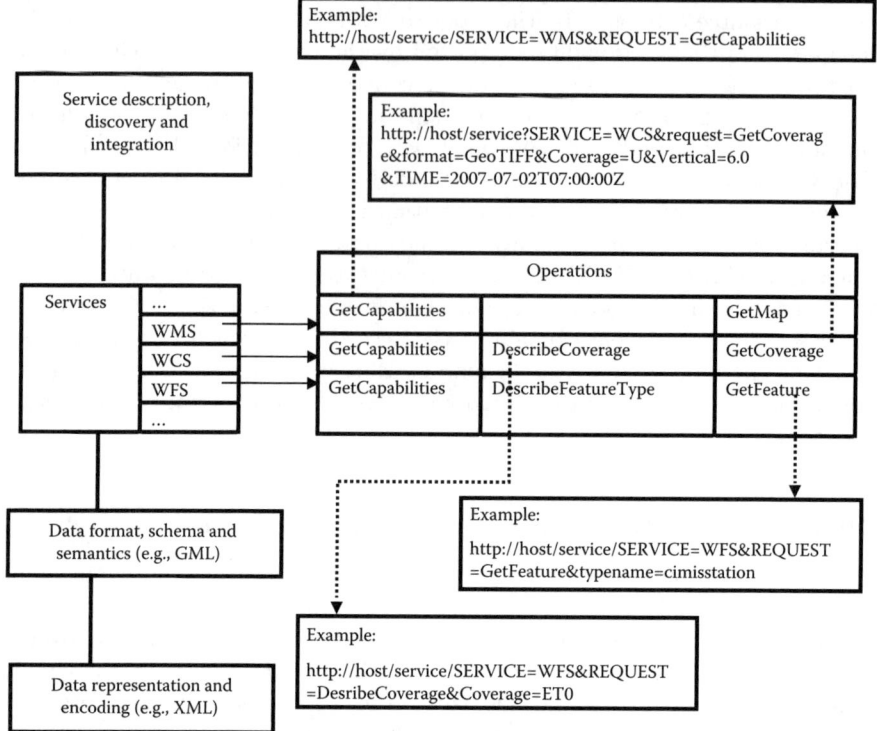

Figure 10.3 OGC Services, Operations, and Example Calls (indicated by dotted lines) for Web Map Service (WMS), Web Coverage Service (WCS), and Web Feature Service (WFS).

three major OGC standard services along with example operations. These services, which cover the two different types of geospatial data (features and coverages) and their visualization, are as follows:

- *Web Feature Service (WFS)*: WFS defines interfaces for querying and retrieving features based on spatial and nonspatial properties of the features. The data is exchanged between a Web Feature Service and the client in the form of GML documents, which in the case of this service encode vector data.
- *Web Coverage Service (WCS)*: WCS defines interface to query and retrieve spatially referenced coverages, i.e., gridded or raster data.
- *Web Map Service (WMS)*: This service produces maps (in the form of digital images) of spatially referenced data (i.e., features or coverages) from a data source managing geographic information. Standard image formats that can be requested by a client include PNG, GIF, GeoTIFF, and JPEG.

Interoperability and Data Integration in the Geosciences 383

In summary, WFS is used for object-based data, and WCS is used for gridded/raster data. Displaying such data and their overlays is done using WMS. To illustrate the functionality of these OGC services, consider the WCS standard as an example. Like the other two services, it defines a mandatory GetCapabilities operation, which allows clients to get WCS server metadata, including an optional list of the offered coverages with some metadata for each coverage. In addition, WCS also defines a mandatory DescribeCoverage operation that allows clients to get more metadata about identified grid coverage(s), including details about the spatial extent of the coverage. A WCS GetCoverage operation requests and returns coverages representing space-time varying phenomena. In general, through a sequence of GetCapabilities → GetMap (WMS), GetCapabilities → DescribeCoverage → GetCoverage (WCS), and GetCapabilities → DescribeFeatureType → GetFeature (WFS), client applications are able to retrieve both metadata and data subsets of interests in a standard way.

The realization of the above services typically occurs in the form of middleware layers that clients can access through the Web. Among the most prominent representatives of such middleware layers are the open source systems GeoServer[71] and MapServer.[72] Either system provides a client with transparent access (using the above OGC services) to diverse types of data stores. That is, these servers can be configured to access geographic data managed in, for example, PostGIS, Shapefiles, or Oracle Spatial, and to provide clients with access to the data through WFS, WMS, and WCS interfaces. In this sense, such a type of middleware layer already realizes an important component to data integration scenarios, namely the transparent and uniform access to the diverse geospatial data sources. For example, using the services, one can request data in a particular (common) coordinate system. Thus, the services help in resolving some data heterogeneity issues.

10.3.5 Sensor Web Enablement

Of particular relevance are the activities recently taken by the OGC Sensor Web Enablement (SWE) program, one of the OGC Web Services initiatives.[73,74] The SWE initiative seeks to provide interoperability between disparate sensors and sensor processing systems by establishing a set of standard protocols to enable a "Sensor Web," by which sensors of all types in the Web are discoverable, accessible, and taskable. The SWE standards allow the determination of the capabilities and quality of measurements from sensors, the retrieval of real-time observations in standard data formats, the specification of tasks to obtain observations of interest, and the asynchronous notification of events and alerts from remote sensors.

SWE components include models and XML Schemas (SensorML, Observations & Measurements, TransducerML) and Web service interfaces (SOS, SPS, SAS, WNS), which are briefly described as follows (see References 73 and 74 for more details):

- SensorML, *Sensor Model Language*: An XML Schema to describe sensors and sensor platforms. SensorML provides a functional description of detectors, actuators, filters, operators, and other sensor systems.
- O&M, *Observations & Measurements*: A specification for encoding observations and measurements from sensors.
- TransducerML, *Transducer Markup Language*: A specification that supports real-time streaming of data to and from transducers and other sensor systems. Besides being used to describe the hardware response characteristics of transducers, TransducerML provides a method for transporting sensor data.
- SOS, *Sensor Observation Service*: This Web service interface is used to request and retrieve metadata information about sensor systems as well as observation data.
- SPS, *Sensor Planning Service*: Using this Web interface, users can control taskable sensor systems and define tasks for the collection of observations and the scheduling of requests.
- SAS, *Sensor Alert Service*: Through this Web service interface, users are able to publish and subscribe to alerts from sensors.
- WNS, *Web Notification Service*. This Web service interface allows the asynchronous interchange of messages between a client and one or more services (e.g., SAS and SPS).

In accordance with the philosophy of Web services in general, and the SWE initiative in particular, data consumers should be concerned only with registries and service interfaces. For example, an SOS provider needs to be "discovered" first through a registry mechanism, which is the OGC Catalog Services (see Section 10.3.3) in the SWE context. Section 10.4 describes specific elements from SensorML, O&M, and SOS that have been included in a prototype to chain data stream processing services. The general sequence of steps to obtain sensor metadata and data is shown in Figure 10.4.

As an SOS service, the provider first provides a *capabilities* document as a response to a GetCapabilities request by a client. This document includes the identification of the provider and the description of the offered services, that is, the available streams in the system, which are organized in the form of *observation offerings*. An offering includes information about the period of time for which observations can be requested, the phenomena being sensed, and the geographic region covered by the observations. A schematic example of a capabilities document is shown in Figure 10.5. (We use a simplified structure style to illustrate XML documents in this section; we use ◇ symbols to indicate relevant XML elements, and example values are shown in cursive.)

Once a client is interested in a particular geospatial data stream, it will submit a DescribeSensor request to the provider. The response is a document describing the sensor that generates the data stream. This response takes the

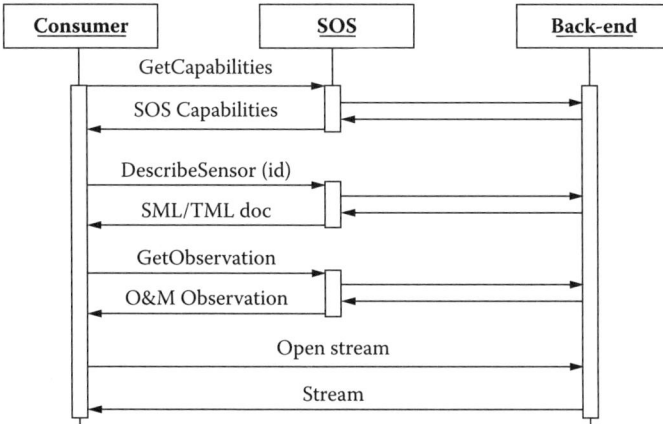

Figure 10.4 SOS sequence diagram for getting metadata and observations, including access to streaming data.

form of SensorML or a TransducerML document. Next, the client will request the actual data from the sensor. This is done by submitting a GetObservation request. The corresponding response is an O&M document containing the observation, either explicitly (inlined in the document), or by providing a hyperlink to the actual data. This is illustrated in the lower part of Figure 10.4

```
◊ sos:Capabilities
    ◊ ServiceIdentification
        ◊ Title, ◊ Abstract, ◊ Keywords
    ◊ ServiceProvider
        ◊ Name, ◊ ContactInfo, ◊ Site
    ◊ Operations
        GetCapabilities, DescribeSensor, GetObservation
    ◊ FilterCapabilities
        ◊ Spatial, ◊ Temporal, ◊ Scalar
    ◊ ObservationOfferings
        visible radiance, near-infrared, mid-infrared
        ◊ boundedBy
            ◊ Envelope: field of view of GOES Imager
        ◊ TimePosition
            ◊ beginPosition: 2003-10-10
            ◊ endPosition: indeterminatePosition
        ◊ Result format: text/xml;subtype="OM"
```

Figure 10.5 Schematic example of an SOS capabilities document with observation offerings exemplified with typical spectral sensors from an environmental satellite (NOAA's GOES satellite).

where the client requests a connection to the data stream directly to the back-end system. We use this mechanism for allowing the access to real-time data in an application scenario, detailed in Section 10.4.

10.3.6 OPeNDAP

We conclude this section by giving a brief overview of a framework that also provides services, mostly in the form of protocols, to manage and access scientific data. While OGC standard components are designed to handle georeferenced data, the Open Source Project for Network Data Access Protocol (OPeNDAP) standards describe the management of multidimensional array data that are not necessarily georeferenced.[75] OPeNDAP includes specifications for encapsulating structured data, annotating the data with attributes and adding semantics that describe the data. In addition to the Distributed Oceanographic Data System (DODS) protocol that allows users to transparently access distributed data across the Internet in a way similar to the Get-Coverage operation in the OGC WCS standard, OPeNDAP has protocols for exchanging metadata. More specifically, the dataset attribute structure (DAS) is used to store attributes for variables in the dataset. The dataset description structure (DDS) is a textual description of the variables and their classes that make up a scientific dataset.

Existing implementations based on OPeNDAP standards provide a convenient framework for retrieving multidimensional scientific data using simple HTTP-GET requests and are widely used by governmental organizations such as NASA and NOAA to serve satellite, weather, and other Earth science data.[76,77] Since there are no coordinate referencing systems involved in OPeNDAP standards, they are best used for datasets with a common underlying coordinate system. However, OPeNDAP services may not be sufficient when datasets with different coordinate systems need to be integrated. In such cases, OGC-based services are more suitable. OGC standards and OPeNDAP standards are not necessarily exclusive. By enhancing multi-dimensional arrays data with proper coordinate systems, it is possible to construct coverages and serve the data using OGC WCS and WMS standards.

10.4 An Example of an Integration and Interoperability Scenario

The concept of service-based geospatial data integration can be demonstrated in many environmental monitoring scenarios. In this section, we consider a particular environmental scenario involving several integration aspects and describe the features an integration framework should in general provide to support such kinds of scenarios.

10.4.1 Environmental Modeling Task—Evapotranspiration

We elaborate on a particular use case involving the integration of real-time evapotranspiration observations and its comparison with estimations from a weather model for accuracy assessment. Evapotranspiration (or ET) is a term used to describe the sum of evaporation and plant transpiration from the Earth's land surface to the atmosphere, which is an important part of the water cycle. ET estimations are used in irrigation scheduling, watershed management, weather forecasting, and the study of long-term effects of land use change and global climate change.[78] A standard reference evapotranspiration, denoted ETo, can be determined by using meteorological measurements, which can be obtained from multiple sources for the same region. Compared with station-observed ETo (point data), the weather model-based ETo (raster data) has a continuous spatial coverage. In general, the model output accuracy needs to be verified against the station observations. Assuming the observed and the predicted data are published as WFS and WCS services, respectively, data integration is needed to retrieve the model predicted data at the station locations and compare the values, possibly after normalization and unit conversions.

The scientific goal in this scenario is the visualization, monitoring, and validation of model-based evapotranspiration for different eco-regions and selected locations in California. This requires the overlay of ETo from the various sources on a single display for visual analysis. We show next a geospatial integration approach that allows the realization of this scenario.

10.4.2 Integration Platform

The overall conceptual architecture for geospatial data integration and interoperability, which consists of data sources, structured repositories, service middleware systems, and client applications, is depicted in Figure 10.6. Data sources include remotely sensed imagery, model outputs, GIS stores, sensor network systems, as well as other service-enabled data providers. The structured data repositories in general refer to data management systems including databases and data stream engines. Service middleware refers to service enabling infrastructures that make the data available to clients by means of Web service standards. Client applications allow users to search, query, and retrieve metadata and data by using the provided Web services (CSW, WFS, SOS, etc.), possibly in combination with more traditional access mechanisms (HTTP, FTP, etc.) with the back-end data repositories.

In general, a service-oriented scientific data integration framework consists of a set of interconnected service-enabled computation nodes. A service enabled computation node consists of both structured data repositories and service middleware. Structured data repositories, such as DBMS, GIS data stores, and data stream engines, store relational data, vector/raster geospatial data, and real-time data streams. They can be connected to physical devices to

388 Scientific Data Management

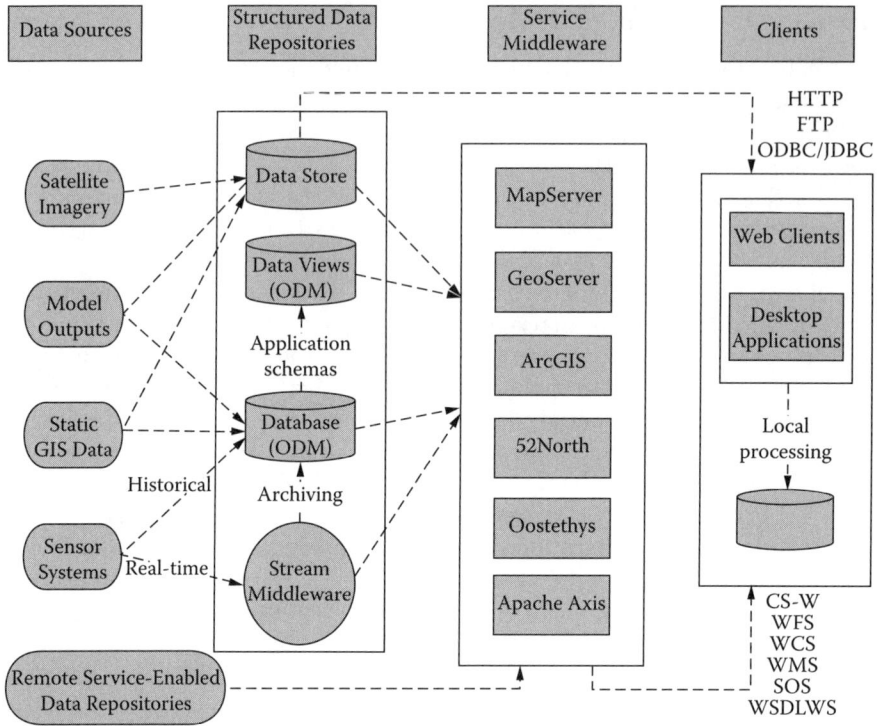

Figure 10.6 Conceptual architecture for geospatial data integration and interoperability.

receive or pull streaming data. Structured data repositories are not necessarily independent; derived products from raw data can be generated and saved as additional data repositories. For example, the daily maximum/average ETo value of an hourly model output (in raster format) can be derived and saved in a GIS. Also monthly ETo average from weather stations can be implemented as either a regular view (named database query) or a materialized view (a physical instantiation of the query result) in a database. In addition, a stream engine (e.g., Ring Buffer Network Bus, RBN,[79]) whose primary functionality is to serve as a middleware for real-time data, can be used by certain client modules to update a database for archiving purposes. Middleware at the service level is responsible for extracting data from structured data repositories and providing it to clients in standard-compliant formats. Several commercial and open source service middleware systems, such as MapServer,[72] GeoServer,[71] THREDDS Data Server (TDS),[80] 52North,[81] and so forth. are currently available. As such, structured data repositories should be formulated to work with the service middleware when possible. For example, PostgreSQL databases storing weather station measurement data can adopt the Community

Observations Data Model (ODM) developed by CUAHSI[82] in formulating their table structures.

We note that the relationship between the structured data repositories, the service middleware, and their services are not necessary one-to-one. For example, both MapServer and GeoServer can connect to the PostgreSQL databases and provide WMS/WFS/WCS services. Similarly, TDS can provide WCS and OPeNDAP services from NetCDF data repositories. We also note that while Web service–based protocols are preferred for geospatial data integration in a Web application environment, more traditional communication protocols among the service-enabled computation nodes, such as distributed databases over TCP/IP, are not precluded. In addition, quite a few service middleware systems have the capability of connecting to remote data repositories either through database interfaces or service interfaces (WFS, WMS, WCS, SOS); thus they can integrate different data sources that are stored in local or remote data repositories and provide new services. While normally such a type of integration is simple and limited in functionality, it could be appropriate in some applications or be used as parts of larger integration tasks. For example, MapServer can be configured to consume remote WFS services and use them along with local data in formulating the structure of a new WMS service.

In the next section, we will illustrate the incorporation of the various components of our scenario including the integration of real-time data.

10.4.3 Overall Integration and Results

As indicated at the beginning of this section, the goal in our scenario is the visualization, monitoring, and validation of model-based evapotranspiration for different eco-regions and selected locations in California. This goal is accomplished as follows. Using the integration tool, the user selects some weather stations to perform a comparison of ETo estimations against observed measurements at the given locations on an hourly basis (see Figure 10.7). The integration platform sends WFS requests to the relevant service endpoints to retrieve the station locations as well as the eco-region data, which are returned as vector features. It then sends a WCS request to the weather model output repository and retrieves the model output for the current time in NetCDF format. In general, the integration platform has to re-project the station geographical coordinates (usually given in latitude/longitude) to the model output projection (which is an equal-area based projection in the case of the WRF model[83]) before the corresponding station locations in the model output grids can be retrieved. The integration platform then loops through the time steps (hours in the figure) to retrieve the ETo predictions at the station locations. The retrieved data can be rendered as charts by the integration platform for all desired stations for visualization purposes.

In real-time, the generated chart for each selected location also includes the observations from ground stations (from the CIMIS network[84] in our

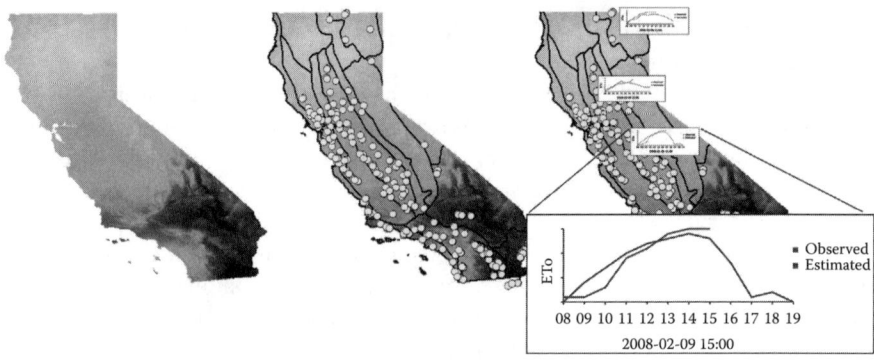

Figure 10.7 (See color insert following page 224.) Main stages in the integration of sensor and model-generated data streams. Left: Estimated ETo map for the current hour; Center: Eco-regions and station locations overlayed; Right: Real-time charts for selected locations including the model prediction for a several-hour period and the actual observed values until current time.

example). For this component, we use sensor web enablement (SWE)-related technologies. As already indicated in Section 10.2, both static and real-time sensor data can be provided through the Sensor Observation Service (SOS) interface; however, here we focus on representative sensor definitions and possible real-time access mechanisms. Following the SOS interface, a provider generates a capabilities document as shown in Figure 10.5. A DescribeSensor request for a particular detector produces a SensorML document describing the instrument that generates the stream. Figure 10.8 is a schematic depiction of part of a SensorML document for ETo measurements from a weather station system.

The response to a GetObservation request is an O&M document including a hyperlink that allows the client to open a connection to the data stream. An example of an O&M observation document is shown in Figure 10.9. A realization architecture would utilize a middleware data stream system as the

```
◇ sml:SensorML
    ◇ sml:Sensor: id = CIMIS_S33_ETO
    ◇ name = CIMIS STATION 33 ETO
        ◇ Identification, ◇ Classification
        ◇ ReferenceFrame
        ◇ Input: name = eto
            ◇ Quantity: urn:ogc:def:phenomenon:eto
        ◇ Output: name = eto
            ◇ Quantity: urn:ogc:def:phenomenon:eto
```

Figure 10.8 Schematic SensorML document describing measured ETo.

> ⋄ om:Observation: id = *CIMIS_S33_ETO*
> ⋄ Description: *Observation with remote streaming result*
> ⋄ name = *CIMIS STATION 33 ETO*
> ⋄ TimePeriod
>   ⋄ beginPosition: *2008-02-10T12:00:00:00*
>   ⋄ endPosition: *future*
> ⋄ Result
>   xlink:href= *"http://comet.ucdavis.edu:9090/CIMIS/?ch=/S33/eto"*
>   xlink:role= *"application/octect-stream"*

Figure 10.9 Example of an O&M observation response.

entry point for all incoming stream data sources. This intermediate component would allow the implementation of various possible connections. An RBNB system[79] can be used as the entry point. In this case, the hyperlink shown in Figure 10.9 points directly to the corresponding channel in the RBNB server. However, various other types of connections are possible, including a TransducerML (TML) stream as a wrapper for the original, native stream, a TML stream as a wrapper for the RBNB stream, and the RBNB stream directly. The capabilities document would advertise the supported connection types so a client application can choose the one it is able to use.

In summary, as shown in Figure 10.7, right, both the prediction and the real-time ETo values can be displayed in chart form next to the respective station locations, and showed in the context of eco-regions (vector data) and current ETo maps (raster data), thus providing users and scientists with a vivid interface to monitor ETo values and easily compare them with weather model prediction over time.

It finally should be noted that the integration platform outlined above provides transparent and uniform access to heterogeneous geospatial data sources. However, in general, certain integration tasks such as resolving structural and semantic heterogeneities (see Section 10.2.4) still need to be explicitly realized at the client side and integration platform. These tasks include matching vector-based objects from two or more sources, selecting respective nonspatial attributes, and resolving general conflation aspects among data to be integrated. A viable approach to support such tasks is through scientific workflows (see Chapter 13), where the logic of conflict-resolving techniques is implemented in the form of actors.

10.5 Conclusions

With the amount of geospatial data growing at unprecedented rates, its effective sharing, exchange, and integration becomes a more critical necessity than ever before. We have seen that this goal involves not only dealing with various types of scientific data, but also integrating the increasing number of data and value-added services that are being deployed by geospatial communities in several important scientific application domains. The primary objective in this context is indeed a high degree of interoperability as a prerequisite for effective data integration and uniform and transparent data access.

In this chapter, we have reviewed emerging data integration requirements particularly in the context of the geosciences, where advancements in sensor and network technologies are placing an immense amount of diverse data at the scientist's disposal. We reviewed integration concepts from basic notions like data formats and metadata standards, to more comprehensive approaches including standards for interoperability and supporting Web-based technologies. We paid special attention to current efforts in the context of geospatial sensor data streams, amply exemplified with the enormous deal of data generated by air and space-borne instruments as well as numerous oceanic and ground sensor networks. With a practical environmental scenario, we illustrated an approach for integration and interoperability involving several of the components discussed in the chapter, which is in fact being developed in the context of the COMET Project.[85] Related projects, such as the Geoscience Network (GEON)[86] and the Science Environment for Ecological Knowledge (SEEK),[87] have also made great progress in building service-oriented architectures and portals that facilitate the efficient access to and integration of diverse geospatial datasets and repositories.

We have illustrated how current technologies, characterized by concerted efforts in standardization, are making the interoperability goal not only better defined but also effectively realizable in critical scientific application scenarios. Although much is still to be accomplished, especially in terms of the specification of ontologies in several areas of the geosciences, the science community can already take advantage of currently available infrastructures and technologies, and start benefiting from the progress underway.

Acknowledgements

This work is in part supported by the National Science Foundation under Awards No. IIS-0326517 and ATM-0619139.

References

[1] A. Szalay and J. Gray. 2020 computing: Science in an exponential world. *Nature*, (440):413–414, 2006.

[2] Sloan Digital Sky Survey (SDSS). http://www.sdss.org/. Accessed on July 10, 2009.

[3] Entrez. http://www.ncbi.nlm.gov/sites/gquery. Accessed on July 10, 2009.

[4] K. A. Brandt. The GDB Human Genome Database: A source of integrated genetic mapping and disease data. *Bulletin of the Medical Library Association*, 81(3), July 1993. http//www.pubmedcentral.nih.gov/articlerender.fcgi?artid=225791. Accessed on October 10th, 2009.

[5] Global Biodiversity Information Facility. http://www.gbif.org/. Accessed on July 10, 2009.

[6] O. A. Bukhres and A. K. Elmagarmid. *Object-Oriented Multidatabase Systems*. Englewood Cliffs, New Jersey: Prentice Hall, 1996.

[7] A. K. Elmagarmid, M. Rusinkiewicz, and A. Sheth. *Management of Heterogeneous & Autonomous Database Systems*. San Francisco: Morgan Kaufman, 1999.

[8] A. P. Sheth and J. A. Larson. Federated database systems for managing distributed, heterogeneous, and autonomous databases. *ACM Computing Surveys*, 22(3):183–236, 1990.

[9] P. Ziegler and K. R. Dittrich. Three decades of data integration—all problems solved? In *Building the Information Society, IFIP 18th World Computer Congress, Topical Sessions*, 3–12, 2004.

[10] A. Y. Halevy, N. Ashish, D. Bitton, M. J. Carey, D. Draper, J. Pollock, A. Rosenthal, and V. Sikka. Enterprise information integration: successes, challenges and controversies. In *Proceedings of the ACM SIGMOD Int. Conference on Management of Data*, 778–787, 2005.

[11] J. J. Bates. Exploratory climate analysis tools for environmental satellite and weather radar data (invited talk). In *Workshop on Management and Processing of Data Streams (MPDS 2003)*, 2003.

[12] Global Earth Observation System of Systems. http://www.epa.gov/geoss/. Accessed on July 10, 2009.

[13] National Spatial Data Infrastructure (NSDI). http://www.fgdc.gov/nsdi/nsdi.html. Accessed on July 10, 2009.

[14] National Ecological Observation Network (NEON). http://www.neoninc.org/. Accessed on July 10, 2009.

[15] The Open Geospatial Consortium (OGC). http://www.opengeospatial.org. Accessed on July 10, 2009.

[16] P. Rigaux, M. Scholl, and A. Voisard. *Spatial Databases with Application to GIS*. Morgan Kaufmann, 2002.

[17] S. Shekhar and S. Chawla. *Spatial Databases: A Tour*. Upper Saddle River, New Jersey: Prentice Hall, June 2002.

[18] J. E. Stern. State Plan Coordinate System of 1983. NOAA Manual NOS NGS 5, Rockville, Maryland, March 1990.

[19] J. C. Iliffe. *Datums and Map Projections: For Remote Sensing, GIS and Surveying.* Whittles Publishing, 2000.

[20] Q. Yang, J. Snyder, and W. Toble. *Map Projection Transformation: Principles and Applications.* CRC, 1999.

[21] P. A. Longley, M. F. Goodchild, D. J. Maguire, and D. W. Rhind. *Geographic Information Systems and Science.* Chichester, Sussex: Wiley, 2005.

[22] ESRI ArcInfo Product Description. http://www.esri.com/software/arcgis/arcinfo. Accessed on July 10, 2009.

[23] T. Ormsby, E. Napoleon, and R. Burke. *Getting to Know ArcGIS Desktop: The Basics of ArcView, ArcEditor, and ArcInfo.* Redlands, California: Esri Press, 2004.

[24] ESRI. *ArcGIS 9.* ESRI Press.

[25] C. Murray. *Oracle Spatial Developer's Guide, 11g Release 1*, 2007.

[26] K. Stolze. SQL/MM Spatial—The Standard to Manage Spatial Data in a Relational Database System. In *Tagungsband der 10. BTW-Konferenz*, LNI 26, 247–264, 2003.

[27] IBM. *DB2Spatial Extender and Geodetic Data Management Feature User's Guide and Reference.* IBM Corp., 2006.

[28] PostGIS. http://postgis.refractions.net/. Accessed on July 10, 2009.

[29] Geographic Resources Analysis Support System (GRASS). http://grass.osgeo.org/. Accessed on July 10, 2009.

[30] M. Neteler and H. Mitasova. *Open Source GIS: A GRASS GIS Approach.* New York: Springer, 2007.

[31] OGC. OpenGIS® Implementation Specification for Geographic information—Simple feature access—Part 1: Common architecture. version 1.2.0. http://www.opengeospatial.org/standards/sfa. Accessed on July 10, 2009.

[32] PROJ.4—Cartographic Projections Library. http://trac.osgeo.org/proj.

[33] GDAL—Geospatial Data Abstraction Library. http://www.gdal.org/. Accessed on July 10, 2009.

[34] ESRI Shapefile Technical Description, White Paper. Technical report, Environmental Systems Research Institute, Inc., 1998.

[35] U.S. Census Bureau. Topologically Integrated Geographic Encoding and Referencing system (TIGER). http://www.census.gov/geo/www/tiger/. Accessed on July 10, 2009.

[36] Network Common Data Form (NetCDF). http://www.unidata.ucar.edu/software/netcdf/. Accessed on July 10, 2009.

[37] Hierachical Data Format (HDF5). http://www.hdfgroup.org/HDF5/. Accessed on July 10, 2009.

[38] GeoTIFF. http://trac.osgeo.org/geotiff. Accessed on July 10, 2009.

[39] R. Lake, D. Burggraf, M. Trninic, and L. Rae. *Geography Mark-Up Language: Foundation for the Geo-Web*. Wiley, 2004.

[40] OGC. OpenGIS® Geography Markup Language (GML) Encoding Specification. version 3.1.1. http://www.opengeospatial.org/standards/gml. Accessed on July 10, 2009.

[41] Federal Geographic Data Commitee (FGDC). http://www.fgdc.gov. Accessed on July 10, 2009.

[42] FGDC. Content Standard for Digital Geospatial Metadata (CSDGM), FGDC-STD-001-1998. http://www.fgdc.gov/metadata/csdgm. Accessed July 10, 2009.

[43] A. M. Ellison, L. J. Osterweil, L. Clarke, J. L. Hadley, A. Wise, E. Boose, D. R. Foster, A. Hanson, D. Jensen, P. Kuzeja, E. Riseman, and H. Schultz. Analytic webs support the synthesis of ecological datasets. *Ecology*, 87(6):1345–1358, 2006.

[44] Ecological Metadata Language (EML). http://knb.ecoinformatics.org/software/eml/. Accessed on July 10, 2009.

[45] M. B. Jones, M. P. Schildhauer, O. Reichman, and S. Bowers. The new bioinformatics: Integrating ecological data from the gene to the biosphere. *Annual Review of Ecology Evolution and Systematics*, 37:519–544, 2006.

[46] Semantic Web for Earth and Environmental Terminology (SWEET). http://sweet.jpl.nasa.gov/. Accessed on July 10, 2009.

[47] A. P. Sheth. Changing focus on interoperability in information systems: From system, syntax, structure to semantics. In *Interoperating Geographic Information Systems*, 5–30. Norwalk, Massachusetts: Kluwer, 1999.

[48] C.-C. Chen, C. A. Knoblock, C. Shahabi. Automatically conflating road vector data with orthoimagery. *GeoInformatica*, 10(4):495–530, 2006.

[49] E. Safra, Y. Kanza, Y. Sagiv, and Y. Doytsher. Efficient integration of road maps. In *14th Int. Symp. on Geographic Information Systems*, 59–66, 2006.

[50] J. M. Ware and C. B. Jones. Matching and aligning features in overlayed coverages. In *Proc. 6th Int. Symp. on Advances in Geographic Information Systems*, 28–33, 1998.

[51] C. Beeri, Y. Doytsher, Y. Kanza, E. Safra, and Y. Sagiv. Finding corresponding objects when integrating several geo-spatial datasets. In *Proc. 13th Int. Symp. on Geographic Information Systems*, 87–96, 2005.

[52] C. Beeri, Y. Kanza, E. Safra, and Y. Sagiv. Object fusion in geographic information systems. In *Proc. 30th Int. Conference on Very Large Data Bases*, 816–827, 2004.

[53] V. Sehgal, L. Getoor, and P. Viechnicki. Entity resolution in geospatial data integration. In *14th Int. Symp. on Geographic Information Systems*, 83–90, 2006.

[54] W. Shi, P. Fisher, and M. F. Goodchild. *Spatial Data Quality*. Boca Raton, Florida: CRC Press, 2002.

[55] IEEE standard computer dictionary: A compilation of IEEE standard computer glossaries, 1990.

[56] I. Foster and C. Kesselman. *The Grid 2: Blueprint for a New Computing Infrastructure (2nd Edition)*. Morgan Kaufmann, 2004.

[57] Global Earth Observation Grid. http://www.geogrid.org/. Accessed on July 10, 2009.

[58] Earth System Grid. http://www.earthsystemgrid.org/. Accessed on July 10, 2009.

[59] GriPhyN. Grid Physics Network. http://www.griphyn.org/. Accessed on July 10, 2009.

[60] I. Foster. Service-oriented science: Scaling e-science impact. In *Proceedings of the 2006 IEEE/WIC/ACM International Conference on Intelligent Agent Technology*, 9–10, 2006.

[61] J. Gray and A. Szalay. The world-wide telescope, an archetype for online science. Technical report msr-tr-2002-75, Microsoft Research, 2002.

[62] OGC Interoperability Institute. http://www.ogcii.org/. Accessed on July 10, 2009.

[63] G. da Rocha Barreto Pinto, S. P. J. Medeiros, J. M. de Souza, J. C. M. Strauch, and C. R. F. Marques. Spatial data integration in a collaborative design framework. *Commun. ACM*, 46(3):86–90, 2003.

[64] J. Lee, Y. Lee, S. Shah, and J. Geller. HIS-KCWATER: context-aware geospatial data and service integration. In *Proceedings of the 2007 ACM Symposium on Applied Computing (SAC)*, 24–29, 2007.

[65] O. Boucelma, M. Essid, and Z. Lacroix. A WFS-based mediation system for GIS interoperability. In *Proc. 10th Int. Symp. on Advances in Geographic Information Systems*, 23–28, 2002.

[66] M. Essid, F.-M. Colonna, O. Boucelma, and A. Bétari. Querying mediated geographic data sources. In *10th Int. Conference on Extending Database Technology*, LNCS 3896, Springer, 1176–1181, 2006.

[67] Y. Lassoued, M. Essid, O. Boucelma, and M. Quafafou. Quality-driven mediation for geographic data. In *Proceedings of the Fifth International Workshop on Quality in Databases, QDB 2007*, 27–38, 2007.

[68] T. Erl. *Service-Oriented Architecture (SOA): Concepts, Technology and Design.* Prentice Hall, 2005.

[69] OGC. Catalogue Service Implementation Specification. http://www.opengeospatial.org/standards/cat. Accessed on July 10, 2009.

[70] J. L. Schattel Jr., A. A. Taylor, P. R. Hershberg, and R. Bunge. Disseminating national weather service digital forecasts using open geospatial standards. In *Proceedings of the 23rd AMS Conference on Interactive Information and Processing Systems for Meteorology, Oceanography, and Hydrology*, page 3B.9, AMS Press, 2007.

[71] GeoServer. http://geoserver.org.

[72] University of Minnesota. Mapserver, http://mapserver.org. Accessed on July 10, 2009.

[73] OGC. OpenGIS® Sensor Web Enablement: Architecture Document. http://www.opengeospatial.org/pt/14140. Accessed on July 10, 2009.

[74] G. Percivall and C. Reed. OGC Sensor Web Enablement Standards. *Sensors & Transducers Journal*, 71(9):698–706, September 2006.

[75] OPeNDAP: Open-source Project for a Network Data Access Protocol. http://www.opendap.org/. Accessed on July 10, 2009.

[76] G. K. Rutledge, J. Alpert, and W. Ebisuzaki. Nomads: A climate and weather model archive at the national oceanic and atmospheric administration. *Bulletin of the Am. Meteorological Society*, 87(3):327–341, 2006.

[77] A. Woolf, K. Haines, and C. Liu. A web service model for climate data access on the grid. *Int. J. High Perform. Comput. Appl.*, 17(3):281–295, 2003.

[78] E. P. Glenn, A. R. Huete, P. L. Nagler, K. K. Hirschboeck, and P. Brown. Integrating remote sensing and ground methods to estimate evapotranspiration. *Crit. Rev. in Plant Sciences*, 26(3):139–168, 2007.

[79] S. Tilak, P. Hubbard, M. Miller, and T. Fountain. The ring buffer network bus (RBNB) dataturbine streaming data middleware for environmental observing systems. In *Proc. 3rd Int. Conf. on e-Science and Grid Computing (e-Science 2007)*, 125–133, IEEE, 2007.

[80] University Corporation for Atmospheric Research. Thematic Real-time Environmental Distributed Data Services. http://www.unidata.ucar.edu/projects/THREDDS/. Accessed on July 10, 2009.

[81] Geospatial Open Source Software GmbH. http://52north.org/. Accessed on July 10, 2009.

[82] Consortium of Universities for the Advancement of Hydrologic Science (CUAHSI). http://www.cuahsi.org/. Accessed on July 10, 2009.

[83] Weather Research & Forecasting Model (WRF). http://www.wrf-model.org Accessed on July 10, 2009.

[84] California Irrigation Management Information System (CIMIS). http://wwwcimis.water.ca.gov/cimis. Accessed on July 10, 2009.

[85] The COMET Project, COast-to-Mountain Environmental Transect. http://comet.cs.ucdavis.edu. Accessed on July 10, 2009.

[86] Geosciences Network (GEON). http://www.geongrid.org/. Accessed on July 10, 2009.

[87] Science Environment for Ecological Knowledge (SEEK). http://seek.ecoinformatics.org/. Accessed on July 10, 2009.

Chapter 11

Analyzing Data Streams in Scientific Applications*

Tore Risch,[1] Samuel Madden,[2] Hari Balakrishan,[2] Lewis Girod,[2] Ryan Newton,[2] Milena Ivanova,[3] Erik Zeitler,[1] Johannes Gehrke,[4] Biswanath Panda,[4] Mirek Riedewald[4]

[1] *Uppsala University, Sweden*
[2] *MIT*
[3] *CWI, The Netherlands*
[4] *Cornell University*

Contents

11.1 Introduction 400
11.2 Stream Processing for Scientific Applications 401
    11.2.1 Stream Processing Background 402
    11.2.2 Scientific Applications 403
    11.2.3 WaveScope 405
        11.2.3.1 The WaveScript Data Model 406
    11.2.4 The WaveScript Language 406
    11.2.5 Distributed Processing 408
    11.2.6 The WaveScope Runtime 409
    11.2.7 Conclusion 409
11.3 Parallelizing High-Volume Scientific Stream Queries 410
    11.3.1 Parallel Stream Queries through
            Data Flow Distribution Templates 411
    11.3.2 Stream Processes in SCSQ 414
11.4 Streaming Function Approximation
     for Scientific Simulations 417
    11.4.1 Survey of Existing Projects 419
    11.4.2 Technology Description 419
        11.4.2.1 Local Models 420
        11.4.2.2 The ISAT Algorithm 420

---

*Section 11.1 was authored by Tore Risch, 11.2 by Samuel Madden, Hari Balakrishan, Lewis Girod, and Ryan Newton, 11.3 by Tore Risch, Milena Ivanova, and Erik Zeitler, and 11.4 by Johannes Gehrke, Biswanath Panda, and Mirek Riedewald.

                11.4.2.3 Indexing Problem 421
                11.4.2.4 An Example: A Binary Tree Index 423
        11.4.3 Deployment Examples 425
        11.4.4 Future Challenges 426
Acknowledgement 426
References 426

11.1 Introduction

Modern scientific instruments such as satellites, on-ground antennas, and simulators collect large volumes of data. For example, instruments monitoring the environment emit streams of environmental sensor readings, particle colliders produce streams of particle collision data, and software telescopes such as LOFAR[33] produce very voluminous digitized radio signals. The measurement data is normally produced as streams rather than formats stored in conventional database tables. A stream has the property that data is ordered in time, and the data volume is potentially unlimited. Scientists perform a wide range of on-line analyses over the data streams. A conventional approach to data management using a relational database management system (DBMS) has the disadvantage that streaming data has to be loaded into a database before it can be queried and analyzed. If the data rate of a stream is too high, it will be impossible for the DBMS to load the streaming data fast enough. This creates backlogs of unanalyzed data, and the high data volume produced by scientific instruments can even be too large to store and process.[2] Furthermore, offline data processing prevents timely analysis of interesting natural events as they occur.

By contrast, data stream management systems (DSMSs) process queries directly over the streams, without preloading them into a database. DSMSs have been a popular research topic in recent years, with a number of commercial and academic projects exploring languages and operators for stream data processing.[3-8] A DSMS reads a stream only once and processes queries over small, moving sections of the streams, called *windows*. The answer to a DSMS query is a stream too. DSMS queries are called *continuous queries* because they are delivering data continuously in near real time. Continuous queries start to deliver data as soon as they are activated and continue to do so until they are deactivated. A continuous query is deactivated either manually or when some condition over the streamed data is fulfilled, for example, it may be deactivated after a specified time interval or when some query condition becomes true.

This chapter describes three aspects of implementing DSMSs for scientific applications:

Section 11.2 describes the programming language WaveScript and the run time environment WaveScope for high-performance processing of streaming scientific data. The system has been applied to several scientific signal processing applications. The importance here is the use of streams as first-class language objects and efficient representations of windows and subwindows over streams. WaveScope provides a typed functional programming language where high-volume signal processing can be specified through the *signal segment* datatype. Several distributed WaveScope nodes can communicate over distributed stream communication channels specified by the user. The section includes a comparison of this approach with DSMSs and discusses the advantage of using a functional programming language for high-volume streams, which traditional DSMSs do not provide.

Section 11.3 describes approaches to specify massively parallel scientific data stream queries. For scalability of stream-processing algorithms it is necessary to provide query language primitives describing how to parallelize a stream computation. It is not always possible to provide fully automated and transparent distribution. Therefore, the query language needs primitives allowing the user to parallelize computations and filters. The second section describes two approaches to specify parallel stream processing: *data flow distribution templates* and *stream processes* in the GSDM (Grid Stream Data Manager) and SCSQ (Super Computer Stream Query processor) systems, respectively. Both systems are based on a functional data model and query language.[9]

Finally, Section 11.4 discusses how to speed up streaming computations of functions by approximate materialization of computed values for scientific simulations. Often, queries over streaming scientific data involve complex computations expressed as functions. These functions may be costly to execute. Therefore, approximate materializations and indexing of their results may speed up the processing by avoiding recomputations.

11.2 Stream Processing for Scientific Applications

DSMSs are ideally suited for *online* scientific data processing applications, because they provide:

1. *Windowed operations:* The ability to segment incoming data streams into *windows*, which can then be sorted or aggregated to compute statistics such as the mean of some element in the stream.

2. *Main memory operation:* The ability to process data without first sending it to disk, decreasing latency and improving throughput.

3. *Specialized streaming operators:* For example, operators that detect out-of-order stream elements or that detect sequences or patterns in streams.

After briefly reviewing traditional DSMSs, we focus on scientific applications and the *WaveScope* scientific stream processing system that has been developed at MIT over the past few years.

11.2.1 Stream Processing Background

The DSMSs[3–8] include a common feature set: a high-level, declarative or graphical programming interface that allows data streams to be filtered according to some set of simple predicates, combined with each other ("joined") to match up related elements by time or values, and "aggregated" to compute simple statistics over groups of tuples.

Traditional applications of DSMSs cover a range of applications, from network monitoring to financial tick stream analysis. For example, suppose a user wants to find times when the price of IBM goes up for three stock ticks in a row. This could be expressed through the following "StreamSQL" query (this example is based on one given at http://www.streambase.com/developers-library-articles-detectingpatterns.htm):

```
CREATE INPUT STREAM InputStream
   (stock string(5), price double, time timestamp);
SELECT s1.time,s1.price AS price1
   FROM PATTERN InputStream AS s1 THEN
             InputStream AS s2 THEN
             InputStream AS s3
   WITHIN 20 TIME
   WHERE s1.stock=s2.stock AND s2.stock=s3.stock AND
           s3.price>s1.price AND s1.price>s2.price  AND
           s1.stock = 'IBM';
```

The first statement defines a stream `InputStream`. The `SELECT` statement then looks for a sequence of three stocks that satisfy the predicates in the `WHERE` clause (consecutively increasing, and all with stock symbol "IBM") and which arrive within 20 seconds of each other, and returns the time and price of the first stock in such a pattern.

Another simple query over this data stream might compute the average price of IBM stocks in the last hour, every minute, assuming one quote arrives pe minute:

```
SELECT time, AVG(price)
   FROM InputStream [SIZE 60 ADVANCE 1] AS s
   WHERE stock = 'IBM'
```

Here, the [] notation indicates that the incoming stream should be "windowed" into batches of 60 tuples, and that each consecutive window should include one new tuple and discard one old tuple.

Typically a streaming query of the sort shown above is converted into a graph of operators, where data flows from input streams ("leaf nodes") to output operators and the user. For example, the AVG query above would consist of three operators: an input operator that reads the incoming data stream and packages it into tuples for processing, a select operator that filters out all non-IBM stocks, and an aggregate operator that computes the average over the IBM stocks.

Streaming systems include a *runtime system* that is responsible for executing the compiled query plan. The main component of any runtime system is a scheduler (or scheduling policy) that dictates the order in which operators run. There are a number of interesting scheduling tradeoffs—for example, a simple scheduling policy is to push each tuple all the way through the query plan before moving on to the next tuple. This is good from the standpoint of data cache locality, as it keeps the same tuple in cache through several operators. It may be inefficient, however, if there is overhead to pass a single tuple from one operator to the next, or if all of the operators don't fit into the instruction cache. An alternative is to batch several tuples together and process them as a group, as proposed in the Aurora system,[10] which reduces per-tuple scheduling overheads and possibly improves instruction cache locality.

Traditional DSMSs are excellent for a variety of applications that need simple pattern matching and filtration over simple data types. As we discuss below, many scientific applications require more sophisticated processing, such as time/frequency domain conversions, signal processing filters, convolution, support for arrays and matries, and so on. In some cases, DSMSs are *extensible*, meaning that they allow users to define their own types and operations over those types, but this extension is typically done in some external language (e.g., C or Java), which introduces several limitations, as discussed below.

11.2.2 Scientific Applications

As noted above, scientific applications often require stream processing. This need is evident in a large number of signal-oriented streaming applications proposed in the sensor network literature (where the predominant applications are scientific in nature), including preventive maintenance of industrial equipment;[11] detection of fractures and ruptures in pipelines,[12] airplane wings (http://www.metisdesign.com/shm.html), or buildings;[13] *in situ* animal behavior studies using acoustic sensing;[14] network traffic analysis;[15] particle detectors in physics experiments; and medical applications such as anomaly detection in electrocardiogram signals.[16] Another important scientific area that requires management of steaming data is geosciences. Examples of multiple sources of streaming data and their integration are discussed in Chapter 10. These target applications use a variety of embedded sensors, each sampling at fine resolution and producing data at rates as high as hundreds of thousands of samples per second.

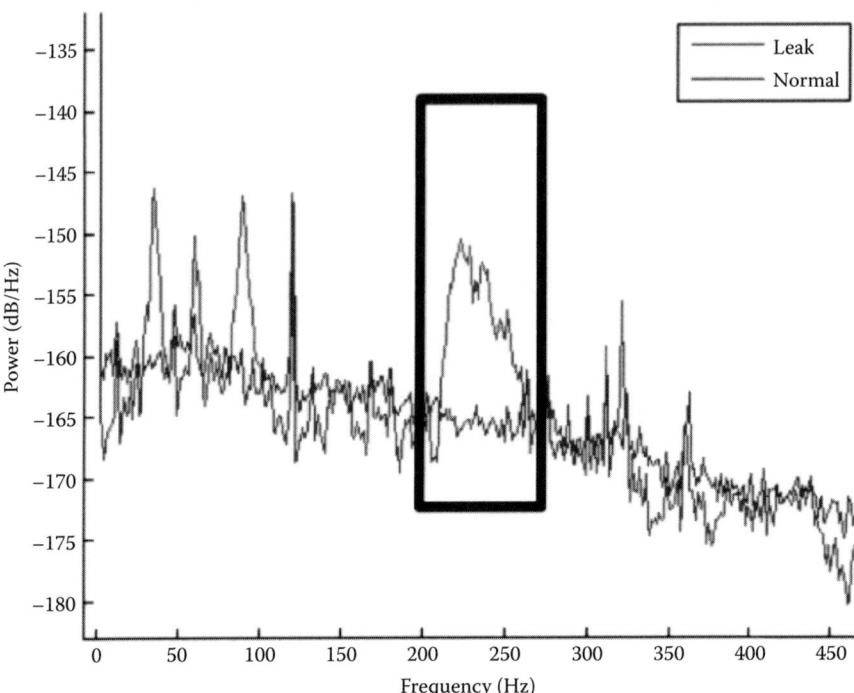

Figure 11.1 (See color insert following page 224.) A leak shows up as additional energy in characteristic frequency bands.

In most applications, processing and analyzing these streaming sensor samples requires nontrivial event-stream and signal-oriented analysis. In many cases, signal processing is application-specific and hence requires some amount of user-defined code. For example, in a pipeline rupture detection application with which we are familiar, the incoming streams of pressure data from sensors on water pipelines, sampled at 600 Hz, are fed into a frequency transform. Ruptures are detected by comparing the energy in certain frequency bands to look for a peak. This is shown in Figure 11.1. Operations like "peak extraction" require user-defined code and are essentially impossible to implement in simple languages like StreamSQL.

As another example, we have been working with biologists in Colorado to build sensor systems that acoustically detect and localize marmots (a kind of rodent endemic to the western United States) by listening for their loud, chirpy calls. These systems consist of several four-microphone arrays that are positioned around an area known to have marmots. Each microphone array "listens" for sound frequencies that are characteristic of marmots and, when it detects such signals, performs a "direction of arrival" analysis to determine the bearing of the marmot. By finding the point where the direction of arrival rays emanating from several microphone arrays intersect ("beam forming"), it is possible to estimate the location of a calling marmot.

Clearly, expressing this kind of signal processing with relational filters and aggregates is not feasible. Conventional DSMSs handle event processing over streaming data, but don't provide a convenient way to write user-defined custom code to handle signal processing operations. In particular, they suffer from an "impedance mismatch," where data must be converted back and forth from its representation in the streaming database to an external language like Java or C++, or even to a separate system like MATLAB. Signal processing operations in these external languages are usually coded in terms of operations on arrays, whereas most DSMSs represent data streams as sequences of tuples. These sequences need to be packed into and unpacked from arrays and be passed back and forth. The conversion overheads imposed by this mismatch also limit the performance of existing stream processing engines when performing signal-processing operations, constraining the applicability of these existing systems to lower-rate domains.

11.2.3 WaveScope

To support these kinds of applications, we have developed a new programming language called WaveScript and a sensor network runtime system called WaveScope.[17] WaveScript is a functional programming language that allows users to compose sensor network programs that perform complex signal and data processing operations on sensor data.

WaveScript includes several noteworthy features. Its data model introduces a new basic data type, the *signal segment*. A signal segment is a sequence of isochronous (i.e., sampled regularly in time) data values (samples) from a signal that can be manipulated as a batch. WaveScript natively supports a set of operations over signal segments. These include various transforms and spectral analyses, filtering, resampling, and decimation operations. Another important feature of WaveScope is that users express both queries and user-defined functions (UDFs) in the same high-level language (WaveScript). This approach avoids the cumbersome "back and forth" of converting data between relational and signal-processing operations. The WaveScript compiler produces a low-level, asynchronous dataflow graph similar to query plans in traditional streaming systems. The runtime engine efficiently executes the query plan over multiprocessor PCs or across networked nodes, using both compiler optimizations and domain-specific rule-based optimizations.

WaveScope allows programs to operate on named data streams, some of which may come from remote nodes. To simplify programming tasks that involve data from many nodes, programmers can group nodes together, creating a single, named stream that represents the union or aggregate of many nodes' data streams.

As with most DSMSs, WaveScope derives a number of benefits from the use of a high-level programming model. Programmers do not need to worry about time synchronization, power management, or the details of networking protocols—they simply express their data processing application using the WaveScript language, and the WaveScope runtime takes care of executing these programs in an efficient manner.

In the remainder of this section, we briefly overview the WaveScope data and programming model, focusing on expressing a few simple example applications.

11.2.3.1 The WaveScript Data Model

The WaveScript data model is designed to *efficiently* support high volumes of isochronous sensor data. Data is represented as streams of tuples in which each tuple in a particular stream is drawn from the same *schema*. Each field in a tuple is either a primitive type (e.g., integer, float, character, string), an array, a set, a tagged union, or a special kind of object called a *signal segment* (SigSeg).

A SigSeg represents a window into a signal (time series) of fixed bit-width values that are regularly spaced in time (isochronous). Hence, a typical signal in WaveScope is a stream of tuples, where each tuple contains a SigSeg object representing a fixed-size window on that signal. A SigSeg object is conceptually similar to an array in that it provides methods to get values of elements in the portion of the signal it contains and determine its overall length.

Although values within a SigSeg are isochronous, the data stream itself may be asynchronous, in the sense that the arrival times of tuples are *not* constrained to arrive regularly in time.

The WaveScope data model treats SigSegs as first-class entities that are transmitted in streams. This is unlike other DSMSs[3–5] that impose windows on individual tuples as a part of the execution of individual operators. By making SigSegs first-class entities, windowing can be done once for a chain of operators, and logical windows passed between operators, rather than being defined by each operator in the data flow, greatly increasing efficiency for high data rate applications.

11.2.4 The WaveScript Language

With WaveScript, developers use a single language to write all aspects of stream processing applications, including queries, subquery-constructors, custom operators, and functions. The WaveScript approach avoids mediating between the main script and user-defined functions defined in an external language, while further allowing *type-safe* construction of queries. In contrast, while SQL is frequently embedded into other languages, there is no compile-time guarantee that such queries are well-formed.

An example WaveScript: Figures 11.2 and 11.3 show a WaveScript for performing marmot detection on a single four-channel microphone array,

Analyzing Data Streams in Scientific Applications

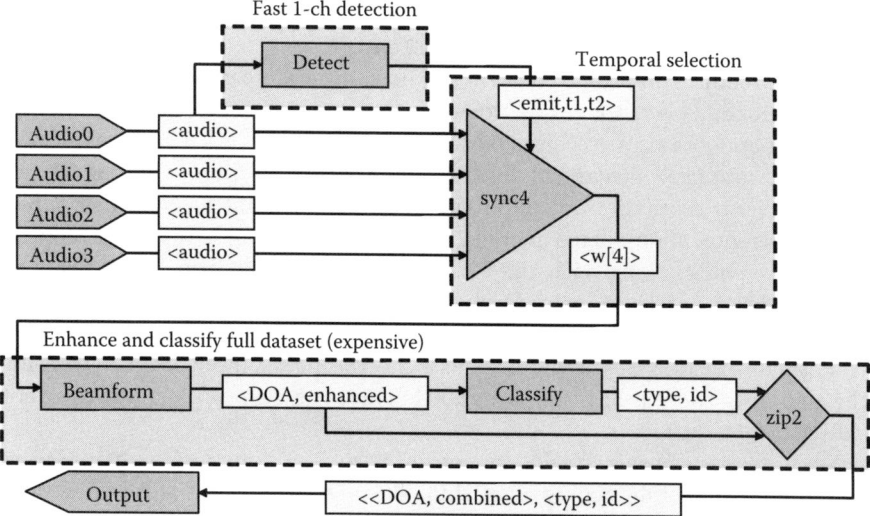

Figure 11.2 Marmot call detection workflow.

first as a workflow diagram and then as the equivalent WaveScript subquery. The `marmot` function uses `detect`, a reusable detection algorithm, to identify the portions of the stream most likely to contain marmot calls, and then extracts those segments and passes them to the rest of the workflow, which enhances and classifies the calls. Several streams are defined: `Ch0..3` are streams of `SigSeg<int16>`, while `control` is a stream of `<bool,time,time>` tuples. Type annotations (e.g., line 2) may be included for clarity, but they are optional. Types are inferred from variable usage using standard techniques[18]:

```
fun marmot(Ch0, Ch1, Ch2, Ch3, net) {
  Ch0 : stream<SigSeg<int16>>
  // Detector on sensor inputs
  control : stream<bool,time,time>
  control = detect(Ch0, marmotScore, <64,192>,
                   <16.0, 0.999, 40, 2400, 48000>);
  // Control stream used to extract data windows.
  windows : stream<SigSeg<int16>[4]>
  windows =
    sync4(filter_lapped(merge(control, net)),
          Ch0, Ch1, Ch2, Ch4);
  // ... and process them: enhance
  beam<doa,enhanced> = beamform(windows, geometry);
  // ... and classify
  marmots = classify(beam.enhanced, marmotSig);
  // Return tuple of control and result streams
  return <control, zip2(beam, marmots)>;
}
```

Figure 11.3 Equivalent WaveScript subquery.

for example, the definition of the `beamform` function implies that the type of `beam` is `Stream<float[360],SigSeg<float>>`.

The `detect` subquery constructor has several parameters, including `marmotScore`, a custom function which computes the energy in frequency bands that are characteristic of marmot calls. This produces a `<bool,time,time>` stream of local marmot call detections that is merged with elements from the `net` input stream (representing a stream of tuples from a remote node). This merged and filtered stream is fed as a control stream to `sync4`, along with the four raw input streams from the audio sensors. `Sync4` aligns the four data streams in time, and "snapshots" synchronized segments of data according to the time ranges specified in the control stream. In this way, `sync4` reduces the volume of input data by passing through only those segments of audio that `detect` suspects contain marmot calls.

Next, the synchronized windows of data from all four audio channels are processed by a beam-forming algorithm. The algorithm computes a direction-of-arrival (DOA) probability distribution and enhances the input signal by combining phase-shifted versions of the four channels according to the most likely direction of arrival. The `beam` function returns a stream of two tuples. We use a special binding syntax, "`beam<doa,enhanced> = ...,`" to give temporary names to the fields of these tuples. That is, `beam.doa` projects a stream of direction-of-arrivals, and `beam.enhanced` contains the enhanced versions of the raw input data. Finally, this enhanced signal is fed into an algorithm that classifies the calls by type (male, female, or juvenile) and, when possible, identifies individuals.

11.2.5 Distributed Processing

WaveScript also includes the ability to specify programs that move data between several nodes, as in the case of the microphone arrays performing marmot detection. The simplest form of distributed processing is as follows:

```
// Send side:
  ToNet(''MyStream'', S);

// Receive side:
  R = FromNet(''MyStream'');
```

Here, the stream S is made visible to other nodes in the network (who are declared in a special configuration file) as the named stream "MyStream." A receiver can read in this stream and process it as though it were produced locally. The WaveScope runtime takes care of ensuring that this data is delivered reliably and in order, and that tuples on this stream are appropriately time synchronized with local data on the receiving node.

Once a WaveScript program has been written, it is fed to the WaveScope compiler, which converts it to a binary that can be run inside of a sensor network by the WaveScope runtime system. Internally, this binary consists of a sequence of operators similar to that used by conventional DSMSs.

11.2.6 The WaveScope Runtime

This compiled binary is fed to the WaveScope Runtime for execution. As with other stream-processing engines described above, operators in compiled WaveScript query plans are run by a scheduler, which picks boxes one at a time and runs them to produce outputs.

A special *timebase manager* is responsible for managing timing information corresponding to signal data. This is a common problem in signal processing applications, since signal processing operators typically process vectors of samples with sequence numbers, leaving the application developer to determine how to interpret those samples temporally. A timebase is a dynamic data structure that represents and maintains a mapping between sample sequence numbers and time units. Examples of timebases include those based on abstract units (such as seconds, hertz, and meters), as well as timebases based on real-world clocks, such as a CPU counter or sample number.

Finally, a *memory manager* is responsible for creation, storage, and management of memory, particularly as is associated with SigSeg objects that represent the majority of signal data in WaveScope. Designing an efficient memory manager is of critical importance, since memory allocation can consume significant processing time during query processing.

These features—the timebase manager, memory manager, and scheduler—simplify the task of the programmer (who no longer has to worry about these details) and allow WaveScope to provide excellent performance. In our initial benchmarks of the marmot application, the current WaveScope system processes 7 million samples per second on a 3 GHz PC, and about 80 K samples per second on a 400 MHz ARM platform (which has a 10x penalty for floating point emulation in addition to a reduced processor speed).

11.2.7 Conclusion

Data stream management systems provide an efficient and powerful way to process data arriving in near real time and are well suited to applications ranging from financial analysis to network monitoring. Scientific applications, however, present an "impedance mismatch" as they often involve large amounts of custom signal processing code and also require support for very high data rates. The WaveScope DSMS, and associated WaveScript language, are designed especially to support such applications, including an integrated language for writing data and signal processing operations, an efficient way to represent batches of high data rate, isochronous signals as SigSegs, mechanisms for transmitting data between networked stream processing nodes, and a very efficient scheduler and memory manager that provide much better throughput than existing stream processing systems.

11.3 Parallelizing High-Volume Scientific Stream Queries

WaveScope provides a complete functional programming language for specifying high-volume stream processing computations. The nodes involved in these computations can communicate using stream communication primitives where the user explicitly specifies data interchange between WaveScope nodes. The purpose of the systems described in this section is to provide primitives to specify massively parallel and distributed computations in a functional query language. The two systems GSDM (Grid Stream Data Manager)[19] and SCSQ (Super Computer Stream Query processor)[21] provide two different ways for parallelizing queries:

- GSDM provides a library of constructors of high-level *data flow distribution templates* to specify parallel execution schemes for functions used in declarative stream queries. GSDM has been applied on signal analysis in space physics applications.
- SCSQ provides declarative parallelization in queries by providing stream processes (SPs) as first-class objects in the query language. SCSQ has been applied on space physics and traffic applications.

Both GSDM and SCSQ are based on a functional data model[9] where declarative queries over streams are expressed in terms of functions.

The motivating application is LOFAR,[33] which is a radio telescope in construction that uses an array of 25,000 omni-directional antenna receivers whose signals are digitized into data streams of very high rate. The LOFAR antenna array will be the largest sensor network in the world. The receivers produce raw data streams that arrive at the central processing facilities at a rate that is too high for the data to be saved on disk. For these data-intensive computations, LOFAR utilizes an IBM BlueGene supercomputer combined with conventional Linux clusters.

High-performance stream processing for this kind of application requires the ability to specify parallel continuous queries (CQs) running on nodes in a heterogeneous hardware environment. To maximize throughput of streams and computations it is important to parallelize CQs into continuous subqueries, each executing as a separate process on some CPU. Often the parallelization method depends on properties of the computation executed by the query, making it impossible to automatically parallelize the execution. The query processing system must therefore provide primitives for customized parallelization of continuous computations.

11.3.1 Parallel Stream Queries through Data Flow Distribution Templates

In GSDM, continuous queries are expressed as functions over stream windows called *stream query functions* (SQFs) specified using a query language. The data flow distribution templates are parameterized descriptions of CQs as distributed compositions of other SQFs together with a logical site assignment for each subquery. For extensibility, a distribution template may be defined in terms of other templates.

For scalable execution of CQs containing expensive SQFs a generic template called PCC is provided for customizable data partitioning parallelism (Figure 11.4). The generic template contains three phases: *partition, compute,* and *combine*. In the partition phase the stream is split into substreams; in the compute phase subqueries are applied in parallel on each substream; and in the combine phase the results of the computations are combined into one stream.

The generic distribution template has been used to define two different, more specialized stream partitioning strategies: query-dependent *window split* (WS) and query-independent *window distribute* (WD). Window split provides application-dependent partition and combine strategies, while window distribute is applicable on any SQF. Window split is favorable, for example, for many numerical algorithms on vectors, such as FFT (Fast Fourier Transform), which scale through user-defined vector partitioning; window distribute provides SQF-independent rerouting of substreams. Both strategies use a pair of nonblocking and order-preserving SQFs to specify the partition and combine phases.

The partition phase in window split is defined by another template, *operator-dependent stream split* (OS-Split) to perform application-dependent splitting of logical windows into smaller ones. An SQF, *operator-dependent stream join* (OS-Join), implements the combine phase. Window split is particularly useful when scaling the logical window size for an SQF with complexity higher than $O(n)$ over the window size. For example, space physics and many signal processing applications require the FFT to be applied on large vector

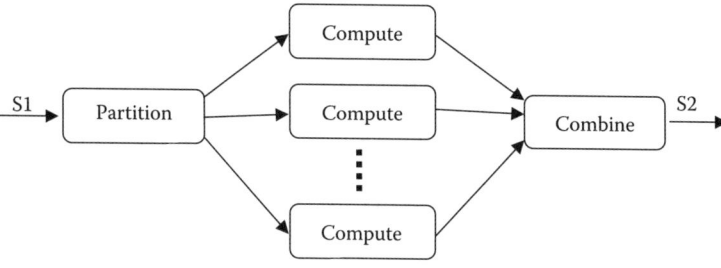

Figure 11.4 The generic dataflow distribution template PCC for partitioning parallelism of expensive stream query functions.

windows, and we use OS-Split and OS-Join to implement an FFT-specific stream partitioning strategy.

As an example of a window distribute strategy, *Round Robin stream partitioning* (RR) is provided where entire logical windows of streams are distributed based on the order in which they arrive. In the combine phase, the result substreams are merged on their order identifier (in our application a time stamp is used). This is an extension of the conventional Round Robin partitioning[22] for data streams. Window distribute by Round Robin does not decrease the size of logical windows, and therefore the compute phase of FFT may run slower than with window split.

Figure 11.5 illustrates the use of window split and window distribute for computing a three-dimensional FFT with the degree of parallelism two.

For example, the window distribute (b) and split (c) strategies are specified with PCC as:

```
PCC(2,"S-Distribute","RRpart","FFT","S-Merge",0.1);
PCC(2,"OS-Split","FFTpart","FFT","OS-Join","FFTcombine");
```

The PCC constructor is parameterized on (1) the degree of parallelism (degree 2); (2) partitioning method (`SDistribute` or `OS-Split`); (3) parameter of the partitioning method (`RRpart` or `FFTPart`); (4) SQF to be computed (FFT); (5) the combining method (`S-Merge` or `OS-Join`); and (6) parameters of the combining method (e.g., 0.1, a time-out).

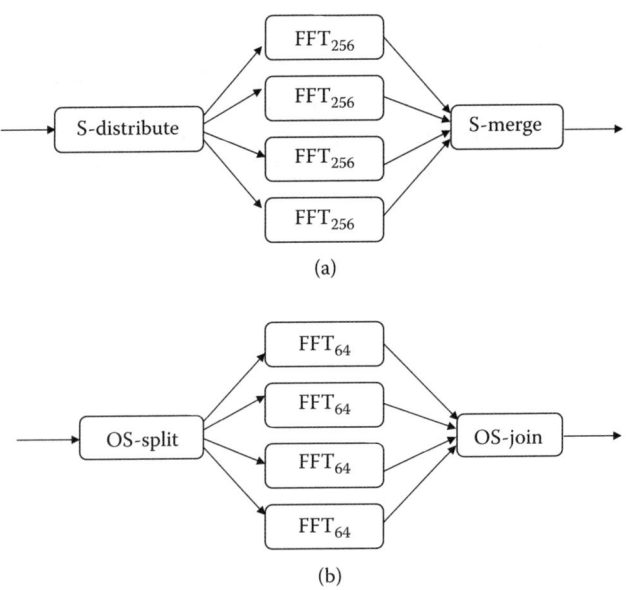

Figure 11.5 (a) Central strategy, (b) Round Robin window distribute strategy, (c) FFT-dependent window split strategy.

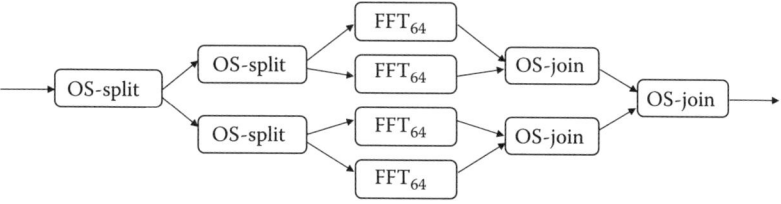

Figure 11.6 Parallel window split strategy with tree partitioning of degree four.

The SQFs `FFTpart` and `FFTcombine` are FFT-specific window split and join functions, respectively, and `RRpart` and `S-Merge` distribute complete windows independent of the SQF to compute.

Complex stream partitioning schemes can be defined by combining distribution templates. For example Figure 11.6 shows a complex tree-shaped distribution scheme containing two levels of window splits and window joins for parallel computation of FFT streams specified by the template:

```
PCC(2,"OS-Split","FFTpart", "PCC",
    {2,"OS-Split","FFTpart","FFT","FFTcombine"},
    "OS-Join", "FFTcombine");
```

Figure 11.7 shows a performance comparison of executing the FFT computation using different distribution templates with degree four in parallelism. The experiments show that both window split and window distribute

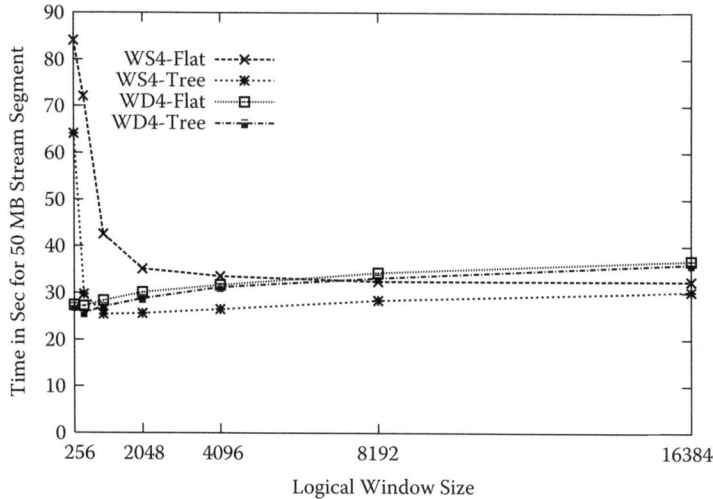

Figure 11.7 FFT performance for parallelism of degree four with various distribution templates.

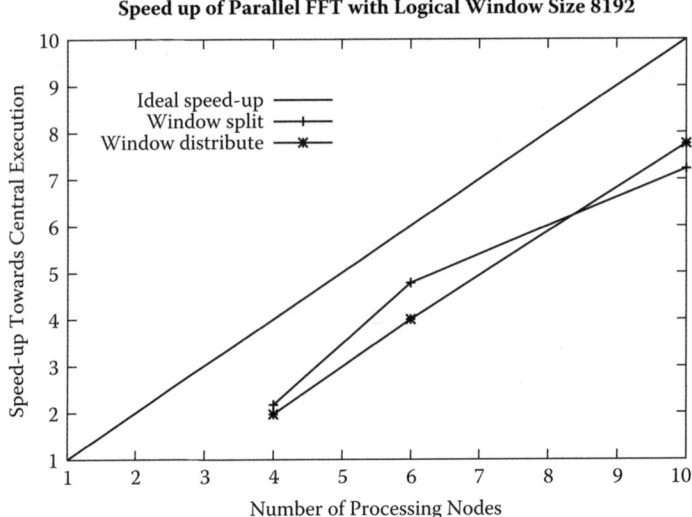

Figure 11.8 FFT speedup for parallelism of degree four with different distribution templates.

strategies have advantages in specific situations. If the continuous query is to be executed with a limited number of compute nodes, so that the load of the compute nodes exceeds the load of the partition and combine phases, the window split is preferable since it utilizes query semantics to achieve a more scalable parallel execution. However, if the system has resources allowing a high degree of parallelism where partition and combine nodes become more loaded than the compute nodes, window distribute may have better performance depending on the cost of partitioning and combining SQFs. This is further illustrated by Figure 11.8, which shows the speed-up of the different partitioning schemes. The window split template provides better speed-up when the number of processors is limited. The tree-shaped window distribute provides the best performance in Figure 11.7. However, it also uses the most compute nodes (8) and does not provide as good speed-up.

11.3.2 Stream Processes in SCSQ

SCSQ enables customized parallelization of continuous queries defined completely in the query language SCSQL (pronounced sis-kel). In SCSQL, parallel computations in queries and views are specified in terms of stream processes (SPs) that are first-class objects in CQs. The CQs can call process construction functions that execute stream subqueries assigned to some CPU. Such queries can be used to define query functions that parallelize computations over streams. The user can specify massively parallel stream computations by defining sets of SPs executing arbitrary subqueries. Properties of the different

CPUs, communication mechanisms, and operating systems substantially influence query execution performance. These properties are stored in a database, which is used by the query optimizer when assigning an SP to a CPU.

For example, the distributed grep MapReduce[23] query using 1,000 parallel grep calls is specified in SCSQL as follows:

```
merge(spv(select grep("pattern", filename(i))
         from Integer i
         where i in iota(1,1000)));
```

Notice, however, that MapReduce is limited to offline processing of stored data with only one possible communication pattern, namely, map and reduce. By contrast, SCSQ enables online processing of streams. Furthermore, arbitrary communication patterns can be expressed using SCSQL. The example above shows that MapReduce processing also can be easily expressed with SCSQL.

To enable easy handling of sets of parallel stream processes, the function spv(s) assigns each continuous subquery in the set of subqueries s to a new stream process on some compute node, and returns a set of handles to the assigned stream processes. The function merge(p) requests elements from each stream process in p. merge() and terminates when (if ever) the last stream process in p terminates.

Splitting of streams is specified by referencing common variables bound to stream processes, as illustrated by the following query function, which implements the Radix2 parallelization of FFT for a stream source named ss.

```
create function radix2(String ss)-> Stream
   as select radixcombine(merge({a,b}))
      from SP a, SP b, SP c
      where a=sp(fft(odd (extract(c))))
        and b=sp(fft(even(extract(c))))
        and c=sp(receiver(ss));
```

The receiver() function returns a stream of 1D arrays of signal data. odd(x) and even(x) obtain odd and even elements from array x, respectively. radixcombine() combines the results from the partial FFT algorithms working in parallel.

The output of an SP is sent to one or more other SPs, which are called subscribers of that SP. The user can control which tuples are sent to which subscriber using a postfilter.[24] The postfilter is expressed in SCSQL, and can be any function that operates on the output stream of its SP. For each output tuple from an SP, the postfilter is called once per subscriber. Hence, the postfilter can transform and filter the output of an SP to determine whether a tuple should be sent to a subscriber.

In the example query above, all elements from c are sent to both a and b. a and b apply the odd() and even() filter functions to extract odd and even elements of the vectors from c. Obviously, the amount of communication from c to a and b can be reduced by 50% if a postfilter is applied in c before its

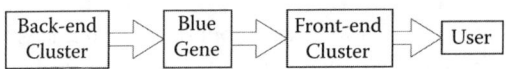

Figure 11.9 Stream data flow in the LOFAR environment.

output is sent to a and b. Using postfilters, the query above is transformed into the following query:

```
create function radix2(String ss) -> Stream
   as select radixcombine(merge({a,b}))
      from SP a, SP b, SP c
      where a=sp(fft(extract(c)))
        and b=sp(fft(extract(c)))
        and c=sp(receiver(ss), #'oddeven');
```

The notation #'oddeven' specifies the object representing the function named oddeven. This function is the postfilter function that extracts odd elements of the vector when sent to a and even elements when sent to b. Postfilter functions have successfully been used in spatial partitioning for parallelization of combinatorial optimization problems.[24]

Figure 11.9 illustrates the stream dataflow in the LOFAR hardware environment. Users interact with SCSQ on a Linux front-end cluster. Another Linux back-end cluster first receives the streams from the sensors where they are preprocessed. Next, the BlueGene processes these streams. The output streams from the BlueGene are then postprocessed in the front-end cluster, and the result stream is finally delivered to the user. Thus, three computer clusters are involved.

Figure 11.10 illustrates a query that is set up for execution in the hardware environment. SCSQ users interact with the client manager, in which they specify CQs using SCSQL. The execution of a CQ forms a directed acyclic graph of running processes (RPs), each executing the subquery specified in one SP.

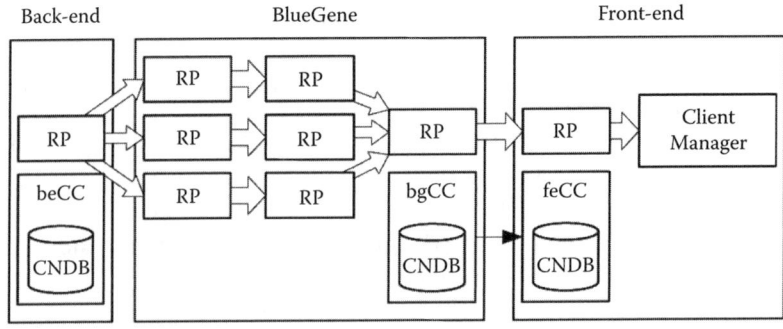

Figure 11.10 Parallel CQ execution in SCSQ. Wide arrows indicate data streams.

When a user submits a CQ, it is optimized and started in the client manager. When the client manager identifies an SP, the subquery of that SP is registered with the coordinator of the cluster where the subquery is to be executed (*feCC, bgCC,* or *beCC* in Figure 11.10). Then, the coordinator starts an RP to execute the subquery. In addition, an RP can dynamically start new RPs by requesting them from the coordinator of the cluster where the new RP is started.

11.4 Streaming Function Approximation for Scientific Simulations

Simulations are one of the most important tools of modern science for studying real world phenomena. The physical laws that govern a phenomenon drive a mathematical model, usually a function defined on a multidimensional space, that is used in simulations as an approximation of reality. In practice scientists face serious computational challenges, because realistic models are expensive to evaluate and simulation runs consist of a large number of model evaluation steps. To make realistic simulations feasible, scientists are often willing to accept approximate results as long as significant improvements in runtime can be achieved. The main idea is to use previously computed function values to construct an approximate model that can be used for future steps instead of the expensive original model.

Speeding up simulations is challenging, because it inherently is a *data stream* processing problem. Like data stream management systems, a simulation engine has to process a stream of data points that describe the current state of the simulated system. Decisions about which previously computed function values to retain and how to leverage them for reducing the number of future function evaluations have to be made in real time and with limited knowledge about future data points. This again is a typical characteristic of data stream applications.

However, despite fitting into the general class of data stream applications, scientific simulations also have unique challenges. These create a novel data stream indexing and data stream model learning problem. Techniques for addressing these challenges are described in the remainder of this chapter.

Example

Consider the simulation of a combustion process, which motivated the line of work discussed in this section. Scientists study how the composition of gases in a combustion chamber changes over time due to chemical reactions. The composition of a gas particle is described by a high-dimensional vector, typically with 10–70 dimensions. The simulation consists of a series of timesteps.

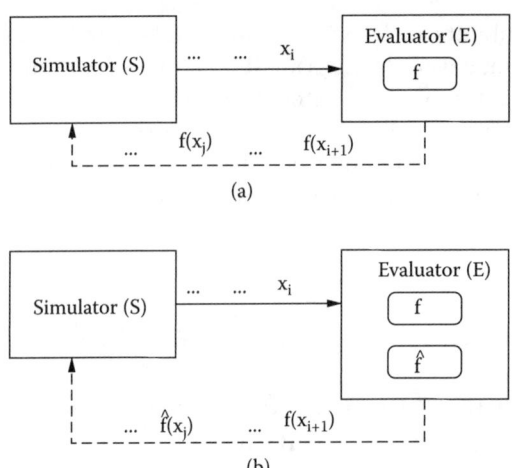

Figure 11.11 Simulation workflow.

During each timestep some particles in the chamber react, causing their compositions to change. This reaction is described by a complex high-dimensional function called the *reaction function*, which, given the current composition vector of a particle and other simulation properties, produces a new composition vector for the particle. Combustion simulations usually require up to 10^8 to 10^{10} reaction function evaluations.

Figure 11.11(a) summarizes the computational workflow. The Simulator (S) generates a sequence of compositions $\{\ldots x_i\}$ at which the reaction function should be evaluated. The Evaluator (E) performs these reaction function evaluations, which are then used by S to generate future compositions. Many scientific simulations have a similar workflow.

Problem Formulation
Scientists often face serious computational challenges in running large-scale scientific simulations. The more realistic the model, the more complex the corresponding mathematical equations and the design of E (Figure 11.11). For example, in combustion simulations E is a differential equation solver that takes in the order of tens of milliseconds for a single reaction function (denoted as f in the figure) evaluation. As a result, even small simulations can run for days; larger, more complex simulations, would run for months or even years.

To make large-scale simulations computationally feasible, scientists trade off accuracy for speed. This is done by modifying the simulation workflow, as shown in Figure 11.11(b). The main idea is to build a computationally inexpensive approximate model ($\hat{f}$) of the original complex model (f). The approximate model $\hat{f}$ is also built and maintained during the course of the simulation, and the Evaluator for each input point $\mathbf{x}$ can either evaluate $\hat{f}(x)$

or use the expensive model to compute $f(x)$ and use $f(x)$ to update $\hat{f}$. Thus, we need to design an Evaluator that ensures that all model evaluations are within an error tolerance (i.e., $\hat{f}$ approximates f within an error threshold) and the total simulation cost is minimized. The total simulation cost includes the cost of evaluating the model and also the cost of building $\hat{f}$.

11.4.1 Survey of Existing Projects

Approximating complex models with simpler and faster ones has been studied in many domains. In the combustion research community, early approaches were offline in the following sense. Function evaluations were collected from simulations and used to learn multivariate polynomials approximating the reaction function.[25] These polynomials were then used later in different simulations, replacing the original reaction function. Recently, more powerful models like neural networks and self-organizing maps have also been used.[26] The offline approaches had only limited success because a single model cannot generalize to a large class of simulations.

In 1997 Pope developed the *In Situ Adaptive Tabulation* (ISAT) Algorithm.[27] ISAT takes an online approach to speeding up combustion simulations. The algorithm caches reaction function evaluations from certain frequently seen regions in the composition space. It then uses the cached values to approximate the reaction function at compositions encountered later on during the simulation. The technique was a major breakthrough in combustion simulation because it enabled scientists to run different simulations without having to first build a model for the reaction function. ISAT was the first algorithm that approached the function approximation problem in combustion simulations according to a stream model, and the algorithm, even today, remains the state of the art in the field.

Several modifications and improvements to ISAT have been proposed. DOLFA[28] and PRISM[29] rely on alternative methods of caching reaction function evaluations. More recently, Panda et al.[30] studied the storage and retrieval problem arising out of caching and reusing reaction function evaluations in ISAT. Their work demonstrates how the streaming nature of simulations creates interesting tradeoffs that can be exploited for significant speed-ups in simulations. This article discusses their major findings and observations.

11.4.2 Technology Description

Even though the ISAT Algorithm was originally proposed in the context of combustion simulations, it can easily be generalized for building approximate models for other high-dimensional functions. This section begins with a discussion on *local models*, which represent the general class of models built by ISAT (Section 11.4.2.1). This is followed by a description of the ISAT Algorithm that uses selective evaluations of the expensive function f to build a local model (Section 11.4.2.2). The streaming nature of the application introduces a new storage and retrieval, and hence indexing, problem in the algorithm. This section then discusses the indexing problem in detail: its challenges and solutions that have been proposed (Sections 11.4.2.3 and 11.4.2.4).

11.4.2.1 Local Models

Local models are used in many applications to approximate complex high-dimensional functions. Given a function $f:\mathbf{R}^m \to \mathbf{R}^n$ a local model defines a set of high-dimensional regions in the function domain: $\mathcal{R}=\{R_1 \ldots R_n | R_i \subseteq \mathbf{R}^m\}$. Each region R_i is associated with a function $\hat{f}_{R_i}:R_i \to \mathbf{R}^n$; such that $\forall \mathbf{x} \in R_i : \|\hat{f}_{R_i}(\mathbf{x}) - f(\mathbf{x})\| \leq \epsilon$, where ϵ is a specified error tolerance in the model and $\|\|$ is an error metric such as the Euclidean distance. Using a local model to evaluate f at some point $\mathbf{x}$ in the function domain first involves finding a region $R \in \mathcal{R}$ that contains $\mathbf{x}$ and then evaluating $\hat{f}_R(\mathbf{x})$ as an approximation to $f(\mathbf{x})$.

11.4.2.2 The ISAT Algorithm

Main algorithm: ISAT is an online algorithm for function approximation. Its pseudocode is shown in Figure 11.12. The algorithm takes as input a query point $\mathbf{x}$ at which the function value must be computed and a search structure I that stores the regions in $\mathcal{R}$. I is empty when the simulation starts. The algorithm first tries to compute the function value at $\mathbf{x}$ using the local model it has built so far (Lines 2 and 3). If that fails, the algorithm computes $f(\mathbf{x})$ using the expensive model (Line 5) and uses $f(\mathbf{x})$ to update existing or add new regions in the current local model (Line 6). The algorithm is online because it does not have access to all query points when it builds the model.

Model updating: ISAT updates the local model using the strategy outlined in Figure 11.13. In general it is difficult to exactly define a region R and an associated $\hat{f}_R$, such that $\hat{f}_R$ approximates f in all parts of R. ISAT uses a two-step process to "discover" regions. It initially starts with a region that is very small and conservative but where it is known that a particular $\hat{f}_R$

ISAT Algorithm
1: **Require:** Query Point $\mathbf{x}$, index structure I
2:     if $\exists \langle R, \hat{f}_R \rangle \in I$ such that $\mathbf{x} \in R$
3:         Compute $\mathbf{y} = \hat{f}(\mathbf{x})$
4:     else
5:         Compute $\mathbf{y} = f(\mathbf{x})$
6:         Update$(I, \mathbf{x}, f(\mathbf{x}))$
7:     end if
8:     return $\mathbf{y}$

Figure 11.12 ISAT algorithm.

	Updating A Local Model
1:	**Require:** I, $\mathbf{x}$, $f(\mathbf{x})$
2:	**if** $\exists \langle R, \hat{f}_R \rangle \in I : \mathbf{x}$ can be included in R
3:	**for all** $\langle R, \hat{f}_R \rangle \in I$
4:	**if** $\mathbf{x}$ can be included in R
5:	Update $\langle R, \hat{f}_R \rangle$ to include $\mathbf{x}$
6:	**end if**
7:	**end for**
8:	**else**
9:	Add new $\langle R, \hat{f}_R \rangle$ to I
10:	**end if**

Figure 11.13 Updating a local model.

approximates f well. It then gradually grows these conservative approximations over time. More specifically the update process first searches index I for regions where $\mathbf{x}$ lies outside the region but $||\hat{f}_R(\mathbf{x}) - f(\mathbf{x})|| \leq \epsilon$. Such regions are grown to contain $\mathbf{x}$ (Lines 2–7). If no existing regions can be grown, a new conservative region centered around $\mathbf{x}$ and the associated $\hat{f}_R$ is added to the local model (Line 9). The Grow Process described is a heuristic that works well in practice for functions that are locally smooth. This assumption holds in combustion and in most other applications.

Instantiation: The original ISAT Algorithm proposed high-dimensional ellipsoids as regions and used a linear model as the function in a region. The linear model is initialized by computing the value of f and estimating the derivative of f at the center of the ellipsoidal region:

$$\hat{f}_R(\mathbf{x}) = f(\mathbf{a}) + f'(\mathbf{x} - \mathbf{a}),$$

where $\mathbf{a}$ is the center of region R and $f'(\mathbf{a})$ is the derivative of f at $\mathbf{a}$.

ISAT performs one of the following basic operations for each query point $\mathbf{x}$. **Retrieve**: Computing the function value at $\mathbf{x}$ using the current local model by searching for a region containing $\mathbf{x}$. **Grow**: Searching for regions that can be grown to contain $\mathbf{x}$ and updating these regions in I. **Add**: Adding a new region (R) and an associated $\hat{f}_R$ into I.

11.4.2.3 Indexing Problem

The indexing problem in function approximation produces a challenging workload for the operations on index I in Figures 11.12 and 11.13. The Retrieve requires the index to support fast lookups. The Grow requires both a fast lookup to find growable ellipsoids and then an efficient update process once an ellipsoid is grown. Finally, an efficient insert operation is required for the Add step. There are two main observations that make this indexing problem different from traditional indexing.[31,32]

1. The regions that are stored in the index are not fixed, but generated by the Add and Grow operations. Due to the streaming nature of the application, past decisions about growing and adding affect the set of regions stored in the index and hence future performance.
2. Traditionally, the performance of index structures has been measured in terms of the cost of search and in some cases update. Since the goal of function approximation is to minimize total simulation cost, function evaluations and region operations must also be accounted for when evaluating the performance of an index.

In the light of these observations a principled analysis of the various costs in the function approximation algorithm leads to the discovery of novel tradeoffs. These tradeoffs produce significant and varying effects on different index structures. The remainder of this section briefly introduces the tradeoffs and the tuning parameters that have been proposed to exploit them. The indexing problem is studied here using the concrete instantiation of the ISAT Algorithm using ellipsoidal regions with linear models. Therefore, regions are often referred to as *ellipsoids* in the rest of the section. It is important to note, however, that the ideas are applicable to other types of regions and associated functions.

Tuning Retrieves. In most high-dimensional index structures the ellipsoid containing a query point is usually not the first ellipsoid examined since an index structure often performs fast approximations during the search before performing an expensive check at each leaf level. The index ends up looking at a number of ellipsoids before finding "the right one." The additional ellipsoids that are examined by the index are called *false positives*. For each false positive, the algorithm pays to search and retrieve the ellipsoid from the index and to check if the ellipsoid contains the query point. In traditional indexing problems, if an object that satisfies the query condition exists in the index, then finding this object during search is mandatory. Therefore, the number of false positives is a fixed property of the index.

However, the function approximation problem provides the flexibility to tune the number of false positives, because we can fall back to evaluating the expensive function if the index search was not successful. The number of false positives can be tuned by limiting the number of ellipsoids examined during the retrieve step. This parameter is denoted by Ellr. Ellr places an upper bound on the number of false positives for a query. Tuning Ellr controls several effects.

- **Effect 1:** Decreasing Ellr reduces the cost of the retrieve operation as fewer ellipsoids are retrieved and examined.
- **Effect 2:** Decreasing Ellr decreases the chance of finding an ellipsoid containing the query point, thereby resulting in more function evaluations.

- **Effect 3:** Misses caused by decreasing Ellr will grow and add other ellipsoids. These grows and adds make regions cover new parts of the domain and also change the overall structure of the index. Both of these affect the probability of retrieves for future queries. This is a more subtle effect unique to this problem.

Tuning Grows and Adds. Just like the Retrieve, the Grow and Add operations can be controlled by the number of ellipsoids examined for growing, denoted as Ellg. Since an Add is performed only if a Grow fails, this parameter controls both the operations. Ellg provides a knob for controlling several effects.

- **Effect 4:** The first part of the grow process involves traversing the index to find ellipsoids that can be grown. Decreasing Ellg reduces the time spent in the traversal.
- **Effect 5:** Decreasing Ellg decreases the the number of ellipsoids examined for the grow and hence the number of ellipsoids actually grown. This results in the following effects.
  - **Effect 5a:** Reducing the number of ellipsoids grown reduces index update costs, which can be significant in high-dimensional indexes.
  - **Effect 5b:** Growing a large number of ellipsoids on each Grow operation covers more parts of the function domain, thereby improving the probability of future retrieves.
  - **Effect 5c:** Growing a large number of ellipsoids on each Grow results in significant overlap among ellipsoids. Overlap among objects being indexed reduces search efficiency in many high-dimensional indexes.
- **Effect 6:** Decreasing Ellg increases the number of Add operations. Creating a new region is more expensive than growing an existing region since it involves initializing the function $\hat{f}_R$ in the new region.

In summary, the two tuning parameters have many different effects on index performance and the cost of the simulation. What makes the problem interesting is that these effects often move in opposite directions. Moreover, tuning affects indexes differently and to varying degrees, which makes it necessary to analyze each index individually.

11.4.2.4 An Example: A Binary Tree Index

The previous section presented a qualitative discussion of the effects that tuning Ellr and Ellg can have on index performance and simulation cost. This section makes these effects more concrete using an example index structure, the Binary Tree. This tree indexes only the centers of the ellipsoids (not the actual ellipsoid) by recursively partitioning the space with cutting planes. Leaf nodes of the tree contain ellipsoids, and nonleaf nodes represent cutting

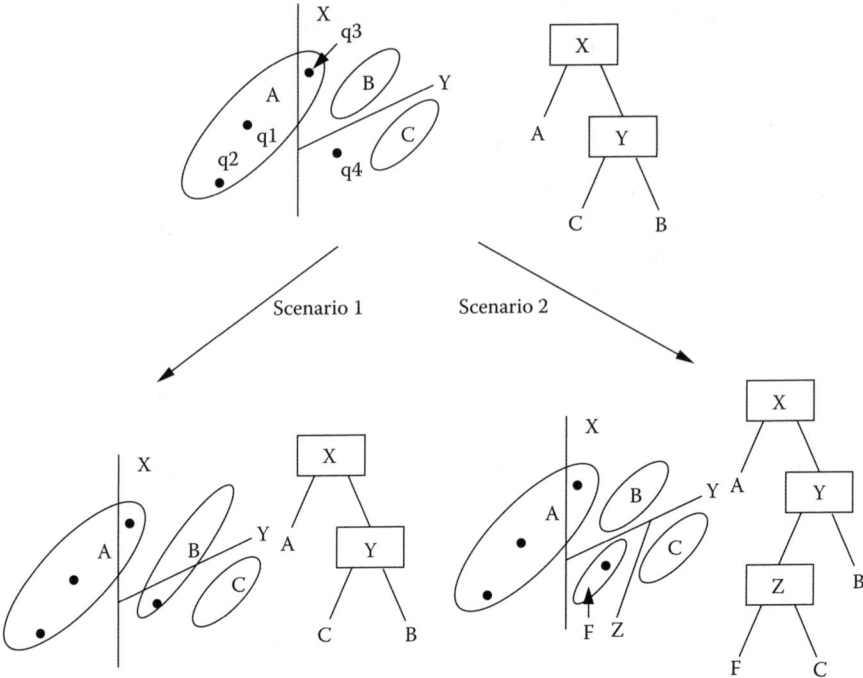

Figure 11.14 Binary Tree.

planes. Figure 11.14 shows an example tree for three ellipsoids *A*, *B*, and *C* and two cutting planes *X* and *Y*.

For now let us focus on the tree in the top part of Figure 11.14 and describe the operations supported by the index.

Retrieve
There are two possible traversals in the index that result in a successful retrieve.

Primary Retrieve. The first, called a Primary Retrieve, is illustrated with query point q_2. The retrieve starts at the root, checking on which side of hyperplane *X* the query point lies. The search continues recursively with the corresponding subtree, the left one in the example. When a leaf node is reached, the ellipsoid in the leaf is checked for the containment of the query point. In the example, *A* contains q_2, therefore, we have a successful Primary Retrieve.

Secondary Retrieve. Since the Binary Tree only indexes centers, ellipsoids can straddle cutting planes; for example, *A* covers volume on both sides of cutting plane *X*. If ellipsoids are straddling planes, then the Primary Retrieve can result in a false negative. For example, q_3 lies to the right of *X* and so the Primary Retrieve fails even though there exists an ellipsoid *A* containing it. To overcome this problem the Binary Tree performs a more expensive Secondary Retrieve if the Primary fails. The main idea of the Secondary Retrieve is to explore the "neighborhood" around the query point by examining nearby subtrees. In the case of q_3, the failed Primary Retrieve ended in leaf B. Nearby subtrees are explored by moving up a level in the tree and exploring the other side of the cutting plane. Specifically, C is examined first (after moving up to Y, C is in the unexplored subtree). Then the search would continue with A (now moving up another level to X and accessing the whole left subtree). This process continues until a containing ellipsoid is found, or Ellr ellipsoids have been examined unsuccessfully.

Update

Scenario 1 (Grow) and Scenario 2 (Add) of Figure 11.14 illustrate the update operations on the index.

Grow. The search for growable ellipsoids proceeds in exactly the same way as a Secondary Retrieve, starting where the failed Primary Retrieve ended. Assume that in the example in Figure 11.14, ellipsoid B can be grown to include q_4, but C and A cannot. After the retrieve failed, the Grow operation first attempts to grow C. Then it continues to examine B, then A (unless Ellg < 3). B is grown to include q_4, as shown on the bottom left (Scenario 1). Growing of B made it straddle hyperplane Y. Hence, for any future query point near q_4 and "below" Y, a Secondary Retrieve is necessary to find containing ellipsoid B, which is "above" Y.

Add. The alternative to growing B is illustrated on the bottom right part of Figure 11.14 (Scenario 2). Assume Ellg = 1; that is, after examining C, the Grow search ends unsuccessfully. Now we add a new ellipsoid F with center q_4 to the index. This is done by replacing leaf C with an inner node Z, which stores the hyperplane that best separates C and F. The Add step requires the expensive computation of f, but it will enable future query points near q_4 to be found by a Primary Retrieve.

Tuning parameter Ellg affects the Binary Tree in its choice of scenario 2 over 1. This choice, that is, performing an Add instead of a Grow operation, reduces false positives for future queries but adds extra cost for the current query. Experiments on real simulation workloads have shown that this tradeoff has a profound influence on the overall simulation cost.[30]

11.4.3 Deployment Examples

The ISAT Algorithm and its optimizations have primarily been applied to combustion simulation workloads. However, the ideas are applicable to any simulation setting where repeated evaluations in a fixed domain of a function that is locally smooth and expensive to compute are required. Preliminary results of applying the techniques to real simulations suggest that a significant reduction in overall simulation time, one or more orders of magnitude, can be achieved by careful index tuning.

11.4.4 Future Challenges

The current technique considered only a limited number of index structures and possible approximation models. Preliminary results suggest that simulation time can be reduced further by allowing the algorithm to adaptively change the regions and the function approximation models used in these regions. This needs to be examined in-depth and for simulations from different scientific domains. Another interesting direction for future work is to study thoroughly the tradeoff between simulation time, error threshold, and approximation quality observed in practice.

Acknowledgement

The research reported in Section 11.3 has been supported by VINNOVA under contract #2001-06074 and by ASTRON.

The research reported in Section 11.4 was performed in collaboration with Stephen Pope and Zhiyun Ren from the Cornell Department of Mechanical Engineering and Paul Chew from the Cornell Department of Computer Science. This research is based upon work supported by the National Science Foundation under grants CBET-0426787, IIS-0621438, IIS-0330201, EF-0427914, and IIS-0612031. Any opinions, findings, conclusions, or recommendations expressed in this material are those of the authors and do not necessarily reflect the views of the National Science Foundation. We thank the Cornell Center for Advanced Computing for support in running the simulations.

References

[1] D. J. Abadi, Y. Ahmad, M. Balazinska, U. Cetintemel, M. Cherniack, J. Hwang, W. Lindner, A. S. Maskey, A. Rasin, E. Ryvkina, N. Tatbul, Y. Xing, and S. Zdonik. The design of the Borealis Stream Processing Engine. In *Second Conference on Innovative Data Systems Research (CIDR 2005)*, 2005.

[2] A. S. Szalay, J. Gray, A. Thakar, P. Z. Kunszt, T. Malik, J. Raddick, C. Stoughton, and J. van den Berg. The SDSS SkyServer: Public access to the Sloan digital sky server data. In *21st ACM SIGMOD International Conference on Management of Data (SIGMOD 2002)*, 2002.

[3] D. Carney, U. Cetintemel, M. Cherniack, C. Convey, S. Lee, G. Seidman, M. Stonebraker, N. Tatbul, and S. Zdonik. Monitoring streams—A new

class of data management applications. In *28th International Conference on Very Large Data Bases (VLDB 2002)*, 2002. 215–226.

[4] R. Motwani, J. Widom, A. Arasu, B. Babcock, S. Babu, M. Datar, G. S. Manku, C. Olston, J. Rosenstein, and R. Varma. Query processing, approximation, and resource management in a data stream management system. In *First Conference on Innovative Data Systems Research (CIDR 2003)*, 2003.

[5] S. Chandrasekaran, O. Cooper, A. Deshpande, M. J. Franklin, J. M. Hellerstein, W. Hong, S. Krishnamurthy, S. R. Madden, V. Raman, F. Reiss, and M. A. Shah. Continuous dataflow processing for an uncertain world. In *First Conference on Innovative Data Systems Research (CIDR 2003)*, 2003.

[6] http://www.streambase.com/. Accessed July 20, 2009.

[7] http://www.coral8.com/. Accessed July 20, 2009.

[8] http://www.aleri.com/. Accessed July 20, 2009.

[9] T. Risch, V. Josifovski, and T. Katchaounov. Functional data integration in a distributed mediator system. In P. Gray, L. Kerschberg, P. King, and A. Poulovassilis (eds.). *Functional Approach to Data Management—Modeling, Analyzing and Integrating Heterogeneous Data*. Heidelberg Germany: Springer, ISBN 3-540-00375-4, 2003.

[10] D. Carney, U. Cetintemel, A. Rasin, S. Zdonik, M. Cherniack, and M. Stonebraker. Operator scheduling in a data stream manager. In *29th International Conference on Very Large Data Bases (VLDB 2003)*, 2003. 838–849.

[11] Intel Corporation. *Mechanical Sound and Vibration Control*. Internal Document.

[12] I. Stoianov, L. Nachman, S. Madden, and T. Tokmouline. PIPENET: A wireless sensor network for pipeline monitoring. In *Sixth International Conference on Information Processing in Sensor Networks (IPSN 2007)*, 2007.

[13] N. Xu and R. Govindan. A wireless seismic sensing array. Poster. Center for Embedded Network Sensing. http://repositories.cdlib.org/cens/Posters/53.

[14] L. Girod, M. Lukac, V. Trifa, and D. Estrin. The design and implementation of a self-calibrating acoustic sensing system. In *4th ACM Conference on Embedded Networked Sensor Systems (SenSys 2006)*, 2006. 71–84.

[15] A. Hussian, J. Heidemann, and C. Papadopoulos. A framework for classifying denial of service attacks. In *ACM SIGCOMM 2003*, Aug. 2003. 99–110.

[16] E. Shih, V. Bychkovsky, and J. Guttag. Pervasive medical monitoring using wireless microsensors. In *Second International Conference on Embedded Networked Sensor Systems (SenSys 2004)*, 2004. Demo.

[17] L. Girod, K. Jamieson, Y. Mei, R. Newton, S. Rost, A. Thiagarajan, H. Balakrishnan, and S. Madden. The case for WaveScope: A signal-oriented data stream management system (position paper). In *Third Conference on Innovative Data Systems Research (CIDR 2007)*, 2007.

[18] R. Milner. A theory of type polymorphism in programming. *Journal of Computer and System Sciences*, 17, pages 348–375, 1978.

[19] M. Ivanova and T. Risch. Customizable parallel execution of scientific stream queries. In *31st International Conference on Very Large Data Bases (VLDB 2005)*, 2005. 157–168.

[20] B. Panda, M. Riedewald, J. Gehrke, and S. B. Pope. High speed function approximation. In *Proceedings of the 7th IEEE International Conference on Data Mining (ICDM 2007)*, Omaha, Nebraslor October 28–31, 2007.

[21] E. Zeitler and T. Risch. Using stream queries to measure communication performance of a parallel computing environment. In *First International Workshop on Distributed Event Processing, Systems and Applications (DEPSA 2007)*, Toronto, Canada, 2007.

[22] M. T. Özsu and P. Valduriez. *Principles of Distributed Database Systems, Second Edition*. Upper Saddle River, New Jersey: Prentice Hall, ISBN 0-13-659707-6, 1999.

[23] J. Dean and S. Ghemawat. MapReduce: Simplified Data Processing on Large Clusters. *Communications of the ACM*, 51(1), pages 107–114, 2008.

[24] G. Gidofalvi, T. B. Pedersen, T. Risch, and E. Zeitler. Highly scalable trip grouping for large-scale collective transportation systems. In *11th International Conference on Extending Database Technology (EDBT 2008)*, Nantes, France, March 25–30, 2008. 678–689.

[25] T. Turanyi. Application of repro-modeling for the reduction of combustion mechanisms. In *25th Symposium on Combustion*, pages 949–955, 1994.

[26] J. Y. Chen, W. Kollmann, and R. W. Dibble. A self-organizing-map approach to chemistry representation in combustion applications. *Combustion Theory and Modelling*, 4(1), pages 61–76, 2000.

[27] S. B. Pope. Computationally efficient implementation of combustion chemistry using *in situ* adaptive tabulation. *Combustion Theory and Modelling*, 1(1), pages 41–63, 1997.

[28] I. Veljkovic, P. Plassmann, and D. C. Haworth. A scientific on line database for efficient function approximation. In *International Conference on Computational Science and Its Applications (ICCSA 2005)*, pages 643–653, 2003.

[29] J. B. Bell, N. J. Brown, M. S. Day, M. Frenklach, J. F. Grcar, R. M. Propp, and S. R. Tonse. Scaling and efficiency of PRISM in adaptive simulations of turbulent premixed flames. In *28th International Combustion Symposium*, 2000.

[30] B. Panda, M. Riedewald, S. B. Pope, J. Gehrke, and L. P. Chew. Indexing for function approximation. In *32nd International Conference on Very Large Data Bases (VLDB 2006)*, 2006.

[31] C. Böhm, S. Berchtold, and D. A. Keim. Searching in high-dimensional spaces: Index structures for improving the performance of multimedia databases. *ACM Computing Surveys*, 33(3), pages 322–373, 2001.

[32] V. Gaede and O. Günther. Multidimensional access methods. *ACM Computing Surveys*, 30(2), pages 170–231, 1998.

[33] http://www.lofar.nl/. Accessed July 20, 2009.

Part V
Scientific Process Management

Chapter 12

Metadata and Provenance Management

Ewa Deelman,[1] Bruce Berriman,[2] Ann Chervenak,[1]
Oscar Corcho,[3] Paul Groth,[1] and Luc Moreau[4]

[1] *University of Southern California Information Science Institute, Marina del Rey, California*
[2] *Caltech, Pasadena, California*
[3] *Universidad Politécnica de Madrid, Madrid, Spain*
[4] *University of Southampton, Southampton, United Kingdom*

Contents

12.1 Metadata and Provenance 434
12.2 Metadata 435
    12.2.1 The Role of Ontologies in Metadata Specification 438
12.3 Provenance 439
12.4 Survey of Existing Approaches 441
    12.4.1 Metadata Schema and Metadata Attributes 441
    12.4.2 Technologies for Storing Metadata Management 443
    12.4.3 Technologies for Provenance Management 444
12.5 Metadata in Scientific Applications 446
    12.5.1 Astronomy 447
    12.5.2 Climate Modeling 449
12.6 Provenance in Scientific Applications 450
    12.6.1 Provenance in Everyday Science 450
    12.6.2 Provenance in Geospatial Applications 451
    12.6.3 Provenance for Oceanographic Applications 453
    12.6.4 End-to-End Provenance for Large-Scale
           Astronomy Applications 454
    12.6.5 Enabling Multidisciplinary and Multiscale Applications
           Using Provenance 456
12.7 Current and Future Challenges 457
Acknowledgments 458
References 459

12.1 Metadata and Provenance

Today data are being collected by a vast number of instruments in every discipline of science. In addition to raw data, new products are created every day as a result of processing existing data and running simulations in order to understand observed data. As the sizes of the datasets grow into the petascale range, and as data are being shared among and across scientific communities, the importance of diligently recoding the meaning of data and the way they were produced increases dramatically.

One can think of *metadata* as data descriptions that assign meaning to the data, and data *provenance* as the information about how data was derived. Both are critical to the ability to interpret a particular data item. Even when the same individual is collecting the data and interpreting them, metadata and provenance are important. However, today, the key drivers for the capture and management of data descriptions are the scientific collaborations that bring collective knowledge and resources to solve a particular problem or explore a research area. Because sharing data in collaborations is essential, these data need to contain enough information for other members of the collaboration to interpret them and then use them for their own research. Metadata and provenance information are also important for the automation of scientific analysis, where software needs to be able to identify the datasets appropriate for a particular analysis and then annotate new, derived data with metadata and provenance information.

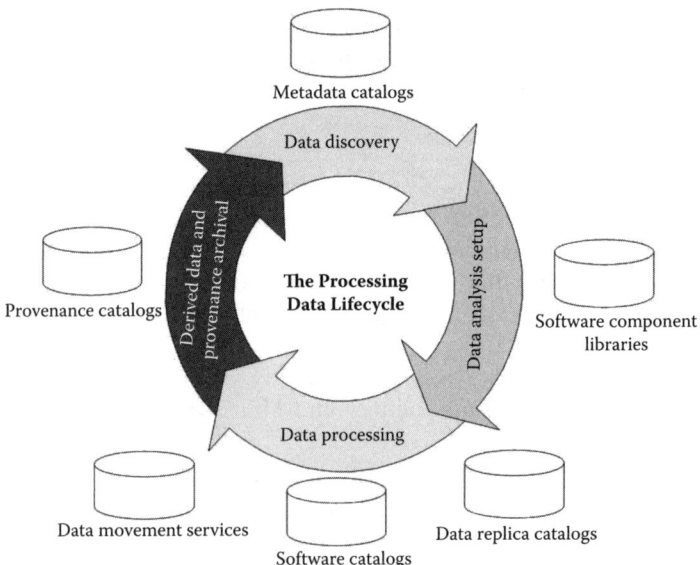

Figure 12.1 The data lifecycle.

Figure 12.1 depicts a generic data lifecycle in the context of a data processing environment where data are first discovered by the user with the help of metadata and provenance catalogs. Next, the user finds available analyses that can be performed on the data, relying on software component libraries that provide *component metadata*, a logical description of the component capabilities. During the data processing phase, data replica information may be entered in replica catalogs (which contain metadata about the data location), data may be transferred between storage and execution sites, and software components may be staged to the execution sites as well. While data are being processed, provenance information can be automatically captured and then stored in a provenance store. The resulting derived data products (both intermediate and final) can also be stored in an archive, with metadata about them stored in a metadata catalog and location information stored in a replica catalog.

12.2 Metadata

From a general point of view, metadata may be defined as "data about data." However, this definition is too broad; hence, other, more specific definitions for this term have been provided in the literature, each of them focusing on different aspects of metadata capture, storage, and use. Probably one of the most comprehensive definitions is from Greenberg,[1] which defines metadata as "structured data about an object that supports functions associated with the designated object." The structure implies a systematic data ordering according to a metadata schema specification; the object can be any entity or form for which contextual data can be recorded; and the associated functions can be activities and behaviors of the object. One of the characteristics of this definition is that it covers the dual function that metadata can have: describing the objects from a logical point of view as well as describing their physical and operational attributes.

We can cite a wide range of objects that metadata can be attached to, such as databases, documents, processes and workflows, instruments and other resources. These objects may be available in different formats. For example, documents may be available electronically in the form of HTML, PDF, Latex, and so forth; on the Web, in a data grid, on a PC hard disk, or on paper in a library, among other formats. At the same time, metadata can also be expressed in a wide range of languages (from natural to formal ones) and with a wide range of vocabularies (from simple ones, based on a set of agreed keywords, to complex ones, with agreed taxonomies and formal axioms). Metadata can be available in different formats, both electronic and on paper, for example, written in a scientist's lab notebook or in the margins of a textbook. Metadata can also be created and maintained using different types of tools, from text editors to metadata generation tools, either manually or automatically.

Given all this variety in representation formats, described resources, approaches for metadata capture, storage, use, and so forth, there is not a commonly agreed taxonomy of types of metadata or types of described resources, but there are different points of view about how metadata can be generated and used. We will now go through some of these points of view, illustrating them with examples.

One of the properties of metadata is that it can be organized in layers; that is, metadata can refer to raw data (e.g., coming from an instrument or being available in a database), to information about the process of obtaining the raw data, or to derived data products. This allows distinguishing different layers (or chains) of metadata: primary, secondary, tertiary, and so forth. As an example, let us consider an application in the satellite imaging domain, such as the one described in Sanchez-Gestido et al.[2] Raw data coming from satellites (e.g., images taken by instruments in the satellite) are sent to the ground stations so that they can be stored and processed. A wide range of metadata can be associated with these data, such as the times when they were obtained and transferred, the instrument used for capturing them, the time period when the image was taken, the position to which it refers, and the like. This is considered as the primary metadata of the images received. Later on, this metadata can be used to check whether all the images that were supposed to be obtained from an instrument in a period of time have actually been obtained or whether there are any gaps, and new metadata can be generated regarding the grouping of pieces of metadata for an instrument, the quality of the results obtained for that time period, statistical summaries, and so forth. This is considered as secondary metadata, because it does not refer to the raw data being described but to the metadata that refer to the analysis, summaries, and observations about the raw data, so that it forms a set of layers or a chain of metadata descriptions. Another common example of this organization of metadata into layers is that of provenance, which is described in the next section.

In all these cases, it is important to determine which type (layer) of metadata we use for searching, querying, and so forth, and which type of metadata we show to users, so that metadata coming from different layers is not merged together and is shown with the appropriate level of detail, as discussed in Hunter.[3]

The organization of metadata into layers also reflects an interesting characteristic of how metadata is used. To some extent, what is considered metadata for one application may be considered data for another. In the previous example in the satellite domain, metadata about the positions of images on the Earth is considered as part of the primary metadata that is captured and stored for the satellite mission application when the information arrives to the ground station. However, the same spatial information would be considered as a data source for other applications, such as a map visualization service (e.g., Google Earth) that positions those resources in a map. In contrast, the dates when the images were taken or the instruments with which they were produced may still be considered as metadata in both cases.

Another aspect of metadata comes into play when new data products are being generated either as a result of analysis or simulation. These derived data are now scientific products and need to be described with appropriate metadata information. In some disciplines such as astronomy (see Section 12.5.1), the community has developed standard file formats that include metadata information in the header of the file. These are often referred to as "self-describing data formats," because each file stored in such a format has all the necessary metadata in its header. Software is then able to read this metadata and to generate new data products with the appropriate headers. One of the difficulties of this approach is to be able to automatically catalog the derived data. In order to do that, some process needs to be able to read the file headers and then extract the information and place it in a metadata catalog. In terms of metadata management, astronomy seems to be ahead of other disciplines, possibly because in addition to the astronomy community, the discipline appeals to many amateurs. As a result, astronomers needed to face the issue of data and metadata publication early on, making their data broadly accessible. Other disciplines of science are still working on the development of metadata standards and data formats. Without those, software cannot generate descriptions of the derived data. Even when the standards are formed within the community, there are often a number of legacy codes that need to be retrofitted (or wrapped) to be able to generate the necessary metadata descriptions as they generate new data products.

Finally, another perspective about metadata is whether a piece of metadata reflects an objective point of view about the resources that are being described or only a subjective point of view about it. In the former case the term *metadata* is usually used, but in the latter the more specific term *annotation* is more commonly used. Annotations are normally produced manually by humans and reflect the point of view of those humans with respect to the objects being described. These annotations are also known as *social annotations*, to reflect the fact that they can be provided by a large number of individuals. They normally consist of sets of tags that are manually attached to the resources being described, without a structured schema to be used for this annotation or a controlled vocabulary to be used as a reference. These types of annotations provide an additional point of view over existing data and metadata, reflecting the common views of a community, which can be extracted from the most common tags used to describe a resource. Flickr[4] or del.icio.us[5] are examples of services used to generate this type of metadata for images and bookmarks, respectively, and are being used in some cases in scientific domains [3].

There are also other types of annotations that are present in the scientific domain. For example, researchers in genetics annotate the Mouse Genome Database[6] with information about the various genes, sequences, and phenotypes. All annotations in the database are supported with experimental evidence, and citations and are curated. The annotations also draw from a standard vocabulary (normally in the form of controlled vocabularies, thesauri, or ontologies, as described in the following subsection) so that they can be consistent. Another example of scientific annotations is in the neuroscience domain, where scientists are able to annotate a number of brain images.[7] The issue in brain imaging is that there are very few automated techniques that can extract the features in an image. Rather, the analysis of the image is often done by a scientist. In some cases, the images need to be classified or annotated based on the functional properties of the brain, information that cannot be automatically extracted. As with other annotations, brain images can be annotated by various individuals, using different terms from an overall vocabulary. An advantage of using a controlled vocabulary for the annotations is that the annotations can be queried and thus data can be discovered based on these annotations.

Annotations can also be used in the context of scientific workflows (see Chapter 13), where workflow components or entire workflows can be annotated so that they can be more readily discovered and evaluated for suitability. The myGrid project[8] has a particular emphasis on bioinformatics workflows composed of services broadly available to the community. These services are annotated with information about their functionality and characteristics. myGrid annotations can be both in a free text form and drawn from a controlled vocabulary.[9]

12.2.1 The Role of Ontologies in Metadata Specification

Together with controlled vocabularies and thesauri, ontologies have become one of the most common means to specify the structure of metadata in scientific applications, such as the previous ones. Ontologies are normally defined as "formal, explicit specifications of shared conceptualizations".[10] A *conceptualization* is an abstract model of some phenomenon in the world derived by having identified the relevant concepts of that phenomenon. *Explicit* means that the type of concepts used, and the constraints on their use, are explicitly defined. *Formal* refers to the fact that the ontology should be machine readable. *Shared* reflects the notion that an ontology captures consensual knowledge; that is, it is not private view for some individual, but accepted by a group.

Ontologies started to be used for this purpose in document metadata annotation approaches in pre-Semantic Web applications like the SHOE project,[11] the (KA)² initiative,[12] and PlanetOnto,[13] among others. Later ontologies have become a commodity for specifying the schema of metadata annotations, not only about Web documents, but also about all types of Web and non-Web resources. The benefits they provide with respect to other artifacts are mainly related to the fact that they capture consensual knowledge (what facilitates interoperability between applications, although consensus is sometimes difficult to achieve, as described in the previous section), and that they can be used to check the consistency of the annotations and to infer new knowledge from existing annotations, since they are expressed in formal languages with a clear logical theory behind.

Not all ontologies have the same degree of formality; neither do they include all the components that could be expressed with formal languages, such as concept taxonomies, formal axioms, disjoint and exhaustive decompositions of concepts, and so forth. Given this fact, ontologies are usually classified either as lightweight or heavyweight.[10] An example of the former would be Dublin Core,[23] which is being widely used to specify simple characteristics of electronic resources, specifying a predefined set of features such as *creator, date, contributor, description, format*, and the like. Examples of the latter would be the ontologies used for workflow annotation in the myGrid project or for product description in the aforementioned satellite imaging application. Lightweight ontologies can be specified in simpler formal ontology languages like RDF Schema,[14] and heavyweight ontologies require more complex languages like OWL.

12.3 Provenance

Provenance is commonly defined as the origin or source or history of derivation of some object. In the context of art, this term carries a more concrete meaning: It denotes the record of ownership of an art object. In this context, such concrete records allow scholars or collectors to verify and ascertain the origin of the work of art, its authenticity, and therefore its price.

This notion of provenance can be transposed to electronic data.[15] If the provenance of data produced by computer systems could be determined as it can for some works of art, then users would be able to understand how documents were assembled, how simulation results were determined, or how analyses were carried out. For scientists, provenance of scientific results would indicate how results were derived, what parameters influenced the derivation, what datasets were used as input to the experiment, and so forth. In other words, provenance of scientific results would help *reproducibility*,[16] a fundamental tenet of the scientific method.

Hence, in the context of computer systems, we define *provenance of a data product* as the process that led to such a data product, where *process* encompasses all the derivations, datasets, parameters, software and hardware components, computational processes, and digital or nondigital artifacts that were involved in deriving and influencing the data product.

Conceptually, such provenance could be extremely large, since potentially it could bring us back to the origin of time. In practice, such level of information is not required by end users, since their needs tend to be limited to specific tasks, such as experiment reproducibility or validation of an analysis.

To support the vision of provenance of electronic data, we make the distinction between *process documentation*, a representation of past processes as they occur inside computer systems, and *provenance queries*, extracting relevant information from process documentation to support users' needs.

Process documentation is collected during execution of processes or workflows and begins to be accumulated well before data are produced, or even before it is known that some dataset is to be produced. Hence, management of such process documentation is different from metadata management. In practice, in a given application context, users may identify commonly asked provenance queries, which can be precomputed, and for which the results are stored and made available.

Similar to the earlier discussion of different metadata layers, we can think of provenance as consisting of descriptions at different levels of abstraction, essentially aimed at different audiences: to support scientific reproducibility, engineering reproducibility, or even deeper understanding of the process that created the derived data (we provide an example of the latter in the context of scientific workflows below). In terms of scientific reproducibility, where scientists want to share and verify their findings with colleagues inside or outside their collaboration, the user may need to know what datasets were used and what type of analysis with what parameters were used. However, in cases where the results need to be reproduced bit by bit, more detailed information about the hardware architecture of the resource, environment variables used, library versions, and the like are needed. Finally, provenance can also be used to analyze the performance of the analyses,[17] where the provenance records are mined to determine the number of tasks executed, their runtime distribution, where the execution took place, and so forth.

In some cases, scientific processes are managed by workflow management systems. These may take in an abstract workflow description and generate an executable workflow. During the mapping the workflow system may modify the executable workflow to the point that it is no longer easy to map between what has been executed and what the user specified.[18] As a result, information about the workflow restructuring process needs to be recorded as well.[19] This information not only allows us to relate the user-created and the executable workflow but is also the foundation for workflow debugging, where the user can trace how the specification they provided evolved into an executable workflow.

In the area of workflow management and provenance, an interesting aspect of workflow creation is the ability to retrace how a particular workflow has been designed, or in other words, to determine the provenance of the workflow creation process. A particularly interesting approach is taken in VisTrails[20,21] where the user is presented with a graphical interface for workflow creation and the system incrementally saves the state of the workflow as it is being designed, modified, or enhanced. As a result the users may retrace their steps in the design process, choose various "flavors" of the same workflow and try and retry different designs. A challenge could be not only to capture the *how* but also the *why* of the design decisions made by the users.

Unlike metadata, much process documentation is relatively easy to produce automatically, especially in the context of workflows, since the workflow system is in charge of setting up the environment for the computation, managing the data, and invoking the analysis steps. Thus, the workflow system is able to capture the information about where the execution took place, what were the parameters and environment variables used by a software component, which data files were used, and so forth. Some of that information may be somewhat obscured, for example, when a configuration file is used instead of placing the parameters on the command line. However, the workflow system can also automatically save information about which configuration file was used. It is also interesting to note that the capabilities of workflow management and provenance management systems are complementary (process execution vs. process documentation), and thus it is possible to integrate workflow management systems with provenance management systems that have been developed independently of each other.[22]

12.4 Survey of Existing Approaches

In this section we describe some of the existing approaches for managing metadata and provenance. We start by giving an example of the various types of attributes that are part of the metadata of scientific data.

12.4.1 Metadata Schema and Metadata Attributes

Metadata attributes that are elements of a metadata schema can encompass a variety of information. Some metadata is application independent, such as the creation time, and author, as described in Dublin Core[23]; and other metadata is application dependent and may include attributes such as the duration of an experiment, temperature of the device, and others. Many applications have expanded the Dublin Core schema to include application-dependent attributes.[24]

Based on experiences with a number of scientific applications, we described nine general types of metadata attributes.[25,26] These descriptions were used in metadata systems developed as part of the Metadata Catalog Service (MCS),[25] the Laser Interferometer Gravitational-Wave Observatory (LIGO) project,[27] the Linked Environments for Atmospheric Discovery (LEAD) project,[28] and others. Below we describe some of the metadata attribute categories.

Logical file metadata: Metadata attributes associated with the logical file include the following. A *logical file name* attribute specifies a name that is unique within the namespace managed by a metadata service. A *data type* attribute describes the data item type, for example, whether the file format is binary, html, XML, and so forth. A *valid* attribute indicates whether a data item is currently valid, allowing users to quickly invalidate logical files, for example, if the file administrator determines that a logical file contains incorrect data. If data files are updated over time, a *version* attribute allows us to distinguish among versions of a logical file. A *collection identifier* attribute allows us to associate a logical file with exactly one logical collection. *Creator* and *last modifier* attributes record the identifications of the logical file's creator and last modifier. Other attributes specify the *creation time* and the *last modification time*. A *master copy* attribute can contain the physical location of the definitive or master copy of the file for use by higher-level data consistency services.

Logical collection metadata: Attributes include the *collection name* and a description of the *collection contents*, which consist of the list of logical files and other logical collections that compose this collection. Each logical file can belong to at most one logical collection. Logical collections may contain other collections, but must form an acyclic collection hierarchy. In addition, the collection metadata includes a *text description* of the collection, information about the *creator* and *modifiers* of the collection, and *audit* information. Finally, there may be a *parent* attribute that records the identifier of the parent logical collection. There may be an arbitrarily deep acyclic hierarchy of logical collections.

Logical view metadata: Attributes include the *logical view name* and *description*; information about the logical files, logical collections, and other logical views that compose this logical view; attributes describing the *creator* and *modifiers* of the view; and *audit* information.

Authorization metadata: Attributes are used to determine the *access permissions* to the data items and metadata.

User metadata: Attributes that describe *writers of metadata*, including contact information. The attributes specify the distinguished *name, description, institution, address, phone,* and *email* information for writers.

User-defined metadata attributes: Extensibility of the schema beyond predefined attributes is provided by allowing users to define new attributes and associate them with logical files, collections, or views. Extensibility is an essential requirement in metadata systems, since each scientific application domain typically produces one or more metadata schemas that capture attributes of interest to that community.

Annotation attributes: Annotations can be attached to logical files, collections, or views. Annotation metadata includes the *identifier* for the object being annotated and the *object type* (logical file, collection or view). The annotation attribute is a string provided by the user. Annotation metadata also includes the distinguished name of the user creating the annotation and a *timestamp* that records when the annotation was created.

Creation and transformation history metadata: These provenance attributes record process information about how a logical file was created and what subsequent transformations were performed on the data. This information may be used to recreate the data item if it ever gets corrupted, or the application may decide to recreate the dataset if the cost of recreating it is less than the cost of retrieval.

External catalog metadata: Because metadata may be spread across multiple heterogeneous catalogs, these attributes can be used to access external catalogs.

12.4.2 Technologies for Storing Metadata Management

Metadata catalogs have utilized a variety of underlying technologies, including relational databases, XML-based databases, grid database services, and RDF triple stores.

Relational databases are well-suited for metadata repositories in application domains that have a well-defined metadata ontology that changes relatively slowly. Relational databases store data in tables and offer good scalability in both the amount of data they can store and the number of simultaneous queries they can support. These databases also support the construction of indexes on particular metadata attributes, which can provide good performance for common queries related to those attributes. Scientific collaborations often rely on open source relational databases such as PostgreSQL[29] and MySQL,[30] but some projects use commercial solutions (Oracle, DB2). Examples of scientific collaborations whose metadata catalogs have used a relational database include the Laser Interferometer Gravitational-Wave Observatory (LIGO) project[31] and the Earth System Grid.[32]

XML-based databases provide the ability to store and query content stored in eXtended Markup Language (XML) format. Although some "native" XML databases store data in XML format, others map XML data to a different format and use a relational or hierarchical database to store the data. XML databases can be queried using a variety of languages, such as XPath and XQuery. Examples of XML databases include the Apache Xindice database[33] and Oracle Berkeley DB XML.[34]

The Resource Description Framework (RDF)[35] supports the representation of graph-based semantic information using a simple data model. An RDF expression is represented by a set of triples, where each triple contains a subject, a predicate, and an object. A triple asserts that a relationship exists between the subject and the object, where the relationship is specified by the predicate. Information about the predicate, subject, and object of the triples may be related to components defined in an existing ontology (which can be implemented in languages like RDF Schema or OWL). This allows defining explicitly the semantics of the objects used in the triples and of the assertions made within these triples. Besides, it allows performing consistency checks and inferring new information from the information provided in the triples.

RDF information can be stored and queried in an RDF triple store. Over time, a growing number of metadata catalogs have made use of RDF to store semantic metadata information. RDF triple stores are often implemented using a relational database or a hierarchical database as a back end. For example, the Jena Semantic Web Toolkit[36] includes functionality to store and query RDF triples using an Oracle Berkeley DB back end or using a relational database (PostgreSQL, MySQL, etc.) via a Java database connectivity (JDBC) interface.[37] Sesame[38] provides a Java framework for storing and querying RDF triples. The storage and inference layer (SAIL) of Sesame interfaces between RDF functions and the API for various databases, including relational and object-oriented databases and other RDF stores.[37] Besides basic querying, these triple stores also implement consistency checking and inference services that exploit the semantics defined in RDF Schema and OWL ontologies.

Another technology used in grid environments to deploy metadata catalogs is the OGSA-DAI (Data Access and Integration) service.[39] The OGSA-DAI middleware provides a grid service interface that exposes data resources such as relational or XML databases. Clients of the OGSA-DAI service can store and query metadata in the back-end database. One example of a metadata catalog that uses the OGSA-DAI service in its deployment is the Metadata Catalog Service (MCS),[40,41] which provides a set of generic metadata attributes that can be extended with application-specific attributes (described in Section 12.4.1). MCS is used to store metadata during workflow execution by the Pegasus workflow management system,[18,42] which in turn is used by a variety of scientific applications, including LIGO,[43] Southern California Earthquake Context (SCEC)[44] and Montage.[45]

Finally, some metadata catalogs are integrated into general-purpose data management systems, such as the Storage Resource Broker (SRB).[46] SRB includes an internal metadata catalog called MCAT.[47] SRB supports a logical namespace that is independent of the physical namespace. Logical files in SRB can also be aggregated into collections. SRB provides various authentication mechanisms to access metadata and data within SRB.

12.4.3 Technologies for Provenance Management

The topic of provenance is the focus of many research communities, including e-Science and grid computing, databases, visualization, digital libraries, Web technologies, and operating systems. Two surveys by Bose and Frew[48] and Simmham[49] provide comprehensive overviews of provenance-related concepts, approaches, technologies, and implementations. In the recent provenance challenge,[50] 16 different systems were used to answer typical provenance queries pertaining to a brain atlas dataset that was produced by a demonstrator workflow in the context of functional magnetic resonance imaging.

Inspired by the summary of contributions in Moreau and Ludaescher,[50] we present key characteristics of provenance systems. Most provenance systems are embedded inside an execution environment, such as a workflow system or an operating system. In such a context, embedded provenance systems can track all the activities of this execution environment and are capable of providing a description of data produced by such environments. We characterize such systems as *integrated environments*, since they offer multiple functionalities, including workflow editing, workflow execution, provenance collection, and provenance querying.[21,51,52] Integrated environments have some benefits, including usability and seamless integration between the different activities. From a provenance viewpoint, there is close semantic integration between the provenance representation and the workflow model, which allows efficient representation to be adopted.[53] The downside of integrated systems is that the tight coupling of components rarely allows for their substitution or use in combination with other useful technologies; such systems therefore have difficulties interoperating with others, a requirement of many large-scale scientific applications.

In contrast to integrated provenance environments, approaches such as Provenance-Aware Service-Oriented Architecture (PASOA)[54,55] and Karma[56] adopt separate, autonomous provenance stores. As execution proceeds, applications produce process documentation that is recorded in a storage system, usually referred to as a *provenance store*. Such systems give the provenance store an important role, since it offers long-term, persistent, secure storage of process documentation. Provenance of data products can be extracted from provenance stores by issuing queries to them. Over time, provenance stores need to be managed to ensure that process documentation remains accessible and usable in the long term. In particular, PASOA has adopted a provenance model that is independent of the technology used for executing the application. PASOA was demonstrated to operate with multiple workflow technologies, including Pegasus,[19] Virtual Data Language (VDL)[57] and Business Process Execution Language (BPEL).[58] This approach that favors open data models and open interfaces allows the scientist to adopt the technologies of their choice to run applications. However, a common provenance model would allow for past executions to be described in a coherent manner, even when multiple technologies are involved.

All provenance systems rely on some form of database management system to store their data, and RDF and SQL stores were the preferred technologies. Associated query languages are used to express provenance queries, but some systems use query templates and query interfaces that are specifically provenance oriented, helping users to express precisely and easily their provenance questions without having to understand the underpinning schemas adopted by the implementations.

Another differentiator between systems is the *granularity of the data* that a provenance management system uses to keep track of the origins of the data. Again, the coupling of the provenance approach to the execution technology can influence the capability of the provenance management system from a granularity viewpoint. For instance, some workflow systems that allow for files to be manipulated by command-line programs such as Pegasus tend to track the provenance of files (and not the data they contain). This capability is sufficient in some cases, but is too coarse-grained in others. Systems such as Kepler,[51] on the other hand, have specific capabilities to track the provenance of collections. Other systems are capable of tracking the origins of programs, such as VisTrails. The PASOA system has been demonstrated to capture provenance for data at multiple levels of granularity (files, file contents, collections, etc.), and its integration with Pegasus showed it could be used to track the change in the workflow produced by the Pegasus workflow compiler.

Systems such as Earth Systems Science Server (ES3)[59] and Provenance-Aware Storage System (PASS)[60] capture events at the level of the operating system, typically reconstructing a provenance representation of files. In such a context, workflow scripts are seen as files whose origin can also be tracked.

The database community has also investigated the concept of provenance. Reusing the terminology introduced in this section, their solutions can generally be regarded as integrated with databases themselves: Given a data product stored in a database, they track the origin of data derivations produced by views and queries.[61] From a granularity viewpoint, provenance attributes can be applied to tables, rows, and even cells. To accommodate activities taking place outside databases, provenance models that support copy-and-paste operations across databases have also been proposed.[62] Such provenance models begin to resemble those for workflows, and research is required to integrate them smoothly.

Internally, provenance systems capture an explicit representation of the flow of data within applications, and the associated processes that are executed. At some level of abstraction all systems in the recent provenance challenge[63] use some graph structure to express all dependencies between data and processes. Such graphs are directed acyclic graphs that indicate from which ancestors processes and data products are derived. Given such a consensus, a specification for an open provenance model is emerging[64] and could potentially become the lingua franca by which provenance systems could exchange information. We illustrate this model over a concrete example in Section 12.6.

12.5 Metadata in Scientific Applications

In this section we present a couple of examples of how scientific applications manage their metadata.

12.5.1 Astronomy

Astronomy has used, for over 25 years, the Flexible Image Transport System (FITS) standard[65] for platform-independent data interchange, archival storage of images and spectra, and all associated metadata. It is endorsed by the U.S. National Aeronautics and Space Administration (NASA) and the International Astronomical Union. By design, it is flexible and extensible and accommodates observations made from telescopes on the ground and from sensors aboard spacecrafts. Briefly, a FITS data file is composed of a fixed logical record length of 2,880 bytes. The file can contain an unlimited number of header records, 80 bytes long, having a "keyword=value" format and written as ASCII strings. These headers describe the organization of the binary data and the format of the contents. The headers are followed by the data themselves, which are represented as binary records. The headers record all the metadata describing the science data. Figure 12.2 depicts an instance of this for the metadata of an image of the galaxy NGC 5584 measured by the Two-Micron All Sky Survey (2MASS).[66]

The complete metadata specification of an astronomical observation not only includes obvious quantities such as the time of the observation, its position and footprint on the sky, and the instrument used to make the observation, but also includes much information custom to particular datasets. FITS therefore was designed to allow astronomers to define keywords as needed. Nevertheless, FITS has predefined reserved keywords to describe metadata common to many observations. For example, the relationship between the pixel coordinates in an image and physical units is defined by the World Coordinate System (WCS),[65] which defines how celestial coordinates and projections are represented in the FITS format as keyword=value pairs. These keywords are listed in sequence in Figure 12.2: They start at CTYPE and end at CDELT.

Tabular data and associated metadata are often represented in FITS format, but the FITS standard is poorly specified for tabular data. Instead, tabular material, whether they are catalogs of sources or catalogs of metadata, are generally stored in relational databases to support efficient searches. Transfer of these data is generally in the form of ASCII files, but XML formats such as VOTable[67] are growing in use. Metadata describing catalog data can be grouped into three types: semantics, which describe science content (units, standards, etc.); logistical, which describe the structure of the table (data types, representation of null values, etc.) and statistical, which summarize the contents (number of sources, ranges of data in each column, etc.). The absence of standard specifications for columns has complicated and confused data discovery and access to tabular data. Work originating at the Centre de Donnees Astronomiques de Strasbourg (CDS)[68] has been embraced by the International Virtual Observatory Alliance (IVOA)[69] as part of an international effort to rectify this problem. When this work is complete, all column descriptors will have a well-defined meaning connected to a hierarchical data model.

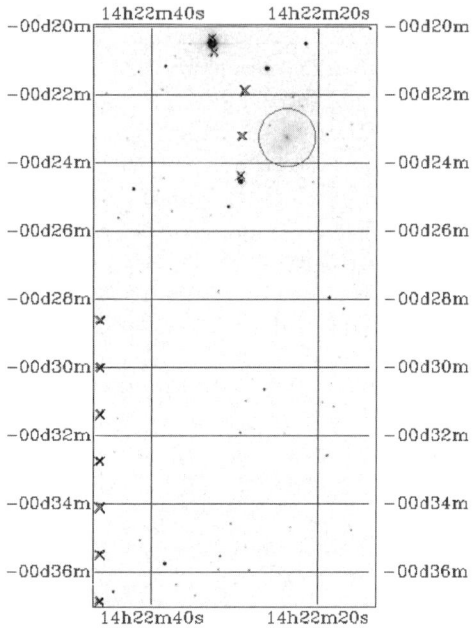

```
ORDATE  = '000503 '        / Observation Ref Date (yymmdd)
DAYNUM  = '1160   '        / Observation Day Num
FN_PRFX = 'j1160059'       / .rdo and .par filename prefix
TYPE    = 'sci   '         / Scan type: dar flt sci cal tst
SCANNO  =            59    / Scan Number
SCANDIR = 'n    '          / Scan Direction: n, s, -
COMMENT                    (O )
STRIP_ID=        301788    / Strip ID (O )
POSITNID= 's001422 '       / Position ID (O )
ORIGIN  = '2MASS  '        / 2MASS Survey Camera
CTYPE1  = 'RA---SIN'       / Orthographic Projection
CTYPE2  = 'DEC--SIN'       / Orthographic Projection
CRPIX1  =          256.5 / Axis 1 Reference Pixel
CRPIX2  =          512.5 / Axis 2 Reference Pixel
CR AL1  =    215.6251831 / RA  at Frame Center, J2000 (deg)
CR AL2  =     -0.4748106667 / Dec at Frame Center, J2000 (deg)
CROTA2  =   1.900065243E-05 / Image Twist +AXIS2 W of N, J2000 (deg)
CDELT1  =   -0.0002777777845 / Axis 1 Pixel Size (degs)
CDELT2  =    0.0002777777845 / Axis 2 Pixel Size (degs)
USXREF  =         -256.5 / U-scan X at Grid (0,0)
USYREF  =         19556. / U-scan Y at Grid (0,0)
```

Figure 12.2 (See Color insert following page 224.) Example of image metadata in astronomy. On the top is an image of the galaxy NGC 5584, shown at the center of the purple circle. The image was measured as part of the Two Micron All Sky Survey (2MASS). The crosses locate the positions of artifacts in the image. At bottom is a sample of the metadata describing the image, written in the form of keyword=value pairs, in compliance with the definition of the Flexible Image Transport System (FITS) in universal use in astronomy.

12.5.2 Climate Modeling

The Earth System Grid (ESG) project[32,70] provides infrastructure to support the next generation of climate modeling research. The ESG allows climate scientists to discover and access important climate modeling datasets, including the Parallel Climate Model (PCM), the Community Climate System Model (CCSM), as well as datasets from the Intergovernmental Panel on Climate Change (IPCC) 4th Assessment Report (AR4).[71]

The original infrastructure for ESG included two data portals, one at the National Center for Atmospheric Research and one at Lawrence Livermore National Laboratory. Each of these ESG portals had an associated metadata catalog that was implemented using a relational database. The next generation of the ESG features a federated architecture, with data nodes at many sites around the world publishing data through several ESG data portals or *gateways*. This data publication will include the extraction and publication of metadata attributes. The bulk of the metadata for this version of the ESG architecture will be stored in a relational database with a different schema from the previous generation of ESG. Metadata for research and discovery will also be harvested into an RDF triple store from the multiple federated ESG gateway sites; this will allow users to execute global searches on the full ESG holdings via the triple store.

The metadata model for the latest generation of the ESG includes metadata classes that describe a climate experiment, model, horizontal grid, standard name, and project as well as a set of data-related classes.[72] The Experiment class includes a specification of the input conditions of a climate model experiment. The Model class describes the configuration of a numerical climate model. The Horizontal Grid class specifies a discretization of the earth's surface that is used in climate models. Standard Names describe scientific quantities or parameters generated by a climate model run, such as air pressure, atmospheric water content, direction of sea ice velocity, and so forth. A Project is an organizational activity that generates datasets. Data objects in the ESG metadata model have one of four types: dataset, file, variable, or aggregation. A *dataset* is a collection of data generated by some activity, such as a project, a simulation run, or a collection of runs. *Files* are usually in the self-describing netCDF[73] data format. A *variable* is a data array with n dimensions, where the array is associated with a dataset. An *aggregation* is a service that provides a view of a dataset as a single netCDF file and that can perform statistical summaries over variable values, such as "monthly means."

There are several ways that ESG users search for and select datasets based on metadata attributes.[74] In one scenario, users can search for datasets based on metadata attributes using a simple Google-style text search over all the metadata associated with ESG datasets. The ESG gateway also presents users with a set of search terms and possible values, as illustrated in Figure 12.3; these terms include the experiment, model, domain, grid, variables, temporal frequency, and dataset. An ESG user may also access an ESG portal using

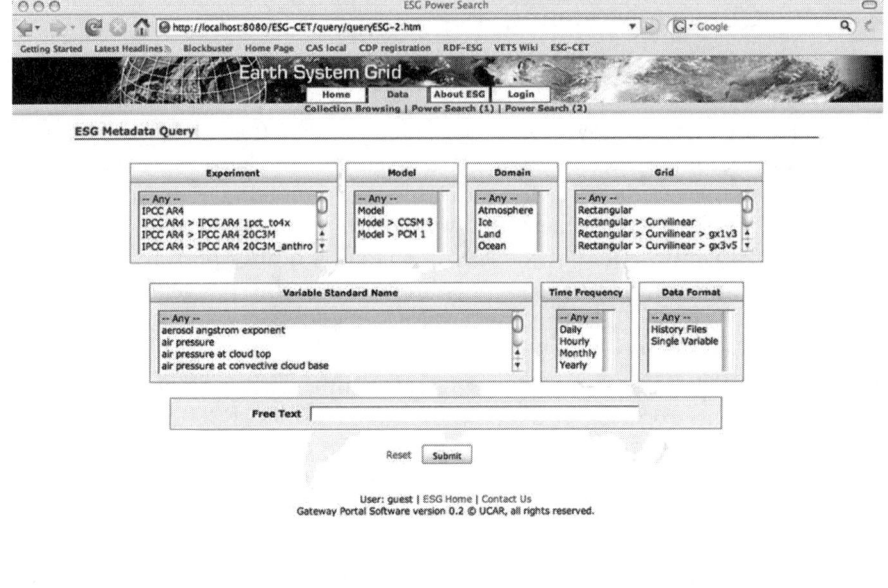

Figure 12.3 Earth System Grid project's interface for metadata queries.

a visualization or analysis tool that provides an API with search capabilities that may include issuing metadata queries.

12.6 Provenance in Scientific Applications

In this section, we detail how scientists use provenance. First, we discuss the technologies that scientists use every day in order to determine provenance. We then look at more advanced systems designed for particular applications where provenance plays a critical role; in particular, we focus on geospatial, oceanographic, and astronomical applications. Finally, we discuss how open data models will facilitate provenance queries in multidisciplinary, multiscale scientific applications.

12.6.1 Provenance in Everyday Science

Lab notebooks are used by scientists every day to record their experimental processes and results. These lab notebooks contain, what we termed previously, *process documentation*. However, determining provenance from physical notebooks is a laborious procedure. Additionally, lab notebooks are ill equipped to capture the data produced by scientific applications. To ameliorate these concerns, scientists have begun to capture their work using electronic systems. While research on electronic notebooks abounds,[75,76] probably the most widely used system for recording lab notebook style information electronically are Wikis.* Wikis provide a system for easily creating, modifying, and collaborating on Web pages. For example, the Bradley Laboratory and Drexel University post the protocols and results for the chemical solubility experiments that they are performing on their Wiki.[77] Similarly, the OpenWetWare project[78] provides a Wiki for the sharing and dissemination of biological protocols. Available protocols range from definitions of how to extract DNA from mouse tissue to setting up microarrays. The project has over 3,000 registered users.[79]

A key piece of functionality is a Wiki's support for a revision history of each page.[80] This revision history provides a coarse view of the provenance of a page. It provides information as to who edited the page, the time when the page was modified, and the difference between the current state of the page and its last revision. For OpenWetWare, the revision history enables the tracking of a protocol's development and thus allows for the creation of community protocols. Indeed, the ability to roughly determine the provenance of a Wiki page is a key enabler to Open Access Science where scientists share drafts of experiments, papers, and preliminary results online for others to comment on and use.[79]

While a Wiki page's coarse provenance is useful, the revision history fails to provide a comprehensive view of the provenance of a page because it does not describe the process by which it was generated. Instead only the difference between pages is known. For example, if a JPEG image added to a Wiki page was produced by converting an image output from an image registration algorithm applied to two other files, the revision history would be unable to inform a scientist that JPEG conversion and image registration were involved in the creation of the JPEG image. Thus, the provenance for the page is incomplete. In many scientific applications, this sort of processing history is required, and hence the provenance technologies discussed previously are needed. We now discuss some of these applications.

12.6.2 Provenance in Geospatial Applications

Some of the first research in provenance was for geospatial systems in particular Geographic Information Systems (GIS).[81] Knowing the provenance of map products is critical in GIS applications because it allows one to determine the quality of those derived map products. In particular, Wang et al.[82] highlights the need for systems to be able to isolate provenance of a specific

---

*The *Oxford English Dictionary* defines *Wiki* as a type of web page designed so that its content can be edited by anyone who accesses it, using a simplified markup language.

spatial region on a map. For example, when computing annual rainfall for the Denver area, the data used as input is the national daily rainfall numbers. When retrieving the provenance of the annual rainfall, it is necessary to have spatial knowledge to know that only the daily rainfall from Denver is used from the input dataset.

A more complex example is given in Frew and Base[83] for tracking the processing of satellite images: Image data covering the western United States and Pacific Ocean is retrieved from National Oceanic and Atmospheric Administration (NOAA) satellites and is sent to a University of California Santa Barbara-operated TeraScan ground station. These Advanced Very High Resolution Radiometer images are processed at the Institute for Computational Earth System Science into two data products for Southern California and the southwestern United States: a sea surface-temperature map and a near-infrared albedo map. Figure 12.4 shows one such sea surface temperature map.

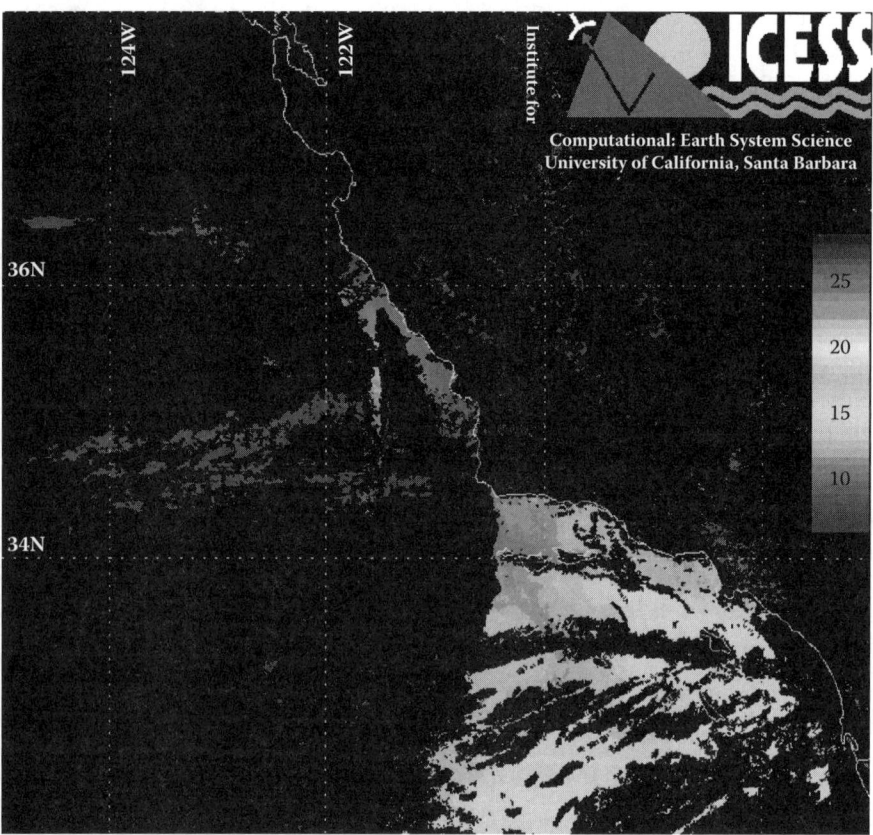

Figure 12.4 (See color insert following page 224.) Map of sea surface temperature taken February 18, 2009, 10:42 p.m. GMT.

To generate these data products, some standard processing steps are applied including calibration, navigation, and registration. These data products are then provided to other scientists, who use it for their own process.[83]

One of the key observations of Frew and Bose[83] is that scientists may use many different command line applications (e.g., Perl scripts, MATLAB, TerraScan tools) to generate new satellite data products. The Earth System Science Workbench (ESSW) system developed at the University of California, Santa Barbara, therefore keeps track of which command line applications are used to modify the data products and records all intermediate data products. Using ESSW, scientists can determine which satellites were used in generating a data product, what version of the calibration algorithm was used, and the number of scan samples used in the original sensor data.

Provenance is also critical in regenerating large-scale data products derived from satellite imagery. An example is how imagery from the Moderate Resolution Imaging Spectroradiometer Satellite is processed and stored at NASA's Goddard Space Flight Center using the MODIS Adaptive Data Processing System (MODAPS). The imagery obtained directly from the satellite is known as Level 0 data. This raw data is irreplaceable and must be archived. However, the initial transformation of this raw data into more useful data called Level 1B data is too large to be archived. Level 1B data includes calibrated data and geo-located radiances for the 36 channels of the satellite.* Level 1B data is then processed and consolidated to create Level 2 data products, which are finally translated into visual images. The Level 2 and final images are the most useful for scientists. Because of its size, Level 1B data is typically kept for 30–60 days and then discarded. To enable this data to be reproduced the MODAPS systems maintains enough process documentation to reproduce the Level 1B data from the raw satellite data. The process documentation includes the algorithms used, their versions, the original source code, a complete description of the processing environment, and even the algorithm design documents themselves. Essentially, provenance enables a *virtual archive* of satellite data, which otherwise would be lost or difficult to recreate.[84]

Provenance for geospatial data is also extremely important for merging data from multiple sources. This aspect was discussed in detail in Chapter 10 in the context of interoperability and data integration in geosciences.

12.6.3 Provenance for Oceanographic Applications

When studying the oceans, scientists require data from multiple data sources—shipboard instruments, buoy-based sensors, underwater vehicles, and permanent stations. These data sources measure everything from the temperature of the ocean to its salinity and the amount of chlorophyll present.

---

*http://modis.gsfc.nasa.gov/data/dataprod/dataproducts.php?MOD_NUMBER=02. Accessed July 20, 2009.

These data sources are combined with published data including satellite imagery in simulations to predict oceanographic features, such as seasonal variations in water levels.[85] Using the current data analysis routines, it is difficult to ascertain how the results of these simulations were produced because the information is spread in log files, scripts, and notes.[85]

To address this problem, the Monterey Bay Aquarium Research Institute has developed their own in-house system, the Shore Side Data System (SSDS), for tracking the provenance of their data products.[86] These data products range from lists of deployed buoys to time series plots of buoy movement. Using SSDS, scientists can access the underlying sensor data, but most importantly, they can track back from derived data products to the metadata of the sensors including their physical location, instrument, and platform. A key part of the system is the ability to automatically populate metadata fields. For example, by understanding that the position of an instrument is caused by the fact that it is located on a mooring platform, the system can traverse the provenance graph to fill in the position metadata for that instrument [86]. It is interesting to note that the metadata produced by SSDS is in the netCDF standard format, which was previously discussed in Chapter 2, Section 2.4. SSDS is an example of a production provenance system as it has been used daily for the past four years for managing ocean observation data.[86]

The focus of SSDS is tracking the provenance of datasets back to the instruments and the associated configuration metadata. A more complex example of provenance in oceanography is given in Howe et al.[85] In this work, the authors present a system that combines sensor data products with simulations to present 3D visualizations of fishery data. Specifically, for the Collaborative Research on Oregon Ocean Salmon Project, they combined data about the location and depth of where salmon were caught in the Northwest of the United States with simulation data about ocean currents to generate visualizations of the depth and distribution of fish when looking at the continental shelf.[85] The key use of provenance here is to enable the scientists to explore the parameter space of a visualization without having to worry about tracking the changes to their visualization pipeline. For example, to see a different perspective on the fish, the scientist may have to reconfigure the model they are using. With the VisTrails[87] system, (described in Chapter 13, Section 13.5) they can easily find the changes they made or go back to other visualization pipelines. This functionality is critical when dealing with these complex oceanographic applications that integrate a variety of simulation techniques and data sources.

12.6.4 End-to-End Provenance for Large-Scale Astronomy Applications

We have seen the need to use provenance to re-create data on demand for satellite imagery, automatically populate metadata fields for oceanographic data, and track the changes in pipelines for visualizing salmon catches.

In this section, we see how provenance enables connecting research results to high-level workflows in an astronomy application.

The application we look at is Montage.[45] Montage produces science-grade mosaics of the sky on demand. This application can be structured as a workflow that takes a number of images, projects them, adjusts their backgrounds, and adds the images together. A mosaic of 6 degrees square would involve processing 1,444 input images, require 8,586 computational steps, and generate 22,850 intermediate data products. Executing the Montage workflow requires potentially numerous distributed resources that may be shared by other users. Because of the complexity of the workflow and the fact that resources often change or fail, it is infeasible for users to define a workflow that is directly executable over these resources. Instead, scientists use "workflow compilers" such as Pegasus[18,42] (see Chapter 13) to generate the executable workflow based on a high-level, resource-independent description of the end-to-end computation (an *abstract workflow*). This approach gives scientists a computation description that is portable across execution platforms and can be mapped to any number of resources. However, the additional workflow mapping also increases the gap between what the user defines and what is actually executed by the system and thus complicates the interpretation of the results: The connection between the scientific results and the original experiment is lost.

To reconnect the scientific results with the experiment, Miles et al.[19] and Miles et al.[88] present a system for tracking the provenance of a mosaic back to the abstract workflow that it was generated from. The system integrates the PASOA[54] and Pegasus systems to answer provenance questions such as what particular input images were retrieved from a specific archive, whether parameters for the re-projections were set correctly, what execution platforms were used, and whether those platforms included processors with a known floating point processing error.

To accomplish this, each stage of the compilation from abstract workflow to executable workflow is tracked in Pegasus. For example, one of Pegasus's features is to select at which sites or platforms each computational step should be executed. During the Pegasus compilation process, this information is stored as process documentation within PASOA's provenance store. Additionally, this information is linked to the subsequent compilation steps such as intermediate data registration, and task clustering. Finally, during the execution of the Pegasus-produced workflow, all execution information is stored and linked to the workflow within the provenance store. Using this process documentation, a provenance graph of the resulting sky mosaic can be generated that leads back to the specific site selected.

The availability of provenance in Montage enables astronomers to take advantage of workflow automation technologies while still retaining all the necessary information to reproduce and verify their results. Outside of Montage, provenance is an underpinning technology that allows for workflow automation, a technology necessary for other large-scale grid-based science applications, such as astrophysics.[89]

12.6.5 Enabling Multidisciplinary and Multiscale Applications Using Provenance

As we have seen, provenance plays an important role in enabling scientific applications. In particular, those applications that use a variety of heterogeneous data sources or computational resources benefit. Scientific problems are increasingly becoming multidisciplinary and multiscale. For example, biomedical applications may combine the results of chemistry simulations of molecular interactions with data about tissues and other organs.[90]

To allow the provenance of data to be determined across boundaries of scale, discipline, and technologies, there is a need for an interoperability layer between systems. One proposed interoperability specification is the Open Provenance Model.[55] This model provides a good outline of what the community developing provenance technologies believes are the core constituents of a provenance graph. We thus use a graphical representation drawn from the Open Provenance Model to illustrate a concrete provenance graph, as shown in Figure 12.5.

In the representation we adopt, nodes of the graph consist of two entities: artifacts and processes. In the context of scientific workflows as considered here, artifacts are immutable pieces of data, whereas processes are transformations that produce and consume artifacts. Artifacts are represented as circles, and processes are denoted by boxes.

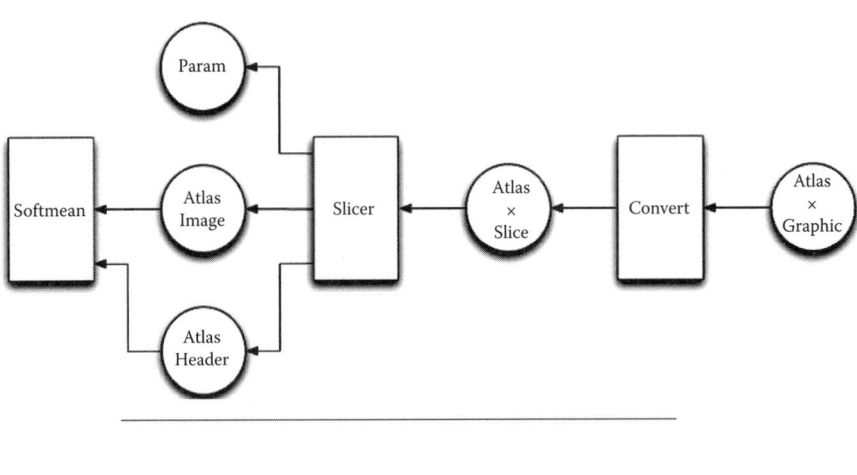

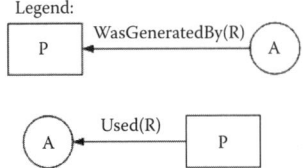

Figure 12.5 Provenance graph for the Provenance Challenge Workflow.

Nodes can be connected by edges expressing causal dependencies between artifacts and processes. The origin of an edge represents an effect, whereas its destination represents a cause: The presence of an edge makes explicit the causal dependency between the effect and its cause. In this presentation, we focus on two types of edges: "wasGeneratedBy" and "used." A "wasGeneratedBy" edge expresses how an artifact was dependent on a process for its generation, whereas a "used" edge indicates that a process relied on some artifacts to be able to complete. An artifact can only be generated by a single process, but it can be used by any number of processes; whereas a process can use and generate any number of artifacts. To be able to distinguish the multiple dependent artifacts a process may rely upon, a notion of role is introduced, allowing the nature of the causal dependency to be characterized explicitly.

Using the above notation, we show a provenance graph generated from the workflow adopted by the provenance challenge,[63] which is inspired by functional MRI (fMRI) workflows to create population-based "brain atlases" from the fMRI Data Center's archive of high-resolution anatomical data.[91] In summary, this workflow produces average images along the axes X, Y, and Z, after aligning each input sample with a reference image. Note that like the other applications discussed, neuroscience applications require provenance.[92]

Figure 12.5 illustrates a subset of the provenance graph that is constructed as the provenance challenge workflow. Such a graph is best read from right to left: The right identifies an artifact, the Atlas X graphic, representing an averaged image along the X axis; all the causal dependencies that led it to be produced appear to its left. Provenance graphs are directed and acyclic, which means that an artifact or a process cannot be (transitively) caused by itself.

Whenever a scientific workflow system executes the fMRI workflow, it would incrementally produce the various elements of that graph (or an equivalent representation), and store them in a repository, usually referred to as provenance store or provenance catalog. Provenance queries can then be issued to extract a subset of the documentation produced, according to the user's needs.

In conclusion, provenance is critical in many scientific applications ranging from neuroscience to astronomy. As scientific applications become increasingly open and integrated across areas, provenance interoperability becomes an important requirement for systems technologies.

12.7 Current and Future Challenges

There are several challenges in the area of metadata and provenance management. They stem mostly from two facts: (1) scientists need to share information about data within their collaborations and with outside colleagues, and (2) the amount of data and related information is growing at unprecedented scales.

As a result of the scale, users need to decide which data to keep (for example, in high-energy physics only selected and already preprocessed collision events are cataloged). When storing provenance, decisions of what to store need to be made as well. Because it is often hard to predict what will be needed in the future, sometimes data and related information are irrevocably lost.

Because of the size of the collaborations and datasets, data, metadata, and provenance information are often not stored at the same location, within the same system. Thus issues of information federation arise. In some sense, the issue for provenance is not as severe as for metadata. Provenance is in some sense inherently distributed, with information about the data coming from different sources, and it also has explicit links (such as those in the provenance graph) that allow one to follow the provenance trail. Additionally, once the process documentation of an item is generated, it will most likely not change since it is a historical record. On the other hand, metadata about data items may change with time, or a piece of data may be found invalid. As a result, metadata requires more effort in the area of consistency management.

The need to share data results in many challenges. First, communities need to agree on metadata standards. Then, these standards need to be followed by data publishers and software systems so that a consistent view of metadata is maintained. When data are shared across communities, mediation between metadata schemas needs to be performed. The challenge for cross-project or cross-community interoperability is not only technical but also social. How does one motivate scientists to provide the necessary metadata about the primary and derived data? What is the incentive to retrofit the codes and publish the data into community repositories?

In general, future work should focus on extensible metadata and provenance systems that follow common standards that are independent of the systems that use them, and can be shared across distributed collaborations. Such systems should support common languages for responding to provenance queries. There is already good progress, but unified metadata and provenance systems for scientific communities are a long way off.

Acknowledgments

Ewa Deelman's work was funded by the National Science Foundation under Cooperative Agreement OCI-0438712 and grant # CCF-0725332. Bruce Berriman is supported by the NASA Multi Mission Archive and by the NASA Exoplanet Science Institute at the Infrared Processing and Analysis Center, operated by the California Institute of Technology in coordination with the Jet Propulsion Laboratory (JPL). Oscar Chorcho's work was funded by the SemsorGrid4Env project (FP7-ICT-223913). The authors would like to thank members of the Earth System Grid for the use of their metadata example.

The Earth System Grid Center for Enabling Technologies is funded by the DOE SciDAC program.

References

[1] J. Greenherg, Metadata and the World Wide Web, *Encyclopedia of Library and Information Science*, Taylor & Francis 2003.

[2] M. Sanchez-Gestido, L. Blanco-Abruna, M. S. Perez-Hernandez, R. Gonzalez-Cabero, A. Gomez-Perez, and O. Corcho, Complex Data-Intensive Systems and Semantic Grid: Applications in Satellite Missions, *Proceedings of the 2nd IEEE International Conference on e-Science and Grid Computing (e-Science 2006), Amsterdam, The Netherlands, December*, 2006.

[3] J. Hunter, Harvesting community tags and annotations to augment institutional repository metadata. Brisbane, Australia: eResearch Australasia, 2007. http://www.eresearch.edu.au/hunter.

[4] Flickr, 2007. http://www.flickr.com/.

[5] delicious, 2007. http://del.icio.us/.

[6] J. A. Blake, J. E. Richardson, C. J. Bult, J. A. Kadin, and J. T. Eppig, The Mouse Genome Database (MGD): the model organism database for the laboratory mouse. *Nucleic Acids Research*, vol. 30, pp. 113–115, 2002.

[7] M. Gertz, K.-U. Sattler, F. Gorin, M. Hogarth, and J. Stone, Annotating scientific images: a concept-based approach, in *14th International Conference on Scientific and Statistical Database Management*, pp. 59–68, 2002.

[8] myGrid, http://www.mygrid.org.uk. Accessed on July 20, 2009.

[9] S. Miles, J. Papay, C. Wroe, P. Lord, C. Goble, and L. Moreau (2004), Semantic Description, Publication and Discovery of Workflows in myGrid, Electronics and Computer Science, University of Southampton. Technical Report ECSTR-IAM04-001, 2004.

[10] R. Studer, V. R. Benjamins, and D. Fensel, Knowledge engineering: principles and methods, *Data & Knowledge Engineering*, vol. 25, pp. 161–197, 1998.

[11] S. Luke, L. Spector, D. Rager, and J. Hendler, Ontology-based Web Agents, *Proceedings of First International Conference on Autonomous Agents*, pp. 59–66, 1997.

[12] V. R. Benjamins, D. Fensel, S. Decker, and A. G. Perez, KA) super (2): building ontologies for the Internet: a mid-term report, *International Journal of Human-Computers Studies*, vol. 51, pp. 687–712, 1999.

[13] J. Domingue and E. Motta, PlanetOnto: from news publishing to integrated knowledge management support, *IEEE Intelligent Systems and Their Applications*, vol. 15, pp. 26–32, 2000.

[14] D. Brickley and R. V. Guha, RDF Vocabulary Description Language 1.0: RDF Schema, W3C Recommendation, February 10, 2004, *B. McBride*, 2004.

[15] L. Moreau, P. Groth, S. Miles, J. Vazquez, J. Ibbotson, S. Jiang, S. Munroe, O. Rana, A. Schreiber, V. Tan, and L. Varga, The Provenance of Electronic Data, *Communications of the ACM*, Vol. 51(4): 52–58, 2008.

[16] Y. Gil, E. Deelman, M. Ellisman, T. Fahringer, G. Fox, D. Gannon, C. Goble, M. Livny, L. Moreau, and J. Myers, Examining the Challenges of Scientific Workflows, *IEEE Computer*, vol. 40, pp. 24–32, 2007.

[17] E. Deelman, S. Callaghan, E. Field, H. Francoeur, R. Graves, N. Gupta, V. Gupta, et al., Managing Large-Scale Workflow Execution from Resource Provisioning to Provenance Tracking: The CyberShake Example, *E-SCIENCE '06: Proceedings of the Second IEEE International Conference on e-Science and Grid Computing*, p. 14, 2006.

[18] E. Deelman, G. Singh, M.-H. Su, J. Blythe, Y. Gil, C. Kesselman, G. Mehta, K. Vahi, G. B. Berriman, J. Good, A. Laity, J. C. Jacob, and D. S. Katz, Pegasus: a framework for mapping complex scientific workflows onto distributed systems, *Scientific Programming Journal*, vol. 13, pp. 219–237, 2005.

[19] S. Miles, E. Deelman, P. Groth, K. Vahi, G. Mehta, and L. Moreau, Connecting scientific data to scientific experiments with provenance, in *e-Science*, 2007.

[20] J. Freire, C. T. Silva, S. P. Callahan, E. Santos, C. E. Scheidegger, and H. T. Vo, Managing Rapidly-Evolving Scientific Workflows, *International Provenance and Annotation Workshop (IPAW'06)*, 2006.

[21] S. P. Callahan, J. Freire, E. Santos, C. E. Scheidegger, C. T. Silva, and H. T. Vo, Managing the Evolution of Dataflows with VisTrails, *IEEE Workshop on Workflow and Data Flow for Scientific Applications (SciFlow 2006)*, 2006.

[22] S. Miles, E. Deelman, P. Groth, K. Vahi, G. Mehta, and L. Moreau, Connecting Scientific Data to Scientific Experiments with Provenance in *Third IEEE International Conference on e-Science and Grid Computing (e-Science 2007)* Bangalore, India, 2007.

[23] Dublin Core Metadata Initiative, http://www.dublincore.org. Accessed October 10th, 2009.

[24] W. D. Robertson, E. M. Leadem, J. Dube, and J. Greenberg, Design and Implementation of the National Institute of Environmental Health Sciences Dublin Core Metadata Schema, *Proceedings of the International Conference on Dublin Core and Metadata Applications 2001 table of contents*, pp. 193–199, 2001.

[25] E. Deelman, G. Singh, M. P. Atkinson, A. Chervenak, N. P. C. Hong, C. Kesselman, S. Patil, L. Pearlman, and M.-H. Su, Grid-Based Metadata Services, in *Statistical and Scientific Database Management (SSDBM)*, Santorini, Greece, 2004.

[26] G. Singh, S. Bharathi, A. Chervenak, E. Deelman, C. Kesselman, M. Manohar, S. Patil, and L. Pearlman, A Metadata Catalog Service for Data Intensive Applications, in *Supercomputing (SC)*, 2003.

[27] LIGO Project, Lightweight Data Replicator, http://www.lsc-group.phys.uwm.edu/LDR/, 2004. http://www.lsc-group.phys.uwm.edu/LDR/.

[28] B. Plale, D. Gannon, J. Alameda, B. Wilhelmson, S. Hampton, A. Rossi, and K. Droegemeier, Active management of scientific data, *Internet Computing, IEEE*, vol. 9, pp. 27–34, 2005.

[29] PostgreSQL Global Development Group, PostgreSQL, http://www.postgresql.org/, 2008.

[30] MySQL AB, MySQL, http:// www.mysql.com 2008.

[31] B. Abbott, et al. (LIGO Scientific Collaboration), Detector description and performance for the first coincidence observations between LIGO and GEO, *Nucl. Instrum. Meth.*, vol. A517, pp. 154–179, 2004.

[32] D. E. Middleton, D. E. Bernholdt, D. Brown, M. Chen, A. L. Chervenak, L. Cinquini, R. Drach, et al., Enabling worldwide access to climate simulation data: the earth system grid (ESG), *Scientific Discovery Through Advanced Computing (SciDAC 2006), Journal of Physics: Conference Series*, vol. 46, pp. 510–514, June 25–29, 2006.

[33] The Apache XML Project, Xindice, http://xml.apache.org/xindice/, 2008.

[34] Oracle Berkeley DB XML, http://www.oracle.com/database/berkeley-db/xml/index.html, 2008.

[35] G. Klyne and J. J. Carroll, Resource Description Framework (RDF): Concepts and Abstract Syntax, W3C Recommendation, http://www.w3.org/TR/rdf-concepts/, 2004.

[36] Jena — A Semantic Web Framework for Java, http://jena.source forge.net/, 2008.

[37] D. Beckett and J. Grant, SWAD-Europe Deliverable 10.2: Mapping Semantic Web Data with RDBMSes, http://www.w3.org/2001/sw/Europe/reports/scalable_rdbms_mapping_report/, 2001.

[38] Sesame, http://sourceforge.net/projects/sesame/, 2008.

[39] University of Edinburgh, Open Grid Services Architecture Data Access and Integration (OGSA-DAI), http://www.ogsadai.org.uk/, 2008. http://www.ogsa-dai.org.uk/.

[40] G. Singh, S. Bharathi, A. Chervenak, E. Deelman, C. Kesselman, M. Manohar, S. Pail, L. Pearlman, A Metadata Catalog Service for Data Intensive Applications, in *SC2003*, 2003.

[41] E. Deelman, G. Singh, M. P. Atkinson, A. Chervenak, N. P. Chue Hong, C. Kesselman, S. Patil, L. Pearlman, M.-H. Su, Grid-Based Metadata Services, in *16th International Conference on Scientific and Statistical Database Management*, 2004.

[42] E. Deelman, G. Mehta, G. Singh, M.-H. Su, and K. Vahi, Pegasus: Mapping Large-Scale Workflows to Distributed Resources, in *Workflows in e-Science*, I. Taylor, E. Deelman, D. Gannon, and M. Shields, Eds. New York: Springer, 2006.

[43] D. A. Brown, P. R. Brady, A. Dietz, J. Cao, B. Johnson, and J. McNabb, A Case Study on the Use of Workflow Technologies for Scientific Analysis: Gravitational Wave Data Analysis, in *Workflows for e-Science*, I. Taylor, E. Deelman, D. Gannon, and M. Shields, Eds. New York: Springer, 2006.

[44] E. Deelman, S. Callaghan, E. Field, H. Francoeur, R. Graves, N. Gupta, V. Gupta, et al., Managing Large-Scale Workflow Execution from Resource Provisioning to Provenance Tracking: The CyberShake Example, in *e-Science*, Amsterdam, The Netherlands, 2006.

[45] G. B. Berriman, E. Deelman, J. Good, J. Jacob, D. S. Katz, C. Kesselman, A. Laity, T. A. Prince, G. Singh, and M.-H. Su, Montage: A Grid Enabled Engine for Delivering Custom Science-Grade Mosaics On Demand, in *SPIE Conference 5487: Astronomical Telescopes*, 2004.

[46] A. Rajasekar, M. Wan, R. Moore, W. Schroeder, G. Kremenek, A. Jagatheesan, C. Cowart, B. Zhu, S. Y. Chen, and R. Olschanowsky, Storage resource broker-managing distributed data in a grid, *Computer Society of India Journal, Special Issue on SAN*, vol. 33(4), pp. 42–54, 2003.

[47] SRB Project, MCAT — A Meta Information Catalog (Version 1.1), http://www.npaci.edu/DICE/SRB/mcat.html. http://www.npaci.edu/DICE/SRB/mcat.html.

[48] R. Bose and J. Frew, Lineage retrieval for scientific data processing: a survey, *ACM Computing Surveys*, vol. 37, pp. 1–28, 2005.

[49] Y. L. Simmhan, B. Plale, and D. Gannon, A survey of data provenance in e-science, *SIGMOD Record*, vol. 34, pp. 31–36, 2005.

[50] L. Moreau and B. Ludaescher, *Journal of Computation and Concurrency: Practice and Experience, Special issue on the First Provenance Challenge*, vol. 20(5) 2007.

[51] B. Ludäscher, I. Altintas, C. Berkley, D. Higgins, E. Jaeger-Frank, M. Jones, E. Lee, J. Tao, and Y. Zhao, Scientific Workflow Management and the Kepler System, *Concurrency and Computation: Practice & Experience, Special Issue on Scientific Workflows*, vol. 18(10) 2005.

[52] T. Oinn, P. Li, D. B. Kell, C. Goble, A. Goderis, M. Greenwood, D. Hull, R. Stevens, D. Turi, and J. Zhao, Taverna/myGrid: Aligning a Workflow System with the Life Sciences Community, in *Workflows in e-Science*, I. Taylor, E. Deelman, D. Gannon, and M. Shields, Eds.: New York, Springer, 2006.

[53] R. S. Barga and L. A. Digiampietri, Automatic capture and efficient storage of e-Science experiment provenance, *Concurrency and Computation: Practice and Experience*, vol. 20(5): 419–429, 2007.

[54] Provenance Aware Service Oriented Architecture, 2006. http://twiki.pasoa.ecs.soton.ac.uk/bin/view/PASOA/WebHome. Accessed on July 20, 2009.

[55] S. Miles, P. Groth, S. Munroe, S. Jiang, T. Assandri, and L. Moreau, Extracting Causal Graphs from an Open Provenance Data Model, *Concurrency and Computation: Practice and Experience*, vol. 20(5), 577–586, 2007.

[56] Y. Simmhan, B. Plale, and D. Gannon. Query capabilities of the Karma provenance framework, *Concurrency & Practice, and Experience*, vol. 20, pp. 441–451, 2007.

[57] B. Clifford, I. Foster, M. Hategan, T. Stef-Praun, M. Wilde, and Y. Zhao, Tracking Provenance in a Virtual Data Grid, *Concurrency and Computation: Practice and Experience*, vol. 20(5): 565–575, 2007.

[58] T. Andrews, F. Curbera, H. Dholakia, Y. Goland, J. Klein, F. Leymann, K. Liu, D. Roller, D. Smith, S. Thatte, I. Trickovic, and S. Weerawarana, Specification: Business Process Execution Language for Web Services Version 1.1, 2003. http://www-106.ibm.com/developerworks/webservices/library/ws-bpel/.

[59] J. Frew, D. Metzger, and P. Slaughter, Automatic capture and reconstruction of computational provenance, *Concurrency and Computation: Practice and Experience*, vol. 20(5): 485–496, 2007.

[60] M. Seltzer, D. A. Holland, U. Braun, and K.-K. Muniswamy-Reddy, Passing the provenance challenge, *Concurrency and Computation: Practice and Experience*, vol. 20(5): 531–540, 2007.

[61] P. Buneman, Why and Where: A Characterization of Data Provenance, http://db.cis.upenn.edu/DL/whywhere.pdf. Accessed October 10th, 2009.

[62] R. Bose, I. Foster, and L. Moreau, Report on the International Provenance and Annotation Workshop:(IPAW'06) May 3–5, 2006, Chicago, *ACM SIGMOD Record*, vol. 35, pp. 51–53, 2006.

[63] L. Moreau, Lud, B. Ascher, I. Altintas, R. S. Barga, S. Bowers, S. Callahan, et al., The First Provenance Challenge, *Concurrency and Computation: Practice and Experience*, vol. 20(5): 409–418, 2007.

[64] L. Moreau, J. Freire, J. Futrelle, R. E. McGrath, J. Myers, and P. Paulson, The Open Provenance Model, University of Southampton 2007.

[65] M. R. Calabretta and E. W. Greisen, Representations of celestial coordinates in FITS, *Arxiv preprint astro-ph/0207413*, 2002.

[66] M. F. Skrutskie, S. E. Schneider, R. Stiening, S. E. Strom, M. D. Weinberg, C. Beichman, T. Chester, R. Cutri, C. Lonsdale, and J. Elias, The Two Micron All Sky Survey (2MASS): Overview and Status, *In The Impact of Large Scale Near-IR Sky Surveys*, eds. F. Garzon et al., p. 25. Dordrecht: Kluwer Academic Publishing Company, 1997.

[67] F. Ochsenbein, R. Williams, C. Davenhall, D. Durand, P. Fernique, R. Hanisch, D. Giaretta, T. McGlynn, A. Szalay, and A. Wicenec, VOTable: Tabular Data for the Virtual Observatory, *Toward an International Virtual Observatory, Proceedings of the ESO/ESA/NASA/NSF Conference* held in Garching, Germany, June 10–14, 2002. Edited by P.J. Quinn, and K.M. Gorski. *ESO Astrophysics Symposia*. Berlin: Springer, 2004, p. 118, 2004.

[68] Centre de Données Astronomiques de Strasbourg, 2007. http://cdsweb.u-strasbg.fr/.

[69] International Virtual Observatory Alliance, 2007. http://www.ivoa.net/.

[70] ESG Project, The Earth System Grid, www.earthsystemgrid.org, 2005. http://www.earthsystemsgrid.org.

[71] IPCC Fourth Assessment Report (AR4), http://www.ipcc.ch/. Accessed July 20, 2009.

[72] B. Drach, and Luca Cinquini, ESG-CET Metadata Model, http://esg-pcmdi.llnl.gov/documents/metadata/ESGCET_metadata_modelV2.doc/view, 2008.

[73] NetCDF (network Common Data Form). http://www.unidata.ucar.edu/software/netcdf/. Accessed July 20, 2009.

[74] B. Drach, Query/Browse Use Cases, http://esg-pcmdi.llnl.gov/documents/metadata-meeting-lbnl-2-12-07/, 2007.

[75] D. Butler, Electronic notebooks A new leaf, *Nature*, vol. 436, pp. 20–21, 2005.

[76] J. Myers, Collaborative electronic notebooks as electronic records: Design issues for the secure electronic laboratory notebook (eln), in *Proceedings of the 2003 International Symposium on Collaborative Technologies and Systems (CTS'03)*, Orlando, Florida, 2003, pp. 13–22.

[77] Usefulchem. http://usefulchem.wikispaces.com. Accessed July 20, 2009.

[78] OpenWetWare. http://openwetware.org/wiki/Main_Page. Accessed on July 20, 2009.

[79] M. Waldrop, Science 2.0, *Scientific American Magazine*, vol. 298, pp. 68–73, 2008.

[80] A. Ebersbach, M. Glaser, and R. Heigl, *Wiki: Web collaboration*: Springer-Verlag New York, Inc., 2008.

[81] D. Lanter, Design of a lineage-based meta-data base for GIS, *Cartography and Geographic Information Science*, vol. 18, pp. 255–261, 1991.

[82] S. Wang, A. Padmanabhan, J. Myers, W. Tang, and Y. Liu, Towards Provenance-Aware Geographic Information Systems, in *Proceedings of the 16th ACM SIGSPATIAL International Conference on Advances in Geographic Information Systems*, 2008.

[83] J. Frew and R. Bose, Earth System Science Workbench: A Data Management Infrastructure for Earth Science Products, in *Proceedings of the 13th International Conference on Scientific and Statistical Database Management (SSDBM 2001)*, 2001, pp. 180–189.

[84] C. Tilmes and A. Fleig, Provenance Tracking in an Earth Science Data Processing System, in *Second International Provenance and Annotation Workshop, IPAW* 2008, p. 221.

[85] B. Howe, P. Lawson, R. Bellinger, E. Anderson, E. Santos, J. Freire, C. Scheidegger, A. Baptista, and C. Silva, End-to-End eScience: Integrating Workflow, Query, Visualization, and Provenance at an Ocean Observatory, in *IEEE Internationational Conference on e-Science*, 2008.

[86] M. McCann and K. Gomes, Oceanographic Data Provenance Tracking with the Shore Side Data System, in *Second International Provenance and Annotation Workshop, IPAW* 2008, p. 309.

[87] C. Scheidegger, D. Koop, E. Santos, H. Vo, S. Callahan, J. Freire, and C. Silva, Tackling the Provenance Challenge One Layer at a Time, *Concurrency and Computation: Practice and Experience*, vol. 20(5): 473–483, 2007.

[88] S. Miles, P. Groth, E. Deelman, K. Vahi, G. Mehta, and L. Moreau, Provenance: The bridge between experiments and data, *Computing in Science & Engineering*, vol. 10, pp. 38–46, 2008.

[89] M. Vouk, I. Altintas, R. Barreto, J. Blondin, Z. Cheng, T. Critchlow, A. Khan, S. Klasky, J. Ligon, and B. Ludaescher, Automation of Network-Based Scientific Workflows, in *Proc. of the IFIP WoCo 9 on Grid-based Problem Solving Environments: Implications for Development and Deployment of Numerical Software*, 2007, p. 35.

[90] S. Krishnan and K. Bhatia, SOAs for scientific applications: experiences and challenges, *Future Generation Computer Systems*, vol. 25, pp. 466–473, 2009.

[91] Functional MRI Research Center. http://www.fmri.org/index.html. Accessed July 20, 2009.

[92] A. MacKenzie-Graham, A. Payan, I. Dinov, J. Van Horn, and A. Toga, Neuroimaging Data Provenance Using the LONI Pipeline Workflow Environment, in *Second International Provenance and Annotation Workshop, IPAW*, 2008, p. 208.

Chapter 13

Scientific Process Automation and Workflow Management

Bertram Ludäscher,[1] Ilkay Altintas,[2] Shawn Bowers,[1] Julian Cummings,[3] Terence Critchlow,[4] Ewa Deelman,[5] David De Roure,[6] Juliana Freire,[10] Carole Goble,[7] Matthew Jones,[8] Scott Klasky,[9] Timothy McPhillips,[1] Norbert Podhorszki,[9] Claudio Silva,[10] Ian Taylor,[11] and Mladen Vouk[12]

[1] *University of California, Davis*
[2] *San Diego Supercomputer Center*
[3] *California Institute of Technology, Pasadena*
[4] *Pacific Northwest National Laboratory*
[5] *USC Information Science Institute, Marina del Rey*
[6] *University of Southampton, United Kingdom*
[7] *The University of Manchester, United Kingdom*
[8] *University of California, Santa Barbara*
[9] *Oak Ridge National Laboratory*
[10] *University of Utah*
[11] *Cardiff University, United Kingdom*
[12] *North Carolina State University*

Contents

13.1 Introduction 468
13.2 Features of Scientific Workflows 470
    13.2.1 The Scientific Workflow Life Cycle 470
    13.2.2 Types of Scientific Workflows 471
    13.2.3 Models of Computation 472
    13.2.4 Benefits of Scientific Workflows 473
13.3 Case Study: Fusion Simulation Management 474
    13.3.1 Overview of the Simulation Monitoring Workflow 475
    13.3.2 Issues in Simulation Management 478
13.4 Grid Workflows and the Scientific Workflow Life Cycle 481
    13.4.1 Workflow Design and Composition 481
    13.4.2 Mapping Workflows to Resources 484
    13.4.3 Workflow Execution 486

13.5 Workflow Provenance and Execution Monitoring 489
      13.5.1 Example Implementation of a Provenance Framework 491
13.6 Workflow Sharing and myExperiment 495
      13.6.1 Workflow Reuse 496
      13.6.2 Social Sharing 496
      13.6.3 Realizing myExperiment 497
13.7 Conclusions and Future Work 499
Acknowledgments 500
References 501

13.1 Introduction

Scientific discoveries in the natural sciences are increasingly data driven and computationally intensive, providing unprecedented data analysis and scientific simulation opportunities. To accelerate scientific discovery through advanced computing and information technology, various research programs have been launched in recent years, for example, the SciDAC program by the Department of Energy[1] and the Cyberinfrastructure initiative by the National Science Foundation,[2] both in the United States. In the UK, the term *e-Science*[3] was coined to describe computationally and data-intensive science, and a large e-Science research program was started there in 2000. With the new opportunities for scientists also come new challenges, for example, managing the enormous amounts of data generated[4] and the increasingly sophisticated but also more complex computing environments provided by cluster computers and distributed grid environments. Scientific workflows aim to address many of these challenges.

    In general terms, a *scientific workflow* is a formal description of a process for accomplishing a scientific objective, usually expressed in terms of tasks and their dependencies.[5–7] Scientific workflows can be used during several different phases of a larger science process, that is, the cycle of hypothesis formation, experiment design, execution, and data analysis.[8,86] Scientific workflows can include steps for the acquisition, integration, reduction, analysis, visualization, and publication (for example, in a shared database) of scientific data. Similar to more conventional business workflows,[9] scientific workflows are composed of individual tasks that are organized at workflow design time and whose execution is orchestrated at runtime according to dataflow and task dependencies as specified by the workflow designer. Workflows are often designed visually, for example, using block diagrams, or textually, using a domain-specific language. From a scientist's perspective, scientific workflows constitute knowledge artifacts or "recipes" that provide a means to automate, document, and make repeatable a scientific process.

The primary task of a *scientific workflow system* is to automate the execution of scientific workflows. Scientific workflow systems may additionally support users in the design, composition, and verification of scientific workflows. They also may include support for monitoring the execution of workflows in real time; recording the processing history of data; planning resource allocation in distributed execution environments; discovering existing workflows and workflow components; recording the lineage of data and evolution of workflows; and generally managing scientific data. Thus, a scientific workflow system primarily serves as a workflow *execution engine*, but may also include features of *problem-solving environments* (PSE).[10]

Wainer et al. describe some of the differences between business (or "office automation") workflows and scientific workflows, stating, "whereas office work is about goals, scientific work is about data".[11] Business workflows are mainly concerned with the modeling of business rules, policies, and case management, and therefore are often control- and activity-oriented. In contrast, to support the work of computational scientists, scientific workflows are mainly concerned with capturing scientific data analysis or simulation processes and the associated management of data and computational resources. While scientific workflow technology and research can inherit and adopt techniques from the field of business workflows, there are several, sometimes subtle, differences ranging from the modeling paradigms used to the underlying computation models employed to execute workflows.[86] For example, scientific workflows are usually dataflow-oriented "analysis pipelines" that often exhibit pipeline parallelism over data streams in addition to supporting the data parallelism and task parallelism common in business workflows.* In some cases (for example, in seismic or geospatial data processing[12]), scientific workflows execute as digital signal processing (DSP) pipelines. In contrast, traditional workflows often deal with case management (for example, insurance claims, mortgage applications), tend to be more control-intensive, and lend themselves to very different models of computation.

In Section 13.2 we introduce basic concepts and describe key characteristics of scientific workflows. In Section 13.3 we provide a detailed case study from a fusion simulation project where scientific workflows are used to manage complex scientific simulations. Section 13.4 describes scientific workflow systems currently in use and in development. Section 13.5 introduces and discusses basic notions of data and workflow provenance in the scientific workflow context, and describes how workflow systems monitor execution and manage provenance. Finally, Section 13.6 describes approaches for enabling workflow reuse, sharing, and collaboration.

---

*In the parallel computing literature, *task parallelism* refers to distributing tasks (processes) across different parallel computing nodes, and *data parallelism* involves distributing data across multiple nodes. *Pipeline parallelism* is a more specific condition that arises whenever multiple processes arranged in a linear sequence execute simultaneously.

13.2 Features of Scientific Workflows

13.2.1 The Scientific Workflow Life Cycle

The various phases and steps associated with developing, deploying, and executing scientific workflows comprise the scientific workflow life cycle. The following phases are largely supported by existing workflow systems using a wide variety of approaches and techniques (cf. Section 13.4).

Workflow Design and Composition. Development of scientific workflows usually starts with gathering requirements from scientists. A specification of the desired workflow functionality is then developed, and an actual workflow is assembled based on this specification. Workflow development differs from general programming in many ways. Most notably, it usually amounts to the composition and configuration of a special-purpose workflow from pre-existing, more general-purpose components, subworkflows, and services. Workflow development thus more closely resembles script- and shell-based programming than conventional application development. During workflow composition, the user (a scientist or workflow developer) either creates a new workflow by modifying an existing one (cf. *workflow evolution*, Section 13.5), or else composes a new workflow from scratch using components and subworkflows obtained from a repository. In contrast to the business workflow world, where standards have been developed over the years (e.g., most recently WS-BPEL 2.0[13]), scientific workflow systems tend to use their own, internal languages and exchange formats (e.g., SCUFL,[14] GPEL,[15] and MOML,[16] among others). Reasons for this diversity include the wide range of computation models used in scientific workflows (Section 13.2.3), and the initial focus of development efforts on scientist-oriented functionality rather than standardization.

Workflow Resource Planning. Once the workflow description is constructed, scientific workflow systems often provide various functions prior to execution. These functions may include workflow validation (e.g., type checking), resource allocation, scheduling, optimization, parameter binding, and data staging. *Workflow mapping* is sometimes used to refer to optimization and scheduling decisions made during this phase. In particular, during workflow design and composition, the target resources to be used for execution are typically not chosen. Workflow mapping then refers to the process of generating an *executable workflow* based on a resource-independent *abstract workflow* description.[17] In some cases, the user performs the mapping directly by selecting appropriate resources (e.g., in Figure 13.2). In other cases, the workflow system automatically performs the mapping. In the latter case, users are allowed to construct workflows at a level of abstraction above that of the target execution environment.

Workflow Execution. Once a workflow is mapped and data has been staged (selected and made available to the workflow system), the workflow can be executed. During execution, a workflow system may record provenance information (data and process history, see Chapter 12 and Section 13.5) as well as provide real-time monitoring and failover functions. Depending on the system, provenance information generally involves the recording of the steps that were invoked during workflow execution, the data consumed and produced by each step, a set of data dependencies stating which data was used to derive other data, the parameter settings used for each step, and so on. If a workflow can change while executing (e.g., due to changing resource availability), the evolution of such a dynamic workflow may be recorded as well in order to support subsequent execution analysis.

Workflow Execution Analysis. After workflow execution, scientists often need to inspect and interpret workflow results. This involves evaluation of data products (*does this result make sense?*), examination of workflow execution traces (*is this how the result should have arisen?*), workflow debugging (*what went wrong here?*), and performance analysis (*why did this take so long?*).

Workflow and Result Sharing. Data and workflow products can be published and shared. As workflows and data products are committed to a shared repository, new iterations of the workflow life cycle can begin.

User Roles. Users of scientific workflow systems can play a number of different roles within the above phases: A *workflow designer* is usually a scientist who develops a new experimental or analytical protocol (or a new variant of an existing method). As mentioned above, a workflow design is often elicited through some form of requirements analysis, and the design and associated requirements can be used by a *workflow engineer* to implement the associated abstract or executable workflow description. A *workflow operator* is a user who executes workflows on the desired inputs. An operator may launch a workflow directly via a scientific workflow system, or indirectly through another application (e.g., within a Web portal), monitor the execution (e.g., via a workflow dashboard), and subsequently validate results based on stored provenance information. The above user roles are not necessarily disjoint; for example, a single person may assume the roles of designer, engineer, and operator. Indeed, scientific workflow systems aim at making workflow design, execution, and result analysis all easier in comparison to traditional script-based approaches to scientific process automation.

13.2.2 Types of Scientific Workflows

Scientific workflows can be used to model and automate scientific processes from many different science domains (for example, particle physics, bioinformatics, ecology, and cosmology, to name a few). Not surprisingly, such workflows can exhibit very different characteristics. For example, workflows might be *exploratory* in nature, starting from ad hoc designs and then requiring frequent changes to the workflow design, parameter settings, etc., to determine which methods and components are most suitable for the particular datasets under investigation. Such exploratory workflow design is common when developing new analysis methods. Conversely, some applications require the development of *production workflows* to be executed on a regular basis with new datasets or simulation parameters (e.g., environmental monitoring and analysis workflows or the fusion simulation workflow in Section 13.3).

Another important distinction has to do with what the workflow components (called *actors* or *tasks*) represent and model. In *science-oriented* workflows, actors model a scientific method or process. In such workflows individual workflow steps generally are meaningful to the scientist, that is, more or less directly correspond to high-level steps of the scientific method being automated. Contrasting with science-oriented workflows are *resource-oriented* workflows. Actors and workflow steps in the latter model require data and resource-handling tasks rather than the science. In such cases, the actual analytical or simulation operations might be "hidden" from the workflow system, and instead the workflow directly handles the "plumbing" tasks such as data movement, data replication, and job management (submit, pause, resume, abort, etc.) The simulation management workflow in Section 13.3 is an example of such a resource-oriented "plumbing workflow."

13.2.3 Models of Computation

Consider a workflow graph W consisting of *actors* (tasks, workflow steps) and *connections* (directed edges) between them.[*] With W we can associate a set of *parameters* $\bar{p}$, input datasets $\bar{x}$, and output datasets $\bar{y}$. A *model of computation* (MoC) M prescribes how to execute the parameterized workflow $W_{\bar{p}}$ on $\bar{x}$ to obtain $\bar{y}$. Therefore, we can view a MoC as a mapping $\mathsf{M}: \mathcal{W} \times \bar{P} \times \bar{X} \to \bar{Y}$, which for any workflow $W \in \mathcal{W}$, parameter settings $\bar{p} \in \bar{P}$, and inputs $\bar{x} \in \bar{X}$ uniquely determine the workflow outputs $\bar{y} \in \bar{Y}$. We denote this by $\bar{y} = \mathsf{M}(W_{\bar{p}}(\bar{x}))$. While most current scientific workflow systems employ a single MoC, the Kepler system,[18] due to its heritage from Ptolemy,[16] supports more than one such MoC: For each each MoC M, there is a corresponding *director* of the same name which implements M.

For example, consider the PN (process network) model of computation. Using the PN director in Kepler, a workflow W executes as a dataflow process network.[19,20] In PN each actor executes as a separate, data-driven process (or thread) which is *continuously* running. Actor connections in PN correspond to unidirectional channels (modeled as unbounded queues) over which ordered token streams are sent, and actors in PN block (wait) only when there are not enough tokens available on the actor's input ports. Process networks naturally support pipeline parallelism as well as task and data parallelism.

---

[*]Here we ignore a number of details: actor *ports*, subworkflows "hidden" within so-called composite actors, and so forth.

In SDF (synchronous data-flow), each actor has fixed token consumption and production rates. In Kepler this allows the SDF director to construct an *actor firing schedule* prior to executing the workflow.[21] This also allows the SDF director to readily execute workflows in a single thread, firing actors one at a time based on the schedule.

Workflows employing the PN and SDF directors in Kepler may include cycles in the workflow graph. We use the term DAG to refer to a model of computation that restricts the workflow graph W to a directed, *acyclic* graph of task dependencies. In DAG each actor node in W is executed only once, and each actor A in W is executed only after all actors A' preceding A (denoted $A' \prec_W A$) in W have *finished* their execution. Note that we make no assumption about whether W is executed sequentially or task parallel; we only require that any DAG-compatible schedule for W satisfy the partial order $\prec_W$ induced by W. A DAG director can obtain all legal schedules for W (i.e., the relation $\prec_W$) via a topological sort of W. Finally, note that the DAG model can easily support task and data parallelism, but not pipeline parallelism.

Another model of computation, extending PN, is COMAD (Collection-Oriented Modeling and Design).[59,89] In this MoC, actors operate on streams of nested data collections (similar to XML data), and can be configured (via XPath-like *scope expressions* and *signatures*) to "pick up" and operate only on relevant parts of the input stream, injecting results back into the output stream for further downstream processing. This MoC can simplify workflow design and reuse when compared with DAG, SDF, and PN workflows.[22]

13.2.4 Benefits of Scientific Workflows

Scientific workflows are designed to help scientists perform effective computational experiments by providing an environment that simplifies (*in silico*) experimental design, implementation, and documentation. The increasing use of scientific workflow environments and systems is due to a number of advantages these systems can offer over alternative approaches.

- Scientific workflows *automate* repetitive tasks, allowing scientists to focus on the science driving the experiment instead of data and process management. For example, automation of parameter studies—where the same process is performed hundreds to thousands of times with different parameter sets—can often be more easily and efficiently achieved than with conventional programming approaches.
- Scientific workflows explicitly *document* the scientific process being performed, which can lead to better communication, collaboration (e.g., sharing of workflows among scientists), and reproducibility of results.
- Scientific workflow systems can be used to *monitor* workflow execution and *record* the provenance of workflow results. Provenance, in particular,

provides a form of documentation that can be used to validate and interpret results produced by (often complex) scientific processes.
- Scientific workflow systems often can *optimize* and then more *efficiently execute* scientific processes, for example, by exposing and exploiting various forms of parallelism inherent in data-driven scientific processes, as well as by employing other techniques for efficient resource management.
- Workflow environments encourage the *reuse* of knowledge artifacts (actors, workflows, etc.) developed when automating a scientific process, both within and across disciplines.

13.3 Case Study: Fusion Simulation Management

We now present a detailed case study to make the previously discussed notions more concrete. We chose a simulation management workflow as our example because it exhibits a number of challenging issues typically not found in other types of scientific workflows. In our terminology, the workflow is a *resource-oriented, production* workflow. The main scientific computations (the fusion simulation) are performed on a remote supercomputer cluster, while the management workflow can be executed on the scientist's desktop. The overall computation managed by the workflow is both *data intensive* and *compute intensive*; involves pipeline parallelism over a stream of data or reference tokens[*]; and is responsible for job management, file transfers, and data archiving. Such workflows have been called "plumbing" workflows due to their focus on explicitly dealing with underlying resources (which the end-user scientist prefers not to deal with). From the scientist's point of view, the primary task is to observe and analyze the simulation results as soon as possible. To understand this challenge, we first describe briefly the physics problems studied in the simulations.

The Center for Plasma Edge Simulation (CPES) is a Deparment of Energey SciDAC project[23] requiring close collaboration between physicists, applied mathematicians, and computer scientists. Together, these researchers have developed a complete plasma fusion simulation, called XGC1,[24] that runs in a high-performance computing environment. The computational scientists use this simulation to study the behavior of hot plasma in a tokamak-type fusion reactor. The central issue under study is as follows. Within a fusion reactor, if the hot edge of the plasma is allowed to contact the reactor wall in an uncontrolled way, it can sputter the wall material into the plasma, degrade or extinguish the fusion burn, and shorten the wall

---

[*]A *reference token* is a "logical pointer" to a data object, for example, a file name.

lifetime to an unacceptable level. Anomalous tokamak plasma transport is thought to be associated with small-scale plasma turbulence. When the heating power to the core plasma is above a certain threshold, a thin plasma layer forms, making the plasma almost free of turbulence; in addition, the central plasma temperature and density rise under these conditions with the added benefit that the fusion probability increases dramatically. However, this layer also triggers large magnetohydrodynamic-type instabilities, which simultaneously destroy the layer (lowering the fusion power in the core) and dump the plasma energy to the material walls. It is currently not fully understood how this layer builds up or how the following crash occurs. The success of next-generation burning plasma experiments is heavily dependent upon solving this problem. Thus, understanding these physical processes is an important area in fusion plasma research (see Cummings et al.[25] for more details).

Due to the complexity of running the simulation (e.g., staging input data, monitoring execution status, and managing result data) as well as to the rapid changes and evolution of the simulation code itself, automating the execution of the simulation via a workflow is crucial for both XGC1 developers and scientists wanting to evaluate XGC1 results. Here, we can distinguish three layers of activity: (1) At the highest level, physicists are interested in understanding (and ultimately taming) nuclear fusion in tokamak-type reactors; (2) in addition to performing the actual experiments, sophisticated, large-scale simulations on supercomputers are used to gain insights into the process (this is a goal of the CPES project); and finally, (3) a simulation management workflow is used to deal with the challenging issues in running the simulations and automating the necessary steps as much as possible.

13.3.1 Overview of the Simulation Monitoring Workflow

The XGC1 simulation outputs one-dimensional diagnostic variables in three NetCDF files. It also outputs three-dimensional data written in a custom binary format for efficient I/O performance. The latter data subsequently is converted to standard formats, such as HDF5, for archiving and analysis. Simple plots are generated from the 1D diagnostic variables, while 2D visualizations (i.e., cross-section slices of the tokamak) are produced from the converted 3D data. The NetCDF, HDF5, and all images produced during the simulation are archived.

Figure 13.1 shows a graphical representation of a CPES simulation monitoring workflow implemented using the Kepler scientific workflow system.[18] After initial preparation steps (e.g., checking if a simulation restart is requested; logging in to all involved machines; creating directories at the processing sites), two independent, concurrently executing pipelines are started for monitoring the XGC1 simulation. The term *monitoring* is used to indicate that for each output step, plots and images are generated in real time, and that these can be visualized on a dashboard, enabling the scientist to observe

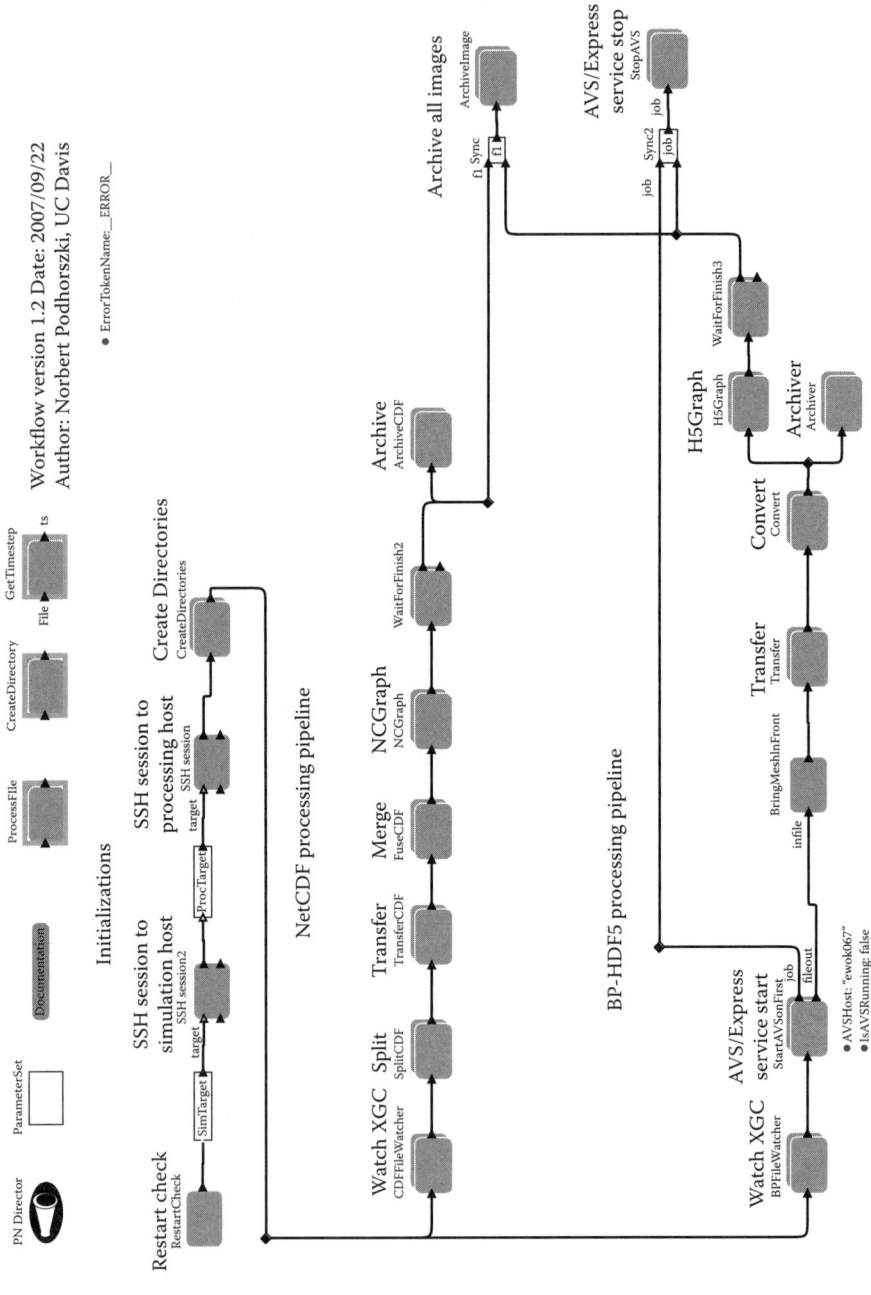

Figure 13.1 (See color insert following page 224.) CPES fusion simulation monitoring workflow (in Kepler).

whether the simulation is progressing correctly. The simulation itself executes on a dedicated supercomputer (*primary cluster*) at Oak Ridge National Laboratory (ORNL), while a *secondary cluster* computer at ORNL is used for on-the-fly analysis of the simulation run on the primary cluster.

The first pipeline (shown in the center of Figure 13.1) performs the **NetCDF file processing** portion of the monitoring workflow. This pipeline starts by checking the availability of NetCDF files. As each such file grows (they are extended after every diagnostic period), the workflow performs *split* (taking the most recent data entry), *transfer*, and *merge* operations on recent data to mirror XGC1's output on the secondary analysis cluster efficiently. Finally, images are generated using *xmgrace*[*] for all variables in the output for each diagnostic time step and placed into a remote directory where the scientist can browse them via the Web-based dashboard application[25] (cf. Section 13.5). The split and merge operations are executed on the login nodes of the primary simulation machine and on the secondary analysis cluster, respectively. To make the plots, however, a job has to be submitted on the secondary cluster for each file in each step. Although one such job is small—lasting only for a couple of seconds—there is almost always one running; this would typically overload the login node of the primary cluster.

The second pipeline (bottom of Figure 13.1) performs the **BP-HDF5 processing**. This pipeline's role is similar to the NetCDF pipeline, but with the following differences. For each step, XGC1 creates new BP files (a custom **b**inary-**p**acked format); hence, there are no split and merge steps when transferring them to the secondary processing site. The BP files are converted to HDF5 using an external code, and then images are created for all 2D slices of the 3D data stored in those files using an AVS/Express[†] dataflow network. For this purpose, the pipeline starts AVS/Express as a remote job on the secondary cluster and then makes image-creation requests to it as a (private) service.

This workflow uses a set of fine-grained *job-control* steps provided by Kepler for calling AVS/Express. The workflow waits until the AVS/Express job is started on its execution host, performs the other tasks while the job is running, and stops the job at the end of the processing. The individual steps in Figure 13.1 are workflows themselves (i.e., subworkflows, or *composite actors* in Kepler terminology), implementing special tasks. One such subworkflow is the archival step in the HDF5 pipeline, which assembles files into large chunks and stores them in a remote mass-storage system. The steps used in this workflow are described further in Podhorszki et al.[26]

---

[*]A tool for **gr**aphing, **a**dvanced **c**omputation and **e**xploration of data; see http://plasma-gate.weizmann.ac.il/Grace/

[†]http://www.avs.com/software/soft_t/avsxps.html

13.3.2 Issues in Simulation Management

In a *compute-intensive* workflow, jobs are typically executed remotely from the workflow execution engine, and thus the output of one job often must be transferred to another host where subsequent jobs are executed. For efficiency reasons, files are directly transmitted between remote sites, while the workflow engine only "sees" reference tokens to the remote files. The XGC1 case study falls into this category of workflows. It also belongs to a category of *data-intensive* workflows in which the data produced during a supercomputing simulation must be processed on the fly and as quickly as possible. This scenario is typical of most scientific simulations that use supercomputers and produce ever-larger amounts of data as the size and speed of the supercomputer clusters continues to increase.

Runtime Decision Support. The typical tasks that a computational scientist performs during and after a simulation run are often tedious to perform manually without automation support. For instance, to maintain high utilization of supercomputing resources, it is essential to be able to detect and halt a divergent simulation. Thus, in most scientific simulations, the status of the computation must be regularly checked to ensure that it is not diverging given the initial input parameters. However, it can be difficult to check the status of an executing simulation because typically the user has to log in to the primary supercomputer cluster (since applications typically write data to local disks) at regular intervals to analyze *diagnostic values* that reveal errors in the input or simulation code. Moreover, simulations typically write out other (more involved) diagnostic data such as physical variables or derivatives of these variables, which must be plotted and analyzed. Although such plots give deeper insight into the current state of the simulation, even more information may be needed for monitoring and runtime decision support, for example, the ability to visually analyze parts of the dataset written out by the simulation. The latter operation usually cannot be done on the supercomputer's login node, however, which is one of the reasons for transferring data to another, secondary, computer such as the scientist's desktop computer or a dedicated visualization computer. Although not described in detail here, the CPES project has automated these various tasks via a separate workflow that greatly reduces the amount of manual work required of users by automatically routing diagnostic information and data, and by displaying the appropriate plots and visualizations on a Web-based dashboard.

Data Archiving. Another important task is the archiving of output data. At present, it is sufficient to archive data after the simulation run. In the near future, however, it is anticipated that the largest simulations will create more data in a single run than can fit onto the disk system of the supercomputers. Therefore, files must be transferred to a remote mass storage system on the fly and then removed from the local disk to make space for more data coming from the simulation. There also is a requirement to create "archival chunks" of an intermediate size; for performance reasons, neither individual files nor the complete simulation output (as a single file) can be sent to the archive system. Thus, the automated solution puts files into appropriately sized chunks while taking care of other requirements, for example, ensuring that all data for one timestep goes into the same chunk.* Finally, recording the *data provenance* of all generated data becomes increasingly important as the size and complexity of the output grows. For example, from an automatically generated diagnostic image, a scientist must be able to easily find the output of the simulation corresponding to the visualization. Tools can greatly help with transferring the relevant data to the scientist's host machine (which could be at a remote site) provided that the above simulation management workflow records the necessary data lineage of all operations.

Pipeline Parallel Processing. An important feature of the Kepler environment is its support for the dataflow process network[19,20] model of computation, implemented via the Process Network (PN) director.[16] Using the PN director, all actors are running continuously in separate threads, waiting for input to be processed immediately. Each pipeline in the above workflow is therefore processing a stream of data items in pipeline-parallel mode. For example, since XGC1 outputs diagnostic data into three NetCDF files at each timestep, plots can be created for one file, a second file is being used in a merge operation, and a third file is being transferred. In a typical production run scenario, XGC1 outputs a new timestep every 30 seconds. The time to get one file through the processing pipeline includes the time for recognizing its presence, the transfer time, and the execution time of the plot generation job on the processing cluster. If the workflow performed only one of these steps at a time (e.g., as prescribed by the SDF director), the simulation would generate files faster than they could be processed. Due to the size of the 3D data in the HDF5 pipeline and the longer transfer time of those files, the situation is similar in this pipeline as well. Finally, the archiving process must obviously work in parallel with the rest of the workflow, since it is a slow process in itself. If the task and pipeline parallelism exhibited by the above workflow is not enough to keep up with the flow of data, one can replicate individual actors on different compute nodes to process multiple data items at the same time. Although the above workflow does not need to do this currently, a more complex production workflow is in use for coupling other codes with the XGC1 predecessor code (such as those described in Section 5), XGC0,[25] where a parameter study has to be executed for each timestep of the simulation, and that study is executed in this parallel mode.

Robustness of Workflows. There are two different but related aspects of robustness that can occur in compute-intensive workflows: What happens if the overall workflow execution fails and stops (e.g., at the workflow engine level), and what happens if an individual task in the workflow fails? For

---

*An additional problem arises when data is generated faster than it can be archived. In this case, an additional workflow step can be inserted that uses an auxiliary disk to queue the data, decoupling the slow archival from the fast data generation.

a workflow responsible for starting and monitoring jobs, both eventualities mean that there is a set of successfully executed jobs (whose results should be salvaged) and a set of not yet executed jobs that cannot be started because they either depend on a failed task, or because the workflow (engine) is no longer running and thus cannot start them. After restarting the workflow, a previously failed task can resume.

The tasks comprising the simulation monitoring workflow represent operations carried out on individual data items (files) as the simulation produces data at each timestep. If some operation during a particular timestep fails (e.g., transfer to another host fails, mass storage is down at archiving time, or a statistic cannot be created — all common failures outside the control of the workflow engine), this should not prevent the workflow from invoking the complete pipeline of operations over the data produced during the next timestep. However, because a downstream actor may be affected by the result of an upstream actor, actors should be prepared for such failures. Two possible solutions are to (1) discard from the token stream that token corresponding to the failed operation, or (2) introduce special "failure tokens" to mark jobs that did not succeed. If we discard the token for the failed operation, downstream actors do not receive a bad task request, and therefore no change to the actor is required to handle them. However, the absence of tokens changes the balance between the consumption and production rates of the actors, and this can lead to difficulties in complex workflow design, for example, if we need to split and merge pipelines. If we replace the token with a failure token, and downstream actors are programmed to simply ignore such failure tokens, the workflow structure remains simpler.*

Resuming Workflow Execution Following a Fault. Pipelined (e.g., PN) workflows are harder to restart than DAG workflows because the current state of the workflow is not as easy to describe and restore; all actors in the workflow graph may be concurrently executing. While the progress of executing a conventional DAG workflow (Section 13.2.3) can be seen as a single "wavefront" progressing from the beginning of the workflow DAG toward the end, in a pipeline-parallel workflow each task can be invoked repeatedly. If the workflow system does not support full restoration of the workflow and actor state (a nearly impossible task when dealing with workflow components outside the control of the engine), the workflow itself has to include some sort of lightweight checkpoint and restart capability.

In the CPES workflow, the solution is to have the remote execution actor — used for executing all of the actual data processing operations along the pipeline — record all successful operations.[26] When restarted, for example,

---

*The first design of CPES workflows was based on approach (1), while for the above reason, an improved design employed the second approach (2). The COMAD model of computation (see Section 13.2.3 and[22]) natively supports mechanisms to tag data, which is an elegant way to achieve variant (2); it can be used to skip over or even bypass data around actors,[27] or perform other forms of exception handling based on tags.

after failure, this actor checks the current tasks to be performed against the set of successful tasks and skips over any that were already executed successfully. In this way, the next actor in the pipeline can immediately start working on it (or skip over it as well). Thus, although the workflow restarts from the very beginning, pushing the initial input tokens back into the pipeline, the actors "fast-forward" to the most recent state prior to the workflow failure, by skipping the tokens corresponding to previously successful tasks. The time spent by the workflow engine in this fast-forward restoration process is negligible compared to the time of actually executing the remote operations.

13.4 Grid Workflows and the Scientific Workflow Life Cycle

The term *grid workflow* applies to workflows that employ distributed (often wide-area) computational resources (often referred to as "the grid"). Like other scientific workflows, grid workflows can be seen as high-level specifications of sets of tasks and the dependencies between them that must be satisfied in order to accomplish a specific goal. The specific goal of grid-enabled workflow systems is to reduce the programming effort required of scientists orchestrating a computational science experiment in a wide-area, distributed system. The vast majority of scientists do not use grid systems in their day-to-day practices, largely because of usability barriers. Workflow systems are beginning to address these usability barriers and to make grid computing far more accessible to general science users. In this section, we focus on the first three stages of the life cycle summarized in Section 13.2.1 — scientific workflow composition, mapping of workflows onto resources, and workflow execution — and describe how several popular grid-enabled workflow systems support these different stages. We present a cross-sectional view of the types of grid workflows that are currently being deployed and compare features provided by Kepler,[18,28] Pegasus,[29,30] Taverna,[14,31,32] Triana,[33,34] and Wings.[35,*]

13.4.1 Workflow Design and Composition

Most e-Science workflow systems provide a graphical tool for composing workflows. For example, Kepler and Triana have sophisticated graphical composition tools for building workflows graphs using a graph or block diagram metaphor, where nodes in the workflow graph represent tasks, and edges represent dataflow dependencies or task precedence. The intent of these graphical

---

*For a high-level overview and attempt at a classification of current systems see[36] and http://www.extreme.indiana.edu/swf-survey/; these include references to other scientific workflow systems, such as Askalon, Karajan, and many others.

composition tools is to simplify for scientists the task of describing workflows. Other, *task-based* systems such as Pegasus focus on the mapping and execution capabilities and leave the higher-level composition tasks to other tools.

Task-level workflow systems focus on resource-level functionality and fault-tolerance, while service-level systems generally provide interfaces to certain classes of services for management and composition. One important factor to the adoption of workflow systems by scientists is the availability of workflow tools and services that scientists can build on in order to create their applications. Such service availability forms part of the composition process since it represents the available tools that can be composed within a system.

Pegasus takes a workflow description in a form of a directed acyclic graph in XML format (DAX). The DAX can be generated using a Java API, any scripting language, or using semantic technologies such as Wings.[35] In some scientific applications, users prefer an interface that simply supports metadata queries while hiding the details of how the underlying systems work. In astronomy, for example, users want simply to retrieve images of an area of the sky of interest to them. In such cases Pegasus is usually integrated into a portal environment, and the user is presented with a Web form for entering desired metadata attributes. Behind the portal a workflow instance is then generated automatically based on the user's input, given to Pegasus for mapping, and then passed to DAGMan[37] for execution. Examples of this approach can be seen in the Montage project (an astronomy application),[38,39] the Telescience portal (a neuroscience application),[40] and the Earthworks portal (an earthquake science application).[41] In all of these applications, Pegasus and DAGMan are being used to run workflows on national infrastructure such as the TeraGrid.

Kepler provides a graphical user interface (GUI) for composing and editing workflows using a hierarchical representation of the workflow graph (see the example in Figure 13.1). Dataflow is indicated by channels represented as edges among the nodes of the graph, and each node represents either an atomic task or a composite task (containing a subworkflow). The user interface provides a semantic-search system across hundreds of different scientific computing components available in the Kepler library. These components cover a wide variety of scientific data processing and modeling activities, such as geospatial data processing, signal processing, statistical algorithms, and data transformations. The semantic search feature[42,43] assists the user in locating components that are relevant to their analysis and modeling tasks. It is also useful when searching the remote Kepler library, allowing users to find components that have been shared by other Kepler users and to share their own components and workflows with others. Kepler workflows can be executed directly from the workflow-composition GUI or saved in an XML representation (MoML) and later passed to Kepler for execution in the absence of the GUI. This feature allows Kepler to be embedded in Web portals and other applications.

Taverna provides a GUI-based desktop application that uses semantic annotations associated with services, employs the use of semantics-enabled helper functions, and uses reasoning techniques to infer service annotations.[44] Over 800 services are described using ontologies[45] expertly annotated by a full-time curator used by clients such as Find-O-Matic, its discovery tool, Feta,[46] which is only available as a plug-in from the Taverna Workflow Workbench. The Bio-Catalogue* project[47] incorporates the experiences of the Taverna Registry and myExperiment (Section 13.6) to build and manage a richly described catalog of Web services in the Life Sciences. The catalog's services have descriptive content capturing functional capabilities curated by experts and by the community through social collaboration; operational content such as quality of service and popularity is automatically curated by monitoring and use analysis. The BioCatalogue is a free-standing component with its own RESTful APIs that can be embedded within and accessed from third-party applications. Developers can incorporate new services through simple actions and can load a pre-existing workflow as a service definition within the service palette, which can then be used as a service instance within the current workflow. Taverna also supports the configuration of the appearance of the graphical representation of workflows, so that a workflow can be suppressed to give higher-level views, for example, to remove details such as data translation (or other "shim") services.†

One of the most powerful aspects of Triana is its GUI. It has evolved in its Java form for over 10 years and contains a number of powerful editing capabilities, wizards for on-the-fly creation of tools, and GUI builders for creating user interfaces. Triana editing capabilities include multilevel grouping for simplifying workflows, cut/copy/paste/undo, ability to edit input/output nodes (to make copies of data and add parameter dependencies, remote controls, or plug-ins), zoom functions, various cabling types, optional inputs, type checking, and so on. Since Triana came from the gravitational-wave field, the system contains a wide-ranging palette of tools (around 400) for the analysis and manipulation of one-dimensional data, which are mostly written in Java (with some in C). Recently, other extensive toolkits have been added for audio analysis, image processing, text editing, for creating retinopathy workflows (i.e., for diabetic retinopathy studies), and even data mining based on configurable Web services to aid in the composition process. See Taylor[48] for a further discussion and description of such applications.

Wings[35] uses rich semantic descriptions of components and workflow templates expressed in terms of domain ontologies and constraints. Wings has a workflow template editor to compose components and their dataflow. The editor assists the user by enforcing the constraints specified for the workflow components. It also assists the user with data selection, to ensure the datasets selected conform to the requirements of the workflow template. With

---
*http://biocatalogue.org
†Shims align or mediate data that is syntactically or semantically closely related but not directly compatible.[49]

this information, Wings generates a workflow instance that specifies the computations (but not where they will take place) and the new data products. For all the new data products, it generates metadata attributes by propagating metadata from the input data through the descriptions and constraints specified for each of the components.

13.4.2 Mapping Workflows to Resources

It is often the case that at the time the workflow is being designed, the target resources are yet to be chosen. *Workflow mapping* refers to the process of generating an executable workflow based on a resource-independent workflow description sometimes called an abstract workflow. In some cases the user performs the mapping directly by selecting the appropriate resources. In other cases, the workflow system performs the mapping.

Depending on the underlying execution model of stand-alone applications, or individual services, different approaches are taken to the mapping process. In the case of service-based workflows, mapping consists of finding and binding to services appropriate for the execution of a high-level functionality. Service-based workflows also can consider quality of service requirements when performing the mapping. In the case of workflows composed of stand-alone applications, the mapping not only involves finding the necessary resources to execute the computations and perform various optimizations, but may also include modifying the original workflow.

Some systems such as Taverna rely on the user to make the choice of resources or services. In the case of Taverna, the user can provide a set of services that match a particular workflow component, so if errors occur, an alternate service can be automatically invoked. The newer versions of Taverna will include late service-binding capabilities.

Kepler, on the other hand, allows the user to specify resource bindings through its distributed computation configuration system. The user designs the workflow in a manner that indicates which components are compute intensive and should be distributed across remote computational resources. The user then is presented with a dialog listing available compute resources, which can include both other Kepler peers and remote Kepler slaves running on computing clusters (see example in Figure 13.2). The user selects which set should be used for the execution, and the Kepler execution engine then determines a schedule for data transfer and execution of jobs based on the execution model used in the abstract workflow model. In addition, Kepler can be used to configure and submit jobs to a variety of other grid-based computing systems, including Griddles,[50] Nimrod,[51] and other systems.

Triana is able to interface to a variety of execution environments using the GAT (Grid Application Toolkit)[34] for task-based workflows and the GAP (Grid Application Prototype) for service-based workflows. In the case of a service-based workflow, a user can provide the information about the services to invoke (or locate them via a repository). Alternatively, a user can create a

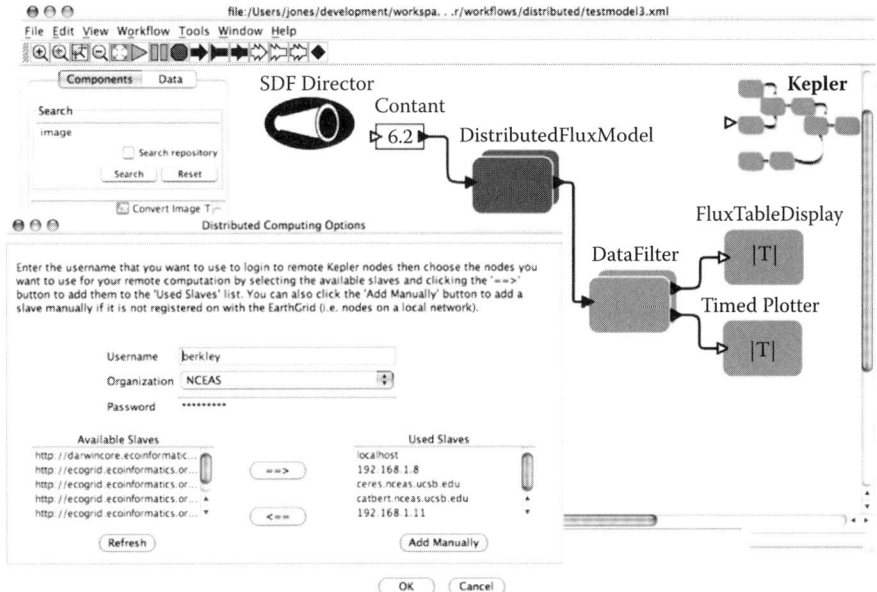

Figure 13.2 (See color insert following page 224.) Kepler supports execution of workflows on remote peer nodes and remote clusters. Users indicate which portions of a workflow should be remotely executed by grouping them in a distributed composite component (shown in blue in the workflow). The user selects from a list of available remote nodes for execution (see dialog), and Kepler calculates a schedule and stages each data token before execution on one of the set of selected remote nodes.

workflow and then map part of the workflow to distributed services through the use of one of the internal scripts, e.g., parallel or pipeline. In this mode, Triana distributes workflows by using (or deploying on the fly) distributed Triana services that can accept a Triana task graph as input. In the case of a task-based workflow, the user can designate portions of the workflow as compute-intensive, and Triana will send the tasks to the available resources for execution. It can, for example, use the GAT interface to the Gridlab GRMS broker[52] to perform the resource selection at runtime. Workflows can also be specified using a number of built-in scripts that can be used to map from a simple workflow specification (e.g., specifying a loop) to multiple distributed resources in order to simplify the orchestration process for distributed rendering. Such scripts can map subworkflows onto available resources by using any of the service-oriented bindings available, e.g., Web Services Resource Framework (WSRF), Web and peer-to-peer (P2P) services using built-in deployment services for each binding.

Workflows specified in DAGMan can be a mixture of concrete and abstract tasks. When DAGMan is interfaced to a Condor task execution system,[37] the abstract tasks can be matched dynamically to Condor resources. The matching is done by the Condor matchmaker, which matches the requirements of an abstract task specified in a Condor classAd* with the resource preferences published in their classAds. We also note that Pegasus uses DAGMan as an execution engine (Figure 13.3). Currently, Pegasus and DAGMan are being integrated into a system, Pegasus-WMS, which provides the user with an end-to-end workflow solution.

Pegasus performs a mapping of the entire workflow, portions of the workflow, or individual tasks onto the available resources. In the simplest case Pegasus chooses the sources of input data (assuming that it is replicated in the environment) and the locations where the tasks are to be executed. Pegasus provides an interface to a user-defined scheduler and includes a number of scheduling algorithms. As with many scheduling algorithms, the quality of the schedule depends on the quality of the information both of the execution time of the tasks and data access as well as the information about the resources. In addition to the basic mapping algorithm, Pegasus can perform the following optimizations: tasks clustering, data reuse, data cleanup, and partitioning. Before the workflow mapping, the original workflow can be partitioned into any number of subworkflows. Each subworkflow is then mapped by Pegasus. The order of the mapping is dictated by the dependencies between the subworkflows. In some cases the subworkflows can be mapped and executed in parallel. The granularity of the partitioning is dictated by how fast the target execution resources are changing. In a dynamic environment, partitions with small numbers of tasks are preferable, so that only a small number of tasks are bound to resources at any one time. On the other hand, in a dedicated execution environment, the entire workflow can be mapped at once. Pegasus can also reuse intermediate data products if they are available and thus possibly reduce the amount of computation that needs to be performed. Pegasus also adds data cleanup nodes to the workflow, which remove the data at the execution sites when they are no longer needed. This often results in a reduce workflow data footprint. Finally, Pegasus can also perform task clustering, treating a set of tasks as one for the purpose of scheduling to a remote location. The execution of the cluster at the remote site can be sequential or parallel (if applicable). Task clustering can be beneficial for fine computational granularity workflows. Pegasus has also been used in conjunction with resource-provisioning techniques to improve the overall workflow performance.[53]

13.4.3 Workflow Execution

In this section, we contrast approaches to workflow execution in Pegasus, Triana, and Kepler. Pegasus can map workflows onto a variety of target

---

*Classified advertisement

Scientific Process Automation and Workflow Management 487

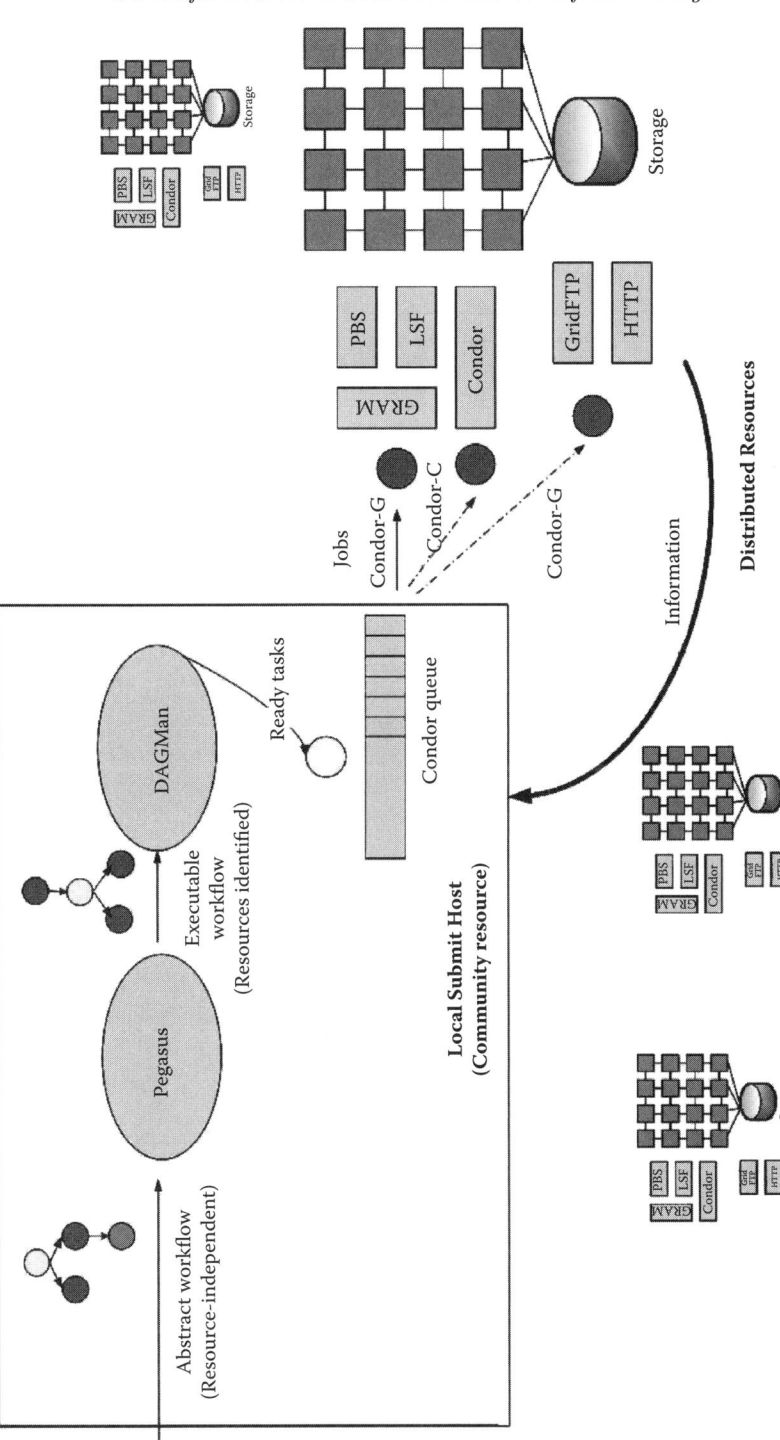

Figure 13.3 An overview of the Pegasus/DAGMan workflow management system. The Pegasus Mapper takes a resource-independent workflow description and maps it onto the available cyberinfrastructure resources. The resulting executable workflow is managed by DAGMan, which uses Condor/Condor-G to send individual workflow tasks from the submit host (a user or community resource) to the cyberinfrastructure: a local machine, a campus cluster, or a Grid.

resources such as those managed by Portable Batch System (PBS), Load Sharing Facility (LSF), Condor,[54] and individual machines. Authentication to remote resources is done via Grid Security Infrastructure (GSI).[55] During workflow execution, Pegasus captures provenance information about the executed tasks. Provenance includes a variety of information including the hosts where tasks executed, task runtimes, environment variables, etc. Pegasus uses the DAGMan workflow engine for execution (Figure 13.3). DAGMan interfaces in turn to a local Condor queue managed by a scheduler daemon. DAGMan uses the scheduler's API and logs to submit, query, and manipulate jobs, and does not directly interact with jobs. DAGMan can also use Condor's grid abilities (Condor-G) to submit jobs to many other batch and grid systems. DAGMan reads the logs of the underlying batch system to follow the status of submitted jobs rather than invoking interactive tools or service APIs. By relying on file-based I/O, DAGMan's implementation can be simpler and more scalable and reliable across many platforms, and therefore more robust. For example, if DAGMan has crashed while the underlying batch system continues to run jobs, DAGMan can recover its state upon restart (by reading logs provided by the batch system) without losing information about the executing workflow. DAGMan workflow management includes not only job submission and monitoring but also job preparation, cleanup, throttling, retry, and other actions necessary to ensure successful workflow execution. DAGMan attempts to overcome or work around as many execution errors as possible; and in the face of errors it cannot overcome, it provides a *Rescue DAG*[*] and allows the user to resolve the problem manually and then resume the workflow from the point where it last left off. This can be thought of as a "checkpointing" of the workflow, just as some batch systems provide checkpointing of jobs.

Triana supports job-level execution through GAT integration, which can make use of job execution components such as GRMS,[52] GRAM[56] or Condor[37] for the actual job submission. It also supports service-level execution through the GAP bindings to Web, WSRF and P2P services. During execution, Triana will identify failures for components and provide feedback to the user if a component fails. Triana does not contain failsafe mechanisms within the system for, e.g., retrying a service, however.

As discussed in Section 13.2.3, execution of a Kepler workflow is managed through an independent component called a *director*, which is in charge of workflow scheduling and execution.[†] A director in Kepler encapsulates a *model of computation* (MoC) and a scheduling algorithm, which allows the same workflow to be executed in different ways depending on which workflow director/MoC is used. Kepler ships with some common MoCs, such as **SDF**, **PN**, and **DDF** (see Section 13.2.3 for more details).

---

[*]http://www.cs.wisc.edu/condor/manual/v7.0/2_10DAGMan_Applications.html
[†]or *orchestration* and *choreography* in Web service parlance

When comparing different workflow execution strategies and approaches, it seems that a "one size fits all" solution is hard, if not impossible, to achieve.* In Pegasus, for example, workflow execution is primarily via Condor/DAGMan, a very mature and reliable platform (with some built-in fault tolerance) for job-oriented scientific workflows that can be expressed as acyclic task dependency graphs. However, there are applications such as scientific workflows over remote data streams (see, e.g., Altintas et al.[12]) which require other models of computation, for example, to express loops and streaming pipeline parallelism. Kepler inherits from Ptolemy[16] a number of such advanced models of computation that can even be combined in different ways.[57] Kepler also adds new models, for example, a data-oriented model of computation called COMAD that results in workflows that are easier to build and understand.[22,59] Triana, on the other hand, is a service-oriented system, supporting a wide variety of different grid and service-oriented execution environments. In the end, user requirements and application constraints have to be taken into account when deciding which execution model or system to choose.

13.5 Workflow Provenance and Execution Monitoring

The absence of detailed provenance information presents difficulties for scientists wishing to share and reproduce results, validate the results of others, and reuse the knowledge involved in analyzing and generating scientific data. In addition, the lack of provenance information may also limit the longevity of data. For example, without sufficient information describing how data products were generated, the value and use of this data may be greatly diminished. Thus, many current scientific workflow systems provide mechanisms for recording the provenance of workflows and their associated data products. This provenance information can be used to answer a number of basic questions posed by scientists related to data, such as: Who created this data product and when? What were the processes used in creating this data product? Which data products were derived from this data product? And were these two data products derived from the same raw (input) data?

The software infrastructure required to accurately and efficiently answer these questions is far from trivial,[60,88] especially in light of the need to make provenance-related software tools usable by domain scientists, who do not necessarily have programming expertise. For instance, while it may be

---

*The different approaches do not necessarily exclude each other, however: for example, Mandal et al.[58] reports on experiences in combining a Kepler frontend with a Pegasus backend, combining features of both systems (but also limiting each system's more general capabilities). While end users typically avoid "system mashups," some interesting insights into the different approaches and capabilities can still be gained by such interoperability experiments.

possible to use session logs generated by software tools to capture provenance information,[61] these logs may not be represented in a format that can be easily queried; and further, they may require sophisticated programming techniques to uncover the information needed to answer the above questions related to data lineage.

Similar to a workflow specification, the provenance of a workflow run is often represented using graph structures in which nodes represent processes and data products, and edges capture the flow of data between processes.[60,62] Some approaches support additional graph representations, for example, by recording explicit process and data dependencies.[63,64] A complete description of provenance models and capture mechanisms is beyond the scope of this chapter (see, for example, Simmhan et al.[55] and Davidson and Freire[60]). Instead, we concentrate on describing one particular scheme of storing provenance information that combines features from Kepler,[18] VisTrails,[65] and Pegasus.[53] We first describe the different types of provenance information considered by these approaches, and then discuss the current implementation of the provenance framework used by the Scientific Data Management (SDM) Center.[66]

Types of Provenance Information. Provenance information related to scientific workflow systems is sometimes divided into three distinct types, or layers[67]: workflow *description*, workflow *evolution*, and workflow *execution*. The workflow description layer consists of the specifications of individual workflows. The workflow evolution layer captures the relationships among a series of workflow specifications that are created in the course of defining an exploratory analysis. Finally, the execution layer stores runtime information about the execution of a workflow. This information may include, for example, the day and time the workflow was run, the execution time of each workflow step, the data provided to and generated by each step, a description of the workflow deployment environment, and so on. There are many ways to store information in each layer. For example, in VisTrails, a "change-based" model is used to represent both the evolution and workflow layers,[54] runtime information is captured by the workflow execution engine and stored in a relational database. The three layers are related by the overall provenance storage infrastructure.

The separation of provenance information into distinct layers can lead to a more normalized representation that avoids storing portions of each layer redundantly. For instance, this is in contrast to provenance approaches that store information about the workflow specification within the execution log, where a module name, the module parameters, and the parameter values are saved for each invocation of a given module. Separating provenance information into distinct layers can also help provenance frameworks become more extensible, for example, by allowing layers to be replaced with new representation approaches or by allowing entirely new layers to be added.[90]

The VisTrails workflow evolution approach captures changes to workflow specifications and displays these changes using a history tree called a *visualization trail*, or *vistrail* for short.[68] As a workflow developer makes modifications to a workflow, the VisTrails system records each change. Instead of storing a set of related workflows, the change-based model stores the operations, or actions, that are applied to the workflows (e.g., the addition or deletion of a module, the addition or deletion of a connection between modules, and the modification of a parameter value). This representation (similar, e.g., to source-code control systems such as Subversion) uses substantially less space than the alternative of explicitly storing each version of a workflow. In addition, VisTrails provides an intuitive interface that can help users to both understand and interact with the version history of a workflow design.[68] This tree-based view (see Figure 13.4) allows a user to return to a previous version, undo changes, compare different workflows, and determine the actions that led to a particular result.

In addition, query languages and user interfaces that can allow users to explore the provenance of workflow runs are also important.[59,64,68,69,87] For example, the ability to query both the specification and provenance of computational tasks enables users to better understand the tasks and their results. In this way, users can identify workflows that are suitable for and can be reused for a given task; identify workflow instances that have been found to contain anomalies; and compare and understand the differences between workflows.[59,68,69] Many existing workflow systems support query and visualization of provenance information associated with the workflow definition and execution layers (e.g., see Moreau et al.[70]).

13.5.1 Example Implementation of a Provenance Framework

Figure 13.5 shows the high-level architecture for the provenance framework employed within the SDM Center. The architecture has been implemented with the goal of supporting scientists as they run large-scale simulations.[71,72] At the heart of this framework is the *provenance store*, which includes one or more databases providing physical storage, as well as various application programming interfaces (APIs) to access and manage provenance information.

The provenance store within the SDM framework captures the following types of information:

- *Process monitoring* information, which includes data transfer rates, file sizes moved, time taken for actor execution and check-pointing, memory usage, process states (initiated, executing, waiting, terminated, aborted), and so forth, This information is useful, for example, to benchmark workflow execution and detect bottlenecks.
- *Data provenance and lineage* information, which links an actor's data output to (1) the specific actor invocation that created the data, (2) the relevant data inputs, and (3) the parameters at the time of invocation. Data provenance allows a scientist to interpret and "debug" analysis results, for example, by stepping through time in the processing history, thus tracing back (intermediate) results to the inputs that

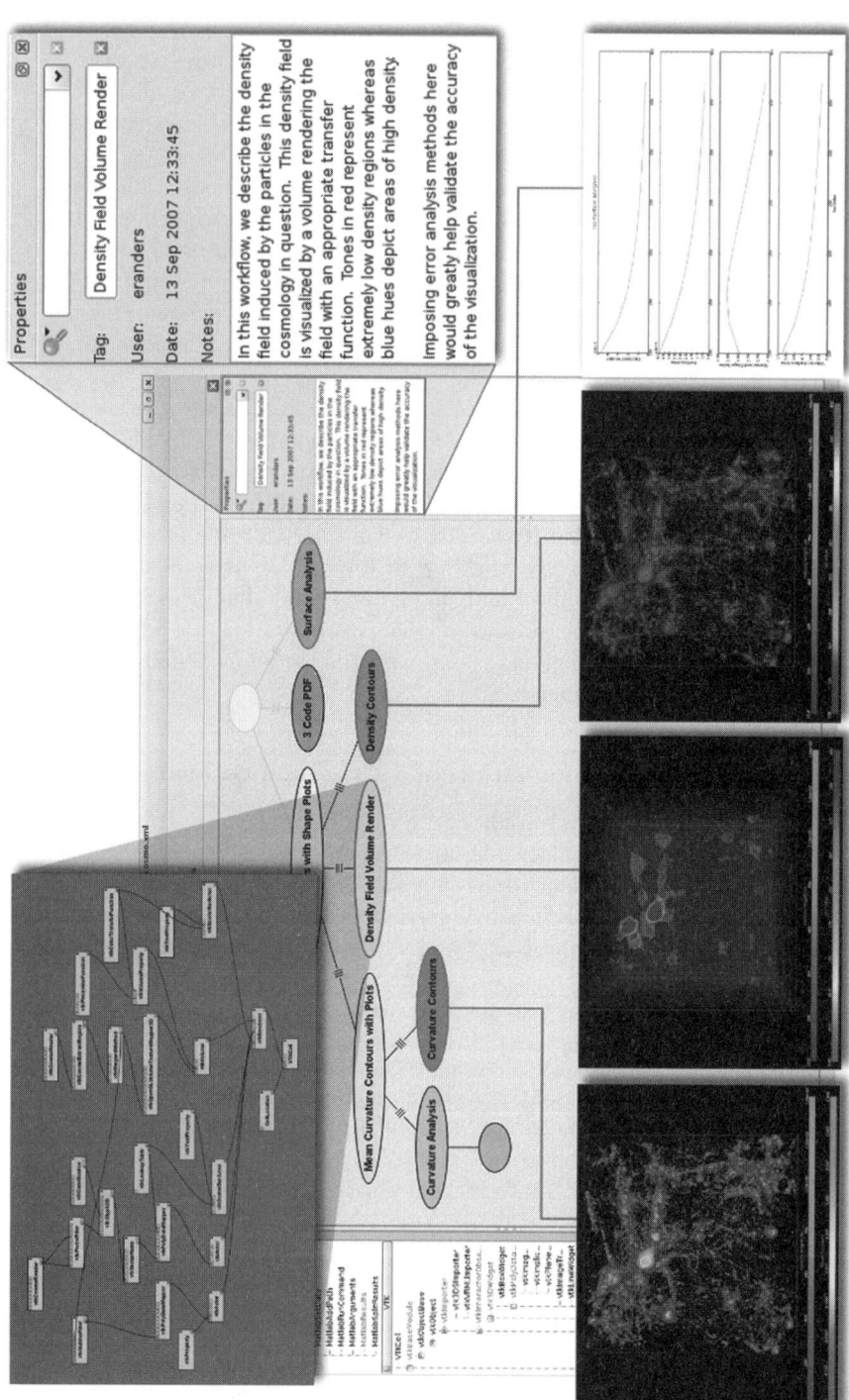

Figure 13.4 (See color insert following page 224.) In VisTrails, workflow evolution provenance is displayed as a history tree, each node representing a workflow that generates a visualization. This tree allows a user to return to previous workflow versions and to be reminded of the actions that led to a particular result. Additional metadata stored with each version includes free-text notes and the user who created it.

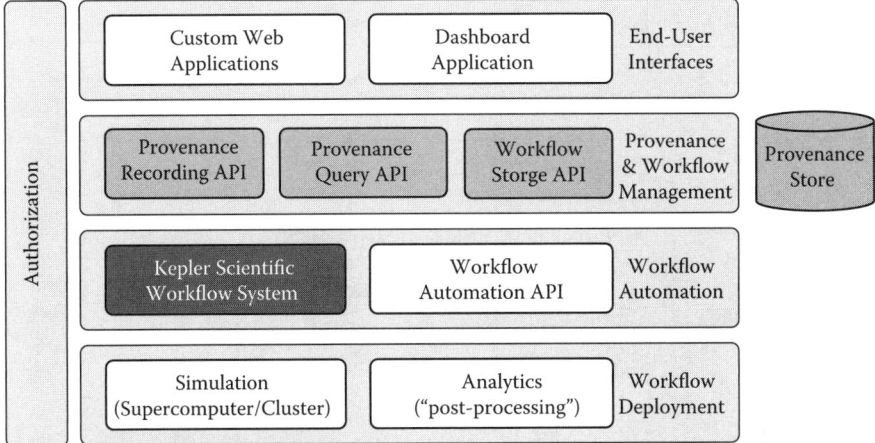

Figure 13.5 Overview of the SDM provenance framework.

created them. Other uses are increased workflow robustness (cf. Section 13.3.2) and improved efficiency upon rerunning only the affected parts after modifying some workflow parameters or inputs.

- *Workflow evolution*, which captures changes over time to the workflow description (including parameter changes). It is particularly important to track workflow evolution as part of exploratory workflow design, when there are many cycles of workflow modifications and workflow runs.

- *System environment* information, which captures data about the system that executes the workflow, and its environment, for example, the machines, operating systems, compiler versions, job queues, and so forth that were used. It is important to capture such information since, in practice, results will depend on the system environment that a workflow executes in.

Kepler has been extended to record and store various forms of provenance information.[64,73,74] Depending on the settings of the Kepler provenance recorder, data may be recorded for all actors in the workflow, or some subset, for example, only top-level composites. The recording API also supports recording of information from components external to Kepler (e.g., from Python or shell scripts that are invoked by an actor).

A provenance query API provides a (read-only) mechanism to retrieve provenance information from the provenance store, for example, a call-back mechanism to notify applications (such as a Web-based workflow monitoring dashboard) during workflow execution. In addition to providing current workflow status, authorized users and applications can query the provenance store about past executions via an SQL interface, thus supporting provenance analytics.

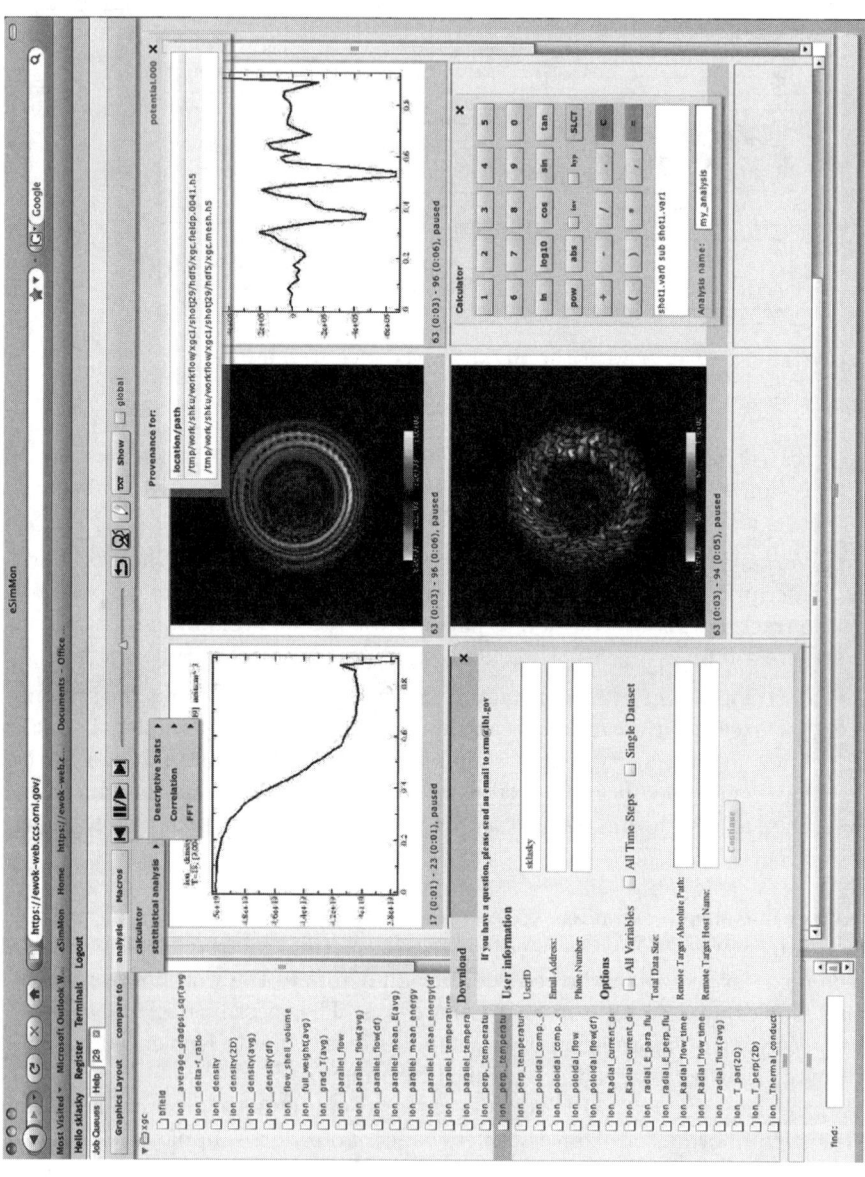

FIGURE 13.6: (See color insert following page 224.) SDM dashboard with on-the-fly visualization.

While provenance information is typically used "post-mortem," that is, after a workflow is run to interpret, validate, or debug results, it can also support runtime execution monitoring.[71,75] The SDM dashboard application supports runtime execution monitoring using the architecture of Figure 13.5. The dashboard is illustrated in Figure 13.6, which shows the on-the-fly visualizations generated by the monitoring workflow described in Section 13.3. Dashboards generally display condensed information about the status of workflow processes, data, the execution environment, and so forth. In addition to providing a high-level overview, such dashboards may also offer a way to navigate into details of runtime progress and provenance trace information, and to show trends in the output data or execution performance.

Other scientific workflow systems have similar capabilities to those described above. For example, Pegasus has been integrated with the PASOA provenance system.[76] Within Triana, provenance information can include the components executed, their parameters, and the datasets that pass through during execution. A data provenance system for Taverna is described in Missier et al.[63]

13.6 Workflow Sharing and myExperiment

Understanding the whole lifecycle of workflow design, prototyping, production, management, publication, and discovery is fundamental to developing systems that support the scientist's work and not just the workflow's execution. Supporting that lifecycle can be the factor that means a workflow approach is adopted or not. Workflow descriptions are not simply digital data objects like many other assets of e-Science, but rather they capture pieces of the scientific process: They are valuable knowledge assets in their own right, capturing valuable know-how that is otherwise often tacit. We can conceive of packages of workflows for certain topics, and of workflow "pattern books," that is, new structures above the level of the individual workflow. Workflows themselves can be the subject of peer review, and can support reproducibility in the scholarly knowledge cycle. We can view them as commodities, as valuable first-class assets in their own right, to be pooled and shared, traded and reused, within communities and across communities. This perspective of the interacting data, services, workflows and their metadata within a scientific environment is a *workflow ecosystem*. Understanding and enabling this ecosystem is the key to unlocking the broader scientific potential of workflow systems.

13.6.1 Workflow Reuse

Workflow reuse is effective at multiple levels: the scientist reuses a workflow with different parameters and data, and may modify the workflow as part of the routine of their daily scientific work; workflows can be shared with other scientists conducting similar work, so they provide a means of codifying, sharing, and thus spreading the workflow designers' practice; and workflows, workflow fragments, and workflow patterns can be reused to support science outside their initial application.

The latter point illustrates the tremendous potential for new scientific advances. An example of this is a workflow used to help identify genes involved in tolerance to trypanosomiasis in east African cattle.[77] The workflow was initially successful because it enabled data to be processed systematically without a need for manual triage. This same workflow was then reused over a new dataset to identify the biological pathways implicated in the ability for mice to expel the *Trichuris muris* parasite (a parasite model of the human parasite *Trichuris trichuria*). This reuse was made easier by the explicit, high-level nature of the workflow that describes the analytical protocol.

Workflows bring challenges too. Realistic workflows require skill to produce, and therefore they can be difficult and expensive to develop. Consequently, workflow developers need development assistance, and prefer not to start from scratch. This is another incentive for reusing workflows. Unfortunately it is easy for the reuse of a workflow to be confined to the project in which it was conceived. In the trypanosomiasis example, the barrier to reuse was how the knowledge about the workflow could be spread to the scientists with the potential need. In this case it was word of mouth within one institution; this barrier needs to be overcome.

Workflow management systems already provide basic sharing mechanisms, through repository stores for workflows developed as part of projects or communities. For example, the Kepler actor repository is an LDAP-based directory for the remote storage, query, and retrieval of actors (processes) and other workflow components. Similarly, the Southern California Earthquake Center (SCEC) uses component and workflow libraries annotated with ontologies.[78] These follow the tradition of cataloging scripting libraries and codes. InforSense's online Customer Hub and workflow library* allow users to share best practices and leverage community knowledge potentially across projects.

13.6.2 Social Sharing

The myExperiment project[†] is taking a more social approach, recognizing the use of workflows by a community of scientists.[79] This acknowledges

---

*http://www.inforsense.com/pdfs/InforSense_WorkflowLibrary_DataSheet.pdf. Accessed on July 20, 2009.

[†]http://www.myexperiment.org/. Accessed on July 20, 2009.

that the lifecycle of the workflows is coupled with the process of science — that the human system of workflow use is coupled to the digital system of workflows. More workflows imply more users and more enactments, which in turn provide scientists with more samples to assist in selecting workflows, to identify best practices, and to learn and build a reputation by sharing workflows within the community.

From the scientist's perspective there are many factors guiding reuse of a workflow, including descriptions of its function and purpose; documentation about the services with which it has been used, with example input and output data, and design explanations; provenance, including its version history and origins; reputation and use within the community; ownership and permissions constraints; quality, whether it is reviewed and still works; and dependencies on other workflows, components, and data types. Workflows also enable us to record the provenance of the data resulting from their enactment, and logs of service invocations from workflow runs can inform later decisions about service use. By binding workflows with this kind of information, a basis is provided for workflows to be trusted, interpreted unambiguously, and reused accurately.

The community perspective brings "network effects." By mining the sharing behavior between users within a community we can provide recommendations for use. By using the structure and interactions between users and workflow tools we can identify what is considered to be of greater value to users. Provenance information helps track down workflows through their use in content syndication and aggregation. By sharing or publishing a workflow, with the appropriate attribution, a scientist can allow their work to be reused with the concomitant spread of their scientific reputation; but even if scientists do not contribute workflows directly, their usage of workflows within the community still adds value to the body of knowledge about those workflows.

13.6.3 Realizing myExperiment

The rise of harnessing the collective intelligence of the Web has dramatically reminded us that it is people who generate and share knowledge and resources, and people who create network effects in communities. Blogs and wikis, shared tagging services, instant messaging, social networks, and semantic descriptions of data relationships are flourishing. Within the scientific community we have many examples, such as OpenWetWare, Connotea, PLoS on Facebook, and so forth.[*]

myExperiment is a virtual research environment to support scientists using workflows by adopting a "Web 2.0 approach." The myExperiment software provides services and a user interface (a social Web site) to address the requirements of the social sharing of workflows. It aims to be a gossip shop to share and discuss workflows and their related scientific objects, regardless of the

---

[*]See corresponding .org Web sites and facebook.com. Accessed on July 20, 2009.

workflow system; a bazaar for sharing, reusing, and repurposing workflows; a gateway to other established environments, for example, depositing into data repositories and journals; and a platform to launch workflows, whatever their system.

In comparison with existing workflow repositories, myExperiment goes the next step: It aims to cross project, community, and product boundaries; it emphasizes social networking around the workflows; it gateways to other environments; and it forms the foundation of a personal or laboratory workbench. It also transcends individual workflow systems, envisaging a multiworkflow environment in which scientists will use whatever workflow is appropriate for their applications — finding workflows and experiments that they can run across multiple systems.

The design of the myExperiment software is completely user-centric. In order to bootstrap the system, both in terms of content and community, the initial user community comprised users of one particular scientific workflow management system — the Taverna workbench. Developed by the myGrid project,[80] Taverna is used extensively across a range of Life Science problems: gene and protein annotation; proteomics, phylogeny, and phenotypical studies; microarray data analysis and medical image analysis; high-throughput screening of chemical compounds and clinical statistical analysis. Importantly, Taverna has been designed to operate in the open wild world of bioinformatics. Rather than large-scale, closed collaborations that own resources, Taverna is used to enable individual scientists to access the many open resources available on the Web. Consequently it has a distributed and decoupled community of users who obtain immediate benefit from sharing workflows through myExperiment.

Released in November 2007, myExperiment was supporting 500 users within 10 weeks and now provides a unique public collection of several hundred workflows from multiple workflow systems.

myExperiment has been designed and built following the mores of Web 2.0 and a set of principles for designing software for adoption by scientists that were established through the Taverna development.[81]

It is a Web-based application built on the Ruby on Rails platform and is not just a single site, like Facebook, YouTube, and others, but rather a software package that can be installed independently and separately in a laboratory, supporting the exchange of content between other Web applications and different installations of myExperiment. It reuses other services as far as possible, and it provides simple APIs so that others can make use of it — to make it easy to bring myExperiment functionality into the scientists' existing environment rather than obliging them to come to myExperiment.

Although initially focused on sharing workflows, myExperiment deals not in workflows or scripts per se but in scientific objects — this allows sharing of documents, presentations, service descriptions, notes, ontologies, plans, and so forth. More generally, myExperiment can be used to glue together heterogeneous collections like distributed experimental data or, for example, packages of workflows — these collections are described as Packs or Research Objects. Hence, rather than a workflow repository, myExperiment can be seen as an aggregator and registry of scientific objects — as its name suggests, it deals in experiments.

13.7 Conclusions and Future Work

Scientific workflows are increasingly being adopted across many natural science and engineering disciplines, spanning all conceivable dimensions and scales, from particle physics and computational chemistry simulations, to bioinformatics analyses, medicine, environmental sciences, engineering, geology, phylogeny, all the way to astronomy and cosmology, to name a few. Not surprisingly, with these rather different domains come different requirements for scientific workflow systems. While some scientific workflows can be conveniently executed on a scientist's laptop or desktop computer, others require significant computational resources, such as compute clusters, possibly distributed over a local or wide area network. In this chapter, we have given an overview of common features of scientific workflows, described the phases of the scientific workflow lifecycle, and provided some background on the different computational models (and other differences) of scientific workflow systems. A detailed case study from plasma fusion simulation was used to take a closer look at the challenges when managing simulation workflows. We also provided an overview of some of the different approaches taken for scientific workflow systems, focusing on workflow composition, resource mapping, and execution. Furthermore, we described approaches for runtime monitoring and provenance management in scientific workflows, and finally discussed workflow sharing and reuse using a "Web 2.0 approach."

The area of scientific workflows is dynamic and growing, as evidenced by many workshops, conferences, and special issues of journals, devoted to the topic (e.g., see Taylor et al.,[6] Gil et al.,[8] and Ludascher and Goble,[82] and Fox and Gannon[83]). Numerous challenges of scientific workflows remain and require future research and development. For example, findings from an NSF-sponsored workshop on scientific workflow challenges are reported in Gil et al.,[8] where they are grouped into *application requirements* (e.g., supporting collaborations, reproducibility of scientific analyses, and flexible system environments), *sharing workflow descriptions* (e.g., how to represent and share different levels of workflow abstractions), *dynamic workflows* (how to support the exploratory and dynamic nature of scientific analyses), and *system-level workflow management* (e.g., how to scale workflows and how to deal with infrastructure constraints). In, Deelman and Chervenak,[84] data-management challenges of data-intensive workflows are presented and organized according to the data lifecycle in a workflow. For example, during workflow creation, effective means are needed to discover data and software tools and to capture workflow evolution.[68] Similarly, during workflow planning and execution, there are numerous challenges, for example, how to efficiently and reliably transfer large amounts of data, or how to deal with distributed, heterogeneous system environments.

Traditionally, in computer science, a core theme is optimization of program runtime and memory usage by developing time- and space-efficient algorithms for the problems at hand. In many application areas, including scientific workflows, "human cycles" are a sometimes neglected resource, which can and should be optimized as well. For example, the use of data and workflow provenance information can be used for traditional purposes (such as optimizing system performance or improving fault-tolerance[74]), but also to enhance the scientist's insights when trying to understand or debug scientific workflow results.[85] Similarly, approaches are needed to facilitate modeling and design of scientific workflows that are easy to use. For example, McPhillips et al.[22] list the following *user-oriented requirements* and provides initial steps toward addressing them: *well-formedness* (facilitate the design of well-formed and valid workflows), *clarity* (facilitate the creation of self-explanatory workflows), *predictability* (make it easy to see what a workflow will do without running it), *recordability* (make it easy to see what a workflow actually did do when it ran), and *reportability* (make it easy to see if a workflow result makes sense scientifically). Clearly, the last two requirements are related to capturing and managing provenance information, a recurring theme in current scientific workflow research. Other research issues mentioned in McPhillips et al.[22] are *reusability* (make it easy to design new workflows from existing ones) and *data modeling* (provide first-class support for modeling scientific data), in addition to the already mentioned *optimization* issues (the system should take responsibility for optimizing performance).

Acknowledgments

Work on Kepler is partially supported by the NSF (under grants IIS-0630033, OCI-0722079, IIS-0612326, DBI-0533368) and the DOE (DE-FC02-ER25809 and DE-FC02-07-ER25811); work on VisTrails is partially supported by the NSF (under grants IIS-0844546, IIS-0751152, IIS-0746500, IIS-0513692, CCF-0401498, EIA-0323604, CNS-0514485, IIS-0534628, CNS-0528201, OISE-0405402, IIS-0905385), the DOE SciDAC2 program (VACET), and IBM Faculty Awards (2005, 2006, 2007, and 2008). Ewa Deelman's work was funded by the NSF under Cooperative Agreement OCI-0438712 and grant # CCF-0725332. We would like to thank Pierre Moualem, Meiyappan Nagappan, and Ustun Yildiz for their support.

References

[1] Scientific Discovery through Advanced Computing (SciDAC), Department of Energy. http://www.scidac.gov/. Accessed July 20, 2009.

[2] National Science Foundation, Office of Cyberinfrastructure. http://www.nsf.gov/dir/index.jsp?org=OCI. Accessed July 20, 2009.

[3] Defining e-Science. http://www.nesc.ac.uk/nesc/define.html. Accessed July 20, 2009.

[4] C. Anderson. The End of Theory: The Data Deluge Makes the Scientific Method Obsolete. *WIRED Magazine*, June 2008.

[5] B. Ludäscher, S. Bowers, and T. McPhillips. Scientific workflows. In T. Özsu and L. Liu, editors, *Encyclopedia of Database Systems*. Heidelberg, Germany: Springer, 2009.

[6] I. Taylor, E. Deelman, D. Gannon, and M. Shields, editors. *Workflows for e-Science: Scientific Workflows for Grids*. Springer, 2007.

[7] E. Deelman, D. Gannon, M. Shields, and I. Taylor. Workflows and e-Science: An overview of workflow system features and capabilities. *Future Generation Computer Systems*, 25(5):528–540, 2009.

[8] Y. Gil, E. Deelman, M. Ellisman, T. Fahringer, G. Fox, D. Gannon, C. Goble, M. Livny, L. Moreau, and J. Myers. Examining the challenges of scientific workflows. *Computer*, 40(12):24–32, 2007.

[9] M. Weske. *Business Process Management: Concepts, Languages, Architectures*. Springer, 2007.

[10] J. R. Rice and R. F. Boisvert. From Scientific Software Libraries to Problem-Solving Environments. *IEEE Computational Science and Engineering*, 3:44–53, 1996.

[11] J. Wainer, M. Weske, G. Vossen, and C. B. Medeiros. Scientific Workflow Systems. In *Proceedings of the NSF Workshop on Workflow and Process Automation in Information Systems: State of the Art and Future Directions*, 1996.

[12] C. Rueda and M. Gertz. Real-Time Integration of Geospatial Raster and Point Data Streams. In B. Ludäscher and N. Mamoulis, editors, *20th Intl. Conf. on Scientific and Statistical Database Management (SSDBM)*, volume 5069 of *LNCS*, pp. 605–611, Hong Kong, China, July 2008. Springer.

[13] D. Jordan and J. Evdemo. Web Services Business Process Execution Language, Version 2.0 (WS-BPEL 2.0), April 11, 2007.

[14] The Taverna Project, http://www.taverna.org.uk. Accessed July 20, 2009.

[15] Y. Wang, C. Hu, and J. Huai. A New Grid Workflow Description Language. In *IEEE International Conference on Services Computing, IEEE SCC*, pages 257–260. IEEE Computer Society, 2005.

[16] C. Brooks, E. A. Lee, X. Liu, S. Neuendorffer, Y. Zhao, and H. Zheng. Heterogeneous Concurrent Modeling and Design in Java (Volume 3: Ptolemy II Domains). Technical Report No. UCB/EECS-2008-37, April 2008. http://www.eecs.berkeley.edu/Pubs/TechRpts/2008/EECS-2008-37.pdf. Accessed on July 20, 2009.

[17] E. Deelman, J. Blythe, Y. Gil, C. Kesselman, G. Mehta, K. Vahi, K. Blackburn, A. Lazzarini, A. Arbree, R. Cavanaugh, and S. Koranda. Mapping Abstract Complex Workflows onto Grid Environments. *Journal of Grid Computing*, 1(1):25–39, 2003.

[18] The Kepler Project, http://www.kepler-project.org. Accessed on July 20, 2009.

[19] G. Kahn. The Semantics of Simple Language for Parallel Programming. In *IFIP Congress*, pp. 471–475, 1974.

[20] E. A. Lee and T. M. Parks. Dataflow Process Networks. In *Proceedings of the IEEE*, pp. 773–799, 1995.

[21] E. A. Lee and E. Matsikoudis. The semantics of dataflow with firing. In G. Huet, G. Plotkin, J.-J. Lévy, and Y. Bertot, editors, *From Semantics to Computer Science: Essays in memory of Gilles Kahn*. Cambridge University Press, March 2008. preprint, http://ptolemy.eecs.berkeley.edu/publications/papers/08/DataflowWithFiring/. Accessed on July 20, 2009.

[22] T. McPhillips, S. Bowers, D. Zinn, and B. Ludäscher. Scientific workflows for mere mortals. *Future Generation Computer Systems*, 25(5):541–551, 2009. http://dx.doi.org/10.1016/j.future.2008.06.013. Accessed on July 20, 2009.

[23] U.S. Department of Energy, The Center for Plasma Edge Simulation Project (CPES). http://www.cims.nyu.edu/cpes. Accessed October 10th, 2009.

[24] S.-H. Ku, C. Chang, M. Adams, J. Cummings, F. Hinton, D. Keyes, S. Klasky, W. Lee, Z. Lin, and S. Parker. Gyrokinetic particle simulation of neoclassical transport in the pedestal/scrape-off region of a Tokamak plasma. *Journal of Physics: Conference Series*, 46:87–91, 2006. Institute of Physics Publishing.

[25] J. Cummings, A. Pankin, N. Podhorszki, G. Park, S. Ku, R. Barreto, S. Klasky, C. Chang, H. Strauss, L. Sugiyama, P. Snyder, D. Pearlstein, B. Ludäscher, G. Bateman, and A. Kritz. Plasma Edge Kinetic-MHD Modeling in Tokamaks Using Kepler Workflow for Code Coupling, Data

Management and Visualization. *Communications in Computational Physics*, 4(3):675–702, 2008.

[26] N. Podhorszki, B. Ludäscher, and S. A. Klasky. Workflow Automation for Processing Plasma Fusion Simulation Data. In *2nd Workshop on Workflows in Support of Large-Scale Science (WORKS)*, pp. 35–44, 2007.

[27] D. Zinn, S. Bowers, T. McPhillips, and B. Ludäscher. X-CSR: Dataflow Optimization for Distributed XML Process Pipelines. In *ICDE*, 2009.

[28] B. Ludäscher, I. Altintas, C. Berkley, D. Higgins, E. Jaeger, M. Jones, E. A. Lee, J. Tao, and Y. Zhao. Scientific workflow management and the Kepler system. *Concurrency and Computation: Practice & Experience*, 18(10):1039–1065, 2006.

[29] The Pegasus Project, http://pegasus.isi.edu. Accessed July 20, 2009.

[30] E. Deelman, J. Blythe, Y. Gil, C. Kesselman, G. Mehta, S. Patil, M.-H. Su, K. Vahi, and M. Livny. Pegasus: Mapping Scientific Workflows onto the Grid. In *European Across Grids Conference*, pp. 11–20, 2004.

[31] T. Oinn, M. Addis, J. Ferris, D. Marvin, M. Senger, M. Greenwood, T. Carver, K. Glover, M. Pocock, A. Wipat, and P. Li. Taverna: a tool for the composition and enactment of bioinformatics workflows. *Bioinformatics*, 20(17), 2004.

[32] T. Oinn, M. Greenwood, M. Addis, M. N. Alpdemir, J. Ferris, K. Glover, C. Goble, A. Goderis, D. Hull, D. Marvin, P. Li, P. Lord, M. R. Pocock, M. Senger, R. Stevens, A. Wipat, and C. Wroe. Taverna: lessons in creating a workflow environment for the life sciences. In Fox and Gannon [32].

[33] The Triana Project, http://www.trianacode.org. Accessed July 20, 2009.

[34] I. J. Taylor, M. S. Shields, I. Wang, and O. F. Rana. Triana applications within grid computing and peer to peer environments. *Journal of Grid Computing*, 1(2):199–217, 2003.

[35] Y. Gil, V. Ratnakar, E. Deelman, G. Mehta, and J. Kim. Wings for Pegasus: Creating Large-Scale Scientific Applications Using Semantic Representations of Computational Workflows. In *AAAI*, pp. 1767–1774, 2007.

[36] J. Yu and R. Buyya. A taxonomy of scientific workflow systems for grid computing. In Ludäscher and Goble [55].

[37] The Condor Team. *Condor Version 7.0.4 Manual*. University of Wisconsin-Madison, 2002. http://www.cs.wisc.edu/condor/manual/v7.0/. Accessed July 21, 2009.

[38] G. B. Berriman, J. C. Good, A. C. Laity, A. Bergou, J. Jacob, D. S. Katz, E. Deelman, C. Kesselman, G. Singh, M. Su, and R. Williams.

Montage: A Grid-Enabled Image Mosaic Service for NVO. *Astronomical Data Analysis Software and Systems, ADASS*, 13, 2003.

[39] D. S. Katz, A. Bergou, G. B. Berriman, G. L. Block, J. Collier, D. W. Curkendall, J. Good, L. Husman, J. C. Jacob, A. C. Laity, P. Li, C. Miller, T. Prince, H. Siegel, and R. Williams. Accessing and Visualizing Scientific Spatiotemporal Data. In *16th Intl. Conf. on Scientific and Statistical Database Management (SSDBM)*, pp. 107–110. IEEE Computer Society, 2004.

[40] A. Lathers, M.-H. Su, A. Kulungowski, A. Lin, G. Mehta, S. Peltier, E. Deelman, and M. Ellisman. Enabling parallel scientific applications with workflow tools. In *Challenges of Large Applications in Distributed Environments, 2006 IEEE*, pp. 55–60, 2006.

[41] J. Muench, H. P. Maechling, Francoeur, D. Okaya, and Y. Cui. SCEC Earthworks Science Gateway: Widening SCEC Community Access to the TeraGrid. In *The First Annual TeraGrid Conference*, 2006.

[42] C. Berkley, S. Bowers, M. Jones, B. Ludäscher, M. Schildhauer, and J. Tao. Incorporating Semantics in Scientific Workflow Authoring. In J. Frew, editor, *17th Intl. Conf. on Scientific and Statistical Database Management (SSDBM)*, pp. 75–78, 2005.

[43] S. Bowers and B. Ludäscher. Actor-Oriented Design of Scientific Workflows. In *24th Intl. Conference on Conceptual Modeling (ER)*, pp. 369–384, Klagenfurt, Austria, October 2005. Springer.

[44] K. Belhajjame, S. M. Embury, N. W. Paton, R. Stevens, and C. A. Goble. Automatic annotation of Web services based on workflow definitions. *ACM Trans. Web*, 2(2):1–34, 2008.

[45] C. Goble, K. Wolstencroft, A. Goderis, D. Hull, J. Zhao, P. Alper, P. Lord, C. Wroe, K. Belhajjame, D. Turi, R. Stevens, and D. D. Roure. Knowledge discovery for in silico experiments with Taverna: Producing and consuming semantics on the Web of science. In C. Baker and K.-H. Cheung, editors, *Semantic Web: Revolutionising Knowledge Discovery in Life Sciences*. New York: Springer Science & Business Media. 2007.

[46] P. Lord, P. Alper, C. Wroe, and C. Goble. Feta: A Light-Weight Architecture for User Oriented Semantic Service Discovery. In *2nd European Semantic Web Conference*, volume 3532 of *LNCS*, pp. 17–31, Crete, 2005. Springer.

[47] C. Goble, R. Stevens, D. Hull, K. Wolstencroft, and R. Lopez. Data curation + process curation = data integration + science. *Briefings in Bioinformatics*, December 2008.

[48] I. Taylor. Triana Generations. In *2nd Intl. Conf. on e-Science and Grid Technologies (e-Science)*, page 143. IEEE Computer Society, 2006.

[49] D. Hull, R. Stevens, P. Lord, C. Wroe, and C. Goble. Treating shimantic Web syndrome with ontologies. In *First Advanced Knowledge Technologies Workshop on Semantic Web Services (AKT-SWS04)*, CEUR-WS.org, Volume 122, 2004.

[50] J. Kommineni and D. Abramson. GriddLeS Enhancements and Building Virtual Applications for the GRID with Legacy Components. In *Advances in Grid Computing (European Grid Conference)*, pp. 961–971, 2005.

[51] D. Abramson, R. Sosic, J. Giddy, and B. Hall. Nimrod: A tool for performing parametised simulations using distributed workstations. In *4th IEEE Symposium on High Performance Distributed Computing*, pp. 112–121, 1995.

[52] GridLab Resource Management System (GRMS), 2005. http://www.gridlab.org/WorkPackages/wp-9/. Accessed July 20, 2009.

[53] G. Singh, M.-H. Su, K. Vahi, E. Deelman, G. B. Berriman, J. Good, D. S. Katz, and G. Mehta. Workflow Task Clustering for Best Effort Systems with Pegasus. In *Mardi Gras Conference*, p. 9, 2008.

[54] R. Wolski, D. Nurmi, and J. Brevik. An Analysis of Availability Distributions in Condor. In *IPDPS*, pp. 1–6, 2007.

[55] Y. Simmhan, B. Plale, and D. Gannon. A Survey of Data Provenance in e-Science. In Ludäscher and Goble, see Reference 82.

[56] Grid Resource Allocation and Management (GRAM), 2008. http://www.globus.org/toolkit/docs/4.2/4.2.0/developer/globusrun-ws.html. Accessed July 20, 2009.

[57] A. Goderis, C. Brooks, I. Altintas, E. A. Lee, and C. A. Goble. Composing Different Models of Computation in Kepler and Ptolemy II. In Y. Shi, G. D. van Albada, J. Dongarra, and P. M. A. Sloot, editors, *International Conference on Computational Science (3)*, volume 4489 of *Lecture Notes in Computer Science*, pp. 182–190. Springer, 2007.

[58] N. Mandal, E. Deelman, G. Mehta, M.-H. Su, and K. Vahi. Integrating existing scientific workflow systems: the Kepler/Pegasus example. In *2nd Workshop on Workflows in Support of Large-Scale Science (WORKS'07)*, pages 21–28, New York, NY, USA, 2007. ACM.

[59] S. Bowers, T. McPhillips, M. Wu, and B. Ludäscher. Project Histories: Managing Data Provenance Across Collection-Oriented Scientific Workflow Runs. In *4th Intl. Workshop on Data Integration in the Life Sciences (DILS)*, University of Pennsylvania, June 2007.

[60] S. B. Davidson and J. Freire. Provenance and Scientific Workflows: Challenges and Opportunities. In *SIGMOD Conference*, pp. 1345–1350, 2008.

[61] B. Ludäscher, N. Podhorszki, I. Altintas, S. Bowers, and T. McPhillips. From computation models to models of provenance: The RWS approach. *Concurrency and Computation: Practice & Experience*, 20(5):507–518, 2008.

[62] S. Bowers, T. M. McPhillips, B. Ludäscher, S. Cohen, and S. B. Davidson. A Model for User-Oriented Data Provenance in Pipelined Scientific Workflows. In L. Moreau and I. T. Foster, editors, *Intl. Provenance and Annotation Workshop (IPAW)*, volume 4145 of *Lecture Notes in Computer Science*, pp. 133–147. Springer, 2006.

[63] P. Missier, K. Belhajjame, J. Zhao, and C. Goble. Data lineage model for Taverna workflows with lightweight annotation requirements. In *Intl. Provenance and Annotation Workshop (IPAW)*, 2008.

[64] S. Bowers, T. McPhillips, S. Riddle, M. Anand, and B. Ludäscher. Kepler/pPOD: Scientific Workflow and Provenance Support for Assembling the Tree of Life. In *Intl. Provenance and Annotation Workshop (IPAW)*, 2008.

[65] The VisTrails Project, http://www.vistrails.org. Accessed July 20, 2009.

[66] A. Shoshani, I. Altintas, A. Choudhary, T. Critchlow, C. Kamath, B. Ludäscher, J. Nieplocha, S. Parker, R. Ross, N. Samatova, and M. Vouk. SDM Center technologies for accelerating scientific discoveries. *Journal of Physics: Conference Series*, 78, 2007.

[67] C. Scheidegger, D. Koop, E. Santos, H. Vo, S. Callahan, J. Freire, and C. Silva. Tackling the provenance challenge one layer at a time. *Concurrency and Computation: Practice & Experience*, 20(5):473–483, 2008.

[68] J. Freire, C. T. Silva, S. P. Callahan, E. Santos, C. E. Scheidegger, and H. T. Vo. Managing Rapidly-Evolving Scientific Workflows. In *Intl. Provenance and Annotation Workshop (IPAW)*, volume 4145 of *Lecture Notes in Computer Science*, pages 10–18. Springer, 2006.

[69] C. E. Scheidegger, H. T. Vo, D. Koop, J. Freire, and C. T. Silva. Querying and creating visualizations by analogy. *IEEE Trans. Vis. Comput. Graph.*, 13(6):1560–1567, 2007.

[70] L. Moreau, et al. The first provenance challenge. *Concurrency and Computation: Practice and Experience—Special Issue on the First Provenance Challenge*, 20(5), 2008.

[71] R. Barreto, T. Critchlow, A. Khan, S. Klasky, L. Kora, J. Ligon, P. Mouallem, M. Nagappan, N. Podhorszki, and M. Vouk. Managing and Monitoring Scientific Workflows through Dashboards. In *Microsoft eScience Workshop*, p. 108, Chapell Hill, NC, 2007.

[72] I. Altintas, G. Chin, D. Crawl, T. Critchlow, D. Koop, J. Ligon, B. Ludäscher, P. Mouallem, M. Nagappan, N. Podhorszki, C. Silva, and

M. Vouk. Provenance in Kepler-based Scientific Workflow Systems. In *Microsoft eScience Workshop*, page 82, 2007. Poster.

[73] I. Altintas, O. Barney, and E. Jaeger-Frank. Provenance Collection Support in the Kepler Scientific Workflow System. In L. Moreau and I. T. Foster, editors, *Intl. Provenance and Annotation Workshop (IPAW)*, volume 4145 of *Lecture Notes in Computer Science*, pp. 118–132. Springer, 2006.

[74] D. Crawl and I. Altintas. A Provenance-Based Fault Tolerance Mechanism for Scientific Workflows. In *Intl. Provenance and Annotation Workshop (IPAW)*, 2008.

[75] S. A. Klasky, M. Beck, V. Bhat, E. Feibush, B. Ludäscher, M. Parashar, A. Shoshani, D. Silver, and M. Vouk. Data management on the fusion computational pipeline. *Journal of Physics: Conference Series*, 16:510–520, 2005.

[76] J. Kim, E. Deelman, Y. Gil, G. Mehta, and V. Ratnakar. Provenance trails in the Wings-Pegasus system. *Concurrency and Computation: Practice & Experience*, 20(5):587–597, 2008.

[77] P. Fisher, C. Hedeler, K. Wolstencroft, H. Hulme, H. Noyes, S. Kemp, R. Stevens, and A. Brass. A systematic strategy for large-scale analysis of genotype phenotype correlations: identification of candidate genes involved in African trypanosomiasis. *Nucleic Acids Res*, 35(16):5625–5633, 2007.

[78] P. Maechling, H. Chalupsky, M. Dougherty, E. Deelman, Y. Gil, S. Gullapalli, V. Gupta, C. Kesselman, J. Kim, G. Mehta, B. Mendenhall, T. A. Russ, G. Singh, M. Spraragen, G. Staples, and K. Vahi. Simplifying Construction of Complex Workflows for Non-Expert Users of the Southern California Earthquake Center Community Modeling Environment. In Ludäscher and Goble [55].

[79] D. D. Roure, C. Goble, and R. Stevens. The design and realisation of the myExperiment virtual research environment for social sharing of workflows. *Future Generation Computer Systems*, 25:561–567, 2009.

[80] R. D. Stevens, A. J. Robinson, and C. A. Goble. myGrid: Personalised Bioinformatics on the Information Grid. In *11th Intl. Conf. on Intelligent Systems for Molecular Biology (ISMB)*, pp. 302–304, 2003.

[81] D. D. Roure and C. Goble. Software design for empowering scientists. *IEEE Software*, 26(1):88–95, January/February 2009.

[82] B. Ludäscher and C. Goble, editors. *ACM SIGMOD Record: Special Issue on Scientific Workflows*, volume 34(3), September 2005.

[83] G. C. Fox and D. Gannon, editors. *Concurrency and Computation: Practice & Experience. Special Issue: Workflow in Grid Systems*, volume 18(10). Hoboken, New Jersey: John Wiley & Sons, 2006.

[84] E. Deelman and A. Chervenak. Data Management Challenges of Data-Intensive Scientific Workflows. In *8th IEEE Intl. Symposium on Cluster Computing and the Grid (CCGRID)*, pp. 687–692. IEEE Computer Society, 2008.

[85] S. B. Davidson, S. C. Boulakia, A. Eyal, B. Ludäscher, T. M. McPhillips, S. Bowers, M. K. Anand, and J. Freire. Provenance in Scientific Workflow Systems. *IEEE Data Eng. Bull.*, 30(4):44–50, 2007.

[86] B. Ludäscher, M. Weske, T. McPhillips, and S. Bowers. Scientific-Workflows: Business as Usual? In U. Dayal, J. Eder, J. Koehler, and H. Reijers, editors, 7th Intl. Conf. on Business Process Management (BPM), LNCS 5701, Ulm, Germany, 2009.

[87] C. Beeri, A. Eyal, S. Kamenkovich, and T. Milo. Querying Business Processes. In Very Large Data Bases (VLDB) conference pp. 343–354, 2006.

[88] J. Freire, D Koop, E. Santos, and C. T. Silva. Provenance for Computational Tasks: A Survey. Computing in Science and Engineering, 10(3): 11–21, 2008.

[89] T. McPhillips, S. Bowers, and B. B. Ludäscher. Collection-Oriented Scientific Workflows for Integrating and Analyzing Biological Data. In 3rd Intl. Workshop on Data Integration in the Life Sciences (DILS), LNCS, European Bioinformatics Institute, Hinxton, UK, July 2006. Springer.

[90] C. E. Scheidegger, D. Koop, E. Santos, H. T. Vo, S. P. Callahan, J. Freire, and C. T. Silva. Tackling the Provenance Challenge one layer at a time. Concurrency and Computation: Practice and Experience, 20 (5): 473–483, 2008.

Conclusions and Future Outlook

Arie Shoshani and Doron Rotem, Editors

In response to the increasing volume and complexity of scientific data, many projects and activities have taken place over the last decade. In this book we have addressed these issues in depth, by describing existing technologies, and how they were used successfully in real example applications. Specifically, we addressed the state of the art of storage technologies, techniques for achieving efficient I/O, and standards for managing storage systems, and large-scale data movement. We also covered methods of efficient in-memory data movement, workflow technology to achieve real-time monitoring, and metadata and provenance management techniques to collect the history of runs. Other topics that were discussed in detail include optimization methods for managing streaming data from sensors and satellites, the integration of diverse geo-science datasets, efficient methods for data searching, analysis and visualization for data explorations, and new types of scientific data management systems.

Some of the main conclusions in the chapters of this book are captured next. In the storage area, while new emerging storage technologies are now introduced, in the near term it is expected that the high performance computing industry will primarily continue to use disks and tapes to store data. A new technology that is gaining importance is the development of SSDs (Solid State Disk Drives). The cost of SSDs has dropped sufficiently to make such devices viable when integrated alongside traditional hard drives in a parallel storage system in order to improve latency of common operations. The main challenges to the continued use of larger numbers of disks with increased capacity are power consumption, recovery from failures, and maintaining data integrity.

Building effective parallel file systems that can take advantage of many thousands of disks will continue to be an important challenging activity. Standards for accessing different storage systems through a uniform interface have already proven viable in simplifying the client applications that use them. Such standards will continue to evolve in the future to permit co-scheduling of compute, storage, and network resources. An important challenge is the development of algorithms and schedulers that can make use of historical data to optimize resource usage, and recover from transient failures. It is predicted that in the future it will not be feasible to store all the raw data generated by large-scale simulations. Consequently, it will be necessary to process much of the data as it is generated before it is stored on disk or archived. Processing of the data as it is being generated can be made on the I/O processors of large supercomputers, or offline on smaller clusters.

Efficient indexing methods for searching and subsetting large datasets are becoming essential in many applications. New emerging indexing methods that take advantage of the fact that scientific data only grows and does not change over time, have proven to be effective in terms of size and performance of the index. New database management systems designed specifically for scientific data are emerging. These include so-called vertical database systems, and systems that support data models for scientific data structures such as arrays. Scientific data analysis will continue to be an art of trying various methods such as filtering, de-noising, dimensionality reduction, and feature extraction. The challenges are to provide facilities where such iterative explorations can take place with minimal effort for the scientists and with good response time. The trend toward parallelizing analysis codes and packages will continue as the volume of data to be analyzed grows. Scientific data visualization is another aspect of data analysis that is indispensible in many applications. Many of the techniques needed for effective real-time interaction with the data, such as multi-resolution analysis, require data structures that can be searched efficiently on parallel machines. Such techniques will continue to evolve, and will most likely be most effective when running on facilities close to where the data is stored. Integration of data from multiple disciplines is already an extremely important problem. Experience in the Geoscience domain has shown that the best chance to succeed is to develop standard data formats and ontologies that various tools can adhere to. In practice, the process of adopting standards is an evolutionary process. Once such standards are developed, data transformations from legacy formats will continue to be applied for some time. Streaming data is becoming a highly challenging problem as the speed and quantity of data generated by sensor devices, experiments, and satellite data increases. Processing the data streams in parallel is an obvious technique, but generating indexes as the data is streaming, as well as approximate summarizations, are effective techniques as well. As more and more data is collected, the metadata associated with it becomes essential. Many datasets have lost their value over time, because the metadata was not adequately collected, or was lost. The challenge is to develop systems that automatically collect metadata on the structural and content information, on the way it was generated, and on its provenance.

Finally, it is evident that workflow management is becoming an essential part of managing the data generation, data processing, and data analysis processes. Many tasks need to be performed soon after the raw data is generated, and workflow systems are needed to perform these in a timely manner. Workflow systems are extremely useful in repetitive tasks, such as running a simulation repeatedly with various parameters, and then generating summaries and graphs while the data is generated in order to monitor progress. There are still many challenges in this relatively new area, including simple tools to design scientific workflows that are easily understood, insuring that the workflows are well-formed, and workflow systems that are fault tolerant. Fault tolerance is important in particular for long-running workflows, so that they can recover even if the machine they are running on temporarily fails.

We see three major emerging trends that may contribute to future developments. The first, energy efficient storage systems, stems from the escalating energy costs of spinning and cooling disk storage systems in data centers. The second, co-location of data and analysis, stems from the impracticality of having to move large volumes of data to scientists' sites. The third, the development of new scientific database management systems, stems from the wish to isolate the scientist from having to deal with various data formats and the details of file and storage systems. Next, we describe each of these trends in more detail.

Energy Efficient Storage Systems

Several studies indicate that disk storage systems and their cooling consume over 30% of the power in scientific data centers. This percentage of disk storage power consumption will continue to increase, as faster and higher capacity disks are deployed with increasing energy costs and as data-intensive applications demand reliable on-line access to data resources. As a result, optimization of energy use in scientific data management has become an important area of research across multiple disciplines such as computer architecture, power management, operations research, and theoretical computer science.

Concern about the amount of energy used by scientific data centers has also led to the introduction of commercial products in the area of energy efficient or "green" data centers. At the system level, a number of integrated storage solutions have emerged, all of which are based on the general principle of spinning down disks when not in use and spinning up disks when they are being accessed. In these systems, disks configured either as RAID sets, or as independent disks, are programmed to be spun down into standby mode after experiencing an interval of time without any activity (idle time). The length of the idle interval, also called *idleness threshold*, can be fixed or determined dynamically based on historical access data. In general, longer idle periods offer more energy saving opportunities. For this reason, several storage vendors are now offering hybrid disks which incorporate an SSD cache where commonly accessed data is maintained, resulting in longer idle periods of the disk. A major research problem is that of determining optimal idleness thresholds that also satisfy quality of service requirements in terms of expected system response time. In recent research works it has been shown that allowing reorganization of disk contents either dynamically or at periodic reorganization points can save much more power as compared with maintaining the disk contents static. Several algorithms for packing files into disks have been developed where it was shown that skewed allocations of files to disks, .i.e., allocations that make few disks very active while the rest are relatively idle, perform much better than balanced file allocations. It has also been observed that traditional caching policies that are aimed only at increased performance are not always optimal in terms of their power consumption. While more research work is required to study power-aware caching policies and prefetching algorithms that can further reduce the power consumption in disk systems, we expect such algorithms to become part of any file system in the future.

Co-Location of Data and Analysis

When the volume of data could fit into a workstation or even a small cluster, copying the data was the common practice, so that it can be analyzed locally. As the volume of data grows, it is becoming increasingly prohibitive to copy hundreds of terabytes or petabytes to scientists' sites. Another consideration is the need to manage the analysis software. As the volume of data grows, many analysis packages do not scale, and a new generation of parallel analysis tools is being developed. The cost of purchasing and managing cluster machines and the parallel analysis software is becoming out of reach for many scientists.

The obvious alternative is "data-side analysis facility" in order to perform the analysis on a shared analysis facility that is co-located with the data. Such a facility, which could be built on a medium size cluster (perhaps a few thousand cores) and include various packages of parallel analysis software. The size of the facility will depend on the number of users expected to be served. However, in addition to managing the use of resources by users, the facility needs to include a way for scientists to express the analysis process. As explained in Chapter 8, the analysis process can be viewed as a pipeline, consisting of steps such as feature extraction, dimensionality reduction, and pattern recognition. Similarly, a visual analysis pipeline, as discussed in Chapter 9, can consist of steps such as filtering, mapping, and rendering. More generally, this process can be an acyclic workflow where flows can branch and join together. Thus, workflow systems need to be deployed on such facilities to facilitate data-side analysis. Further, metadata and provenance capabilities need to be supported as well.

There are already some limited examples where the analysis scripts are sent to facilities near the data to be executed. However, a general purpose facility needs to provide a way for the user to specify the analysis components to be used by the workflow, the inputs needed for each step, and the output produced. This facility should insulate the user from unnecessary details of where the data is stored and from interacting with the workflow system directly. The analysis facility should take care of the storage of the data, execution of the workflow, translation of formats between steps to match each step's input and output requirements, and generation of products for the user to track and monitor progress of the analysis process. This suggests the concept of an "abstract workflow" that is mapped into "concrete (or executable) workflows" as described in Chapter 13. We expect such facilities to become the norm in conducting scientific analysis.

In general, data-side analysis facilities will not eliminate replicating data. One would expect that important data, which large communities share, will be replicated to multiple sites, each providing its own data-side analysis facility. For example, climate modeling data generated by long runs on supercomputers will most likely be mirrored to multiple sites worldwide. Similarly, it is expected that some subsets of data will still be moved to scientists' sites, as cost of cluster hardware continues to fall, and networking speed grows. As cloud computing and storage grows in use, it is expected that data-side analysis will be offered on cloud facilities as well.

Scientific Database Management Systems

Historically, the concept of separating the logical organization of the data from its physical organization dominated the development of database management systems (DBMSs). This is referred to as "physical data independence". The logical organization referred to "what" is the structure of the data, and the physical organization referred to "how" the data is stored and organized on physical media, including memory and disk. In order to access the data faster, different types of storage organization and indexes were invented, which did not affect the logical organization of the data. Such concepts brought about the use of DBMSs in many areas, especially in business and commercial application domains.

The dominant DBMS system today is still the relational database system. Its simple data model of representing data as tables, where rows represent instances of objects (such as people, books, etc.), and columns represent properties (or attributes) of the objects, made it very attractive to many applications. However, by and large, relational database systems have not been used extensively by scientific applications. Instead, most large scientific datasets are stored as files in specific standard file formats, such as NetCDF and HDF5. There are several reasons for this state of affairs. First, it is the desire of scientists to exchange data by simply sending each other files. By agreeing on a standard file format for some communities, and even including the metadata in the header of the files, the files became "self-describing". Second, there is a need to write data into and read data from file systems efficiently. Focusing on files rather than a database systems, simplified the task. Today, there are several popular parallel file systems that are extensively used with supercomputers and large clusters, including Lustre, PVFS, and GPFS, described in Chapter 2. Furthermore, the libraries for reading and writing the specialized file formats are being adapted to take advantage of these file systems. The third, and perhaps the most important reason, is that the logical data model presented by existing database system, especially the relational data model, is not appropriate for most scientific data. Much of the scientific data is multidimensional, such as space-time data for representing simulations of natural phenomena, or array data representing multivariate data. Furthermore, some data uses grids that are not regular, such as geodesic data, or toroidal meshes, and have different data models.

Yet, the concept of physical data independence is very attractive for scientific data as well. It is very attractive to a scientist to be concerned only with the "abstract data model", without dealing with files, datasets, file formats, and the file systems involved. As data volumes grow, having to deal with I/O bottlenecks, matching the application code to the type of data formats and file systems used, is increasing the overhead to the scientists, taking away productivity in doing their science. Recent activities, such as SciDB described in Chapter 7 have taken the approach that the data model of the system should represent multidimensional structures, and other structures that match the scientific domains. The goal is to have all the data stored in such scientific database systems, eliminating the burden on the scientist to deal with the concepts of files, file formats, and various physical organization considerations. Rather, the performance requirements will be specified, and accordingly the database system will choose the most appropriate storage structures and indexes for accessing the data, and the underlying file structures to hold the data. Assuming that such scientific database management systems will become operational over time and will mature, it is still unclear whether scientists will abandon their practice of storing data in specialized file formats.

Index

A

Acceptors, 138
Accord programming infrastructure, 172–174
ACID properties, 186, 188–189, 260
Adaptable IO System (ADIOS), 154, 155–166, 176
　applications interface, 157–158, 163
　code-coupling framework, 156–157, 158
　DataTap, 154, 159–166
　data workspaces, 169
　event-processing architecture, 168–169
　in-transit services, 168–171
　metadata and, 155–156
　XML structure, 157
Adaptive data-streaming, 170–171
Ad hoc queries, 190, 240
Advanced technology attachment (ATA) disk systems, 11, 41
　serial ATA (SATA) standards, 15, 26, 41
ALDS, 242
ALICE experiment, 91
Analytic I/O, 152
Annotation
　metadata management and, 437–438
　workflow management and, 483
Apache Hadoop project, 220
Apache Xindice database, 443
Append-only scientific datasets, 186, 189, 193, 267
Application programming interfaces (APIs), 55, See also POSIX I/O standard
　Adaptable IO System (ADIOS), 157–158, 163
　HDF5, 61–62
　HDF5_FastQuery, 352–354, 355
　high-performance visualization and analytics, 347–355
　myExperiment Web-based application, 498
　parallel file systems, 55–63
　parallel NetCDF, 61

　provenance store, 491–493
　storage resource management access, 79
Approximating streaming functions, 416–425
ArcGIS products, 374
Archival storage sustainability, 31
Archiving simulation data, 478–479
Argonne Leadership Computing Facility (ALCF), 51
Array-based datasets, 200, See also HDF5; NetCDF
　mapping into MonetDB BAT, 262
　SciDB and Extremely Large Databases, 268–272
Array blocking, 257
Array files, 200, 219
Astronomical metadata, 437, 447–448, 455
Astrophysics supernova code, 158
ATLAS, 91
Attribute value clustering, 240
Authorization enforcement, 108–109
AutoMate, 173
Autonomic data streaming infrastructure, 172–175
AVS/Express dataflow network, 477

B

BANG-file, 203
Batch-Aware Distributed File System (BADFS), 118
Battery-backed RAM, 66
Beauty experiment, 91
Berkeley Storage Manager (BeStMan), 78, 80, 82
Binary association tables (BATs), 242, 257
　MonetDB and BAT Algebra, 257–261
　MonetDB space optimizations, 249
　processing data models and query languages, 262–263
Binary tree, 340–342, 423–425
Binning, 215–216
Bio-Catalogue project, 483

515

Bioinformatics
  high-throughput quantitative
      proteomics, 308–315
  protein function prediction, 291–299
  social sharing of workflows, 498
  workflow reuse, 496
  workflow services, 438
Biological data analysis, 306–308
BI-TCP, 127
Bitmap indexing, 188, 191, 208–218, 226
  applications, 210
  binning, 215–216
  compression, 210–213
  encoding, 213–215
  FastBit, 209, 219, 222–223, 226, 352, 359, 360
  implementations, 217–218
  query performance and, 222–224, 226
Bit packing, 248, 249
BLAST, 293
B-list structure, 249–250
Block-based storage systems, 12, *See also* Disk storage
  parallel file systems, 40–43
Blocks, 12, 37, 189
  alignment issues, 15
  lock granularity, 38
  SCSI fixed-block format, 40–41
Blue Gene, 410, 416
Blue Gene/P, 50–53
Blu-ray, 21
Brain imaging data annotation, 438
Brain imaging data provenance, 457
B-Tree, 191, 194–195, 209, 241, 266
  B+-Tree, 61, 195
  b-list structure, 249–250
Buckets, 196, 272
Buffering techniques for accessing metadata, 241, 250–251
Bulk modulus, 303–304
Business-orientated data integration, 370
Business Process Execution Language (BPEL), 445
Business workflows, 469
BYNET, 221
Bypass, 133
Byte-aligned bitmap code (BBC), 210, 218

C

Cache-conscious data structures, 243, 265
Caching
  database performance issues, 244
  false sharing, 38
  parallel file systems, 38, 42
  pinning, 81, 85–86
  POSIX I/O and, 58
  vertical data organization and, 238
  X100 vectorized query processor, 246
Cantor, 239, 243, 246, 249, 251, 252
CASTOR, 78, 81, 92, 99
Catalog and registry services, 380–381
Center for Plasma Edge Simulation (CPES), 474–481
Centre de Donnees Astronomiques de Strasbourg (CDS), 447
CERN large hadron collider, *See* Large Hadron Collider (LHC)
Chalcogenide, 28
CHARISMA project, 56
Checksum, 25
ChemBank, 285
ChemDB, 285
Chemical compound classification algorithms, 290–291
Cheminformatics data analysis, 285–291
  compound classification algorithms, 290–291
  databases, 285
  data mining and modeling trends, 286–287
  descriptor-based chemical representations, 287–288
  indirect similarity measures, 288–290
  scaffold hopping, 289
  structure-activity relationship modeling, 285–287
  tools, 285–286
Chimera, 158
Chirp, 134–136
Chromium dioxide-based magnetic media, 7
Chunking, 189, 198–200
ClassAd job description language, 119
Client failures, 39–40, 48
Climate modeling, 75, 188, 306
  Community Climate System Model, 107, 449
  Earth System Grid (ESG), 82, 106–107, 449
  metadata management and, 449–450
Cluster analysis, principal components analysis (PCA), 300–301
Clustered/nonclustered index, 190
Clustering of attribute values, 240, 241–242
Co-allocation of resources, 137–141

Coercivity, 7
Collective communication calls, 59
Column-based database systems, *See* Vertical database systems
Column-based data organization, 184–185, 219, 237, 240, 264–265, *See also* Vertical data organization
Column cardinality, 213, 215
COMAD, 489
Combustion simulation, 417–419, 425
COMET Project, 392
Common data form language (CDL), 60
Common neighborhood, 296
Community Climate System Model (CCSM), 107, 449
Community Observations Data Model (ODM), 388–389
Compact discs (CDs), 20
Complex analytics, 237–239
Complexity of scientific data, 282
Compound TCP, 127
Compressed, fully transposed, ordered file (CFTOF), 250–252
Compression, 210–213, 240, 247–250
 database performance issues, 247–248
 dictionary schemes, 248
 I/O bandwidth conservation and, 247
 lazy decompression, 241, 248
 querying fully transposed files, 251–252
 query performance and, 222–224
 run-length encoding, 240
 sorted data, 247–248
 transactional, write-oriented database design, 238
 vertical data organization and, 239
Compression-aware equi-join operations, 252
Computational and network resource coallocation, 117, 136–141
Computational grid, 80–81
Condor, 133, 488, 489
Conflation, 378
Congestion avoidance, 125–126, 127
Congestion window, 126, 127
Connotea, 497
Content Standard for Digital Geospatial Metadata (CSDGM), 376
Coordinate reference system (CRS), 374
Crystallographic databases, 299, 305
C-Store, 219, 221, 247, 251, 255–256
Curse of dimensionality, 207–208
Custom data processing hardware, 221
Cyberinfrastructure initiative, 468
Cyclic redundancy check (CRC), 25
Cylinders, 12

D

DAGMan, 482, 485–486, 488, 489
DART (decoupled and asynchronous remote transfers), 154, 166–167
Data analysis, 75, 283–285, 510, *See also* Scientific data analysis
 analytic I/O, 152
 co-location of data and analysis, 512–513
 dimension reduction, 299–303
 high-throughput data movement and, 152
 Large Hadron Collider (LHC) applications, 95–96
 MapReduce parallel analysis system, 220–221
 multiresolution data layout, 336–345
 special-purpose systems, 219–220
 visual, *See* Visual data analysis
Database administrator (DBA), 266
Database management systems (DBMSs), 184, 510, *See also* Query processing
 bitmap index implementations, 210, 217–218
 compute and memory bound, 243–244
 data compression and performance, 247–248, *See also* Compression
 data mining benchmarks, 245
 data overwriting, 271
 data warehouse benchmarks, 253
 hardware architecture and performance, 243–245
 high-performance scalability, 253–254
 indexing, *See* Index methods
 physical data independence, 513–514
 query processing, *See* Query processing
 read-optimized, 240, 248, 263–266
 relational model, 184–185, 236–237, 513, *See also* Relational database systems
 SciDB and Extremely Large Databases, 268–272
 scientific application domains, 186
 scientific data characteristics, 185–189
 streaming data and, 400, *See also* Data stream management systems
 traditional (relational) DBMS limitations, 187–188
 transactional, write-oriented design, 237–239

vertical (column-based) systems, *See* Vertical database systems
workload analysis and online tuning, 266
Data compression, *See* Compression
Data consistency and coherence, 37–38
Data-driven scheduling, 117, 118, *See also* Stork data scheduler
Data grid, 81
Data integration, 370, 510
  business applications, 370
  geospatial data, *See* Geospatial data integration
  interoperability issues, 371, 378, 379–380
  physical and logical, 377
  schema-matching approach, 377
  service-oriented architectures, 387–391
Data integrity, 25
Data lifecycle, 434–435
Data mining benchmark, 245
Data mining in materials science, 301–305
Data movement, high-throughput, *See* High-throughput data movement
Data movement scheduling, 119–125
Data organization, 184–185, 189, 218, 236–237, *See also* Database management systems
  access methods, 189
  column-based, 184–185, 219, 237, 240, *See also* Vertical data organization
  data processing systems, 219–221
  indexing, *See* Index methods
  parallel systems, 218, *See also* Parallel data management systems
  partitioning, 242
  row-based, 184, 237, *See also* Horizontal data organization
Data overwriting, 271
Data placement job types, 120
Data postprocessing, 74–75
Data processing, 74–75, 283
  column-based systems, 219
  hardware, 221
  Large Hadron Collider (LHC) applications, 94–95
  missing values problem, 299
  noise reduction, 284, 295
  parallelized production-level visualization tool, 328–332
  protein function prediction, 295–297
  scientific data analysis process, 283–285
  simulation data reconstruction, 236
Data processing systems, 219–221
  custom hardware, 221
  index utility, 221–222
  MapReduce parallel analysis system, 220–221
  special-purpose data analysis systems, 219–220
Data provenance, *See* Provenance
Data replication service (DRS), 118
Data-side analysis facility, 512–513
Data storage, *See* Storage technology; *specific technologies*
Data storage resource management, *See* Storage resource management
Data streaming, 399–400, 510, *See also* Data stream management systems
  autonomic services for wide-area and in-transit data, 171–172
  autonomic streaming infrastructure, 172–175
  geospatial data integration, 372
  IQ-Paths, 170–171
  QoS management, 174–175
  storage resource management functionality, 86–87
Data stream management systems (DSMSs), 399–402
  distributed processing, 408–409
  extensible, 403
  high-volume query parallelization, 410–417
  parallelized query processing in SCSQ, 412–417
  runtime systems, 403
  scientific applications, 403–405
  traditional applications, 402
  WaveScope and WaveScript, 401, 405–409
DataTap, 154, 159–166
Data transfer management for distributed systems, *See* Distributed computing environment
Data warehousing, 210, 219, 239
  benchmark, 253
  scalability, 253–254
  SkyServer and MonetDB, 263–267
Data workspaces, 169
DAX, 482
DB2, 374
dCache, 81, 92, 98–99
Decomposition storage model (DSM), 242
  MonetDB and, 254, 256
  pivot algorithm, 242–243
Decoupled and asynchronous remote transfers, *See* DART

Index

Delta, 242
Descriptor space, 287
Dictionary schemes for data compression, 248
Differential file, 255
Digital signal processing (DSP) pipelines, 469
Digital versatile discs (DVDs), 21
DIII-D, 152, 154
Dimensionality, curse of, 207–208
Dimension reduction, 299–303
DIP database, 295–296, 297
Direct access indexing, 192
  multidimensional schemes, 203
DirectFlow, 45, 53
DirectorBlades, 44, 53
Disk arrays, 41
Disk failures, 41, 63–66
Disk Pool Manager (DPM), 99
Disk storage, 11–16
  ATA systems, 11
  bandwidth growth projections, 65
  energy-efficient systems, 511–512
  enterprise-level versus commodity technology, 15–16
  fundamentals, 12–13
  history, 11–12
  hybrid disk drives, 66
  latencies, 13, 66
  MAID, 15, 26
  parallel data storage, 35–40, *See also* Parallel file systems
  performance characteristics, 13–15
  power consumption, 15, 31
  SCSI fixed-block format, 40–41
  solid-state devices, 66, 509
  storage capacity-delivered bandwidth gap, 30
  trends, 11, 16, 30, 65–66
  virtual tape libraries, 24
  visualization of remote data, 130
Distributed computing environment, 115–117, *See also* Grid computing
  coscheduling compute and network resources, 117, 136–141
  efficiency and reliability issues, 116–117
  Enabling Grids for E-sciencE (EGEE), 91, 104–105
  e-science, 468
  Fibre Channel, 17, 128
  grid-enabled workflow design, 481–484
  grid-enabled workflow mapping, 484–486

GridFTP service, 97, 98, 100, 107, 118, 129, 133
high-level data management tools, 118
Highly Available Resource Co-allocator (HARC), 137–141
high-performance wide area data transfers, 117, 125–131
  remote I/O, 117, 131–136
  scheduling data movement, 117, 119–125, *See also* Stork data scheduler
  storage resource management functionality, 80–91, 108–109, *See also* Storage resource management
  storage space management, 89–90, 108, 121–123
  WaveScript and data stream management, 408–409, *See also* Data stream management systems
Distributed Oceanographic Data System (DODS), 386
DOLFA, 419
DrugBank, 285
DSM, *See* Decomposition storage model
Dublin Core, 439, 441
Dynamic random access memory (DRAM), database performance issues, 244
Dynamic storage management, 73–77, 80
  hierarchical storage management (HRM), 78
  metadata management and, 78
  network storage (NeST) approach, 77
  requirements, 75–76
  SRM approach, 76–91, *See also* Storage resource management
  storage elements, 97–100
  storage resource broker approach, 76–77
  Worldwide LHC Computing Grid (WLCG) infrastructure, 91–103, *See also* Worldwide LHC Computing Grid

E

Earth System Grid (ESG), 82, 106–107, 449
Earth System Science Workbench (ESSW), 453
Earth Systems Science Server (ES3), 446
Earthworks, 482
Ecological Metadata Language (EML), 376–377

Electronically erasable programmable read-only memory (EEPROM), 26
Electronic notebooks, 451
Embarrassingly parallel visualization operations, 330
Enabling Grids for E-sciencE (EGEE), 91, 104–105
Encoded Vector Index, 218
Energy-efficient storage systems, 511–512
EnLIGHTened Computing Project, 139
EnSight, 328
Enstore, 78, 81, 92
Enterprise-class disk technology, 15–16
Enterprise-class tape technology, 18–19
Enterprise information integration, 370
Entity resolution techniques, 378
Entrez Genome Database, 370
Equality encoding, 213
Equi-join optimization, 252
E-science, 468, See also Distributed computing environment
  Enabling Grids for E-sciencE (EGEE), 91, 104–105
  workflow design and composition, 481–484
EUROFAN project, 292
Evapotranspiration data integration, 387–391
EVPath, 168–169
Exact-match query, 190
Exodus, 329
eXplicit Control Protocol (XCP), 128
Extremely Large Databases (XLDB), 268–272

F

Facebook, 497, 498
FastBit, 209, 219, 226, 352
  HDF5_FastQuery, 352–354
  network traffic analysis visualization implementation, 360
  query-driven visualization implementation, 359
  query performance and, 222–223
Fast Fourier transform (FFT), 411–414
FAST TCP, 127
Fault tolerance
  Panasas storage system, 46
  parallel file systems, 39–40, 48, 63–66
  parallel virtual file system, 48, 52–53
  projections model, 65
  RAID and, 39, 41
  workflow execution and, 480–481

Feature detection and analysis, 358
Feature extraction, 283
Features, object-based models, 373
Feature vectors, 196–197
Federal Geographic Data Committee (FGDC), 376
Ferromagnetic media, 5, 7
Feta, 483
Fibre Channel, 17, 128
Field-based data, 378
Field-based geospatial models, 373
  file formats, 375
Field effect transistor (FET), 26
File caching, See Caching
File sharing
  false sharing, 38
  storage resource management functionality, 87
File streaming, storage resource management, 86–87, 105–107
Fill words, 211
Find-O-Matic, 483
First-come, first served (FCFS) scheduling, 120
FITS, 189, 219, 447
FLASH memory, 11, 26–28, 31
Flexible Image Transport System (FITS), 189, 219, 447
Floating-point data, 215
Frame of reference encoding (FOR), 248
  FOR-delta, 248, 249
FTP, 133, See also GridFTP service
Fully transposed storage model, See Decomposition storage model
FunctionalFlow algorithm, 297
Function approximation, 419–425
Fusion simulation, 152, 153
  DART and, 167
  data archiving, 478–479
  gyrokinetic toroidal code (GTC), 153, 163, 164, 167
  kinetic and simulation code coupling, 154
  monitoring and visualization, 153–154, 163, 475–477
  wide-area data-streaming, 172
  workflow services, 469, 474–481

G

GenBank, 292
Gene expression datasets, 293
Generalized multi-protocol label switching (GMPLS), 139

Index

General parallel file system (GPFS), *See* GPFS
Genome sequencing, 291, 292
Geographic information systems (GIS), 372, 374–375, 451–452
Geographic Resources Analysis Support System (GRASS), 374–375
Geography Markup Language (GML), 375–376, 381
Geoscience Network (GEON), 392
Geosciences applications, 369–372, *See also* Geospatial data
GeoServer, 383, 388
Geospatial data, 369, 371
    data management systems and formats, 374–375
    metadata standards, 376–377
    models and representations, 373–374
    provenance, 451–453
Geospatial data integration, 371–372
    approaches, 377–379
    catalog and registry services, 380–381
    conflation, 378
    evapotranspiration case study, 387–391
    GML application schemas, 375–376
    interoperability issues, 371, 378, 379–380
    real-time streaming data, 372
    resolving spatial conflicts, 378–379
    service-oriented architectures, 380, 387–391
    Web services, 371–372, 380, 381–383
GeoTIFF, 375
GFarm, 118
Ghost data, 330, 335
Giant magnetoresistance (GMR), 8, 28
gLite data management client tools, 93
Global Biodiversity Information Facility, 370
Global Lambda Grid Workshop, 139
Globus Grid Security Infrastructure, 135
Google file system (GFS), 39
GPFS (general parallel file system), 14, 15, 38, 41–43
    Enabling Grids for E-sciencE (EGEE), 104
    fault tolerance, 39–40, 48
    RAID configurations, 41–42
    storage resource management functionality, 82
Gravitational wave workflows, 483
Gray-Code encoding, 205
Green data centers, 511
Grid Application Toolkit (GAT), 484–485

Grid Collector, 218, 224–226
Grid computing, *See also* Distributed computing environment
    Earth System Grid (ESG), 82, 106–107
    Enabling Grids for E-sciencE (EGEE), 91, 104–105
    Large Hadron Collider Computing Grid, 91–103
Griddles, 484
Grid-enabled workflow systems, 481–484
    workflow execution, 486–489
    workflow mapping, 484–486
Grid-File, 203, 204
GridFTP service, 97, 98, 100, 107, 118, 129, 133
Grid Security Infrastructure (GSI), 488
Grid Stream Data Manager (GSDM), 410, 411
Group Transport Protocol (GTP), 128
GSDM (Grid Stream Data Manager), 410, 411
GTC (fusion simulation code), 153, 158, 163, 164
GUR, 137
Gyrokinetic toroidal code (GTC) fusion simulation, 153, 158, 163, 164

H

H5Part, 345–352, 354–355
Hardware architecture and DBMS performance, 243–245
Hashing-based indexing schemes, 196
    multidimensional schemes, 203
    radix cluster algorithm, 243, 255
h-confidence, 296, 297
HDF4, 61, 200
HDF5, 55, 60, 513
    Adaptable IO System (ADIOS) and, 157, 163
    array files, 200
    data organization method, 189
    data repartitioning during reads, 329
    H5Part, 345–352, 354–355
    high-level I/O library, 61–62
    high-performance visualization and analytics, 345–352
    parallel file systems, 188
    raster and gridded data format, 375
    special-purpose data analysis system, 219
    XGC1 fusion simulation data format, 475, 477

HDF5_FastQuery, 352–354
Heap, 189
Hierarchical data format version 5, *See* HDF5
Hierarchical indexing, out-of-core visual data processing, 337–345
Hierarchical storage management (HSM), 78
Hierarchical Triangular Mesh, 205
Hierarchical Z-order data layout, 337
High-definition DVD (HD-DVD), 21
High-level I/O libraries, 60
Highly Available Resource Co-allocator (HARC), 137–141
High-performance computing (HPC), 188
  shared-disk, 254
  shared-memory, 253–254
  shared-nothing, 254
High-performance computing system (HPCS) failures, 63–66
High-performance DBMS scalability, 253–254
High-performance storage system (HPSS), 31–32, 80, 81, *See also* Mass storage systems
  fusion simulation data movement, 153
  WLCG infrastructure, 92
HighSpeed TCP, 127
High-throughput data movement, 151–155, 166–167, 170–171, 188, *See also* Parallel file systems
  ADIOS, 154, 155–166, 176, *See also* Adaptable IO System
  analytic I/O, 152
  autonomic data streaming infrastructure, 172–175
  autonomic services using IQ-Paths, 170–171
  code coupling, 154, 156–157, 158
  DataTap, 154, 159–166
  fusion simulation example, 152, 153
  high-performance data capture, 155
  high-speed data extraction using DART, 154, 166–167
  in-transit services, 162, 168–170
  online extraction and transfer service effectiveness, 176
  parallel systems, *See* Parallel file systems
  QoS management at in-transit nodes, 175–176
  wide-area data streaming, 171–172

High-throughput quantitative proteomics, 308–315
High-throughput screening (HTS), 291
Hilbert-order encoding, 205
Holographic storage, 21, 28, 32
Horizontal data organization, 184
  column-wise storage performance vs., 264–265
H-TCP, 127
HTTP, 133
Huffman encoding, 248
Human cycles, 500
Human Genome Database, 370
Human genome sequencing project, 291
Hybrid disk drives, 66
Hybrid index schemes, 192, 203–207
Hypercliques, 297
Hyperslab reads, 329, 353

I

IBM 3340, 12
IBM 350 RAMAC, 11–12
IBM bitmap index implementations, 218–219
IBM Blue Gene deployments, 50–53, 410, 416
IBM GPFS, *See* GPFS
Idleness threshold, 511
Indexed sequential access method (ISAM) index, 194–195
Indexing problem for function approximation, 421–425
Index methods, 188, 189–193, 510
  access methods, 189
  binary tree, 340–342, 423–425
  bitmap, 188, 191, 208–218, 226, *See also* Bitmap indexing
  B-tree, 191, 194–195, 209, 241, 249, 266
  curse of dimensionality, 207–208
  data compression, *See* Compression
  factors affecting choice of method, 189–191
  future trends, 226
  Grid Collector, 218, 224–226
  hashing schemes, 196, 255
  hybrid methods, 192, 203–207
  inverted, 191
  making smart iterators, 224–226
  metrics, 191–192
  MonetDB support, 266
  multidimensional direct access, 203

multidimensional schemes, 192, 196–207
out-of-core visualization, 337–345
quaternary/hierarchical triangular mesh, 205–207
query performance and, 222–224, 226
query types and, 190–191
sequential scan access, 189, 193, 267
single-attribute schemes, 192, 193–196
specialized data management systems, 221–222
spherical surfaces, 205, 208
taxonomy of, 192–193
tessellation and chunking, 198–200
transactional, write-oriented database design, 238
tree-structured approaches, 192, 194–196, 200–203
Index size constraints, 191
Index types, 190
Indirect similarity measures, 288–290
InfiniBand, 128, 162, 164
Information management system (IMS), 185
Input/output (I/O) systems and issues, 4
access patterns, 55–57
Adaptable IO System (ADIOS), 154, 155–166
analytic I/O, 152
caching and, 58
collective communication calls, 59
data compression and performance, 247–248, *See also* Compression
disk device performance, 13–15
high-level libraries, 60–63
high-performance computing issues, 188, *See also* High-throughput data movement
I/O graphs, 160
layers, 55
MPI-IO, 55, 58–59, 67–68
parallel file systems, 55–63, 67–68
POSIX, *See* POSIX I/O standard
production-level, parallel visualization tool, 328–329
remote access, 117, 131–136
Stork data scheduler, 119
In Situ Adaptive Tabulation (ISAT) algorithm, 419–425
Integrated device electronics (IDE), 15
Integration of data, *See* Data integration
Intel iPSC/860, 56
Intergovernmental Panel on Climate Change (IPCC), 107, 449

International Virtual Observatory Alliance (IVOA), 447
Internet Engineering Task Force (IETF), 129
Interoperability, 378
geospatial data integration and, 371, 378, 379–380
OGC Interoperability Institute, 380
OGC Sensor Web Enablement (SWE) program, 383, 390
ontologies and, 438
provenance management and, 456–457
service-oriented science, 379
Interval encoding, 213
Interval tree, 197
In-transit data services, 162, 168–170
QoS management, 174–175
Inverted indexes, 191
I/O graphs, 160
I/O servers, 36–37
IQ-Paths, 170–171
Iron oxide-based magnetic media, 7
ISAT Algorithm, 419–425
ITER, 152, 153

J

Jaccard similarity, 296
Jaguar system, 65
JASMine, 81
Java database connectivity (JDBC) interface, 444
Jena Semantic Web Toolkit, 444
Job description language, 119
Job scheduling techniques, 120–125
Join index, 242–243, 262
Join operator
compression-aware optimization, 252
MonetDB and BAT Algebra, 259–260

K

(KA)2 initiative, 438
K-D-Tree, 200
Kepler, 446, 479, 481, 482, 484, 486, 488, 489, 493, 496
Kerberos, 135
Kernel-based learning algorithms, 287
k-nearest-neighbor graph, 289
Kx Systems, 219

L

Laboratory notebooks, 450–451
LambdaStream, 128

Index

Large Hadron Collider (LHC), 82
   data processing and reconstruction, 94–95
   data storage requirements, 97–104
   experiments, 91
   storage classes, 101–103
   WLCG infrastructure, 91–103, *See also* Worldwide LHC Computing Grid
Laser Interferometer Gravitational-Wave Observatory (LIGO) project, 441
Lazy decompression, 241, 248
LEAD project, 441
Lead Sharing Facility (LSF), 488
Lebesgue space-filling curve, 340–342
Lightweight data replicator (LDR), 118
Lightweight object storage facility (LWSF), 162
LIGO project, 441
Linear tape-open (LTO) technology, 19
Linked Environments for Atmospheric Discovery (LEAD) project, 441
Linux OS, 53
Linux clusters, 44–45
Literal words, 211
LoadLeveler, 139
Local models, 419–420
Locking, 37–38, 42–43, 50, 58, 239
LOFAR, 400, 410
Logarithmic binning, 215
Logical integration of data, 377
Logistical networking, 118
LONI, 141
Loop pipelining, 246, 257
Los Alamos National Laboratory, Roadrunner supercomputer, 43, 53–55
LSD-Tree, 200
L-Store, 80
LTO (linear tape-open) tape technology, 19
Lustre, 38, 39, 47, 49–50, 80, 165

M

M3D, 172
Magnesium diboride (MgB2) superconductor, 301
Magnetic storage, 5–11
   giant magnetoresistance, 8
   magnetoresistive reading, 8, 28
   media, 5, 7
   perpendicular recording, 8–9
   spintronics, 8, 11, 28
   superparamagnetism, 7
   tape media, *See* Tape storage technology
   trends and limits, 9–11
Magnetoresistive Random access memory (MRAM), 8, 28
MAID (Massive Arrays of Idle Disks), 15, 26
Map projections, 374
MapReduce parallel analysis system, 220–221
MapServer, 383, 388–389
Marmot detection and localization, 404–405, 406–408
Massive Arrays of Idle Disks (MAID), 15, 26
Mass storage systems (MSSs), *See also* Dynamic storage management; Parallel data storage; Storage technology
   file sizes, 77–78
   hierarchical storage management (HRM), 78
   high-performance storage systems, 31–32, 80, 81
   metadata management and, 78
   performance and reliability goals, 77
   storage resource management functionality, 78–91, 105–107
   sustainability paradox, 78
   WLCG infrastructure, 92, *See also* Worldwide LHC Computing Grid
Materials informatics data analysis, 299–305
MDDR, 285
Memory access latency, database performance issues, 243–245
Message digest algorithm #5 (MD5), 25
Metadata, 433–439, 458
   Adaptable IO System (ADIOS) and, 155–156
   analytic I/O, 152
   annotation and, 437–438
   buffering techniques, 241, 250–251
   data lifecycle, 434–435
   definition, 435
   Ecological Metadata Language (EML), 376–377
   FITS standard, 447
   geospatial metadata standards, 376–377
   layer organization, 436
   mass storage systems and, 78
   ontologies and, 438–439
   parallel data storage systems, 37

schema and attributes, 441–443
scientific applications, 446–450
scientific data management requirements, 187
self-describing data formats, 437
technologies for storage, 443–444
visualization tool optimization, 332–335
Metadatabase (MDB) buffer load, 250–251
Metadata Catalog Service (MCS), 441, 444
Metadata intent locking method, 50
Metadata managers, Panasas distributed storage system, 44–47
Metadata servers, 49
Midrange tape, 19
Millipede, 30, 32
Missing data problem, 299
Moab, 137, 139
MODAPS, 453
Models of computation, 472–473
Moderate Resolution Imaging Spectroradiometer Satellite, 453
Modis Adaptive Data Processing System (MODAPS), 453
Molecular manipulation storage technology, 30, 32
MonetDB, 219, 221, 243, 244–245, 249, 254–255
  BAT Algebra, 257–261
  benefits of vertical organization, 261–263
  decomposition storage model, 254, 256
  design principles, 256–257
  improvements and extensions, 261, 264–265
  index support, 266
  non-optimal performing applications, 267
  processing data models and query languages, 262–263
  reduced redundancy and storage needs, 266
  SkyServer data warehouse and, 263–267
  tuple reconstruction joins, 267
  weaknesses, 261
Montage, 455
Monterey Bay Aquarium Research Institute, 454
Morton mapping, 204–205
Mouse Genome Database, 437
MPI-IO, 55, 58–59, 62, 67–68, 133, 162, 351–352
MRAM (magnetoresistive RAM), 8, 28
MS SQL server, 210
Multidimensional extendible hashing, 203

Multidimensional index schemes, 192, 196–207
Multidisciplinary and multiscale applications and provenance, 456–457
Multilevel indexing schemes, 194, 214, *See also* Multidimensional index schemes
Multilevel queue priority scheduling, 121
Multiresolution data layout, 336–345, 364
  hierarchical indexing for out-of-core data access, 337–345
  space-filling curves, 204–205, 337, 340–342, 364
myExperiment project, 496–499
myGrid project, 438–439, 498
MySQL, 187, 443

N

Nanotube devices, 30
N-ary Storage Model (NSM), 242, 260
National Grid Service (NGS), 141
National Weather Service, 381
Neighborhood-based similarity measure, 296
NEON, 371
NetCDF, 60, 513
  Adaptable IO System (ADIOS) and, 157
  array files, 200
  data organization method, 189
  Earth System Grid, 449
  geospatial data integration implementation, 389
  parallel NetCDF, 55, 60–61, 163, 188
  raster and gridded data format, 375
  Shore Side Data System metadata, 454
  special-purpose data analysis system, 219
  XGC1 fusion simulation files, 475, 477
Netezza, 221
Network common data form, *See* NetCDF
Network failures, 40
Network file system (NFS), *See* NFS
Network storage (NeST) project, 77, 121
Network traffic analysis, 360–363
Neuroimaging data annotation, 438
Neuroimaging data provenance, 457
NFS, 53, 133
NFSv3, 39
NFSv4, 104–105
Nimrod, 484
Noise reduction, 284, 295

Nordic Data Grid Facility (NDGF), 91
Notebooks, 450–451
NSTX, 152, 154
Nuclear weapons certification, 53

O

Oak Ridge National Laboratory fusion simulation, 153
Oak Ridge National Laboratory Jaguar system, 65
Object-based geospatial models, 373
  file formats, 375
Object-based storage systems, 43–50
  Lustre, 49–50
  PanFS, 43–47
  PVFS, 47–49, See also Parallel virtual file system
Objectivity, 187
Object Storage Device (OSD) protocol, 44
Object-storage servers, 49
Oceanographic data provenance, 453–454
OceanStore, 118
OGC Interoperability Institute, 380
OGSA-DAI, 444
On-line analytic processing (OLAP), 210, 222
Online database tuning, 266
Online function approximation algorithm, 419–425
Ontologies and metadata specification, 438–439
OPeNDAP, 386
Open Geospatial Consortium (OGC), 371, 381
  Catalog Service for the Web implementation standard, 380
  Sensor Web Enablement (SWE) program, 383–384
Open Grid Forum (OGF), 82
Open Provenance Model, 456–457
Open Science Grid (OSG), 91
Open-source Project for a Network Data Access Protocol (OPeNDAP), 118, 386
OpenWetWare project, 451, 497
Optical storage, 20–21
  CDs, 20
  DVDs, 21
  Fibre Channel, 17, 128
ORACLE, 187, 217–218
Oracle Berkeley DB, 443, 444
Oracle Spatial, 383

Order-preserving bin-based clustering (OrBiC), 216
Orthogonal-range query, 190
Out-of-core visual data processing, 337–345
OWL, 439, 444

P

Panasas storage system, 43–47
  supercomputer deployment, 53–55
PanFS, 43–47
  Roadrunner deployment, 53–55
Parallel data management systems, 218
  MapReduce parallel analysis system, 220–221
Parallel data storage, 35–40, 42, See also Mass storage systems; Parallel file systems
  access patterns, 55–57
  data consistency, 37
  disk failures and, 63–66
  extreme-scale devices, 66–68
  fault tolerance, 39–40, 63–66
  file locking, 37–38, 42–43, 50, 58
  hybrid devices, 66
  lowering access latencies, 66
  Panasas system, 43–47
  RAID and, See RAID
  solid-state devices, 66
Parallel file systems, 36, 40, 188, 509, 514
  applications interface, 55–63
  block-based storage systems, 40–43
  caching and, 38, 42
  IBM's GPFS, See GPFS
  I/O issues, 42, 55–63, 67–68
  I/O layers, 55
  Lustre, 49–50, See also Lustre
  MapReduce parallel analysis system, 220
  object-based storage systems, 43–50
  PanFS, 43–47
  PanFS, Roadrunner deployment, 53–55
  parallel HDF5 for visualization applications, 350–352
  PVFS, 39, 47–49, See also Parallel virtual file system
  PVFS, IBM Blue Gene/P deployment, 50–53
Parallelized stream query processing, 410–417
Parallel NetCDF, 55, 60–61, 163, 188
Parallel Ocean Program (POP), 107
Parallel processing, ROOT applications, 220

Index

Parallel **R**, 306–311
Parallel virtual file system (PVFS), 39, 47–49
  fault tolerance, 48, 52–53
  IBM Blue Gene/P deployment, 50–53
Parallel visualization tools, 327–336, *See also* Production-level, parallel visualization tool
Paralogous verification method (PVM), 295–296
ParaView, 328
Parrot, 104, 134–136
Partial least squares (PLS) technique, 303, 304
Partial-match query, 190
Partial orthogonal-range query, 190
Particle-in-cell (PIC) technique, 153, 172
Partitioning, 242
PASOA, 445–446, 455, 495
Pathfinder project, 262
PATRICIA index, 196
Pattern recognition, 283, 284, 285
Paxos Commit protocol, 137–138
PBIO, 162, 164
Pegasus, 445, 446, 455, 481–482, 486, 488, 489, 495
Perpendicular recording technology, 8–9
Petascale computing issues, 4
Petascale Data Storage Institute (PDSI), 63–64
PGOS algorithm, 170–171
Phase-change memory, 28
Phase diagrams, 300
PhEDEx, 93
Physical data independence, 513–514
Physical integration of data, 377
Pinning, 81, 85–86
Pipeline, 329–330
Pipeline parallel processing, 469, 479
Pipeline Pilot, 286
Pivot algorithm, 242–243
PlanetOnto, 438
PLoS on Facebook, 497
Plumbing workflows, 474–481
Portable Batch System (PBS), 488
Portable operating system interface for unix (POSIX) standard, *See* POSIX I/O standard
POSIX I/O standard, 13–14
  caching and, 58
  Enabling Grids for E-sciencE (EGEE), 104
  Lustre and, 49–50
  mass storage systems and, 77
  parallel file systems and, 42, 55, 57–58, 68
PostGIS, 374–375, 383
PostgreSQL, 210, 253, 374, 388–389, 443
Power consumption, 15, 31
  energy-efficient storage systems, 511–512
  supercomputer, 53
Primary index, 190
Primary key, 190, 196
Primary metadata, 436
Principal components analysis (PCA), 300–301, 304–305, 310–312
PRISM, 419
Process documentation, 440–441
Production-level, parallel visualization tool, 327–336, 363
  data models and semantics, 334–335
  data processing, 328–332
  embarrassingly parallel operations, 330
  file formats, 345–355
  I/O infrastructure, 328–329
  metadata and performance optimization, 332–335
  remote framebuffer protocol, 331–332
  rendering, 330–331, 335
  SDM-related concerns, 330–331
  VisIt architecture and operation, 335–336
ProRata, 309–311
Protein function prediction, 291–299
  biological data and analysis techniques, 294
  databases, 293
  dataset preprocessing, 295–297
Protein interaction data, 293, 295
Proteomics data analysis, 308–315
Provenance, 271, 433–434, 439–441
  data lifecycle, 434–435
  definition, 439
  integrated management environments, 445
  interoperability issues, 456–457
  lab notebooks, 450–451
  management technologies, 444–446
  multidisciplinary and multiscale applications, 456–457
  process documentation, 440–441
  query language, 271
  scientific applications, 450–455
  scientific insights and, 500
  simulation management, 479
  storage decisions, 458

types or layers, 490
VisTrails, 440, 446, 454, 490–491
workflow services and, 471, 479, 489–495
Provenance-Aware Service-Oriented Architecture (PASOA), 445–446, 455, 495
Provenance-Aware Storage System (PASS), 446
Provenance store, 445, 491–493
PSockets, 129
PubChem, 285, 288
Push-caching, 136
PVFS, *See* Parallel virtual file system
PVM (paralogous verification method), 295–296
Pyramid tree, 208

Q

Quality of service (QoS) management, 174–175
Quantitative shotgun proteomics, 309–315
Quantitative structure-activity relationships (QSAR), 285, 303–304
Quaternary/hierarchical Triangular Mesh, 205
Query-driven visualization (QDV), 358–363
Query processing, 183–185, *See also* Database management systems; *specific systems applications*
  access methods, 189
  ACID properties and, 186, 188–189, 260
  binary association tables (BATs) and, 262–263
  compressed, fully transposed files, 251–252
  data compression, *See* Compression
  data processing systems, 219–221
  data repartitioning during reads, 329
  datastream management systems, 400
  DSM parallel strategy, 242–243
  feature vectors, 196–197
  future trends, 226
  HDF5_FastQuery API, 352–354, 355
  horizontal vs. vertical data organization, 238–239
  indexes and performance, 222–224, 226
  indexing, *See* Bitmap indexing; Index methods
  parallelizing high-volume stream queries, 410–417

provenance, 271, *See also* Provenance
  query types and index method selection, 190–191
  radix cluster algorithm, 243, 255
  read-optimized database systems, 240
  smart iterators, 224–226
  special-purpose data analysis systems, 219–222
  vectorized operations, 241, 245–247
  version management vs. locking, 239
  workload analysis and online tuning, 266
Query types, 190–191
Quorum techniques, 40, 46, 52

R

R, 306–311
Radix cluster algorithms, 243, 255
RAID (redundant arrays of independent disks), 12–13, 22–23, 77
  data consistency control, 37–38
  error correction, 23, 31
  fault tolerance, 39, 41
  GPFS and, 41–42
  levels, 22–23, 41
  MAID, 15, 26
  Panasas distributed storage system implementation, 46–47
RAIT (redundant arrays of inexpensive tape), 24
RAMAC, 11–12
Random access memory (RAM)
  battery-backed, 66
  database performance issues, 244
  inefficiencies, 262
  magnetoresistive (MRAM), 8, 28
  X100 vectorized query processor, 246
Random scheduling, 121
RAPID, 242
Raster and gridded data formats, 375
RDF Schema, 439, 444, *See also* Resource Description Framework (RDF)
RDMA, *See* Remote direct memory access
Read-only scientific datasets, 186, 189
Read-optimized databases, 240, 248
  SkyServer data warehouse, 263–266
Realtime Transmission Protocol, 130
Redundant arrays of independent disks, *See* RAID
Redundant arrays of inexpensive tape (RAIT), 24

Index

Relational database systems, 184–185, 236–237, 513
   CPU cycle costs, 256
   metadata management and, 443
   transactional, write-oriented database design, 237–239
Reliable Blast UDP (RBUDP), 128, 129
Reliable file transfer service (RFT), 118
Remote data visualization, 130
Remote direct memory access (RDMA), 154, 162
   DART, 166–167
   DataTap, 154, 159–166
   online extraction and transfer service effectiveness, 176
Remote framebuffer (RFB) protocol, 331–332
Remote input/output, 117, 131–136
Rendering, 330–331, 335
Resource Description Framework (RDF), 210
   metadata storage, 443–444
Retinopathy workflows, 483
Ring Buffer Network Bus (RBNB), 388, 391
RM, 242
Roadrunner, 43, 53–55
ROMIO, 59
ROOT, 186, 187, 188, 219–220
   H5Part reader, 356
   parallel processing applications, 220
Round Robin stream partitioning, 412
Row-based data organization, 184, 237, *See also* Horizontal data organization
   appropriate applications for, 267
   column-based storage performance vs., 264–265
   MapReduce parallel analysis system, 220–221
R-trees, 198, 200–203, 272
Ruby on Rails, 498
Run-length encoding, 251–252
Run-length encoding (RLE), 240, 248, 249
Runtime systems, 403

S

SAF, 329
Sapphire project, 283
Satellite-generated data, provenance management, 453
Satellite-generated data integration, *See* Geospatial data integration
Scaffold hopping, 289
Scalability architectures for DBMSs, 253–254
Scalable I/O Initiative Applications Group, 56
Scalable TCP, 127
Scheduling data movement, 119–125
SciDAC program, 468, 474
SciDB, 268–272, 514
Science Environment for Ecological Knowledge (SEEK), 392
Scientific data analysis, 281–284, 510, *See also* Data analysis; *specific applications*
   cheminformatics data, 285–291
   dimension reduction, 299–303
   high-throughput quantitative proteomics, 308–315
   materials informatics, 299–305
   parallel R applications, 306–311
   process of, 283–285
   protein function prediction, 291–299
   structure-activity relationship modeling, 285–287
Scientific data characteristics
   complexity, 282
   data lifecycle, 434–435
   floating-point values, 215
   provenance, *See* Provenance
   uncertainty, 271
Scientific data integration, *See* Data integration
Scientific data management application domains, 186
Scientific data processing and analysis stages, 74–75, *See also* Data analysis; Data processing
Scientific dataset characteristics, 185–189
   append-only, 186, 187, 189
   dimensionality, 187
   read-only, 186, 189
   size, 187
Scientific process automation, 467, *See also* Workflow management
Scientific visualization, 326, *See also* Visualization
Scientific workflows, 467–473, 499–500, *See also* Workflow management
   benefits, 473–474
   business workflows vs., 469
   fusion simulation, 153, 474–481

grid-enabled system design and composition, 481–484
human cycles and optimization, 500
life cycle, 470–471
models of computation, 472–473
pipelines, 469
plumbing workflows, 474
social sharing, 496–499
types, 471–472
user-oriented requirements, 500
workflow execution, 486–489
workflow reuse, 496
workflow sharing, 495–499
SCREEN, 286
SCSI (small computer system interface)
fixed-block disk format, 40–41
Object Storage Device (OSD) protocol, 44
SCSI tape drives, 17, 24
SCSQ (Super Computer Stream Query) processor, 410, 412–417
SCTP (stream control transmission protocol), 127
SDM-related visualization issues, 330–331
Secondary index, 190
Secondary metadata, 436
Sectors, disk, 12
Security issues, remote I/O, 133–134
Segment tree, 197
Self-describing data formats, 437
Semantic Web applications, 253
Semantic Web for Earth and Environmental Terminology, 377
Semantic Web Toolkit, 444
Semi-Quadcode (SQC), 205
Sensor network runtime system (WaveScope), 405–409, *See also* Data stream management systems
Sensor Observation Service (SOS), 390
Sensor Web Enablement (SWE) initiative, 383–384, 390
Sequential data access, 189, 193, 262, 267
Serial ATA (SATA) standards, 15, 26, 41
Server failure, 39, 52–53
Service-oriented data integration framework, 380, 387–391
Service-oriented science, 379
Shapefiles, 375, 383
Shared-disk architecture, 254
Shared file management, 81
Shared-memory computer systems, 253–254
Shared-nothing, 254

SHOE project, 438
Shore Side Data System (SSDS), 454
Shortest, 120
Shortest job first (SJF) scheduling, 121
Silo, 329
Similarity measures, 288–290, 296
Simulation codes, 153–154
  code coupling issues, 154, 158, 348
  parallelized production-level visualization tool, 327–336, *See also* Production-level, parallel visualization tool
Simulation data movement, *See* High-throughput data movement
Simulation management workflows, 153, 474–481
Simulation monitoring, 153–154, 163, 326, *See also* Visualization
Simulation streaming function approximations, 416–425
Single-attribute index schemes, 192, 193–196
SkyServer data warehouse, 263–267
Small computer system interface (SCSI), *See* SCSI
Smart iterators, 224–226
Snippet processing units (SPUs), 221
Social annotation, 437
Social sharing of workflows, 496–499
Solid-state storage devices, 66, 509
Sony Playstation 3, 53
Sorted data compression, 247–248
Sort ordering of attribute values, 240
Southern California Earthquake Center (SCEC), 496
Space allocation authorization enforcement, 108–109
Space-filling curves, 204–205, 337, 340, 364
SPARQL, 263
Sparse Relational Array Mapping (SRAM) project, 262
Spatial representation system (SRS), 374
Spherical indexing, 205, 208
Spindle, 12
Spintronics, 8, 11, 28
SQL, 184
  binary association tables (BATs) and, 262
  bitmap index support, 210
  MySQL, 187, 443
  PostgreSQL, 210, 253, 374, 388–389, 443
  SkyServer and MonetDB, 264
  StreamSQL queries, 402

SRM, *See* Storage resource management
SR-Tree, 203
STAR experiments, 105–107, 186, 224
State Plane Coordinate System, 374
Stones, 168–169
StorageBlades, 44, 53
Storage classes, CERN Large Hadron Collider data, 101–103
Storage elements (SEs), 97–100
Storage resource broker (SRB), 76–77, 118, 444
Storage resource management (SRM), 76–80, 108, 118
  authorization enforcement issue, 108–109
  directory management, 90
  Earth System Grid (ESG), 82, 106–107
  Enabling Grids for E-sciencE (EGEE), 91, 104–105
  file movement, 84–87, 105–107
  file sharing, 87
  file streaming, 86–87, 105–107
  file types and expiration mode, 87–88
  functionality, 80–91
  getting/putting file requests, 88–89
  grid environment, 80–81
  hierarchical storage management systems, 78
  high-energy physics applications, 82
  logic, 82
  logical file names, 84, 93
  mass storage system interfaces, 78–80
  performance estimation, 108
  redundant request handling, 124–125
  shared file management, 81
  solid-state devices, 66, 509
  space allocation, 89–90, 108, 121–123
  specifications (SRM v1.1 or SRMv2.2), 81–82, 98–99, 101
  storage types, 89
  transfer protocol negotiation, 84
Storage resource manager (StoRM), 78, 80, 82, 104
Storage technology, 3–5, 509, *See also* Dynamic storage management; Mass storage systems; Parallel data storage; *specific technologies*
  common interface requirements, 75
  composite devices, 22–25, *See also* RAID
  data consistency and coherence, 37–38
  data integrity, 25
  direct molecular manipulation, 30, 32
  disk, *See* Disk storage
  energy-efficient systems, 511–512
  FLASH, 26–28, 31
  holographic, 21, 28, 32
  magnetic storage fundamentals, 5–11
  MAID, 15, 26
  MRAM, 8, 28
  optical, 20–21
  parallel systems, 35–40, *See also* Parallel file systems
  phase-change approaches, 28
  power consumption, 15, 31
  solid-state devices, 66
  storage density trends, 5
  tape, 16–20, 24
  trends, 30–32
Stork data scheduler, 117, 119–125
  data placement job types, 120
  I/O control, 119
  job scheduling techniques, 120–125
  storage space management, 121–123
StoRM (storage resource manager), 78, 80, 82, 104
Stream control transmission protocol (SCTP), 127
Streaming data, *See* Data streaming
Streaming function approximation, 417–425
Streaming sensor systems, *See* Data stream management systems
Stream partitioning, 411
Stream processing, *See* Data stream management systems
Strength reduction, 257
Structure-activity relationship (SAR) modeling, 285, 287, 291
  materials informatics, 303
  performance of descriptors, 288
Suffix tree, 195–196
Supercomputers
  parallel file system deployments, Blue Gene/P, 50–53
  parallel file system deployments, Roadrunner, 53–55
  power consumption, 53
Super Computer Stream Query (SCSQ) processor, 410, 412–417
Superconductors, 301
Supernova code, 158
Superparamagnetism, 7
Support vector machines (SVMs), 290–291
Sustainability paradox, 78
SWISS-PROT, 292, 293

Sybase, 187
Sybase IQ, 219, 237–238
SYBYL, 286

T

Tanimoto coefficient, 289
Tape libraries, 18–19
    virtual, 24
Tape storage technology, 16–20
    commodity-class, 19
    enterprise-class, 18–19
    media, 17
    RAIT, 24
    tape marks, 17–18
    trends and limits, 11, 20, 32
Taverna, 481, 482–484, 495, 498
TCP (Transmission Control Protocol), 125–127
TCP Vegas, 127
TCP Westwood, 127
Teradata, 221
TeraGrid, 141
Tessellation, 198–200
Thermochemical databases, 299, 305
Thinking Machines CM-5, 56
THREDDS Data Server (TDS), 388
TIGER, 375
Tiger Shark file system, 41
TOD, 242
Tokamak fusion reactors, 152, 153, 167, 474–475, *See also* Fusion simulation
Topological analysis, 337
Topologically Integrated Geographic Encoding and Referencing (TIGER) format, 375
Toroidal LHC apparatus (ATLAS), 91
TPC-H data warehouse benchmark, 253
Transactional, write-oriented database design, 237–239
TransducerML (TML), 391
Transmission Control Protocol (TCP), 125–126
    alternative transport protocols, 127
Transposed files, 240, 241–242
    querying compressed, fully transposed files, 251–252
Transposition of data, 75
Tree-structured indexing, 192, 194–196, 200–203
Tree-structured multidimensional indexing, 200–203
Triana, 481, 483–486, 488, 489, 495
Trie index, 195–196

Tuple identifiers (TIDs), 242
    compressed lists in vertical databases, 241
    MonetDB and BAT Algebra, 259
    MonetDB virtual TIDs, 249
Tuple reconstruction joins, 267

U

UDP (user datagram protocol), 127
UDT, 128, 129
Uncertain data, 271
Universal Transversal Mercator (UTM), 374
Unix I/O interface, 134–136
Unsupervised learning, 301
User datagram protocol (UDP), 127

V

VA File approach, 208
Van der Waals interaction, 30
VDL, 445
Vector data file formats, 375, 378
Vectorized dataflow network architecture, 241, 246–247
Vectorized data stream operations, 241, 245–247
Vertica, 219, 221, 253
Vertical database systems, 235–240, 510, *See also* C-Store; MonetDB
    advantages of sequential vs. random data access, 262
    architectural principles, 241
    benchmark studies, 253
    binary association tables (BATs), 242, 257
    buffering techniques for accessing metadata, 241, 250–251
    cache-consciousness, 243, 265
    design rules and user needs, 237–239
    DSM, *See* Decomposition storage model
    join operator optimization, 252
    performance advantages, 240–241
    querying compressed, fully transposed files, 251–252
    transposed files, 240, 241–242
    vectorized operations and network architecture, 241, 245–247
Vertical data organization, 185, 237–239, *See also* Column-based data organization
    architectural opportunities, 240–241
    assessing benefits of, 261–263

attribute clustering, 241–242
data processing systems, 219
non-optimal performing applications, 267
page size and, 238
row-wise storage performance vs., 264–265
version management vs. locking, 239
Vertical partitioning, 242
Videoconferencing applications, 130
Virtual network computing (VNC), visualization applications, 331–332
Virtual shared disk (VSD), 41
Virtual tape libraries (VTLs), 24
VisIt, 328, 334, 335–336
VisTrails, 440, 446, 454, 490–491
Visual data analysis, 325–326, 510, *See also* Visualization
   file formats, 345–355
   H5Part, 345–352, 354–355
   HDF5_FastQuery, 352–354, 355
   hierarchical indexing for out-of-core data access, 337–345
   multiresolution data layout, 336–345
   network traffic analysis example, 360–363
   parallel I/O, 350–352
   query-driven visualization, 358–363
Visual information overload, 359
Visualization, 325–327, 363
   embarrassingly parallel operations, 330
   file formats for high-performance, 345–355
   ghost data, 330, 335
   in-transit manipulation of data, 170
   parallelized production-level tool, 327–336, *See also* Production-level, parallel visualization tool
   remote data, 130
   remote framebuffer protocol, 331–332
   rendering, 330–331, 335
   SDM-related concerns, 330–331
   simulation monitoring, 153–154, 163, 475–477
   3-D grid slicing experiment, 344
   virtual network computing, 331–332
ViSUS, 329

W

WaveScope, 401, 405–409
WaveScript, 405–409
Web 2.0 approach, 497, 499

Web Coverage Service (WCS), 382–383
Web Feature Service (WFS), 381–383
Web Map Service (WMS), 382–383
Web services
   geospatial, 371–372, 380, 381–383
   myExperiment workflow sharing project, 496–499
Wide area data transfers, 117, 125–131, *See also* Distributed computing environment
   alternative transport protocols, 127–128
   autonomic services, 171–172
   CERN Large Hadron Collider grid, *See* Worldwide LHC Computing Grid
   TCP and, 125–126
   transport protocol evaluation, 129
Wikis, 451
Winchester disk technology, 12
Windowed operations, 401
Wings, 481, 483–484
Word-aligned hybrid (WAH) code, 210
Workflow compilers, 455
Workflow ecosystem, 495
Workflow management, 440–441, 455, 467, 510, *See also* Scientific workflows
   annotation and ontologies, 483
   fault tolerance, 480–481
   grid-enabled workflow design and composition, 481–484
   grid-enabled workflow mapping, 484–486
   job scheduling techniques, 120–125
   myExperiment project, 496–499
   provenance management, 471, 479, 489–495
   scientific annotations, 438
   scientific workflow life cycle, 470–471
   simulation management, 153, 474–481
   Stork data scheduler, 119
   workflow execution, 486–489
   workflow robustness, 479–480
Workflow mapping, 470, 484–486
Workflow reuse, 496
Workflow sharing, 495–499
Workload analysis, 266
World Coordinate System (WCS), 447
Worldwide LHC Computing Grid (WLCG), 91–103
   calibration study, 96–97
   chaotic analysis, 97
   data management services and clients, 92–93
   data reconstruction, 94–95
   deployment working group, 103

general computing and data model, 93–94
mainstream analysis, 95–96
SRM client support, 98–99, 101
storage classes, 101–103
storage requirements, 97–104
tiers model, 92
WorldWide Telescope, 379
WS-BPEL, 470

X

X100, 246
XChange, 162
XGC0, 154
XGC1, 153, 474–481, *See also* Fusion simulation
XML-based databases, 443
XML-Schema, 376
XQuery, 262
XROOTD, 80, 93
X-Tree, 208
XUnion, 133

Y

YouTube, 498

Z

Zeolite database, 304–305
Zettabyte file system (ZFS), 25
Zonal recording, 12
Z-order mapping, 204–205, 337, 338, 340–342